T0329240

Reproductive Technologies in Animals

Reproductive Technologies in Animals

Edited by

Giorgio A. Presicce
ARSIAL, Rome, Italy

ACADEMIC PRESS
An imprint of Elsevier

ELSEVIER

British Library Cataloguing-in-Publication Data
A catalogue record for this book is available from the British Library

Library of Congress Cataloging-in-Publication Data
A catalog record for this book is available from the Library of Congress

ISBN: 978-0-12-817107-3

For Information on all Academic Press publications
visit our website at https://www.elsevier.com/books-and-journals

Publisher: Charlotte Cockle
Acquisitions Editor: Anna Valutkevich
Editorial Project Manager: Devlin Person
Production Project Manager: Joy Christel Neumarin Honest Thangiah
Cover Designer: Victoria Pearson

Typeset by MPS Limited, Chennai, India

Contents

List of contributors

David Arney Institute of Veterinary Medicine and Animal Sciences, Estonian University of Life Sciences, Tartu, Estonia

Judit Barna National Centre for Biodiversity and Gene Conservation (NCBGC), Institute for Farm Animal Gene Conservation, Gödöllő, Hungary

Giuseppe Campanile Department of Veterinary Medicine and Animal Production, University of Naples "Federico II,"Napoli, Italy

Sebastián Cánovas Department of Physiology, University of Murcia, Murcia, Spain

Jose Cibelli Michigan State University, East Lansing, MI, United States

Daguia Zambe John Clotaire College of Veterinary Medicine, Shaanxi Centre of Stem Cells Engineering & Technology, Northwest A&F University, Yangling, China; Laboratory of Biological and Agronomic Sciences for Development, Faculty of Science, University of Bangui, Bangui, Central African Republic

M. Colombo Department of Health, Animal Science and Food Safety "Carlo Cantoni" - Università degli Studi di Milano, Milan, Italy

Pilar Coy Department of Physiology, University of Murcia, Murcia, Spain

Elizabeth G. Crichton Camel Reproduction Center, Dubai, UAE

Giuseppe De Rosa Department of Agricultural Sciences, University of Naples Federico II, Portici, Italy

J. Farías Chemical Engineering Department, Faculty of Engineering and Science, University of La Frontera, Temuco, Chile

E. Figueroa Nucleus of Research in Food Production, Department of Biological and Chemical Sciences, Faculty of Natural Resources, Catholic University of Temuco, Temuco, Chile

Giovanni Formato Istituto Zooprofilattico Sperimentale del Lazio e della Toscana, Roma, Italy

Joaquín Gadea Department of Physiology, University of Murcia, Murcia, Spain

Julie Gard Department of Clinical Sciences, Auburn University College of Veterinary Medicine, Auburn, AL, United States

Wiebke Garrels Institute for Laboratory Animal Science and Central Animal Facility, Hannover Medical School, Hannover, Germany

Bianca Gasparrini Department of Veterinary Medicine and Animal Production, University of Naples "Federico II,"Napoli, Italy

Muren Herrid International Livestock Center, Queensland, Australia

Jinlian Hua College of Veterinary Medicine, Shaanxi Centre of Stem Cells Engineering & Technology, Northwest A&F University, Yangling, China

Dolors Izquierdo Department of Animal and Food Sciences, Faculty of Veterinary, University Autonomus of Barcelona, Barcelona, Spain

Takehito Kaneko Division of Fundamental and Applied Sciences, Graduate School of Science and Engineering, Iwate University, Morioka, Japan

S. Ledda Department of Veterinary Medicine, University of Sassari, Sassari, Italy

Krisztina Liptói National Centre for Biodiversity and Gene Conservation (NCBGC), Institute for Farm Animal Gene Conservation, Gödöllő, Hungary

G.C. Luvoni Department of Health, Animal Science and Food Safety "Carlo Cantoni" - Università degli Studi di Milano, Milan, Italy

Clara M. Malo Camel Reproduction Center, Dubai, UAE

Gabriela F. Mastromonaco Reproductive Sciences, Toronto Zoo, Toronto, ON, Canada

Carmen Matás Department of Physiology, University of Murcia, Murcia, Spain

O. Merino Center of Excellence of Biotechnology in Reproduction (BIOREN-CEBIOR), Faculty of Medicine, University of La Frontera, Temuco, Chile

M.G. Morselli Department of Health, Animal Science and Food Safety "Carlo Cantoni" - Università degli Studi di Milano, Milan, Italy

Ajda Moškrič Agricultural Institute of Slovenia, Ljubljana, Slovenia

Daniel Mota-Rojas Department of Animal Production and Agriculture, Metropolitan Autonomous University *(UAM)*−Campus Xochimilco (UAM), México City, México

S. Naitana Department of Veterinary Medicine, University of Sassari, Sassari, Italy

Fabio Napolitano School of Agriculture, Forest, Food and Environmental Sciences, University of Basilicata, Potenza, Italy

Gianluca Neglia Department of Veterinary Medicine and Animal Production, University of Naples "Federico II,"Napoli, Italy

Maria-Teresa Paramio Department of Animal and Food Sciences, Faculty of Veterinary, University Autonomus of Barcelona, Barcelona, Spain

Eszter Patakiné Várkonyi National Centre for Biodiversity and Gene Conservation (NCBGC), Institute for Farm Animal Gene Conservation, Gödöllő, Hungary

Janez Prešern Agricultural Institute of Slovenia, Ljubljana, Slovenia

Giorgio A. Presicce ARSIAL, Rome, Italy

J. Richard Pursley Michigan State University, East Lansing, MI, United States

J. Risopatrón Center of Excellence of Biotechnology in Reproduction (BIOREN-CEBIOR), Faculty of Medicine, University of La Frontera, Temuco, Chile

Raquel Romar Department of Physiology, University of Murcia, Murcia, Spain

Angela Salzano Department of Veterinary Medicine and Animal Production, University of Naples "Federico II,"Napoli, Italy

L. Sandoval Nucleus of Research in Food Production, Department of Agricultural and Aquaculture Sciences, Faculty of Natural Resources, Catholic University of Temuco, Temuco, Chile

Marina Sansinena Laboratory of Animal Reproduction and Biotechnology, School of Engineering and Agricultural Sciences, Universidad Católica Argentina, Buenos Aires, Argentina; National Research and Technology Council, CONICET, Buenos Aires, Argentina; Department of Veterinary Clinical Sciences, School of Veterinary Medicine, Louisiana State University, Baton Rouge, LA, United States; College of Agricultural and Veterinary Sciences, Universidad del Salvador, Buenos Aires, Argentina

Julian A. Skidmore Camel Reproduction Center, Dubai, UAE

Maja Ivana Smodiš Škerl Agricultural Institute of Slovenia, Ljubljana, Slovenia

Nucharin Songsasen Center for Species Survival, Smithsonian Conservation Biology Institute, Front Royal, VA, United States

Sandra Soto-Heras Department of Animal and Food Sciences, Faculty of Veterinary, University Autonomus of Barcelona, Barcelona, Spain

I. Valdebenito Nucleus of Research in Food Production, Department of Agricultural and Aquaculture Sciences, Faculty of Natural Resources, Catholic University of Temuco, Temuco, Chile

Jane L. Vaughan Cria Genesis, Victoria, Australia

Barbara Végi National Centre for Biodiversity and Gene Conservation (NCBGC), Institute for Farm Animal Gene Conservation, Gödöllő, Hungary

Nisar A. Wani Reproductive Biotechnology Centre, Dubai, UAE

Yudong Wei College of Veterinary Medicine, Shaanxi Centre of Stem Cells Engineering & Technology, Northwest A&F University, Yangling, China

Shihua Yang College of Veterinary Medicine, South China Agricultural University, Guangzhou, P.R. China

Luigi Zicarelli Department of Veterinary Medicine and Animal Production, University of Naples "Federico II,"Napoli, Italy

Chapter 1

Reproductive technologies in cattle

J. Richard Pursley and Jose Cibelli
Michigan State University, East Lansing, MI, United States

1.1 Introduction

Reproductive technologies are critical for genetic modification and management efficiencies that directly impact sustainability of dairy and beef herds. These technologies provide ways to improve individual production and health traits as well as manipulate production cycles to maximize herd performance parameters.

This chapter describes current reproductive technologies from whole animal to cloning in cattle. Section 1 focuses on recent technologies that have transformed reproduction in cattle industries. Technologies that control ovarian function and timing of ovulation for the purpose of timed-artificial insemination (AI) are discussed in depth. These technologies succinctly control the time of ovulation and can improve fertility. The increases in reproductive efficiency from these technologies have had major positive impacts on profit of herds that utilize them. In addition, new electronic technologies that enhance detection of estrus and new ways to diagnose early pregnancies in cattle are discussed. Lastly, this section will introduce new ways to control ovarian development during ovarian stimulation with follicle-stimulating hormone (FSH).

Section 2 discusses bovine cloning and emerging reproductive technologies to improve cattle genomic parameters. It includes a brief description on the origins of cloning cattle, and the current techniques implemented commercially. We also discuss new technologies that, coupled with somatic cell nuclear transfer (SCNT), have the potential to exponentially increase genetic selection in a very short period of time.

Section 1: Increasing the efficiency of pregnancy production utilizing technologies that control ovarian function in cattle

1.1.1 Controlling ovarian development to enhance pregnancy production following artificial insemination

1.1.1.1 Physiological basis for ovarian manipulation of the estrous cycle

Technologies that manipulate ovarian structures in systematic ways are important for a number of reasons. The primary purpose of manipulating ovarian function is to synchronize timing of estrus and/or ovulation for preparation to AI. Additionally, synchronizing estrus and/or ovulation is important for recipients of embryos in addition to donor cows being prepared for superovulation or ovum pickup (OPU).

Follicles and corpora lutea (CL) growth and function during an estrous cycle are very well understood. Follicles grow in waves following estrus [1]. A cohort of antral follicles begins growing due to an FSH surge that occurs simultaneously with the luteinizing hormone (LH) surge at the onset of estrus. The largest of these follicles continues to grow and develop after the remainder of the follicles in the wave become atretic. This follicle has been termed the dominant follicle (DF) because the inhibin and estradiol that it produces inhibits the release of FSH, thus dominating the ovary as the only growing follicle until it either becomes atretic or, in the case of most second- or third-wave DFs, ovulates. FSH will surge once lack of inhibitory substances, inhibin and estradiol, are no longer being produced from the DF [2–4]. This natural FSH surge will induce a new follicular wave. Once the DF becomes atretic or ovulates a new wave of follicles starts the process again.

Reproductive Technologies in Animals. DOI: https://doi.org/10.1016/B978-0-12-817107-3.00001-1

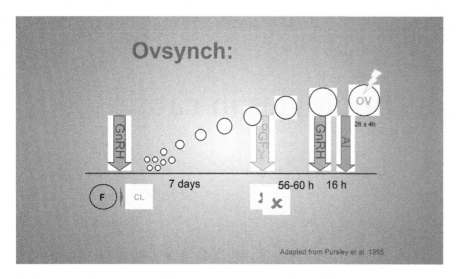

FIGURE 1.1 Description of how Ovsynch controls follicle and corpora lutea development in cattle to synchronize the timing of ovulation in an 8-hour period. *Adapted from Pursley JR, Mee MO, Wiltbank MC. Synchronization of ovulation in dairy cows using PGF2alpha and GnRH. Theriogenology 1995;44:915–23. doi:10.1016/0093-691x(95) 00279-h.*

Acquisition of LH receptors in granulosa cells of the DF occurs when the DF deviates in growth from the other follicles in the wave [5]. This critical change in functionality to a follicle with LH receptors makes it possible to manipulate this follicle with LH, human chorionic gonadotropin (hCG), or an exogenously induced LH surge using gonadotropin-releasing hormone (GnRH). Understanding this mechanism opened doors to novel ways to manipulate ovarian function and control time to estrus and ovulation. For example, gaining a greater understanding of these concepts led to the discovery of Ovsynch (Fig. 1.1 [6]).

Historically, four types of hormones have been utilized to manipulate ovarian function: GnRH type products including the direct impacts of LH or hCG, prostaglandin F2-alpha (PG), progesterone (P4), and pharmaceutical estrogens. The earliest technology derived from one of these hormones was the systematic use of PG to induce and schedule estrus in cattle [3,7]. This technology enhanced the percentage of cows detected in estrus and provided a management tool to organize labor more efficiently.

1.1.1.2 Importance of inducing a new follicular wave

The most important step in assuring success of timed-AI technologies is the induction of a new follicular wave. It is a critical step in manipulating ovarian development to control time of ovulation to allow for timed-AI. A new follicular wave can be pharmacologically induced following ovulation with an endogenous or exogenous surge of LH, estrogen-induced atresia of functional follicles, or ablation of functional follicles with ultrasound-guided aspiration [8].

Asynchrony of follicle development relative to luteolysis and induction of ovulation is likely to occur if a new follicular wave is not induced in timed-AI technologies. DFs can become atretic prior to PG induced luteolysis. This would result in an untimely new wave of follicles near the time of the final induction of ovulation. Thus follicles from this untimely new wave may lack sufficient maturity to respond to the final LH surge and ovulate. Delayed time to estrus and ovulation would occur 3–4 days later.

GnRH products induce an LH surge from the anterior pituitary. Functional DFs generally ovulate approximately 28 hours following an LH surge even in the presence of progesterone. An FSH surge occurs at the same time as the LH surge, and stimulates a new wave of follicular development. Numbers of follicles in this new wave are positively associated with circulating levels of anti-Müllerian hormone [9].

1.1.2 Ovsynch technologies

Ovsynch technologies utilize GnRH and PG in timely fashions to synchronize ovulation during a period of low progesterone imitating as closely as possible the physiological environment of a natural estrus [6]. Ovsynch technologies control follicle and CL development in order to synchronize ovulation amongst groups of cattle for fixed timed-AI [10,11]. Most countries have approved GnRH and PG products available for use. GnRH products are decapeptides that may or may not have a hydrophobic peptide for a longer half-life [12]. Most PG products utilized in Ovsynch technologies contain either cloprostenol sodium or dinoprost tromethamine. Ovsynch-based technologies generally include Ovsynch,

Double Ovsynch, G6G, and Presynch/Ovsynch protocols. Ovsynch may include the use of a progesterone-releasing device during the period between the GnRH and PG.

1.1.2.1   The basis for development of Ovsynch-based technologies

Embryonic survival continues to be a critical problem in lactating dairy cows, limiting thus profitability and sustainability of dairy farms [13]. The inability of the lactating dairy cow to foster complete embryonic and fetal development is escalating due to greater genetic pressure on milk production. Pregnancies per AI at 32 days following a detected estrus are approximately 30% in Holstein cows compared to 60% in virgin dairy heifers with similar genetic makeup [14].

Remarkable changes in follicular dynamics occur in dairy cattle during transition from nulliparous to primiparous and multiparous. Size of the ovulatory follicle, length of follicular dominance, number of double ovulations (twinning rates), and ovarian cysts [15] are greater in cows versus nulliparous heifers. Nulliparous heifers have approximately double the progesterone concentrations during mid-luteal phase of the estrous cycle compared to cows. The considerable decrease in fertility and increase in twinning rates transitioning from nulliparous heifer to cow negatively affects farm profit. "Fertility" programs (Fig. 1.2) were developed in the past 15 years to increase pregnancies per AI and reduce the chance for twinning compared to Ovsynch alone or natural estrus [16–18]. Fertility programs utilize pre-Ovsynch hormonal manipulations to increase levels of progesterone and manipulate the age and size of ovulatory follicles in cows treated with Ovsynch [19].

1.1.2.2   First gonadotropin-releasing hormone of Ovsynch

Fertility program technologies in dairy cows are designed to "presynchronize," so Ovsynch can be implemented on day 6 or 7 of the estrous cycle. Ovsynch is based on three treatments. The first treatment is GnRH. The intent of the GnRH-induced LH surge is to cause ovulation of the day 6 or 7 DF(s). Cows on day 6 or 7 of the estrous cycle have >95% chance of ovulating the DF. Ovulation of the DF induces the subsequent emergence of a new follicular wave ~1.5 days later [6] followed by the growth and development of both a new DF and an accessory CL during the next 7 days. The new DF has approximately a 95% chance of remaining functional during the 7 days leading up to the PG and then continuing on to ovulation following the final GnRH-induced LH surge [17].

Cows that have not been presynchronized with a fertility program and receive Ovsynch at a random stage of the estrous cycle have approximately a 30% chance to be in early stages of follicular development (first or second wave) when granulosa cells have not yet acquired sufficient LH receptors to respond to the GnRH-induced LH surge [5]. In this case, the first GnRH does not induce ovulation and the potential DF continues to grow, deviate from subordinates, and develop as a DF. This DF has approximately an 80% chance of remaining functional prior to the PG-induced luteolysis and subsequent increase in LH pulsatility that allows further development. If this DF remains functional until the time of PG, it will continue to develop into a pre-ovulatory follicle and has a 97% chance of ovulation following the final injection of GnRH [17]. However, this follicle could be as much as 12 days from emergence. Thus fertility of this follicle could be compromised due to the antral age of the follicle with an oocyte developing under insufficient levels of progesterone. Conversely, approximately 20% of these follicles become atretic prior to the PG. If this happens, a new wave will develop generally just prior to the PG and the new potential DF will emerge, deviate from subordinates, and become a DF. But this follicle will likely not have deviated from subordinate follicles and will not have acquired

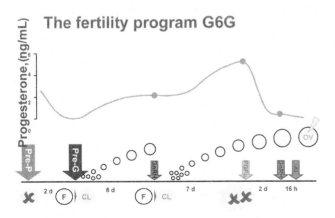

FIGURE 1.2 The fertility program G6G presynchronizes most cows to start Ovsynch on day 6 of the estrous cycle. Cows on day 6 of the cycle have a 95% chance of ovulating a first-wave DF, inducing an accessory CL, and starting a new follicular wave (Bello et al., 2006).

LH receptors prior to the final GnRH-induced LH surge. In this scenario, cows will likely have natural estrus that may or may not be detected 3−4 days after the timed-AI as the new DF develops under basal concentrations of progesterone into an ovulatory follicle. Conception rates from the timed-AI in this case would be near 0%. Thus it is critical to control ovulation of a DF in response to the first GnRH of Ovsynch to not only induce an *accessory CL* to increase P4 but to control the age of the DF to control ovulation to the final GnRH-induced LH surge and avoid the asynchrony of ovulation just described. To ensure that cows respond to the first GnRH-induced LH surge of Ovsynch, cows must be on day 6 or 7 of the estrous cycle [10]. This is the value of fertility programs that presynchronize cows to day 6 or 7 of the estrous cycle at the time of the first GnRH of Ovsynch.

1.1.2.3   Prostaglandin F2-alpha of Ovsynch

PG is administered to induce luteolysis, thus enabling the DF of the new follicular wave to develop into a preovulatory follicle. Unfortunately, lactating dairy cows have approximately an 80% chance of complete luteolysis prior to the final GnRH following a single dose of PG [20]. Even though cows have a high likelihood of responding to the final GnRH and ovulating, data indicate that chances of a pregnancy were only 5% in cows that do not decrease to <0.5 ng/mL P4 at the time of the final GnRH [21]. Studies indicate that a second PG 24 hours later resolves this problem [22].

When Ovsynch is initiated late in the estrous cycle, there is a high likelihood that CL may undergo natural luteolysis prior to the PG. If this happens, cows may have a natural estrus and ovulate early. In this case, if cows are not detected in estrus at this time, usually around the time of PG, conception rates are significantly less due to the asynchrony of AI and ovulation, that is, timed-AI may occur well after ovulation.

1.1.2.4   Final gonadotropin-releasing hormone of Ovsynch

This additional GnRH treatment is administered 48−60 hours after the PG to induce a preovulatory LH surge, trigger ovulation of the functional DF 24−32 hours later [6], and release the oocyte to be fertilized following AI. The chance of ovulation to this treatment is >95% even if luteolysis is not complete prior to this injection [20]. As mentioned above, cows can have a synchronized ovulation but still not have a chance to become pregnant due to incomplete or prolonged time to luteolysis.

1.1.3   Synchronization technologies for cows diagnosed not-pregnant, anovular, or cystic

Reducing the time between inseminations in cows diagnosed not-pregnant is essential to control herd calving interval. Cows that become pregnant later in lactation (>150 days in milk) are at greater risk for poorer body condition at the next calving, and subsequently more health issues and reduced fertility during the next lactation [23]. Resynchronization technologies can be initiated before or after pregnancy diagnosis [24,25]. Generally, these protocols are shorter and simpler to implement and maintain compliance compared with first-service programs. Pregnancies per AI achieved are often less than first-service programs. These technologies take into consideration the importance of limiting time between inseminations to ensure pregnancy occurs earlier in lactation.

The most common program for re-synchronization is Ovsynch (Fig. 1.1). But, programs can be implemented based on veterinarian diagnoses of ovarian status at time of a negative pregnancy diagnosis. Cows that have a CL at time of a negative pregnancy diagnosis should be treated with Ovsynch or Ovsynch plus an intra-vaginal progesterone-releasing insert. Cows that do not have CL, including cows with large cysts, should be administered the GnRH 7 days ahead of the start of the Ovsynch program. This would create a new CL and start a new follicular wave similar to fertility programs so that cows can start Ovsynch on day 6 or 7 of the new cycle. This program can be used as a treatment for anovular and cystic ovaries with acceptable pregnancy rates per AI. Pregnancy rates can be increased using fertility and resynchronization technologies if management can be highly accurate utilizing estrus detection prior to pregnancy diagnoses.

1.1.4   Detection of estrus technologies

Electronic technologies that assist management in detection of estrus are key tools that can improve reproductive management of herds. Many of these technologies have additional features that monitor other cow health parameters. These technologies are highly efficient at determining estrus. Most data report these technologies are effective at finding 70%

of eligible cows in estrus during a 21-day period [26]. These technologies are very helpful in determining the percentage of cows that are anovular (not cycling). This type of information can lead to making herd management changes that increase the percentage of cows that have ovulated and are cycling during the breeding period. The downsides to these technologies include the inability to induce cyclicity in anovular cows, and in addition they do not enhance fertility of cows when compared to fertility programs.

1.1.5   Technologies to evaluate pregnancy status

Accurate determination of pregnancy at the earliest possible time following AI or embryo transfer followed with re-examinations to detect losses are critical management interventions that are necessary to control calving intervals. Ultrasound technologies have been used for decades to determine pregnancy, but newer ultrasound technologies with color Doppler may become essential for predictions of pregnancy loss. Advantages of ultrasound technologies are the ability to determine a heartbeat, evaluate embryonic or fetal age and health, and evaluate ovarian function and uterine diseases in nonpregnant cattle. Other advantages occur when the veterinarian is present at time of pregnancy diagnosis and can evaluate the health status of the whole animal.

A newer technology that is highly accurate is the blood test for pregnancy-associated glycoproteins and pregnancy-specific protein-B [27]. These proteins are produced from binucleate cells of the trophectoderm of the conceptus. Detection of these proteins in blood is direct evidence of a recent viable pregnancy. These large proteins may remain elevated in circulation following embryonic or fetal death for 1−2 weeks allowing for false positives to occur in these animals. New ways to utilize this technology are being reported in the literature and hold promise for detection of pregnancy just following uterine attachment of the conceptus [28].

1.1.6   Embryo transfer technologies

Producers can make faster genetic progress and at the same time enhance reproductive measures utilizing embryo transfer technologies. Currently, embryos for transfer are produced through in vivo and in vitro means. In vivo embryos are collected following superovulation. In vitro produced embryos are made via oocytes collected from either oocyte pickup (OPU) or ovaries collected from abattoirs. Yet, follicle development during FSH stimulation of ovaries clearly influences outcomes from these technologies.

1.1.6.1   In vivo embryo production through follicle stimulating hormone-stimulated ovulations

In in vivo production of embryos, numbers of embryos collected following superovulation are variable and highly correlated, as expected, with the number of ovulations as measured from number of corpora lutea at time of embryo retrieval [29,30]. Thus embryo production is dependent upon the ovulation of a large cohort of mature FSH-stimulated follicles. FSH-stimulated follicles must grow to ovulatory size (approximately ≥ 10 mm in diameter) and be capable of responding to an LH surge to ensure ovulation of these follicles [31]. Initiating FSH treatment at the onset of a new follicular wave increases consistency of ovulation rates. This can be accomplished in a laboratory setting with ultrasound-guided follicle ablation of dominant and mid-sized follicles. Because the use of ultrasound-guided aspiration can be limited due to lack of equipment or logistics (time), superovulation protocols are generally initiated between day 8 and 12 of the estrous cycle to coincide with emergence of the second follicular wave. Unfortunately, the onset of the second follicular wave is quite variable in cows and can range from day 8 to 14 of the cycle. Thus the FSH treatment period could begin several days before or after the second follicular wave. In these cases, it is not clear how the DF or the largest follicles in a new follicular wave may affect the FSH-stimulated new wave of follicles or superovulation outcome. Data suggest that if FSH is administered at the time of a DF, a new wave of follicles will be induced in response to the FSH. This DF may stay functional after induced luteolysis, and if so may induce an LH surge before the FSH-induced follicles have sufficient time mature enough to acquire LH receptors and ovulate. Once an LH surge occurs, the growing but not yet "mature" follicles will become atretic (cell death). In this scenario, the cow or heifer would likely show signs of estrus following FSH stimulation but would have just one or two ovulations.

1.1.6.2   New ideas to increase ovulations in follicle stimulating hormone−treated cattle

Progesterone-releasing devices have been utilized in superovulation studies [29], but in most cases the removal of progesterone coincided with timing of induced luteolysis on the third day of FSH stimulation. Yet, maintaining a very low level of progesterone device in the vagina until the final treatment with FSH appears to prolong the time of the LH surge long enough to allow a greater number of superstimulated follicles to reach the ovulatory pool of follicles. In this case, the low-level progesterone device would allow more follicles to continue to grow, mature, and eventually ovulate due to the attenuation of an LH surge from subluteal circulating levels of progesterone. So, as in the previous example, if a DF is still functional after luteolysis, this follicle could not cause the positive feedback mechanism that induces an LH surge due to the presence of the low level of progesterone and would allow more follicles to continue to grow and reach the ovulatory pool of follicles prior to the removal of the progesterone device. It is important to note that circulating concentrations of progesterone must be low to have this effect. Only progesterone devices used for approximately 14 days that are cleaned and sanitized should be considered for use in this case [32].

1.1.7 Section 1—Summary

The reproductive technologies outlined in the section can completely control physiological processes in cattle to allow for fixed-time AI, greater fertility, timely pregnancy diagnoses, precise estrus information, and a greater chance for embryo production in FSH-stimulated donor cows. Control of reproductive processes allows for precise management of calving intervals and length of lactations of dairy cows. These management tools increase meat and milk production of cattle herds without increasing cattle numbers through greater control of reproduction.

Section 2: Bovine cloning and emerging reproductive technologies to improve cattle genomic parameters

1.1.8 Brief history of cattle cloning

Making genetically identical cattle using embryonic cells as donors has been possible since the 1970s [33]. This was done using embryo splitting, in which a fertilized embryo between the eight-cell and the late morula stage is split into two under a microscope [34]. The halved embryos are then placed in two emptied zona pellucida before transferring them to recipient cows. In 1987, the birth of cloned calves using nuclear transfer, in which a cell from a preimplantation embryo is introduced into an enucleated oocyte, was reported [33]. The potential to generate revenue from this new bovine assisted reproductive technique (ART) enticed investors to create companies based on such promise [35]. The idea was to produce multiple individuals from a single embryo, also called multigenerational embryo cloning [36,37]. In theory, one could take a 4- 8- or 16-cell fertilized embryo and generate 4, 8, or 16 cloned embryos that, in turn, could be re-cloned multiple times to generate an unlimited number of embryos [38]. In practice, though, this did not work as planned. The two main unaccounted roadblocks were (1) wide variation in the in vivo development of the clones depending on the generation, that is, second, third, or fourth round, and (2) the absence of a reliable cryopreservation protocol that would allow for proper storage and distribution of embryos [39,40]. Another roadblock not appreciated at the time that compromised the developmental potential of cloned embryos was our poor understanding of the need to coordinate the cell cycle of the donor nucleus with that of the oocyte [41].

Fast forward 10 years and Dolly, the cloned sheep, was born from a differentiated epithelial cell [42]. Soon after, multiple laboratories started working on translating the breakthrough into various species. In 1998, the birth of the first transgenic calves generated from fetal fibroblasts was published, and soon after that, Tsunoda's laboratory showed that it was feasible to clone cows from adult somatic cells [43,44].

1.1.9 Reprogramming somatic cells into embryos

In the field of SCNT, "reprogramming" is defined as the process of transforming the nucleus of a somatic cell into the nucleus of a preimplantation embryo. The factors that influence the efficiency of this man-made process can be classified as technical and biological ones.

The technical factors include the expertise of the person performing the procedure, the experimental protocol, and the tools and reagents used. We should also add the knowledge and skills of the farm crew managing the recipient heifers and performing embryo transfers.

Details of the most common protocols used to clone cows have been described as elsewhere [45–47]. For the purposes of this chapter, we will focus on the SCNT procedure that our laboratory has successfully implemented to generate cloned calves repeatedly [44,47–49].

We isolate dermal fibroblasts from a small, full-thickness skin biopsy of 3 mm in diameter taken from the donor animal. We culture the epidermis and dermis using fibroblast culture medium at 38.5°C in 5% CO_2 in air. Provided that the tissue is attached to the culture plate, somatic cells start to proliferate within a week. We usually first observe a mix of epithelial and fibroblast cells. However, after two to three passages, only dermal fibroblasts remain. At passage 4, we perform karyotypic analysis that confirms that they have normal ploidy and freeze and store the rest of the cells until needed. Approximately a week before the SCNT procedure is scheduled, we thaw the cells and culture them under the same culture conditions described above. The day that SCNT is done, fibroblasts are enzymatically dissociated and resuspended as single cells in culture media until used. This section of the protocol—the preparation of the cell before fusion/injection with the oocyte—varies between laboratories depending on the specific cell cycle stage the somatic cell is synchronized [50].

We isolate the recipient oocytes from ovaries recovered from the slaughterhouse. Immediately after we remove the oocytes from the reproductive tract, we rinse them with phosphate-buffered saline and maintain them at room temperature during transportation to the laboratory. Upon arrival, we aspirate the immature oocytes at the germinal vesicle stage with their granulosa cells from follicles that have a diameter between 2 and 8 mm [51]. We perform a brief selection of aspirated oocytes to eliminate those that do not have granulosa cells attached to them, otherwise called denuded [52]. Alternatively, others have reported using oocytes isolated in vivo from the ovaries using ovum pickup and matured in the laboratory [53]. Regardless of the source of oocytes, before we can use them for SCNT, they have to resume meiosis and reach metaphase II, the point at which an oocyte is ready to be fertilized. To accomplish this, we place the immature oocytes in media containing FSH and LH and aseptically incubate them at 38.5°C with 5% CO_2 in air for a period of 17−20 hours. We then take the oocytes with their expanded cumulus cells and place them in a solution of hyaluronidase, we vigorously pipette or vortex them until they are devoid of any somatic cell attached to its zone pellucida. If, after this process, we see some oocytes that still have granulosa cells attached to them, we assume they are not properly matured, and we discard them. Denuded—or "naked"—oocytes are placed in embryo culture medium at 38.5°C and stained with a fluorescent, nontoxic, DNA intercalator such as Hoechst or a similar compound [47]. For the process of oocyte enucleation and cell transfer, we use an inverted microscope with 20X objective. Brief exposures of ultraviolet light allow us to confirm that the eggs have extruded the second polar body and remove the maternal chromosomes arranged in metaphase using a glass needle. The same needle is subsequently used to deliver the somatic cells underneath the zona pellucida, in close contact with the oocyte's cell membrane [47].

We fuse the somatic cell with the oocyte by incubating the eggs with inactivated Sendai virus; alternatively, we use electrical current [54,55].

From this point forward, the eggs have a diploid nucleus (in G1 or G0 stage of the cell cycle) and a cytosol arrested at metaphase stage.

After a period of incubation of 2−3 hours, the somatic nucleus starts to remodel, and while the typical mitotic spindle of a cell at metaphase is not observed, the nuclear envelope disappears, and its chromatin condenses [56].

For the eggs to divide, though, we must induce the resumption of the first cell cycle like the sperm would do, but without letting the oocyte extrude the second polar body. We accomplish this by incubating the eggs in a HEPES-buffered solution containing ionomycin, a cell-permeable calcium ionophore, that triggers intracellular calcium release. Subsequently, we place the eggs for four hours into a solution containing compounds that can lower maturation-promoting factors. From this point onward, embryos are treated as conventional-fertilized embryos until they are transferred into recipient cows. Phenotypically, these embryos are indistinguishable from fertilized ones [47].

1.1.10 Biological factors of bovine cloning with somatic cells

One-cell embryos, produced by fertilization, are generated using sperm and eggs, two highly specialized cells that start their developmental process months in advance, during postimplantation embryonic development. Briefly, primordial germ cells (PGCs), originated in the embryonic epiblast, reach the genital ridge—developing gonads—when the fetus is 35 days old [57]. PGCs continue to divide, and start differentiation into oogonia in the female and spermatogonia in male. Final maturation into fertile eggs and sperm occurs, on average, at 9 months after birth, depending on factors such as breed and nutrition [58−61]. Throughout that time, the genomic DNA of the eggs and sperm undergo extensive epigenetic reprogramming, including the erasure and reestablishment of gene imprinting. Basically, it takes months to prepare the genome of a gamete to form a zygote that will develop into a healthy calf [62,63]. One-cell embryos, created by SCNT, have no oocyte's or sperm's genomic DNA. Instead, they have a nucleus carrying the epigenetic signature of a somatic cell. When taking into account the molecular processes required for gametogenesis, it is surprising that SCNT works at all.

Skin fibroblasts are easy to isolate and culture in vitro, making them the cell of choice for SCNT. Notwithstanding, a fibroblast's function is to provide structure to tissues, extracellular matrix, and participate in wound healing. It stands to reason that the epigenetic state of a fibroblast's DNA is not prepared to turn, in one cell division, into a fully functional zygote. However, this does happen. When data from multiple laboratories is compiled, the chances of obtaining a healthy clone calf hover between 10% and 15%. That is, for every 7%–10% recipient cows that received a cloned embryo, one calf will be born healthy. This efficiency pales in comparison to pregnancy rates by embryo transfer using in vitro fertilized embryos that can reach up to 60%, with a rate of healthy calves born healthy in the 40%–50% range, over the total number of cows transferred [64].

There are reasons to be optimistic about the prospects of having a bovine cloning efficiency that is the same as in vitro fertilization embryos. Recent experimental results in other mammalian species have provided a better understanding of the epigenetic changes that take place in the cloned embryos. Molecules responsible for a somatic nucleus "resistance to reprogramming" have been identified. These are histones and histone-residues accountable for conveying a somatic cell its phenotypic identity and provide an opportunity to manipulate these targets to increase cell plasticity.

Wakayama's laboratory conducted one of the first studies in 2006, where they demonstrated that the addition of Trichostatin A in the culture medium of mouse-cloned embryos for only 10 hours after oocyte activation increased the efficiency of production of cloned pups up to five times [65,66].

Trichostatin A (TSA) is a potent histone deacetylase inhibitor, targeting a large number of proteins, including H3K4, H3K5, and H3K9 histones [67–69]. Subsequent studies in bovine indicate that TSA, when used after cell fusion, and before the first embryo cleavage, improves embryo quality and survival of cloned calves [69,70].

Another relevant histone repressive mark, H3K9me3, was found to be responsible for the main failures of reprogramming in mouse-cloned embryos. Matoba and colleagues elegantly demonstrated this in 2014 [71]. They showed that the efficiency of mouse SCNT could be increased when mRNA for Kdm4d, the enzyme responsible for removing methyl residues from lysine 9, was injected in cloned embryos at the one-cell stage. These results have been replicated in vitro with human- and bovine-cloned embryos [72].

These are just a handful of studies that still require further validation by other laboratories. Nonetheless, encouraging data continue to pour in on new histone targets that when modified, SCNT efficiency is improved.

1.1.11 Need for more basic research to improve the efficiency of bovine somatic cell nuclear transfer

Synchronized recipient cows that received a cloned bovine embryo have a slightly lower pregnancy rate and a higher incidence of abortion [73]. Pregnancies are lost either in the first trimester or near parturition [73]. Incidences of perinatal deaths are also above normal [74–76]. The signs and symptoms present in abnormal calves are collectively called "large calf syndrome (LCS)," that was initially reported in embryos that were in vitro fertilized and cultured in the presence of serum or with coculture of cells [77].

A pregnant cow carrying a fetus that is developing LCS displays a progressively large abdominal cavity to the point that, in the most severe cases, compromises the life of the dam [73,78].

We still do not fully understand the pathogenesis of LCS. The working hypothesis is that it starts with a defective trophoblast differentiation in the early embryo, leading to a dysfunctional placenta, which in turn compromises placenta's blood circulation triggering edema, fluid retention, hydroallantois, and placental necrosis. Parts of the placenta that are still functional develop larger cotyledons to compensate for the ones dying. A dysfunctional placenta negatively impacts the fetal circulatory system, that is, the fetal heart must work harder to maintain normal blood circulation. There is an increase in fetal pulmonary blood pressure, which contributes to pulmonary edema, heart hypertrophia, and dilated cardiomyopathy [75,79]. Liver steatosis and enlarged kidney are reported as well [80].

Newborn cloned calves that had a faulty placenta during pregnancy are born with an abnormally sizeable umbilical cord that can reach up to three times its average diameter. Such overgrowth interferes with the physiological occlusion of umbilical blood vessels after parturition [73,80,81]. The recommendation is to surgically cut the cord and tie the vein and artery permanently.

Independent from placentation failures, some cloned fetuses are much larger than average. They show excessive bone development, forcing the leg joints to bend inside the constrained space of the uterus. Upon birth, cloned must be aided to stand on their own, and in the worst cases, euthanized.

Despite these associated pathologies observed in some cloned fetuses and calves, the vast majority of cloned calves born alive are healthy, requiring minimal support during the first days of their life [82]. Once they reach 3 months of age, they have a health- and life-span equivalent to that of animals produced by fertilization, a phenomenon also reported in sheep [80,83].

Producing clone cattle at a higher efficiency will: (1) lower the cost of making cloned calves, (2) mitigate some animal welfare concerns, (3) accelerate genetic progress, and (4) increase food security.

The economics of producing cloned calves at efficiencies of 10%−15% restricts the use of cloning to selected groups of breeders that sell genetically superior cows and bulls that are then used as founder animals. At the moment of this writing, the cost of cloning cattle is 20,000 American dollars per animal. For commercial farmers that would want to make genetic copies of animals for food production purposes, cloning is untenable.

There are valid concerns regarding the welfare of cloned calves and the recipient dam. The most recent evidence indicates that, while the efficiency of cloning remains low, there is a fewer incidence of hydrops in recipient cows and neonatal death of cloned calves [84]. Nonetheless, the origin of these problems remains poorly understood and may still occur, something that could be construed as an iatrogenic phenomenon to take into account before broadly deploying this technology [85,86].

Genetically modified (GM) organisms, such as plants and prokaryotic species, are an essential component of our food supply, especially in developing countries where food sources are scarce. There is still a need, however, to generate enough animal protein to feed the estimated 2 billion undernourished people by 2050, as stated in the United Nations Zero Hunger Challenge (www.zerohungerchallenge.org). In November 2019, AquAdvantage became the first GM animal—Atlantic salmon—to be approved by the Food and Drug Administration for human consumption. This is interpreted as a sign that regulatory agencies, and the public at large, are becoming more aware of the need for more aggressive strategies to cope with the expected food shortages. GM cattle as a source of food should be considered a valid option. Transgenic calves produced using SCNT were first reported in 1998. More recently, SCNT, coupled with precise gene-editing tools, was used to generate prion-free and polled calves [87,88]. It is reasonable to envision genetically engineered cattle that have increased production traits without compromising the welfare of the animal and that can be raised in an environmentally sustainable manner.

1.1.12   Emerging assisted reproductive techniques in cattle

The broad implementation of reliable genomic markers for the selection of genetically superior animals is the main factor that contributed to the improvement of production traits over the last decade. In the past, the genetic merit of an animal—mostly males—was determined when the production traits of its progeny were measured, a process that would take 4−5 years. With the implementation of genetic selection based on DNA markers, the genomic estimated breeding value (GEBV) of an animal can be determined at birth, shortening the generation interval by 3 years [89,90]. The critical issue is to rapidly generate new genotypes from founders with the desired GEBV. In practice, the selection of an individual that has high GEBV can be made as soon as diploid cells are isolated from a calf, a fetus, or an embryo. Once these cells are determined to have a superior genotype, it is possible to produce a new founder animal by SCNT, shortening the generation interval even further. Nonetheless, at present, we still need a whole animal to produce the gametes for a calf, fetus, or embryo, to start the process, for now that is.

Experiments in mice have demonstrated that functional gametes can be obtained by differentiating embryonic and PGCs into sperm and eggs, embryos such produced, when transferred to a recipient female, have generated viable offspring [91−95]. In vitro produced bovine gametes, coupled with SCNT, could have an extraordinary impact on bovine genetic selection. Producers would have, at their disposal, a method to create unique genomes over multiple generations without the need to make an animal.

1.1.13   Section 2—Summary

Bovine cloning is an ART currently used to replicate the genome of genetically superior animals. At present, cloned cattle are used as "gamete donors" for the generation of fertilized embryos. The entry barrier for commercial producers is its cost, a consequence of the low efficiency of the procedure as measured by the number of healthy clones born over the number of recipient cows transferred. Since first reported two decades ago, the efficiency of cloning cattle using somatic cells have been in the low teens. Recent reports describing the molecular mechanisms of cellular reprogramming have unveiled new genes and gene products that can increase the plasticity of the donor cells. These new findings give credence to the notion that shortly, embryos produced using SCNT could reach the same developmental competence as those generated using in vitro fertilization, allowing for the use of bovine cloning in commercial herds.

References

[1] Pierson RA, Ginther OJ. Ultrasonography of the bovine ovary. Theriogenology 1984;21:495−504. Available from: https://doi.org/10.1016/0093-691x(84)90411-4.

[2] Ireland JL, Ireland JJ. Changes in expression of inhibin/activin a, bA and bB subunit messenger ribonucleic acids following increases in size and during different stages of differentiation or atresia of non-ovulatory follicles in cows. Biol Reprod 1994;50:492.

[3] King BF, Britt JH, Esbenshade KL, Flowers WL, Ireland JJ. Evidence for a local role of inhibin or inhibin subunits in compensatory ovarian hypertrophy. Reproduction 1995;104:291−5. Available from: https://doi.org/10.1530/jrf.0.1040291.

[4] Ireland JJ, Fogwell RL, Oxender WD, Ames K, Cowley JL. Production of estradiol by each ovary during the estrous cycle of cows. J Anim Sci 1984;59:764−71. Available from: https://doi.org/10.2527/jas1984.593764x.

[5] Xu Z, Garverick HA, Smith GW, Smith MF, Hamilton SA, Youngquist RS. Expression of follicle-stimulating hormone and luteinizing hormone receptor messenger ribonucleic acids in bovine follicles during the first follicular wave. Biol Reprod 1995;53:951−7. Available from: https://doi.org/10.1095/biolreprod53.4.951.

[6] Pursley JR, Mee MO, Wiltbank MC. Synchronization of ovulation in dairy cows using PGF2alpha and GnRH. Theriogenology 1995;44:915−23. Available from: https://doi.org/10.1016/0093-691x(95)00279-h.

[7] Momont HW, Sequin BE. Prostaglandin therapy and the postpartum cow. In: the bovine proceedings; 1985.

[8] Silcox RW, Powell KL, Kiser TE. Ability of dominant follicles to respond to exogenous GnRH administration is dependent on their stage of development. J Anim Sci 1993;71:219.

[9] Ireland JL, Scheetz D, Jimenez-Krassel F, Themmen AP, Ward F, Lonergan P, et al. Antral follicle count reliably predicts number of morphologically healthy oocytes and follicles in ovaries of young adult cattle. Biol Reprod 2008;79(6):1219−25.

[10] Pursley JR, Kosorok MR, Wiltbank MC. Reproductive management of lactating dairy cows using synchronization of ovulation. J Dairy Sci 1997;80:301−6. Available from: https://doi.org/10.3168/jds.s0022-0302(97)75938-1.

[11] Pursley JR, Silcox RW, Wiltbank MC. Effect of time of artificial insemination on pregnancy rates, calving rates, pregnancy loss, and gender ratio after synchronization of ovulation in lactating dairy cows. J Dairy Sci 1998;81:2139−44. Available from: https://doi.org/10.3168/jds.s0022-0302(98)75790-x.

[12] Twagiramungu H, Guilbault LA, Proulx J, Dufour JJ. Synchronization of estrus and fertility in beef cattle with two injections of buserelin and prostaglandin. Theriogenology 1992;38:1131−44. Available from: https://doi.org/10.1016/0093-691x(92)90126-c.

[13] Martins JPN, Wang D, Mu N, Rossi GF, Martini AP, Martins VR, et al. Level of circulating concentrations of progesterone during ovulatory follicle development affects timing of pregnancy loss in lactating dairy cows. J Dairy Sci 2018;101:10505−25. Available from: https://doi.org/10.3168/jds.2018-14410.

[14] Sartori R, Haughian JM, Shaver RD, Rosa GJM, Wiltbank MC. Comparison of ovarian function and circulating steroids in estrous cycles of Holstein heifers and lactating cows. J Dairy Sci 2004;87:905−70. Available from: https://doi.org/10.3168/jds.s0022-0302(04)73235-x.

[15] Kesler DJ, Garverick HA. Ovarian cysts in dairy cattle: a review. J Anim Sci 1982;55:1147−59. Available from: https://doi.org/10.2527/jas1982.5551147x.

[16] Moreira F, Orlandi C, Risco CA, Mattos R, Lopes F, et al. Effects of presynchronization and bovine somatotropin on pregnancy rates to a timed artificial insemination protocol in lactating dairy cows. J Dairy Sci 2001;84:1646−59. Available from: https://doi.org/10.3168/jds.s0022-0302(01)74600-0.

[17] Bello NM, Steibel JP, Pursley JR. Optimizing ovulation to first GnRH improved outcomes to each hormonal injection of ovsynch in lactating dairy cows. J Dairy Sci 2006;89:3413−24. Available from: https://doi.org/10.3168/jds.s0022-0302(06)72378-5.

[18] Souza AH, Ayres H, Ferreira RM, Wiltbank MC. A new presynchronization system (Double-Ovsynch) increases fertility at first postpartum timed AI in lactating dairy cows. Theriogenology 2008;70:208−15. Available from: https://doi.org/10.1016/j.theriogenology.2008.03.014.

[19] Pursley JR, Martins JPN. Impact of circulating concentrations of progesterone and antral age of the ovulatory follicle on fertility of high-producing lactating dairy cows. Reprod Fertil Dev 2012;24:267−71. Available from: https://doi.org/10.1071/rd11917.

[20] Martins JPN, Policelli RK, Neuder LM, Raphael W, Pursley JR. Effects of cloprostenol sodium at final prostaglandin F2alpha of Ovsynch on complete luteolysis and pregnancy per artificial insemination in lactating dairy cows. J Dairy Sci 2011;94:2815−24. Available from: https://doi.org/10.3168/jds.2010-3652.

[21] Martins JPN, Policelli RK, Pursley JR. Luteolytic effects of cloprostenol sodium in lactating dairy cows treated with G6G/Ovsynch. J Dairy Sci 2011;94:2806−14.

[22] Brusveen DJ, Souza AH, Wiltbank MC. Effects of additional prostaglandin F2alpha and estradiol-17beta during Ovsynch in lactating dairy cows. J Dairy Sci 2009;92:1412−22. Available from: https://doi.org/10.3168/jds.2008-1289.

[23] Middleton EL, Minela T, Pursley JR. The high-fertility cycle: how timely pregnancies in one lactation may lead to less body condition loss, fewer health issues, greater fertility, and reduced early pregnancy losses in the next lactation. J Dairy Sci 2019;102:5577−87. Available from: https://doi.org/10.3168/jds.2018-15828.

[24] Fricke PM, Caraviello DZ, Weigel KA, Welle ML. Fertility of dairy cows after resynchronization of ovulation at three intervals following first timed insemination. J Dairy Sci 2003;86:3941−50. Available from: https://doi.org/10.3168/jds.s0022-0302(03)74003-x.

[25] Giordano JO, Wiltbank MC, Guenther JN, Pawlisch R, Bas S, Cunha AP, et al. Increased fertility in lactating dairy cows resynchronized with Double-Ovsynch compared with Ovsynch initiated 32 d after timed artificial insemination. J Dairy Sci 2012;95:639−53. Available from: https://doi.org/10.3168/jds.2011-4418.

[26] Valenza A, Giordano JO, Lopes G, Vincenti L, Amundson MC, Fricke PM. Assessment of an accelerometer system for detection of estrus and treatment with gonadotropin-releasing hormone at the time of insemination in lactating dairy cows. J Dairy Sci 2012;95:7115−27. Available from: https://doi.org/10.3168/jds.2012-5639.

[27] Sasser RG. Detection of pregnancy by radioimmunoassay of a novel pregnancy-specific protein in serum of cows and a profile of serum concentrations during gestation. Biol Reprod 1986;35:936−42. Available from: https://doi.org/10.1095/biolreprod35.4.936.

[28] Middleton EL, Pursley JR. Short communication: blood samples before and after embryonic attachment accurately determine non-pregnant lactating dairy cows at 24 d post-artificial insemination using a commercially available assay for pregnancy-specific protein B. J Dairy Sci 2019;102:7570−5. Available from: https://doi.org/10.3168/jds.2018-15961.

[29] Bo GA, Mapletoft RJ. Historical perspectives and recent research on superovulation in cattle. Theriogenology 2014;81:38−48. Available from: https://doi.org/10.1016/j.theriogenology.2013.09.020.

[30] Sartori R, Suárez-Fernández CA, Monson RL, Guenther JN, Rosa GJM, et al. Improvement in recovery of embryos/ova using a shallow uterine horn flushing technique in superovulated Holstein heifers. Theriogenology 2003;60:1319−30.

[31] Sartori R, Fricke PM, Ferreira JCP, Ginther OJ, Wiltbank MC. Follicular deviation and acquisition of ovulatory capacity in bovine follicles. Biol Reprod 2001;65:1403−9. Available from: https://doi.org/10.1095/biolreprod65.5.1403.

[32] Channa AA, Martins JPN, Jimenez-Krassel F, Pursley JR. Effect of immersion of CIDR in ethanol on steady state circulating concentrations of progesterone and follicular dynamics in Holstein heifers and non-lactating cows. Livest Sci 2017;198:191−4.

[33] Prather RS, et al. Nuclear transplantation in the bovine embryo: assessment of donor nuclei and recipient oocyte. Biol Reprod 1987;37:859−66.

[34] Moore SG, Hasler JF. A 100-year review: reproductive technologies in dairy science. J Dairy Sci 2017;100:10314−31.

[35] Marx J. Cloning sheep and cattle embryos. Science 1988;239:463−4.

[36] Takano H, Kozai C, Shimizu S, Kato Y, Tsunoda Y. Cloning of bovine embryos by multiple nuclear transfer. Theriogenology 1997;47:1365−73.

[37] Stice SL, Keefer CL. Multiple generational bovine embryo cloning. Biol Reprod 1993;48:715−19.

[38] Ectors FJ, et al. Viability of cloned bovine embryos after one or two cycles of nuclear transfer and in vitro culture. Theriogenology 1995;44:925−33.

[39] Seidel GE. Production of genetically identical sets of mammals: cloning? J Exp Zool 1983;228:347−54.

[40] Lehn-Jensen H, Willadsen SM. Deep-freezing of cow 'half' and 'quarter' embryos. Theriogenology 1983;19:49−54.

[41] Campbell KH, Loi P, Otaegui PJ, Wilmut I. Cell cycle co-ordination in embryo cloning by nuclear transfer. Rev Reprod 1996;1:40−6.

[42] Wilmut I, Schnieke AE, McWhir J, Kind AJ, Campbell KH. Viable offspring derived from fetal and adult mammalian cells. Nature 1997;385:810−13.

[43] Kato Y, et al. Eight calves cloned from somatic cells of a single adult. Science 1998;282:2095−8.

[44] Cibelli JB, et al. Transgenic bovine chimeric offspring produced from somatic cell-derived stem-like cells. Nat Biotechnol 1998;16:642−6.

[45] Cibelli JB, et al. Cloned transgenic calves produced from nonquiescent fetal fibroblasts. Science 1998;280:1256−8.

[46] Lonergan P, Fair T. Maturation of oocytes in vitro. Annu Rev Anim Biosci 2016;4:255−68.

[47] Ross PJ, Cibelli JB. Bovine somatic cell nuclear transfer. Methods Mol Biol 2010;636:155−77.

[48] Lanza RP, et al. Cloned cattle can be healthy and normal. Science 2001;294:1893−4.

[49] Lanza RP, et al. Extension of cell life-span and telomere length in animals cloned from senescent somatic cells. Science 2000;288:665−9.

[50] Akagi S, Matsukawa K, Takahashi S. Factors affecting the development of somatic cell nuclear transfer embryos in cattle. J Reprod Dev 2014;60:329−35.

[51] Matos DG, Furnus CC, Moses DF, Baldassarre H. Effect of cysteamine on glutathione level and developmental capacity of bovine oocyte matured in vitro. Mol Reprod Dev 1995;42:432−6.

[52] Lonergan P, Monaghan P, Rizos D, Boland MP, Gordon I. Effect of follicle size on bovine oocyte quality and developmental competence following maturation, fertilization, and culture in vitro. Mol Reprod Dev 1994;37:48−53.

[53] Brüggerhoff K, et al. Bovine somatic cell nuclear transfer using recipient oocytes recovered by ovum pick-up: effect of maternal lineage of oocyte donors. Biol Reprod 2002;66:367−73.

[54] Galli C, Lagutina I, Vassiliev I, Duchi R, Lazzari G. Comparison of microinjection (piezo-electric) and cell fusion for nuclear transfer success with different cell types in cattle. Cloning Stem Cells 2002;4:189−96.

[55] Song BS, et al. Inactivated Sendai-virus-mediated fusion improves early development of cloned bovine embryos by avoiding endoplasmic-reticulum-stress-associated apoptosis. Reprod Fertil Dev 2011;23:826−36.

[56] Choi J, Kim C, Park C, Yang B, Cheong H. Effect of activation time on the nuclear remodeling and in vitro development of nuclear transfer embryos derived from bovine somatic cells. Mol Reprod Dev 2004;69:289−95.

[57] Cherny R, et al. Strategies for the isolation and characterization of bovine embryonic stem cells. Reprod Fertil Dev 1994;6:569−75.

[58] Desjardins C, Hafs HD. Maturation of bovine female genitalia from birth through puberty. J Anim Sci 1969;28:502−7.

[59] Wolfe MW, et al. Effect of selection for growth traits on age and weight at puberty in bovine females. J Anim Sci 1990;68:1595−602.

[60] Salamone DF, Damiani P, Fissore RA, Robl JM, Duby RT. Biochemical and developmental evidence that ooplasmic maturation of prepubertal bovine oocytes is compromised. Biol Reprod 2001;64:1761−8.

[61] Presicce GA, et al. Age and hormonal dependence of acquisition of oocyte competence for embryogenesis in prepubertal calves. Biol Reprod 1997;56:386−92.

[62] Kaneda M, Akagi S, Watanabe S, Nagai T. Comparison of DNA methylation levels of repetitive loci during bovine development. BMC Proc 2011;5:S3.

[63] Li E. Chromatin modification and epigenetic reprogramming in mammalian development. Nat Rev Genet 2002;3:662–73.

[64] Bousquet D, et al. In vitro embryo production in the cow: an effective alternative to the conventional embryo production approach. Theriogenology 1999;51:59–70.

[65] Kishigami S, et al. Significant improvement of mouse cloning technique by treatment with trichostatin A after somatic nuclear transfer. Biochem Biophys Res Commun 2006;340:183–9.

[66] Tsuji N, Kobayashi M, Nagashima K, Wakisaka Y, Koizumi K. A new antifungal antibiotic, trichostatin. J Antibiot. 2006;29:1–6.

[67] Bui HT, et al. Effect of trichostatin A on chromatin remodeling, histone modifications, DNA replication, and transcriptional activity in cloned mouse embryos. Biol Reprod 2010;83:454–63.

[68] Lee MJ, et al. Trichostatin A promotes the development of bovine somatic cell nuclear transfer embryos. J Reprod Dev 2011;57:34–42.

[69] Iager AE, et al. Trichostatin A improves histone acetylation in bovine somatic cell nuclear transfer early embryos. Cloning Stem Cells 2008;10:371–80.

[70] Wang YS, et al. Production of cloned calves by combination treatment of both donor cells and early cloned embryos with 5-aza-2/-deoxycytidine and trichostatin A. Theriogenology 2011;75:819–25.

[71] Matoba S, et al. Embryonic development following somatic cell nuclear transfer impeded by persisting histone methylation. Cell 2014;159:884–95.

[72] Zhou C, et al. H3K27me3 is an epigenetic barrier while KDM6A overexpression improves nuclear reprogramming efficiency. FASEB J 2019;33:4638–52.

[73] Chavatte-Palmer P, et al. Review: placental perturbations induce the developmental abnormalities often observed in bovine somatic cell nuclear transfer. Placenta 2012;33:S99–104.

[74] Batchelder CA, et al. Perinatal physiology in cloned and normal calves physical and clinical characteristics. Cloning Stem Cells 2007;9:63–82.

[75] Hill JR, Edwards JF, Sawyer N, Blackwell C, Cibelli JB. Placental anomalies in a viable cloned calf. Cloning 2001;3:83–8.

[76] Hill JR, et al. Clinical and pathologic features of cloned transgenic calves and fetuses (13 case studies). Theriogenology 1999;51:1451–65.

[77] Behboodi E, et al. Birth of large calves that developed from in vitro-derived bovine embryos. Theriogenology 1995;44:227–32.

[78] Heyman Y, et al. Frequency and occurrence of late-gestation losses from cattle cloned embryos. Biol Reprod 2002;66:6–13.

[79] Hill JR, et al. Clinical and pathologic features of cloned transgenic calves and fetuses (13 case studies). Theriogenology 1999;51:1451–65.

[80] Chavatte-Palmer P, et al. Review: health status of cloned cattle at different ages. Cloning Stem Cells 2004;6:94–100.

[81] Kohan-Ghadr HR, et al. Ultrasonographic and histological characterization of the placenta of somatic nuclear transfer-derived pregnancies in dairy cattle. Theriogenology 2008;69:218–30.

[82] Cibelli JB, Campbell KH, Seidel GE, West MD, Lanza RP. The health profile of cloned animals. Nat Biotechnol 2002;20:13–14.

[83] Sinclair KD, et al. Healthy ageing of cloned sheep. Nat Commun 2016;7:12359.

[84] Duranthon V, Chavatte-Palmer P. Long term effects of ART: what do animals tell us? Mol Reprod Dev 2018;85:348–68.

[85] Pascalev AK. We and they: Animal welfare in the era of advanced agricultural biotechnology. Livest Sci 2006;103:208–20.

[86] Ormandy EH, Dale J, Griffin G. Genetic engineering of animals: ethical issues, including welfare concerns. Can Vet J 2011;52:544–50.

[87] Richt JA, et al. Production of cattle lacking prion protein. Nat Biotechnol 2007;25:132–8.

[88] Carlson DF, et al. Production of hornless dairy cattle from genome-edited cell lines. Nat Biotechnol 2016;34:479–81.

[89] Hayes BJ, Bowman PJ, Chamberlain AJ, Goddard ME. Invited review: genomic selection in dairy cattle: progress and challenges. J Dairy Sci 2009;92:433–43.

[90] Schaeffer LR. Strategy for applying genome-wide selection in dairy cattle. J Anim Breed Genet 2006;123:218–23.

[91] Sato T, Katagiri K, Kubota Y, Ogawa T. In vitro sperm production from mouse spermatogonial stem cell lines using an organ culture method. Nat Protoc 2013;8:2098–104.

[92] Hübner K, et al. Derivation of oocytes from mouse embryonic stem cells. Science 2003;300:1251–6.

[93] Toyooka Y, Tsunekawa N, Akasu R, Noce T. Embryonic stem cells can form germ cells in vitro. Proc Natl Acad Sci USA 2003;100:11457–62.

[94] Thomas CD, et al. Derivation of embryonic germ cells and male gametes from embryonic stem cells. Nature 2004;427:148–54.

[95] Sato T, et al. In vitro production of fertile sperm from murine spermatogonial stem cell lines. Nat Commun 2011;2:472.

Chapter 2

Assisted reproductive biotechnologies in the horse

Marina Sansinena[1,2,3,4]

[1]Laboratory of Animal Reproduction and Biotechnology, School of Engineering and Agricultural Sciences, Universidad Católica Argentina, Buenos Aires, Argentina, [2]National Research and Technology Council, CONICET, Buenos Aires, Argentina, [3]Department of Veterinary Clinical Sciences, School of Veterinary Medicine, Louisiana State University, Baton Rouge, LA, United States, [4]College of Agricultural and Veterinary Sciences, Universidad del Salvador, Buenos Aires, Argentina

2.1   Introduction

Assisted reproductive technologies (ARTs) have been defined as the group of procedures aimed at the treatment of infertility. A Word Health Organization's recent update described ARTs as all treatments or procedures that include the in vitro handling of oocytes, sperm, and embryos for the purpose of establishing a pregnancy. This includes in vitro fertilization (IVF) and embryo transfer (ET), gamete intrafallopian transfer, zygote intrafallopian transfer, tubal ET, gamete and embryo cryopreservation, oocyte and embryo donation, and gestational surrogacy. In humans, the current definition of ARTs does not include artificial insemination (AI) [1]. Similarly, animal reproductive biotechnologies have been collectively described as the various procedures and techniques involving the laboratory handling of gametes and embryos [2].

According to a review conducted by Allen in 2005 [3], the first undocumented application of an assisted reproductive biotechnology in the horse can be traced to the year 1322 when an Arab chief attempted to obtain the genetics of a neighboring stallion by recovering semen from the vagina of a mated mare, diluting it in camel's milk, and then using it to inseminate several of his own mares. It is believed that this early attempt of AI was rewarded with the birth of a live foal [4]. Since then, many biotechnologies have been applied successfully in the horse, although it is important to distinguish between valuable, proof-of-concept achievements from repeatable procedures that can be adapted to commercial settings.

2.2   Oocyte recovery for assisted reproductive biotechnologies

Fertilization is defined as the process of the union between the male and female gametes. In the mare, as in all domestic species, fertilization occurs in the ampulla of the oviduct and involves a complex cascade of events that conclude with the union of the male and female pronuclei (syngamy), chromosomal recombination, and formation of a zygote [5]. Some in vitro reproductive technologies such as IVF require the exteriorization of female gametes in a way that cells retain their viability and future developmental competence. In mares, this oocyte recovery technique must be repeatable and should not jeopardize health and future fertility of the donor.

2.2.1   Oocyte recovery from live mares

The method of ovum or oocyte recovery was initially developed in humans using a laparoscopic technique [6,7]; however, this method is considered traumatic by today's standards because it requires patients to undergo general anesthesia and endotracheal intubation [8]. The first report of a new, less invasive, ultrasound-guided approach to oocyte collection was reported in Scandinavia by Lentz et al. in 1981 [9], followed by Dellenbach et al. in 1985 [10] and Wikland et al. in 1985 [11]. This novel technique quickly replaced the laparoscopic procedure, revolutionizing the field of reproductive medicine and becoming the basis for human IVF to date. The method, which only requires mild sedation and analgesia, relies on the visualization of ovarian follicles by ultrasonography through the vaginal wall, penetration of

Reproductive Technologies in Animals. DOI: https://doi.org/10.1016/B978-0-12-817107-3.00002-3

follicles with an adapted ultrasound grip containing a needle guide, and the application of negative pressure within the punctured follicle resulting in the aspiration of the follicular fluid and the extraction of the oocyte.

In the mare, the first successful ultrasound-guided follicle aspiration or ovum pickup (OPU) was reported by Brück in 1992 [12], although it took over a decade for the technique to gain enough repeatability and reliability to be applied to the commercial practice. Several factors have contributed to the delayed demand for OPU in equine clinical settings; however, the more consistent results obtained in the last 5 years combined with the treatment of mare infertility have renewed the interest in oocyte recovery from donor mares. The technique has now evolved to become increasingly common in equine veterinary practice, mainly in combination with the production of equine embryos by intracytoplasmic sperm injection (ICSI) or for the treatment of infertility of mares with oviductal pathologies. OPU has also been used in recent years for the recovery of oocytes for somatic cell nuclear transfer (SCNT) or cloning, although this technique relies mostly on abattoir-derived equine oocytes as the main source of cytoplast for the transfer of a donor cell.

In domestic animals, the procedure of OPU was first described in cattle by Pieterse et al. in 1988 [13] at the University of Utrecht, The Netherlands. The system is a combination of three main components: (1) an ultrasound and appropriate transducer with the correct depth penetration for the visualization of ovarian follicles through the vaginal wall, (2) an adapted grip for the transducer and needle guide system, and (3) an aspiration pump (Fig. 2.1).

The grip that contains the transducer and the needle guide is frequently constructed of a hard polymer in two specular, fitting halves and should be compatible with the ultrasound and transducer combination to be utilized by the operator (Figs. 2.2 and 2.3). It should have a length and weight that will permit the transducer to be manipulated vaginally in close contact with the ovaries. In addition to the transducer, the grip will accept an echogenic-tip needle (sometimes accompanied by a needle guide) that will be used to puncture the follicle once visualized. This needle is connected to a vacuum pump using sterile Teflon or silicon tubing. When the pump is activated, it produces a negative pressure at the needle tip and the aspirated content is collected into a sterile container.

The mare's unique histological features of the follicular wall in relation to the cumulus-oocyte complex (COC) attachment needed some adaptations from the original techniques described for humans and cattle. In humans, cattle, and other domestic species, the COCs are loosely attached to the theca layer and the application of negative pressure or vacuum is sufficient for removal. The equine attachment within the follicle is characterized by a firm and persistent connection to the thecal wall; the strength of this attachment is more evident in immature oocytes that have yet to undergo resumption of meiosis and expansion of cumulus cells in response to gonadotropins [14]. Therefore, after numerous unsuccessful attempts, it became evident that vacuum pressure alone was not sufficient for the extraction of the equine oocyte. Finally, a combined approach was reported for the mare in which follicles were alternatively collapsed and perfused with isotonic aspiration medium. By gently massaging the follicle *per rectum*, repeating the aspiration/perfusion cycle, and scrapping the follicular wall with the aspiration needle cause the oocyte to eventually dislodge from the follicular wall and result in a successful recovery [15].

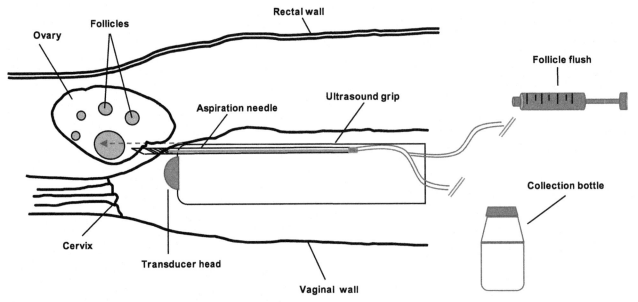

FIGURE 2.1 Schematic representation of OPU setup and components.

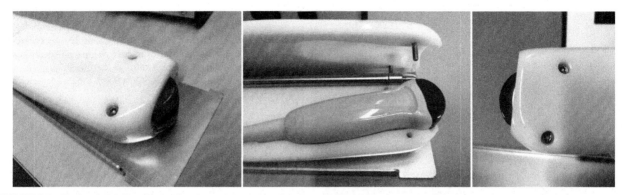

FIGURE 2.2   Equine OPU grip with details of transducer guide, showing two fitting halves and microconvex transducer head. *Photograph: Courtesy Universidad del Salvador, School of Veterinary Medicine, Buenos Aires, Argentina.*

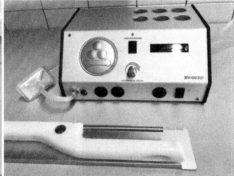

FIGURE 2.3 Equine OPU components featuring ultrasound and microconvex transducer, grip, and aspiration pump. *Photograph: Courtesy Universidad del Salvador, School of Veterinary Medicine, Buenos Aires, Argentina.*

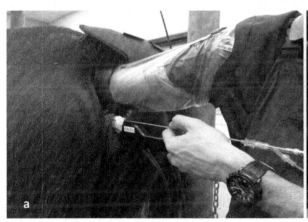
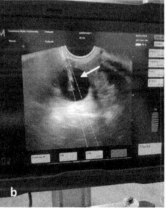

FIGURE 2.4 (A) OPU aspiration needle insertion into guide and (B) close-up of ultrasound monitor, arrow showing a punctured follicle, echogenic needle tip, and software needle-assist tracking lines. *Photograph: Courtesy C. Pinto, Louisiana State University School of Veterinary Medicine, Department of Veterinary Clinical Sciences.*

For the procedure, the OPU handle is inserted into the vagina and oriented to the left or right of the cervix, while the operator simultaneously manipulates the ovary per rectum to ensure close contact and good echogenic visualization of follicles (Fig. 2.4A). Nowadays, the software on most modern ultrasounds has the capability of producing a reference line, which should coincide with the direction of the needle penetration; this feature aids in the puncture of the follicle (Fig. 2.4B). Once aligned, the operator advances the needle through the vaginal wall and the aspiration vacuum is applied using a foot pedal. For best results, a 60 cm, 12-g double lumen needle for simultaneous aspiration/perfusion is recommended for the mare, although aspirations with single lumen, disposable needles have been reported. The needle's inner inlet, connected to the vacuum pump and sterile collection vessel, is used for aspiration at a suction rate of approximately 20 mL/minute; this rate must be carefully monitored and adjusted according to oocyte maturation status and individual mare characteristics, as it can result in undesired oocyte damage and denudation of the cumulus investments [15]. The follicle is then flushed with aspiration or commercial ET medium, sometimes supplemented with

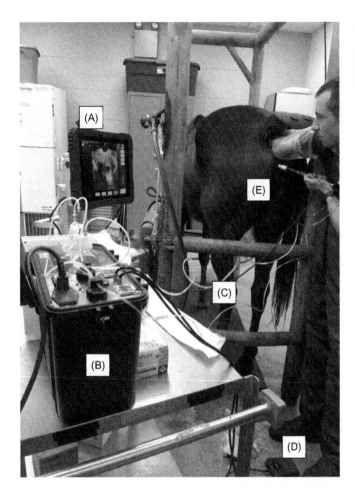

FIGURE 2.5 OPU procedure, indicating system components: (a) ultrasound; (b) aspiration pump (Minitube, dual aspiration/flush system) with heated collection and flush medium bottles; (c) aspiration/flush lines; (d) foot pedal; and (e) transducer grip and needle guide. *Photograph: Courtesy C. Pinto, Louisiana State University School of Veterinary Medicine, Department of Veterinary Clinical Sciences.*

heparin in order to reduce blood clot formation and facilitate isolation of oocytes. The components of the OPU system are shown in Fig. 2.5.

There are basically two approaches for the recovery of oocytes in the mare: the aspiration of multiple, immature oocytes from small and medium-sized follicles, or the aspiration of a single mature oocyte from a large, preovulatory follicle (Fig. 2.6); both approaches have been successfully applied to clinical settings.

There are several advantages to the aspiration of immature oocytes from small to medium-sized follicles. Mares undergoing these aspirations do not require such intense monitoring of follicular growth, as opposed to those mares with a human chorionic gonadotropin–stimulated, preovulatory follicle. Also, because it is more common to find several small to medium-sized follicles in one individual mare, the probability of recovering oocytes is increased. In addition, aspiration of small to medium-sized follicles is adaptable to working with several mares in a single OPU session. Recent reports indicate that, with an established team and system, an average recovery rate of 50%–70% from immature follicles can be expected [16–20], although individual OPU variability can range between 20% and 100% [15]. On the other hand, oocytes recovered from these types of follicles are often immature and require additional in vitro maturation (IVM) in gonadotropin-enriched medium in order to reach metaphase-II stage and be able to be fertilized [15].

Harvesting in vivo–matured oocytes from a preovulatory follicle has also been used in clinical programs, particularly for the recovery of oocytes from infertile or subfertile mares and their transfer into the oviduct of a previously inseminated recipient. High oocyte recovery rates (>70%) have been reported from large preovulatory follicles [21,22]; partly because these oocytes have already been exposed to gonadotropins within the follicle, are surrounded by expanded, hyaluronan-rich cumulus cells and are mostly detached from the follicular wall. These in vivo–matured oocytes are normally recovered at metaphase-II stage and do not require additional IVM. Consequently, they are considered to have better cytoplasmic maturity and developmental potential than immature oocytes. The main downside of recovering oocytes from large follicles is that mares have to be continuously monitored and normally only one of these follicles will be present in the ovary per cycle.

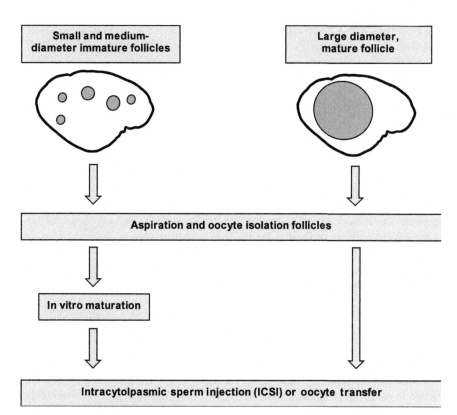

FIGURE 2.6 Schematic approach for oocyte collection and maturation from different follicular populations.

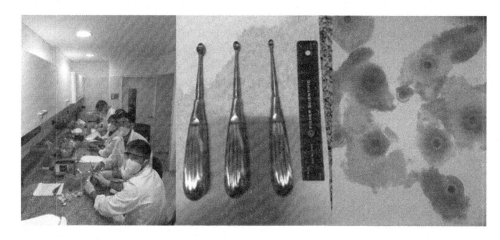

FIGURE 2.7 Postmortem equine ovary processing. *Photograph: M. Sansinena, Laboratory of Animal Reproduction & Biotechnology, Catholic University of Argentina.*

2.2.2   Oocyte recovery postmortem

The aspiration of oocytes postmortem may be the only option for rescuing the genetics of a mare that dies unexpectedly. In this case, ovaries can be removed from the deceased mare and shipped to the laboratory where they can be processed for oocyte recovery [23]. In order to prevent temperature fluctuations that compromise oocyte viability, it is recommended that ovaries be immediately shipped at room temperature in a Styrofoam box or Equitainer and allowed to passively cool during transport. The time period between the death of the mare and the ovaries reaching the processing facility is a critical factor affecting results [23].

Once ovaries reach the laboratory, they are normally rinsed with phosphate-buffered saline and antibiotics in order to reduce surface contamination. The best oocyte recovery rates (approximately six oocytes per ovary) are obtained when follicles are individually processed (Fig. 2.7), sliced with a scalpel blade and follicle walls are scraped either with bone curettes or the scalpel blade [24,25]. Material isolated from scrapping is then rinsed into 50-mL sterile tubes containing oocyte holding medium and antibiotics. Once all follicles are processed, the contents of the tube are allowed to settle to the bottom and then aspirated and transferred to a search Petri dish with a grid. Oocytes are then isolated and classified under a stereo-microscope (50 ×), rinsed several times, and then placed alternatively into oocyte holding or IVM medium. Other oocyte recovery procedures such as syringe flushing + follicular wall scrapping with needle [26], needle scraping + vacuum + rinsing [27], or individual follicle dissection and slicing in separated Petri dishes [27] have been reported.

2.3 Oocyte in vitro maturation

Although there is still some debate about the nuclear status of ovulated oocytes in mares, the general consensus is that horse oocytes, such as those of most mammals, are ovulated after completion of the first meiotic division (metaphase II) and formation of the first polar body [28]. The oocytes recovered from immature follicles have yet to be stimulated by gonadotropins during the last stages of follicular maturation and are therefore found at the germinal vesicle stage. Therefore when conducting OPU from immature nonpreovulatory follicles, the recovered oocytes must still undergo additional IVM in order to acquire competence (both nuclear and cytoplasmic) to successfully undergo fertilization. One of the main requirements for oocytes to undergo successful IVM is for them to be surrounded by several layers of compact cumulus cells. These cells have an active role in nuclear maturation and have paracrine effects within the culture environment [29]. Therefore it is important for the vacuum applied during OPU to be forceful enough to cause the COC to dislodge from the wall, yet gentle enough for the oocyte to remain with intact cumulus as opposed to be partially or completely denuded.

Embryo development is highly dependent on initial oocyte quality. The first successful report of IVM of horse oocytes was produced by Fulka and Okolski in 1981 [30]. IVM is generally conducted in Earl's M199 medium supplemented with gonadotropins (mostly follicle-stimulating hormone, although luteinizing hormone has also been added), growth factors (such as epidermal growth factor and Insulin-like growth factors, IGFs), antibiotics, and fetal serum [31]. In addition, commercially defined formulations specifically tailored to the equine requirements are slowly but steadily becoming available. IVM is conducted in water-jacketed incubators at 38.2°C, 5% CO_2, and air, or alternatively 5% CO_2, 5% O_2, and 90% N_2 mixtures (Fig. 2.8). Results for oocyte IVM vary greatly between laboratories, ranging from 50% to 70% maturation rates [24,25,32]. In 2016, Claes et al. [33] reported the results from a large commercial herd of 700 donor mares with OPU intervals of 2−4 weeks and continuous monitoring of follicular dynamics, where OPU oocyte recovery and in vitro oocyte maturation rates were 67% and 64%, respectively.

The proper length of oocyte IVM is still an active area of research in the equine species. The population of recovered equine oocytes is often heterogeneous, particularly when oocytes are collected from small and medium-sized follicles. Two distinct populations have been reported: oocytes recovered with compact layers of cumulus (Cp) and those presenting an expanded cumulus (Ex) [34]. These oocyte populations have been shown to have distinctive differences in their nuclear configuration; while Cp oocytes tend to have a diffuse distribution within the germinal vesicle, Ex oocytes mostly have their chromatin condensed in one dense mass [35]. In addition, cytoplasmic mRNA accumulation and the required IVM period for them to reach metaphase-II stage differ. Although a range of maturation times has been reported (from 26 to 36 hours), oocytes having compact cumuli (Cp) in general have lower developmental competence and require more hours to mature than those oocytes with expanded (Ex) cumuli [36].

Few veterinary practices include a high complexity laboratory and technicians needed for equine ICSI; therefore oocytes must typically be shipped to another facility. The conditioning and shipping of mature oocytes (recovered from dominant follicles) are different from immature oocytes (recovered from small and medium follicles) [37]. Oocytes recovered from dominant follicles require shipment at equine body temperature because they are in the final stages of

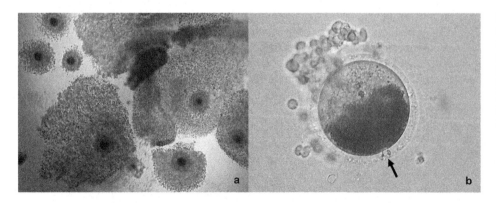

FIGURE 2.8 (A) Equine oocytes after 26 hours of in vitro maturation. (B) Mature equine oocyte after cumulus cell stripping (arrow indicates position of first polar body). *Photograph: M. Sansinena, Laboratory of Animal Reproduction and Biotechnology, Catholic University of Argentina.*

acquiring nuclear maturity. They must be shipped in battery-operated, portable heated incubators (preferably in M199 or oocyte maturation medium). In these oocytes, ICSI should be conducted within a short interval from aspiration, since they are actively metabolizing and reaching metaphase-II stage. On the other hand, one interesting aspect of immature equine oocytes recovered from small and medium-sized follicles is that they are at germinal vesicle stage and can be held at room temperature in conditions that inhibit nuclear maturation while allowing cytoplasmic acquisition of competence [23]. Equine immature oocytes have been held in holding medium (a mixture of 40% Earl's: 40% Hank's based M199, 20% fetal bovine serum, and 25 μg/mL gentamicin) at room temperature (approximately 22°C−25°C) for up to 18 hours and then transferred to IVM with no detrimental effects on maturation rates or subsequent embryo development [38,39]. This apparent flexibility of the equine immature oocyte has been a useful tool for commercial practitioners, particularly when OPU is conducted at farms geographically separated from the in vitro laboratory.

2.4 Oocyte cryopreservation

The female gamete or oocyte contributes with 50% to the genetic makeup of the new embryo; therefore the preservation of the mare germplasm is extremely important to equine ART. The ability to systematically collect and preserve oocytes from valuable mares, as well as the possibility to rescue and preserve germplasm in the case a mare dies unexpectedly, is all necessary biotechnologies associated with oocyte collection and in vitro embryo production. Unfortunately, the oocyte has proven to be one of the hardest cells to cryopreserve. Its large volume to surface area determines that equilibrium, freezing protocols using relatively low concentrations of cryoprotectants (5%−10%) and slow cooling rates (0.25°C−5°C) are mostly unsuccessful, resulting in intracellular ice crystal formation and oocyte death by lysis. An alternative cryopreservation method, vitrification, was initially proposed by Luyet in 1937 [40] and later successfully applied to cryopreservation of oocytes in the mouse [41]; this is currently the procedure of choice for oocyte banking.

Vitrification is a cryopreservation method that results in an amorphous, vitreous, or glass-like structure without the formation of ice crystals. For this phenomenon to occur the sample must undergo extremely high cooling rates (> 100.000° C/minute). This process requires for oocytes to be exposed to high concentrations to permeating (DMSO, ethylene glycol, propanediol, etc.) and nonpermeating cryoprotectants (sucrose, trehalose, galactose), which lower the freezing point of the cytoplasm and cause oocyte dehydration (Fig. 2.9).

The oocyte, exposed to sequential vitrification solutions with increasing cryoprotectant concentrations, is then loaded with minimal volume (< 1 μL) of surrounding medium onto supports or carriers such as thin-walled straws (open-pulled straws) and polypropylene strips (cryotops) and finally plunged directly into liquid nitrogen. The high viscosity of the cryoprotectant solution, in combination with the minimal loading volume, results in ultrarapid cooling rate and sample vitrification when plunged into −196°C. In humans, the IVF first baby born from a vitrified oocyte was reported by

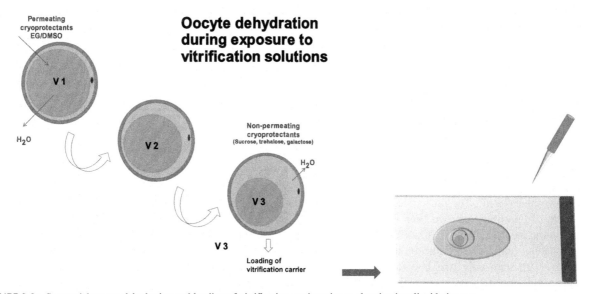

FIGURE 2.9 Sequential oocyte dehydration and loading of vitrification carrier prior to plunging into liquid nitrogen.

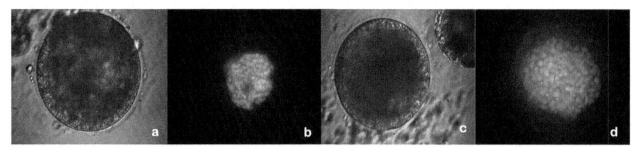

FIGURE 2.10 Equine ICSI blastocysts and nuclei DNA staining: (A and B) from nonvitrified oocytes; (C and D) from vitrified oocytes. *Photograph: M. Sansinena, Laboratory of Animal Reproduction and Biotechnology, Catholic University of Argentina.*

Kuleshova et al. in 1999 [42]. Since then, the procedure has gone from experimental to clinical application and is now an integral part of human reproductive medicine. Oocyte vitrification is currently used not only for fertility preservation in the event of chemotherapy, but also as a systematic approach to elective fertility preservation in women.

In the horse, the first successful oocyte vitrification was reported by Maclellan et al. [43]. In this study, oocytes collected by OPU from stimulated folliclles were vitrified, warmed, and surgically transferred to the oviduct of inseminated recipients. This approach resulted in the birth of a healthy foal (named "Vitreous") and a filly (named "Ethyl"). More recently, Ortíz-Escribano et al. [44] reported the birth of a foal (named "VICSI") from vitrified oocytes fertilized by ICSI. Although these live births represent a significant accomplishment, the efficiency of the procedure still remains extremely low.

There are multiple factors associated with the poor results with equine oocytes. Current scientific evidence indicates the oocyte's cytoskeleton is deeply affected during vitrification. A functional and healthy meiotic spindle is fundamental for the correct alignment of the chromosomes and migration of chromatids toward the poles. If the microtubule assembly is damaged, then chromosomal imbalance and embryo aneuploidy are likely to occur. In the horse, Ducheyne et al. in 2019 [45] reported severe chromosomal abnormalities and aberrant meiotic spindles in equine oocytes undergoing vitrification. Mitochondria, which are associated with the cytoskeleton, have also been shown to be affected. Because mitochondria play a key role in cell's energy reserves and oxidative status, vitrification has been associated with more reactive oxygen species and apoptosis; this was reported recently by Clérico et al. in 2018 [25].

Information about the best nuclear status (germinal vesicle vs metaphase-II) prior to equine oocyte vitrification is limited. In humans, most vitrification is conducted in metaphase-II oocytes; this approach has not been successful in the horse. Instead, immature equine oocytes seem to be more tolerant to cryodamage. Vitrification of immature oocytes, followed by warming, maturation, and ICSI, has resulted in moderate rates of maturation and embryo production (1%−15% blastocysts per injected oocyte) [46−49]. Promising results were recently obtained in our laboratory using a short incubation, DMSO/EG/trehalose protocol. In this study, blastocyst rates from vitrified oocytes were similar to control, nonvitrified oocytes (18%) [25]. However, slower development was observed in blastocysts from vitrified oocytes, indicating that embryo viability could be compromised (Fig. 2.10). Therefore until there is more information about the health of foals derived from vitrified oocytes, the technique should be considered in the experimental phase and not ready for widespread equine clinical practice.

2.5 In vitro embryo production

2.5.1 In vitro fertilization

The process of IVF is a complex sequence of events initiated by the penetration of a metaphase-II, mature oocyte by a capacitated sperm cell. In humans, success was reported by Steptoe and Edwards with the birth of a healthy baby girl, Louise Brown in 1978 [6]. The procedure, which is now routinely used in the field of human reproductive medicine for the treatment of infertility, has also been applied successfully to domestic animals such as cattle, sheep, goats, pigs, and buffaloes, and in the case of cattle, it has become a cornerstone technology for commercial in vitro embryo production.

Unfortunately, unlike most domestic animal species, traditional IVF by gamete coincubation has been unsuccessful in the horse [50], with only two live foals ever reported [51,52]. Although the reasons for failure remain unclear, authors have indicated in vitro zona pellucida (ZP) hardening, inadequate sperm capacitation, and overall fertilization failure as factors affecting success [53−55].

2.5.2   Intracytoplasmic sperm injection

In order to bypass the ZP and circumvent IVF failures, several microfertilization techniques have been explored over the last decades. Although techniques such as zona drilling or renting, subzonal sperm injection (SUZI), and partial zona dissection have been reported in humans and other species [56−59], no equine offspring has ever been reported using these procedures. In 1992 a new microfertilization technique, ICSI was described by Palermo et al. for the treatment of male factor infertility [60]. This technique was proposed in humans as a treatment in cases of repeated, failed fertilization attempts with conventional IVF cycles and involves the direct injection of one single sperm cell into the ooplasm of a mature oocyte (Fig. 2.11).

The first equine pregnancies and live births by ICSI were reported in 1996 from abattoir-derived oocytes [61] and OPU-derived oocytes from nonpregnant mares [62,63], as well as pregnant mares [64,65]. In those early days, partly because in vitro culture conditions for equine embryos were still suboptimal, ICSI pregnancies were only established after surgical transfer of early-stage (2−4 cell) embryos into the oviduct. The establishment of a successful ICSI/ in vitro embryo production system for the horse demanded that embryos be cultured to later stages of development (blastocyst stage) and nonsurgical transfer into the recipient's uterus. In the early years, the lack of a reliable culture system, poor cleavage (<50%), and low blastocyst rates (<10%) significantly slowed down the progress and cooled enthusiasm among equine practitioners. The development of Dulbecco's modified Eagle medium/Hams F-12 culture medium capable of supporting ICSI blastocyst rates >35% was a major breakthrough in the transition of this procedure from the research laboratory to clinical practitioner settings [66,67]. More recently Galli et al. in 2014 [18] reported 0.6 blastocysts/OPU cycle in a commercial OPU-IVP program, indicating that OPU-IVP could become a competitive technology to commercial ET. Currently, blastocyst rates of 20% per injected oocyte and over 1 blastocyst produced per OPU cycle (with overnight shipping of oocytes at 20°C) are encouraging results that justify the current surge in interest in equine OPU-IVP [18].

The ICSI technique consists of the injection of a single sperm directly into the ooplasm of a mature oocyte using a microscopic needle. This procedure can bypass many barriers of fertilization including the process of sperm penetration through cumulus-corona cells, ZP, and oolemma during fertilization by directly depositing the sperm into the ooplasm. The ICSI procedure requires complex equipment in order to visualize and micromanipulate oocytes (100−120 μm diameter) and sperm cells (4−5 μm diameter). The micromanipulator system, which is used to physically interact with a live specimen under a microscope, enables a precision movement that cannot be achieved by the unaided human hand. It basically consists of an inverted microscope (normally 20−40 ×), fitted with micromanipulation joysticks and microinjectors; these can be hydraulically or pneumatically driven. The micromanipulator holders are fitted with thin-walled borosilicate glass injection (5−7 μm) and holding micropipettes (80−90 μm) (Figs. 2.12−2.14). The selection of

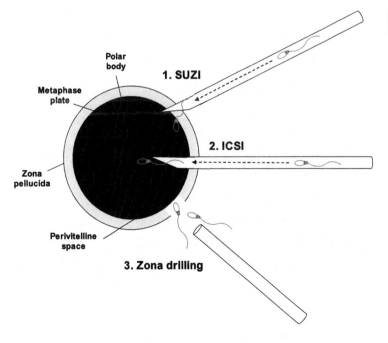

FIGURE 2.11 Schematic representation of microfertilization techniques SUZI, ICSI, and zona drilling.

Polar body

Metaphase plate

1. SUZI

2. ICSI

Zona pellucida

Perivitelline space

3. Zona drilling

the injection needle is a critical aspect for the success of the procedure; it must be beveled and sharp enough to perforate both the ZP and the oolemma; the outer and inner diameter (OD and ID) should be adequate to draw the sperm cell without aspirating excess medium.

For the ICSI procedure, the oocyte is secured into position by applying negative pressure through the holding pipette and rotating the oocyte in order for the technician to select the injection site. It is generally preferred to locate the first polar body at the 6 or 12 o'clock position, in order to minimize the probability of damaging nuclear material contained in the metaphase-II spindle. A single sperm cell is selected based on motility and strict morphology criteria; it is then immobilized by a quick strike of the injection pipette (this mechanical sperm membrane damage has been shown to trigger sperm capacitation and acrosome reaction) and then injected into the ooplasm of the oocyte (Fig. 2.15).

Immobilization of motile sperm cells seems to be necessary prior to ICSI, because it induces membrane permeabilization that in turn releases a cytosolic factor that causes Ca^{2+} oscillations and oocyte activation [68]. For this, a sperm cell is immobilized at the edge of the sperm micromanipulation droplet (which normally contains polyvinyl pyrrolidone or other polymer of large molecular weight) by touching its tail with the needle. The sperm tail is then crushed by sliding the injection pipette over the mid-piece of the tail until a kink is seen and it is then aspirated tail-first into the injection pipette (Fig. 2.15). Then, the membrane surrounding the ooplasm (oolemma) must be carefully stretched and ruptured in order for the sperm cell to be properly injected. This can be achieved by gentle suction of ooplasm into the injection pipette until a noticeable change in membrane resistance is detected. Alternatively, it can be accomplished by

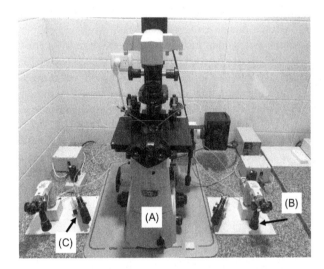

FIGURE 2.12 Micromanipulation station used for ICSI: (a) inverted microscope, (b) micromanipulation joysticks, and (c) microinjectors. *Photograph: M. Sansinena, Laboratory of Animal Reproduction and Biotechnology, Catholic University of Argentina.*

FIGURE 2.13 Close-up of ICSI microtools: (a) holding (90 μm) and (b) injection (5−7 μm). *Photograph: M. Sansinena, Laboratory of Animal Reproduction and Biotechnology, Catholic University of Argentina.*

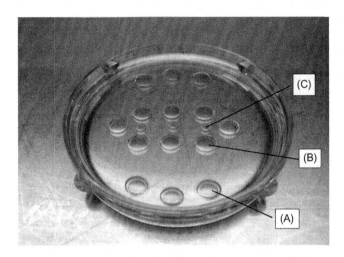

FIGURE 2.14 Petri dish showing microdroplet arrangement for ICSI: (a) rinse droplets, (b) micromanipulation droplets containing oocytes, and (c) sperm immobilization droplets. *Photograph: M. Sansinena, Laboratory of Animal Reproduction and Biotechnology, Catholic University of Argentina.*

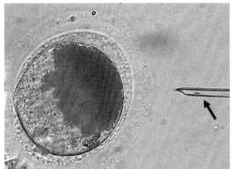
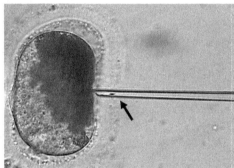

FIGURE 2.15 Intracytoplasmic sperm injection in the horse. Arrow indicates position of sperm cell within injection pipette. *Photograph: M. Sansinena, Laboratory of Animal Reproduction and Biotechnology, Catholic University of Argentina.*

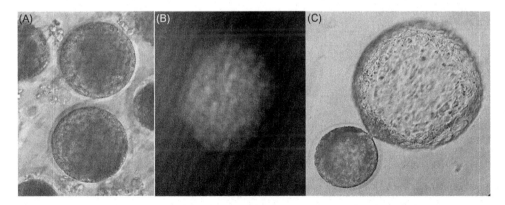

FIGURE 2.16 Equine blastocysts produced by ICSI. (A) Day 7 blastocysts, (B) Hoechst 33342 staining showing nuclear DNA, and (C) hatched equine blastocyst. *Photograph: M. Sansinena, Laboratory of Animal Reproduction and Biotechnology, Catholic University of Argentina.*

more sophisticated equipment such as the "Piezo drill." This piezoelectric injection device produces a stabbing, punctuate movement to puncture the oolemma with minimal distortion of the oocyte, which reduces the chance of damage. Although piezo injection is not mandatory for the procedure of ICSI in the horse, some authors have reported that its use improves oocyte activation and fertilization rates [69].

Success rates of ICSI vary greatly among laboratories; an average of 50%−60% oocyte maturation rate and 20%−35% blastocyst production can be expected [25,66,67] (Fig. 2.16). Irrespective of their developmental kinetics, it is generally recommended to consider all in vitro, ICSI blastocysts as day-6, in vivo embryos [70,71]. Therefore the recommendation for ICSI blastocysts is for them to be transferred to recipient mares that are at days 4−6 postovulation.

2.6 Embryo cryopreservation

In the horse, as in many mammalian species, blastocoel cavity formation initiates at the compact morula stage and continues to grow in volume thereafter. The equine in vivo–derived embryo undergoes an exponential increase in volume after 7 days postfertilization, resulting in a major complexity to obtain small size embryos (less than 300 μm) without compromising the recovery rate. There is evidence that equine embryos equal or smaller than 300 μm in diameter can be cryopreserved by freezing or vitrified with similar protocols to those applied to other domestic species. In order to obtain small diameter in vivo embryos, it is necessary to attempt recovery early after insemination, generally on day 6. The problem of this approach is that the recovery rates on day 6 are considerably lower than flushes conducted on days 7 and 8. Therefore most ET practitioners prefer to flush donor mares on those days, obtaining embryos that are well over the 300 μm cut-off [72].

Cryopreservation of expanded equine blastocysts (> 300 μm) was initially unsuccessful and associated with low pregnancy rates [73,74]. A breakthrough in cryopreservation of large embryos was due to the blastocoel collapse prior to vitrification; this reduction of embryonic volume resulted in improved survival and implantation rates [75–77]. Although this discovery has in some way enabled the preservation of large in vivo embryos, pregnancy rates are still low. In addition, a major limitation is that it requires access to an expensive micromanipulator system, which prevents this approach to be practical for clinical practice. Recently, our group produced pregnancies from vitrified, > 900 μm-diameter equine embryos collapsed using microneedle and forced pipetting. Embryo blastocoel was collapsed either using the traditional micromanipulator (Fig. 2.17) or collapsed in a modified microwell for embryo immobilization using a penta-point, 32-g microneedle followed by forced pipetting into a small glass pipette for additional fluid leakage (Fig. 2.18). Both groups resulted in 50% pregnancy rates after transfer; these encouraging results indicate that a simpler approach may be feasible for on-farm use.

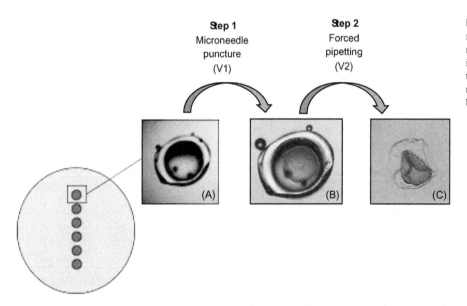

FIGURE 2.17 Blastocoel collapse using microneedle and forced pipetting technique. (A) Expanded blastocyst placement in microwell prior to microneedle puncture, (B) embryo immediately after microneedle puncture, and (C) additional blastocoel collapse after forced pipetting.

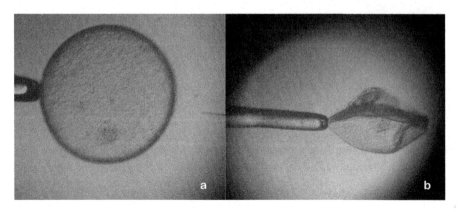

FIGURE 2.18 Artificial blastocoel collapse of expanded equine embryo blastocyst by micromanipulator-assisted technique that resulted in a pregnancy. (A) Prior to blastocoel aspiration. (B) Same embryo, after blastocoel collapse. *Photograph: M. Sansinena, Laboratory of Animal Reproduction and Biotechnology, Catholic University of Argentina.*

2.7   Biopsy and embryo sexing

The publication of the complete human genome has resulted in significant contributions toward the early identification of disorders and gene-associated mutations. Additionally, human embryo pregenetic testing (PGT) has been applied to the identification of traits and chromosome-related abnormalities in early embryonic life. In the horse, the complete genome sequence was first reported by Wade et al. in 2009 [78]; this advance paved the road for early embryonic screening in the equine species.

In some equestrian disciplines such as polo, the ability to determine the sex of the embryo and therefore select the sex of the offspring (females are preferred, in the case of polo) poses an extremely high added value. In 2010, Choi et al. [79] showed it was possible to maintain equine embryos overnight in holding medium prior to biopsy and also to keep them viable for 4–6 hours after micromanipulation prior to transfer into a recipient. This was a promising milestone toward being able to ship embryos to a satellite laboratory for PGT (where micromanipulation for biopsy would take place) and then immediately send those embryos back to the farm for transfer. Although this strategy was successful, it was still not ideal for large-scale programs. Analysis by polymerase chain reaction (PCR) of embryo biopsies took more than 24 hours, which meant that many of the embryos were confirmed of the undesired sex after they had been transferred.

Several cell types or structures have been biopsied as sources for DNA amplification and PCR analysis. These structures include first and second polar bodies from oocytes, blastomeres from cleavage-stage embryos and inner cell mass, trophectoderm cells, or blastocoel fluid from large blastocysts. The aspiration of blastocoel cavity and analysis of the fluid was first reported for human embryos by Tobler et al. in 2015 [80]. This approach, combined with the ability to cryopreserve the biopsied embryos, had the potential to be particularly useful if applied to equine embryos, because it would permit to dissociate the embryo biopsy/PCR gender determination from the actual ET. In addition, because large equine blastocysts can only be cryopreserved (vitrified) successfully if the blastocoel cavity is collapsed, it could potentially eliminate the limitation of long PCR turn-around time and the lag period for ET. In 2015, Herrera et al. [81] reported the presence of DNA in blastocoel fluid from equine expanded blastocysts and survival of embryos vitrified at the time of blastocoel biopsy. Although results were promising, the cost of the technology, in combination with the logistics and the many steps involved have, in part, prevented this technology from being applied in a commercial setting.

Research is still ongoing in the field of equine preimplantation genetic diagnosis for gender determination, considering that it could have a significant economic impact on equine commercial programs. In addition, screening and selection for performance traits (such as a "speed gene") are also areas of active research. It is important to notice that the genomic field is advancing very rapidly also, with commercially available microarrays for comparative genomic hybridization and single nucleotide polymorphism being applied to the genomic selection of domestic animals.

2.8   Somatic cell nuclear transfer (cloning)

The birth of cloned animals from differentiated adult somatic cells provided the evidence that mammalian development has a far greater developmental plasticity than imagined [82]. SCNT encompasses the removal of nuclear DNA from a mature oocyte and the insertion of a donor cell nucleus derived from a somatic cell. The donor nucleus is then subjected to a complete "reprogramming" by undetermined factors located inside the ooplasm, which enable the complete set of instructions that were once turned off in the differentiated donor nucleus to become active and commence development, not as another somatic cell, but as a one-cell embryo.

Prior to the horse, SCNT had been successfully accomplished in several domestic and wild animal species. The mule was the first equid to be cloned [83] and shortly afterward a group in Italy announced the birth of the first cloned horse, Prometea [84]. The birth of Prometea was the product of arduous work that involved 841 injected oocytes, 22 blastocysts cultured, 17 embryos transferred, 4 pregnancies (two were lost around day 30, and one was lost after 6 months) and one foal carried to term. Since this first cloned filly was reported, approximately 100 live foals had been reported in the scientific literature [85–87]. In addition, commercial cloning is now available in North America (United States), South America (Argentina, Brazil, and Colombia), Europe (Italy), and Australia; it is estimated these companies have by now produced over 375 cloned horses [88].

There are basically three approaches to equine cloning. The first micromanipulation or "traditional" approach follows the procedure originally outlined by Willadsen to produce sheep clones from embryonic donor cells [89] and later replicated by Wilmut et al. [82] to produce Dolly from adult epithelial fibroblasts [82]. The technique consists of the removal of nuclear DNA from a mature, metaphase-II equine oocyte (enucleation) and the injection of a donor cell by

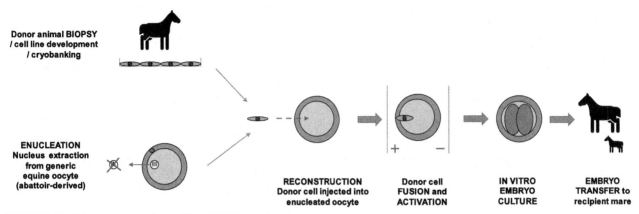

FIGURE 2.19   Diagram representing traditional SCNT procedure.

micromanipulation, followed by cell incorporation by electrofusion, couplet activation, in vitro embryo culture, and nonsurgical transfer to a synchronized recipient (Fig. 2.19). A second approach is the "zona-free" method, originally reported by Vatja et al.in 2008 for cattle [90] and later adapted by Lagutina et al. for the horse [91]. This method involves the removal of the oocyte's ZP, the extraction of nuclear DNA with a blunt pipette and, instead of injected, the donor cell is adhered to the enucleated oocyte using phytohemagglutinin; this step is also followed by electrofusion, activation, and in vitro embryo culture. A third procedure is similar to the traditional method, but involves the activation of the oocyte prior to its enucleation and reconstruction with a synchronized donor cell [87].

The initial reported pregnancy rates for equine clones were highly variable, ranging from 9% to 100%. Recent results appear to have become more consistent; Choi et al. in 2015 [92] reported a total of 4 live foals/15 total pregnancies (27%), whereas Olivera et al. in 2016 [93] reported 22 live foals/78 pregnancies (28%) suggesting that an average of three to four pregnancies are needed to produce one live cloned foal [88]. The scientific information on the health of cloned foals is limited, in part because most clones are currently being produced by private commercial companies. Johnson et al. [94] reported on the health of 14 cloned foals; two died 2 weeks postpartum and the remaining foals exhibited problems such as neonatal maladjustment syndrome, enlarged umbilical cord, and/or front leg contracture, all of which were resolved with treatment. Therefore the current recommendation is for clones to be born in facilities with available neonatal care, including supplemental oxygen, colostrum administration, and prophylactic antibiotic therapy.

There are striking differences among the levels of acceptance, regulatory and registration requirements for equine cloning among countries, as well as individual breed associations/registries, ranging from prohibition (Jockey Club of North America, International Studbooks, American Quarter Horse Association, and European Commission) to open advocacy (Argentine Polo Pony Breeder's Association) [88]. In spite of this, cloned horses have been reported for several breeds and sporting types including Haflinger, Arabian, Argentine criollo, Polo, Warmblood, and others such as Colombian paso, Pasofino, American Cutting Horse, Mangalarga, and Campolina, as indicated by various commercial cloning companies [88].

Equine cloning is such a powerful biotechnology that it has captured the collective imagination of enthusiasts and producers. Unprecedented possibilities such as the cloning of a champion gelding to be used as stallion, the replication of valuable individuals in a particular equestrian discipline, the cloning of deceased animals, or even having multiple cloned horses competing in the same event are now a reality. As an example, in a recent edition of the prestigious Palermo Polo Championship in Buenos Aires, Argentina, polo player Adolfo Cambiaso rode six clones of the same mare in different chukkers of one single match. This has resulted in the local press coining the phrase "The Clone Era" for the game of high polo. Argentina, often referred to as the Polo Capital of the World, has seen in its most recent 133rd International Livestock and Agriculture Expo a total of six clones out of a total of 134 horses auctioned in the Argentine Polo Pony breed category; all of the auctioned clones coming from animals which, at some point, had participated in the Palermo Polo Open. In spite of this, the application of cloning to polo discipline is not without controversy. Many experts, particularly geneticists and breeding specialists, continue to warn about the potentially negative impact of the massive application of the cloning technology to a particular breed, with the reduction of the breed's diversity and dramatic increase in the levels of endogamy/inbreeding being among top concerns [95].

Public perception plays a big role in the acceptance and demand of cloning as an ART. Because cloning is still a very inefficient process, numerous oocytes are needed in order to produce sufficient embryos and establish pregnancies that will ultimately result in a viable foal. This means that cloning companies are largely dependent on their ability to access abattoir-derived ovaries and therefore currently operate in countries that do not have regional restrictions regarding equine slaughter. Because of public perception and animal rights activists' pressures, equine slaughter for meat production is a very volatile operation; this may ultimately have an impact on the sustainability of cloning as a commercial business. Although some groups have attempted to replace (in part or totally) abattoir-derived oocytes with OPU-derived ones, the fact that the current efficiency of equine cloning remains very low (current cloning efficiency is about 10%−12% blastocyst rate based on number of enucleated oocytes) has determined that the total replacement of abattoir-derived oocytes is not feasible under the current conditions. Finally, a recent poll conducted during one of the workshop sessions at the last International Symposium of Equine Reproduction (ISER) in 2018 in Cambridge, United Kingdom, indicated that about one half of the attendance (composed mostly of researchers and equine reproduction practitioners) expressed reservations about the value of equine cloning as an assisted reproduction procedure. In this poll, practitioners expressed concerns related to public perception and pressure (particularly EU), reduction of diversity, and narrowing of gene pool base by widespread application of the technology. This indicates that, in spite of the relative success, press releases, and commercial visibility of equine cloning, the value of the procedure as a genetic propagation tool remains, among practitioners at least, controversial.

2.9   Future challenges

Reproductive biotechnologies in the horse have come a long way since the early attempts of AI by Arabian chiefs. As technologies evolve and procedures improve, focus will likely shift toward a more integrated approach; in which the management of mare and stallion infertility will closely follow those trends seen today in human reproductive medicine. This suggests a new and exciting challenge regarding the academic learning and continuing education of the modern equine reproduction researcher and practitioner, incorporating proven technologies while continuously developing and improving new methods.

Acknowledgments

The author would like to thank Ph.D. candidate Gabriel Clérico for providing pictures of equine ICSI embryos produced as part of his research and the School of Veterinary Medicine, Universidad del Salvador for providing pictures of ovum pickup equipment.

References

[1]   Zegers-Hochschild F, Adamson GD, de Mouzon J, Ishihara O, Mansour R, Nygren K, et al. The international committee for monitoring assisted reproductive technology (ICMART) and the world health organization (WHO) revised glossary on ART terminology, 2009. Hum Reprod 2009;24(11):2683−7.

[2]   Hafez YM. Assisted reproductive technologies in farm animals. Egypt: ICMALPS, Alexandria University; 2015.

[3]   Allen WR. The development and application of the modern reproductive technologies to horse breeding. Reprod Domest Anim 2005;40 (4):310−29.

[4]   Bowen JM. Artificial insemination in the horse. Equine Vet J 1969;1:98−110.

[5]   Hafez E.S., Hafez B. Reproductive cycles. In: Reproduction in farm animals. 2000. p. 55−67.

[6]   Steptoe PC, Edwards RG. Birth after re-implantation of a human embryo. Lancet 1978;2:366.

[7]   Lopata A, Johnston IWH, Hoult IJ, Speirs AI. Pregnancy following intrauterine implantation of an embryo obtained by in vitro fertilization of a preovulatory egg. Fertil Steril 1980;33:117.

[8]   Mettler L, Seki M, Baukloh V, Semm K. Human ovum recovery via operative laparoscopy and in vitro fertilization. Fertil Steril 1982;38 (1):30−7.

[9]   Lenz S, Laurritsen JG, Kjellow M. Collection of human oocytes for in vitro fertilization by ultrasonically guided follicular puncture. Lancet 1981;1:1163−4.

[10]   Dellenbach P, Nisand I, Moreau L, Feger B, Plumere C, Gerlinger P. Transvaginal sonographically controlled follicle puncture for oocyte retrieval. Fertil Steril 1985;44(5):656−62.

[11]   Wikland M, Enk L, Hamberger L. Transvesical and transvaginal approaches for the aspiration of follicles by use of ultrasound. Ann N Y Acad Sci 1985;442(1):182−94.

[12]   Brück I, Raun K, Synnestvedt B, Greve T. Follicle aspiration in the mare using a transvaginal ultrasound-guided technique. Equine Vet J 1992;24(1):58−9.

[13]   Pieterse MC, Kappen KA, Kruip TA, Taverne MA. Aspiration of bovine oocytes during transvaginal ultrasound scanning of the ovaries. Theriogenology. 1988;30(4):751−62.

[14]   Hawley LR, Enders AC, Hinrichs K. Comparison of equine and bovine oocyte-cumulus morphology within the ovarian follicle. Biol Reprod. 1995. p. 243−52.

[15] Bols PE, Stout TA. Transvaginal ultrasound-guided oocyte retrieval (OPU: ovum pick-up) in cows and mares. Animal Biotechnology 1. Cham: Springer; 2018. p. 209—33.

[16] Jacobson CC, Choi YH, Hayden SS, Hinrichs K. Recovery of mare oocytes on a fixed biweekly schedule, and resulting blastocyst formation after intracytoplasmic sperm injection. Theriogenology. 2010;73(8):1116—26.

[17] Galli C, Colleoni S, Duchi R, Lagutina I, Lazzari G. Equine assisted reproduction and embryo technologies. Anim Reprod 2013;10:334—43.

[18] Galli C, Duchi R, Colleoni S, Lagutina I, Lazzari G. Ovum pick up, intracytoplasmic sperm injection and somatic cell nuclear transfer in cattle, buffalo and horses: from the research laboratory to clinical practice. Theriogenology. 2014;81(1):138—51.

[19] Kanitz W, Becker F, Alm H, Torner H. Ultrasound-guided follicular aspiration in mares. Biol Reprod 1995;52:225—31 (monograph_series1).

[20] Hinrichs K. The equine oocyte: factors affecting meiotic and developmental competence. Mol Reprod Dev 2010;77(8):651—61.

[21] Carnevale EM, Maclellan LJ. Collection, evaluation, and use of oocytes in equine assisted reproduction. Vet Clin Equine Pract 2006;22 (3):843—56.

[22] Foss R, Ortis H, Hinrichs K. Effect of potential oocyte transport protocols on blastocyst rates after intracytoplasmic sperm injection in the horse. Equine Vet J 2013;45:39—43.

[23] Ribeiro BI, Love LB, Choi YH, Hinrichs K. Transport of equine ovaries for assisted reproduction. Anim Reprod Sci 2008;108(1-2):171—9.

[24] Reggio B.C., Sansinena M., Cochran R.A., Guitreau A., Carter J.A., Denniston R.S., et al. Nuclear transfer embryos in the horse. In: Proceedings of the fifth international symposium on equine embryo transfer, 2001. p. 45—46.

[25] Clérico G, Rodriguez MB, Taminelli G, Butteri A, Veronesi JC, Fernández S, et al. Vitrification of immature oocytes for the production of equine embryos by ICSI: protocol effect on maturation, embryo development, mitochondrial distribution and functionality. J Equine Vet Sci 2018;66:192—3.

[26] Gambini A, Jarazo J, Olivera R, Salamone DF. Equine cloning: in vitro and in vivo development of aggregated embryos. Biol Reprod 2012;87 (1) 15-1.

[27] Ortiz-Escribano N, Bogado Pascottini O, Woelders H, Vandenberghe L, De Schauwer C, Govaere J, et al. An improved vitrification protocol for equine immature oocytes, resulting in a first live foal. Equine Vet J 2018;50(3):391—7.

[28] King WA, Bezard J, Bousquet D, Palmer E, Betteridge KJ. The meiotic stage of preovulatory oocytes in mares. Genome. 1987;29(4):679—82.

[29] Hawley LR, Enders AC, Hinrichs K. Comparison of equine and bovine oocyte-cumulus morphology within the ovarian follicle. Biol Reprod 1995;52:243—52 (monograph_series1).

[30] Fulka J, Okolski A. Culture of horse oocytes in vitro. Reproduction. 1981;61(1):213—15.

[31] Hinrichs K, Schmidt AL, Friedman PP, Selgrath JP, Martin MG. In vitro maturation of horse oocytes: characterization of chromatin configuration using fluorescence microscopy. Biol Reprod 1993;48(2):363—70.

[32] Galli C, Colleoni S, Duchi R, Lagutina I, Lazzari G. Developmental competence of equine oocytes and embryos obtained by in vitro procedures ranging from in vitro maturation and ICSI to embryo culture, cryopreservation and somatic cell nuclear transfer. Anim Reprod Sci 2007;98 (1—2):39—55.

[33] Claes A, Galli C, Colleoni S, Necchi D, Lazzari G, Deelen C, et al. Factors influencing oocyte recovery and *in vitro* production of equine embryos in a commercial OPU/ICSI program. J Equine Vet Sci 2016;41:68.

[34] Goudet G, Bezard J, Duchamp G, Gerard N, Palmer E. Equine oocyte competence for nuclear and cytoplasmic in vitro maturation: effect of follicle size and hormonal environment. Biol Reprod 1997;57(2):232—45.

[35] Hinrichs K, Schmidt AL. Meiotic competence in horse oocytes: interactions among chromatin configuration, follicle size, cumulus morphology, and season. Biol Reprod 2000;62(5):1402—8.

[36] Del Campo MR, Donoso X, Parrish JJ, Ginther OJ. Selection of follicles, preculture oocyte evaluation, and duration of culture for in vitro maturation of equine oocytes. Theriogenology 1995;43(7):1141—53.

[37] Galli C, Colleoni S, Claes A, Beitsma M, Deelen C, Necchi D, et al. Overnight shipping of equine oocytes from remote locations to an ART laboratory enables access to the flexibility of ovum pick up-ICSI and embryo cryopreservation technologies. J Equine Vet Sci 2016;41:82.

[38] Choi YH, Love LB, Varner DD, Hinrichs K. Holding immature equine oocytes in the absence of meiotic inhibitors: effect on germinal vesicle chromatin and blastocyst development after intracytoplasmic sperm injection. Theriogenology. 2006;66(4):955—63.

[39] Sansinena MJ, Reggio B, Hylan D, Klumpp A, Paccamonti D, Lyle SK, et al. Effect of different pre-maturation treatments on in vitro maturation of equine oocytes and subsequent competence for nuclear transfer. Transporting gametes and embryos 2004;23.

[40] Luyet BJ. On the growth of the ice phase in aqueous colloids. Proc R Soc Lond Ser B Biol Sci 1957;147(929):434—51.

[41] Rall WF, Fahy GM. Ice-free cryopreservation of mouse embryos at − 196° C by vitrification. Nature. 1985;313(6003):573.

[42] Kuleshova L, Gianaroli L, Magli C, Ferraretti A, Trounson A. Birth following vitrification of a small number of human oocytes: case report. Hum Reprod 1999;14(12):3077—9.

[43] Maclellan LJ, Carnevale EM, Da Silva MC, Scoggin CF, Bruemmer JE, Squires EL. Pregnancies from vitrified equine oocytes collected from super-stimulated and non-stimulated mares. Theriogenology. 2002;58(5):911—19.

[44] Ortiz-Escribano N, Bogado Pascottini O, Woelders H, Vandenberghe L, De Schauwer C, Govaere J, et al. An improved vitrification protocol for equine immature oocytes, resulting in a first live foal. Equine Vet J 2018;50(3):391—7.

[45] Ducheyne K, Rizzo M, Beitsma M, Deelen C, Daels P, Stout T, et al. Vitrifying equine oocytes at the germinal vesicle stage disturbs spindle morphology and chromosome alignment. J Equine Vet Sci 2018;66:178.

[46] Hurtt AE, Landim-Alvarenga F, Scidel Jr GE, Squires EL. Vitrification of immature and mature equine and bovine oocytes in an ethylene glycol, ficoll and sucrose solution using open-pulled straws. Theriogenology. 2000;54(1):119—28.

[47] Tharasanit T, Colenbrander B, Stout TA. Effect of maturation stage at cryopreservation on post-thaw cytoskeleton quality and fertilizability of equine oocytes. Mol Reprod Dev 2006;73(5):627−37.

[48] Campos-Chillon LF, Suh TK, Barcelo-Fimbres M, Seidel Jr GE, Carnevale EM. Vitrification of early-stage bovine and equine embryos. Theriogenology. 2009;71(2):349−54.

[49] Canesin HS, Brom-de-Luna JG, Choi YH, Ortiz I, Diaw M, Hinrichs K. Blastocyst development after intracytoplasmic sperm injection of equine oocytes vitrified at the germinal-vesicle stage. Cryobiology. 2017;75:52−9.

[50] Dell'Aquila ME, Cho YS, Minoia P, Traina V, Fusco S, Lacalandra GM, et al. Intracytoplasmic sperm injection (ICSI) versus conventional IVF on abattoir-derived and in vitro-matured equine oocytes. Theriogenology 1997;47(6):1139−56.

[51] Palmer E, Bezard J, Magistrini M, Duchamp G. In vitro fertilization in the horse. A retrospective study. J Reprod Fertil Suppl 1991;44:375−84.

[52] Bezard J. In vitro fertilization in the mare. Rec Med Vet 1992;168:993−1003.

[53] Dell'Aquila ME, De Felici M, Massari S, Maritato F, Minoia P. Effects of fetuin on zona pellucida hardening and fertilizability of equine oocytes matured in vitro. Biol Reprod 1999;61(2):533−40.

[54] Hinrichs K, Love CC, Brinsko SP, Choi YH, Varner DD. In vitro fertilization of in vitro-matured equine oocytes: effect of maturation medium, duration of maturation, and sperm calcium ionophore treatment, and comparison with rates of fertilization in vivo after oviductal transfer. Biol Reprod 2002;67(1):256−62.

[55] Leemans B, Gadella BM, Stout TA, De Schauwer C, Nelis H, Hoogewijs M, et al. Why doesn't conventional IVF work in the horse? The equine oviduct as a microenvironment for capacitation/fertilization. Reproduction 2016;152:233−45.

[56] Gordon JW, Grunfeld L, Garrisi GJ, Talansky BE, Richards C, Laufer N. Fertilization of human oocytes by sperm from infertile males after zona pellucida drilling. Fertil Steril 1988;50(1):68−73.

[57] Ng SC, Bongso A, Chang SI, Sathananthan H, Ratnam S. Transfer of human sperm into the perivitelline space of human oocytes after zona-drilling or zona-puncture. Fertil Steril 1989;52(1):73−8.

[58] Cohen J. Assisted hatching of human embryos. J In Vitro Fert Embryo Transf 1991;8(4):179−90.

[59] Cohen J, Malter H, Wright G, Kort H, Massey J, Mitchell D. Partial zona dissection of human oocytes when failure of zona pellucida penetration is anticipated. Hum Reprod 1989;4(4):435−42.

[60] Palermo G, Joris H, Devroey P, Van Steirteghem AC. Pregnancies after intracytoplasmic injection of single spermatozoon into an oocyte. Lancet 1992;340(8810):17−18.

[61] Squires EL, Wilson JM, Kato H, Blaszczyk A. A pregnancy after intracytoplasmic sperm injection into equine oocytes matured in vitro. Theriogenology. 1996;1(45):306.

[62] Meintjes M, Graff KJ, Paccamonti D, Eilts BE, Cochran R, Sullivan M, et al. In vitro development and embryo transfer of sperm-injected oocytes derived from pregnant mares. Theriogenology. 1996;1(45):304.

[63] McKinnon AO, Lacham-Kaplan O, Trounson AO. Pregnancies produced from fertile and infertile stallions by intracytoplasmic sperm injection (ICSI) of single frozen-thawed spermatozoa into in vivo matured mare oocytes. J Reprod Fertil Suppl 2000;56:513−17.

[64] Cochran R, Meintjes M, Reggio B, Hylan D, Carter J, Pinto C, et al. Live foals produced from sperm-injected oocytes derived from pregnant mares. J Equine Vet Sci 1998;18(11):736−40.

[65] Cochran R, Meintjes M, Reggio B, Hylan D, Carter J, Pinto C, et al. Production of live foals from sperm-injected oocytes harvested from pregnant mares. J Reprod Fertil Suppl 2000;56:503−12.

[66] Choi YH, Love LB, Varner DD, Hinrichs K. Factors affecting developmental competence of equine oocytes after intracytoplasmic sperm injection. Reproduction. 2004;127(2):187−94.

[67] Hinrichs K. In vitro production of equine embryos: state of the art. Reprod Domest Anim 2010;45:3−8.

[68] Homa ST, Swann K. Fertilization and early embryology: a cytosolic sperm factor triggers calcium oscillations and membrane hyperpolarizations in human oocytes. Hum Reprod 1994;9(12):2356−61.

[69] Foss R, Ortis H, Loncar K. Effect of artificial activation of equine oocytes on cleavage and blastocyst production following ICSI. J Equine Vet Sci 2018;66.

[70] Claes A, Cuervo-Arango J, van den Broek J, Galli C, Colleoni S, Lazzari G, et al. Factors affecting the likelihood of pregnancy and embryonic loss after transfer of cryopreserved in vitro produced equine embryos. Equine Vet J 2019;51(4):446−50.

[71] Cuervo-Arango J, Claes A, Stout T. Effect of embryo-recipient synchrony on post-ET survival of in vivo and in vitro-produced equine embryos. J Equine Vet Sci 2018;66:163−4.

[72] Seidel Jr GE. Cryopreservation of equine embryos. Vet Clin North Am Equine Pract 1996;12(1):85−99.

[73] Slade NP, Takeda T, Squires EL, Elsden RP, Seidel Jr GE. A new procedure for the cryopreservation of equine embryos. Theriogenology. 1985;24(1):45−58.

[74] Barfield JP, McCue PM, Squires EL, Seidel Jr GE. Effect of dehydration prior to cryopreservation of large equine embryos. Cryobiology. 2009;59(1):36−41.

[75] Choi YH, Hartman DL, Bliss SB, Hayden SS, Blanchard TL, Hinrichs K. High pregnancy rates after transfer of large equine blastocysts collapsed via micromanipulation before vitrification. Reproduction, Fertil Dev 2009;22(1):203.

[76] Choi YH, Velez IC, Riera FL, Roldan JE, Hartman DL, Bliss SB, et al. Successful cryopreservation of expanded equine blastocysts. Theriogenology. 2011;76(1):143−52.

[77] Stout TA. Cryopreservation of equine embryos: current state-of-the-art. Reprod Domest Anim 2012;47:84−9.

[78] Wade CM, Giulotto E, Sigurdsson S, Zoli M, Gnerre S, Imsland F, et al. Genome sequence, comparative analysis, and population genetics of the domestic horse. Science. 2009;326(5954):865−7.

[79] Choi YH, Gustafson-Seabury A, Velez IC, Hartman DL, Bliss S, Riera FL, et al. Viability of equine embryos after puncture of the capsule and biopsy for preimplantation genetic diagnosis. Reproduction. 2010;140(6):893.

[80] Tobler KJ, Zhao Y, Ross R, Benner AT, Xu X, Du L, et al. Blastocoel fluid from differentiated blastocysts harbors embryonic genomic material capable of a whole-genome deoxyribonucleic acid amplification and comprehensive chromosome microarray analysis. Fertil Steril 2015;104 (2):418−25.

[81] Herrera C, Morikawa MI, Castex CB, Pinto MR, Ortega N, Fanti T, et al. Blastocele fluid from in vitro− and in vivo−produced equine embryos contains nuclear DNA. Theriogenology. 2015;83(3):415−20.

[82] Wilmut I, Schnieke AE, McWhir J, Kind AJ, Campbell KH. Viable offspring derived from fetal and adult mammalian cells. Nature. 1997;385 (6619):810.

[83] Woods GL, White KL, Vanderwall DK, Li GP, Aston KI, Bunch TD, et al. A mule cloned from fetal cells by nuclear transfer. Science. 2003;301(5636):1063.

[84] Galli C, Lagutina I, Crotti G, Colleoni S, Turini P, Ponderato N, et al. Pregnancy: a cloned horse born to its dam twin. Nature. 2003;424 (6949):635.

[85] Hinrichs K, Choi YH, Varner DD, Hartman DL. Production of cloned horse foals using roscovitine-treated donor cells and activation with sperm extract and/or ionomycin. Reproduction. 2007;134(2):319−25.

[86] Gambini A, Jarazo J, Olivera R, Salamone DF. Equine cloning: in vitro and in vivo development of aggregated embryos. Biol Reprod 2012;87(1):15.

[87] Maserati M, Mutto A. In vitro production of equine embryos and cloning: today's status. J Equine Veterinary Sci 2016;41:42−50.

[88] Gambini A, Maserati M. A journey through horse cloning. Reprod Fertil Dev 2018;30(1):8−17.

[89] Willadsen SM. Cloning of sheep and cow embryos. Genome. 1989;31(2):956−62.

[90] Vajta G, Lewis IM, Hyttel P, Thouas GA, Trounson AO. Somatic cell cloning without micromanipulators. Cloning. 2001;3(2):89−95.

[91] Lagutina I, Lazzari G, Duchi R, Turini P, Tessaro I, Brunetti D, et al. Comparative aspects of somatic cell nuclear transfer with conventional and zona-free method in cattle, horse, pig and sheep. Theriogenology. 2007;67(1):90−8.

[92] Choi YH, Velez IC, Macías-García B, Hinrichs K. Timing factors affecting blastocyst development in equine somatic cell nuclear transfer. Cell Reprogram 2015;17(2):124−30.

[93] Olivera R, Moro LN, Jordan R, Pallarols N, Guglielminetti A, Luzzani C, et al. Bone marrow mesenchymal stem cells as nuclear donors improve viability and health of cloned horses. Stem Cell Cloning Adv Appl 2018;11:13.

[94] Johnson AK, Clark-Price SC, Choi YH, Hartman DL, Hinrichs K. Physical and clinicopathologic findings in foals derived by use of somatic cell nuclear transfer: 14 cases (2004−2008). J Am Vet Med Assoc 2010;236(9):983−90.

[95] Lopez Frias FJ, Torres CR. The ethics of cloning horses in polo. Int J. Appl Philos 2019;.

Chapter 3

Reproductive technologies in sheep

S. Naitana and S. Ledda

Department of Veterinary Medicine, University of Sassari, Sassari, Italy

3.1 Introduction

Sheep breeding has been growing significantly over the last decades, playing an important role in the improvement of economical and social activity in many countries. The steady increase in the number of animal breeds and their geographic distribution is due to the ability of small ruminants to live and produce in unfavorable environments, compared to other animal species.

More recently, the increase in sheep breeding has been supported by the development and improvement of reproductive technologies. Reproduction control and manipulation represent one of the most effective tools for the genetic progression in this productive species [1].

However, assisted reproductive technologies (ARTs) have not reached wide application, and while induction of estrus and synchronization and artificial insemination (AI) are currently used in breeding programs, many others such as superovulation and embryo transfer (ET), in vitro embryo production (IVEP), embryo cryopreservation, and embryo manipulation are not yet extensively applied [2]. For example, multiple ovulation embryo transfer (MOET) in sheep is, in many cases, almost restricted to a few countries and, in most instances, is still at an experimental stage. The success of this technique is affected by several limitations responsible for unpredictable overall results. This is also the case of more advanced reproductive biotechnologies such as transgenesis and gene editing for the production of animal models for human diseases and for the improvement or modification of valuable animal products.

The main limitation to a wider application is due to sheep reproductive seasonality, with a long, naturally occurring period of anestrus, a variability of response to superovulatory treatments, and the need of surgery for the collection and transfer of gametes and embryos (reviewed by Cognie et al.[3]). These unpredictable results, combined with the high costs of pharmacology stimulation and treatment, have limited the large-scale use of MOET programs in sheep and, up to now, the technique, even if considered affordable, is not fully applicable in large-scale systems.

More recently, the new prospects offered by in vitro embryo production, by repeated ovum pickup from live females and by juvenile breeding, are possibly suggesting an alternative system to the MOET program and may be more extensively used, moving from an exclusively in lab research to the field context. Recent improvements in embryo production and freezing technologies could lead to a wider propagation of valuable genes in small ruminant populations in the future and be used to form flocks without the risk of diseases. In addition, they can offer a substantial contribution in the preservation of endangered species or breeds. The new era of gene editing could offer new perspectives in sheep breeding, but the application of these innovative techniques needs to involve specialized professional roles and is limited by the relative high cost of embryo manipulation and of molecular biology analysis [4].

The aim of this manuscript is to provide an overview of some recent developments in ART in small ruminants and the latest information regarding estrus synchronization methods, follicular wave synchronization, and/or ovulation induction techniques during superovulatory treatments in ewes, embryo collection and transfer. The possibilities offered by the generation of in vitro−produced embryos from selected adult or juvenile donors will be also discussed in terms of accelerating the genetic progression of selected choice animals. Recent progress in embryo manipulation and embryo cryopreservation will be also reported.

Reproductive Technologies in Animals. DOI: https://doi.org/10.1016/B978-0-12-817107-3.00003-5

3.2 Seasonal control and induction of gonadal activity

Reproductive function in mammals falls under the hypothalamic control, which integrates and elaborates the various signals deriving from internal and external environment including genetic condition, age, seasonal photoperiod, social interaction, nutritional aspect, and breeding system. The final output of these factors activates the hypothalamic-pituitary-gonadal axis by the alternating secretion of gonadotrophin-releasing hormone (GnRH) in the median eminence [5] or gonadotrophin-inhibiting hormone (GnIH) in the dorsomedial hypothalamic area [6]. Both neuropeptides regulate the feedback actions of follicle-stimulating hormone (FSH) and luteinizing hormone (LH) of the pituitary and the gonadal steroids such as progesterone (P4) and estradiol-17β (E2) [7].

It has been suggested that the basic signal to stimulate a GnRH pulse is initiated by an interactive network of neurons (KNDy) in the arcuate nucleus of the hypothalamus (ARC) containing kisspeptin, neurokinin B, and an endogenous opioid peptide, the dynorphin [8]. Neurokinin B neurons, in sheep, stimulate kisspeptin release acting on the secretory terminals in the median eminence of GnRH [9]. A few minutes later dynorphin neurons, expressing the kappa opioid receptors, activate the interruption of kisspeptin release from KNDy neuron in order to end GnRH pulses (see review [10]).

In temperate latitudes, sheep have a seasonal reproductive activity from late summer to late autumn, leading to lambing 5 months later, when high forage availability and favorable temperatures can improve the chances of offspring survival [11]. Photic information of day length annual variations is translated into an endocrine message by the nocturnal secretion of melatonin by the pineal gland [12]. Melatonin acts indirectly, through its receptors (MT1) present in the pars tuberalis of the hypothalamus [13], as a potent seasonal control on kisspeptin stimulation to GnRH secretion and dynorphin inhibition release neurons to start the reproductive activity [14]. During periods characterized by reduced daylight, melatonin inhibits GnRH in sheep, resulting in a development of competent oocytes and, to a minor extent, in increased sperm production and quality [15].

From the acquired knowledge of reproductive physiology, several approaches have been established in order to mimic the seasonal physiological reproductive conditions and gonadal control. The aim is firstly to improve the economical value of sheep breeding and secondly to reduce the environmental impact of sheep breeding.

3.2.1 Progesterone and progestagen treatment

During the year, sheep alternate their reproductive season (short daylight periods) with periods characterized by reduced reproductive activity (long photoperiod). In the follicular phase, P4 inhibits GnRH pulse frequency stimulating the dynorphin release from KNDy neurons, while, E2 inhibits GnRH pulse amplitude by inhibiting kisspeptin secretion onto GnRH neurons [16].

Exogenous administration of progesterone's other more potent synthetic analogs (progestagens) can be used to mimic the luteal phase during the physiological cycle and consequently (control) the dynamics of follicular growth and ovulation. Thus estrous synchronization can be successfully achieved by inserting vaginal polyurethane sponges impregnated with progestagens for 12−14 days, in both breeding and nonbreeding seasons [17]. This protocol of synchronization is still largely used thanks to its low cost. However, the composite material of these implants may cause an accumulation of microorganisms, responsible for vaginal microbiota modifications and potential infections and disease [18].

Another way to limit the incidence of vaginal inflammation is to reduce the duration of intravaginal device treatments. Martinez-Ros et al. [19] found different alteration rates of the vaginal pH in females treated with sponges for 14 days (80%) and 7 days (15%). This short synchronization method is not usually adopted during the breeding season due to the presence of the physiological corpus luteum (CL) after ovulation. Furthermore, progesterone treatment over 7 days during the nonbreeding season resulted in a high proportion of prepubertal-ewes ovulating and a pregnancy rate similar to the standard treatment protocol of 12 days [20].

A more suitable synchronization protocol can result from the use of controlled internal drug release (CIDR), consisting of a Y-shaped inert silicone elastomer usually containing 300 mg of natural progesterone. CIDR showed similar rates in ovine estrous synchronization to flugestone acetate without altering the pH of the vagina and increasing embryonic survival and pregnancy rates thanks to the high progesterone levels [21].

3.2.2 Prostaglandin

Prostaglandin $F_{2\alpha}$ (PG), thanks to its rapid metabolization in lungs and limited accumulation in tissues [22], represents a valid alternative to progestagens and CIDR for ovarian synchronization (see review [23]). Furthermore, the absence of residues encounters a favorable public opinion in advanced countries that are asking for safe and high-quality food as well as respect for animal welfare [24]. Natural PG, or more active synthetic analogs such as D-cloprostenol, administered at 10 mg and 100 µg doses, respectively, between day 5 and 14 of the estrous cycle, induce rapid and complete luteal regression within 24 hours [25]. In temperate latitudes, PGs are injected for estrous synchronization only in the presence of functional CL. This method of synchronization offers several practical advantages such as the administration by a simple intramuscular injection, a cheap approach, and no environmental contamination compared to progestagen intravaginal treatment [23], although other disadvantages can limit its extensive application. When PG treatment is performed between day 3 and 6 after estrous [26], the ovulating follicles are surrounded by few granulosa cells and consequently P4 levels are low, reducing pregnancy rates in treated ewes [27]. This treatment also decreases sperm movements to the fertilization site due to the alteration of myometrial contraction and vaginal mucus [28]. A great variability of the interval from PG treatment to the onset of ovulation due to follicle dynamics during the estrous cycle has been also reported [29,30]. Usually, a double PG treatment at an interval of 9−12 days is performed for estrous synchronization during the breeding season, but this protocol does not result in an adequate tight synchrony of estrous and/or ovulation [30].

Recently, it has been shown that ewes treated during the breeding season with a long interval between PG injections (15 or 16 days) had a better reproductive outcome after cervical AI with fresh semen than at mid-interval (12 or 13 days) [31]. More research needs to be done before regularly applying PG synchronization in AI programs under field conditions.

3.2.3 Melatonin

Melatonin implants (18 mg released over 100 days) are used in temperate regions to advance the reproductive period in anestrus small ruminants [11,32]. Melatonin is considered a luteotrophic agent [33,34] and in sheep living in temperate regions, it prolongs the ovarian activity in spring, increases estrous behavior and advances puberty [35].

How the presence of melatonin can positively influence ovarian activity in sheep has been clearly demonstrated by pinealectomy experiments. After the removal of the pineal gland, ewes significantly extended the length of their breeding season (310 days) compared to controls (217 days), had fewer follicular waves during the oestrous cycle, a reduced mean area of the CL, and a lower steroidogenic activity, with a consequent decrease of P4 levels in the plasma [34]. In rams, the pineal hormone increased sperm motility and fertility rates, improved blastocyst development [36] and mating efficiency, and reduced spermatozoa apoptosis [37,38]. The positive effect of melatonin treatment during the nonbreeding season on ram spermatozoa could derive from the modification of seminal plasma (SP) composition in terms of hormonal profile and the activity of some antioxidant enzymes [39]. Rams in nonbreeding season treated with melatonin for 100 days showed changes of several metabolites in their SP, which can affect the fertilizing potential of sperm [40] and increase the number of sperm cells with intact acrosome [41]. Short-term treatment by subcutaneously injection of melatonin 36 hours after CIDR withdrawal [42] at different concentrations (0, 5, and 10 mg/ewe), significantly promoted the number and quality of embryos in donors, increased the pregnancy rate and litter size after ET. Melatonin implants inserted 3 months prior to the superovulatory treatment in aged high-prolificacy ewes [43] did not improve the superovulation rate, but improved the quality of recovered embryos with a better viability compared to controls and an increase in the number of blastocysts.

3.3 Artificial insemination

The application of AI in sheep is limited by two main factors: the poor quality of frozen-thawed ram semen and the anatomy of the cervix. The cervix presents several folds (ranging between 4 and 7), which are variable among females and pose an obstacle to transcervical passage of conventional catheters for AI [44]. During AI, semen is usually released at the entrance of the cervix through the vagina canal (cervical insemination) and this kind of insemination results in unsatisfactory conception rates (40%−60%) [45].

Different approaches have been proposed and tested to improve the passage through the ovine cervix. The cervix fixation, such as the Guelph System, obtained positive results [46] but when the time employed to inseminate the ewes (number of ewes per hour) was associated with pregnancy rates, the efficiency, due to the long time required and intense animal handling, was considerate lower when compared to cervical or laparoscopic techniques [47]. Another approach consists in the use of hormonal treatments to enhance the dilation of the cervical canal, mimicking the pathway that involves the oxytocin-mediated synthesis of PGE2 enhanced by gonadotropins and estrogens [48].

Recently, to overcome this obstacle, Pau et al. [49] proposed an alternative technique for the transcervical intrauterine deposition of semen based on the surgical removal of the cervical folds immediately after birth. This procedure should be used in the near future with promising results.

As cervical insemination did not reach satisfactory results, the possibility to deposit semen doses into the uterine lumen through the uterine wall by laparoscopy (laparoscopic insemination) was considered and performed in sheep [50]. This is the preferred technique when frozen semen is used, since fertility of cryopreserved sperm used in cervical insemination is extremely poor (15%−30%) and is not available to be used in the field [51].

Previous studies have shown that intrauterine insemination by laparoscopy results in a greater pregnancy rate than cervical insemination also when using fresh semen [52].

3.4 Superovulation

The female reproductive potential in monotocous species such as sheep is very low. MOET represents an important tool to improve and spread the genetic transmission of superior females, exchange germplasm (Fig. 3.1) between isolated geographical regions, avoid the risk of wildlife species extinction, and limit the risk of disease transmission [3].

However, this promising technique cannot be applied on a large scale [53] due to the absence of repeatable and standardized ovarian stimulation and embryo recovery protocols. The results are highly variable in terms of number of embryos recovered from single donors within the same farm, and seasonal and breed effects and physiological condition of both donor and recipient animals [54,55]. Superovulation requires the synchronization of donors starting 2 days before the end of a short (7 days) or a long (12 days) progestagen/progesterone treatment. Exogenous FSH is usually administered over 3−4 days every 12 hours at decreasing doses [56]. The gonadotropin treatment improves follicle growth and prevents early atresia of antral follicles, stimulating the growth of multiple follicles that can ovulate oocytes prone to be fertilized [57]. It has been suggested [58] that some drawbacks may result from the administration of gonadotrophin containing high levels of LH, which negatively may affect the superovulatory response. More purified preparations with low LH concentration are now commercially available, but they are still too expensive to be extensively adopted in breeding programs [59].

Recent results have shown that the main causes of variability in ovarian response, following pharmacological stimulation, are related to the systemic endocrinological conditions and to local ovarian factors, including anti-Mullerian hormone (AMH) levels and paracrine molecules such as bone morphometric protein (BMP15) [60−64]. The presence of dominant follicles, secreting estradiol and inhibin A, determines an inhibition on the growth of subordinate gonadotropin-dependent follicles in sheep [65]. Ablation of dominant follicles [66] before commencing the

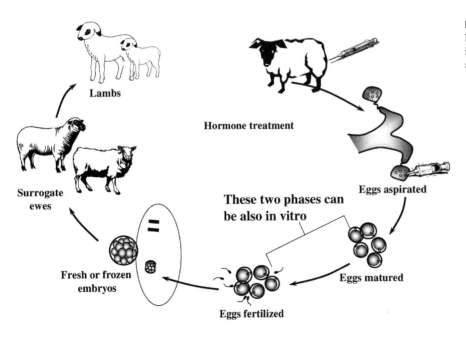

FIGURE 3.1 Workflow of multiple ovulation and embryo transfer (MOET) and in vitro embryo production (IVEP) in adult sheep.

Lambs

Hormone treatment

Eggs aspirated

Surrogate ewes

These two phases can be also in vitro

Eggs matured

Fresh or frozen embryos

Eggs fertilized

superovulation treatment increases the mobilization of primordial follicles in the ovarian pool reserve and the number of ovulations and viable embryos [64].

It has been suggested that AMH blood levels could represent the potential follicular pool reserve in the ovary [67] and could be used as a predictive indicator of ovarian response to FSH in stimulated ewes. Moreover, the ovarian evaluation of the follicle population via ultrasonography and follicle blood flow by Doppler ultrasonography may be used to estimate the potential response to the superovulation treatment, avoiding unnecessary economic costs, stress to the animals, and waste of time of personnel involved [16,68].

3.5   Embryo collection and transfer

In MOET programs, the embryos recovered at morula and blastocyst stages, 7/8 days after mating or AI, can be transferred in synchronized recipients of low genetic value, improving selective breeding programs [69]. A similar approach has been successfully used to preserve wildlife species such as mouflons [70]. Embryo collection in small ruminants is usually performed by surgical technique, although the same donor cannot be subjected to too many surgical procedures, as it will result in a severe decrease of embryo recovery rate due to the formation of scar tissue and/or surgical adhesions [71]. Alternatively, laparoscopic embryo collection and transfer has been proposed to reduce postsurgical adhesions that would limit the reuse of the same donor, but this technique is very expensive and requires highly skilled personnel [72].

Another method to perform ET called "semilaparoscopy technique," consists in the exteriorization of the final part of the uterine horn with the relative ovary by a clamp, under laparoscopic visualization. This method permits to count the CLs and to easily transfer the embryos by tomcat catheter in synchronized ewes [73].

Recently, the collection and transfer of embryos derived from superovulated ewes by the introduction of a catheter through the cervix [71] has been suggested as alternative technique to surgical and laparoscopic methods. The introduction of a catheter is relatively easy in goat, while in sheep the passage of the catheter is more complex due to the different anatomical structure of the cervix, making the transcervical pathway more difficult [44]. However, the insertion of the catheter through the cervical folds can be facilitated by cervix dilatation obtained by a local administration of estradiol and PGs E2. More recently it has been suggested to use a combined administration of D-cloprostenol, oxytocin, and estradiol benzoate in ewes synchronized with a 7-day protocol, in order to induce cervical dilatation that would facilitate the catheter passage and permit uterine flushing over a short time [71]. Although significant improvements in embryo recovery by laparoscopic and transcervical embryo flushing techniques have been achieved, they are not as widespread as the laparotomy embryo collection method.

Pregnancy rates are not only influenced by the methods of transfer employed, but also by other factors such as season, breed, age and condition of recipients, and stage and quality of embryos [74] (Fig. 3.2). Some researchers reported that blastocyst stage had a higher survival rate than morula stage in sheep [75,76].

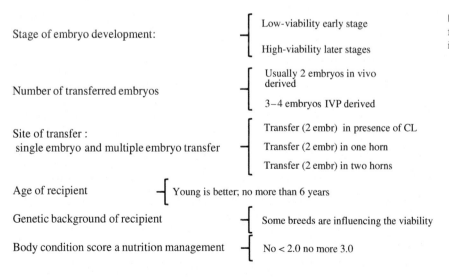

Stage of embryo development:
- Low-viability early stage
- High-viability later stages

Number of transferred embryos
- Usually 2 embryos in vivo derived
- 3–4 embryos IVP derived

Site of transfer :
single embryo and multiple embryo transfer
- Transfer (2 embr) in presence of CL
- Transfer (2 embr) in one horn
- Transfer (2 embr) in two horns

Age of recipient — Young is better; no more than 6 years

Genetic background of recipient — Some breeds are influencing the viability

Body condition score a nutrition management — No < 2.0 no more 3.0

FIGURE 3.2 Factors that can be considered for embryo transfer (ET) of in vivo and in vitro derived embryos in sheep.

3.6 Cryopreservation

The purposes of cryopreservation of germplasm samples are to temporally halt physiological activity. The cryoprocedures should be resulted in the highest post-warming viability, minimizing the cell injuries. Storage of oocytes, sperm, and embryos in cryobanks in order to keep the genetic resources and animal fertility by ART could provide a new opportunity to increase selection programs among breeds, protect population integrity of local ecotypes, and preserve endangered wildlife species [77]. However, gametes and embryos are chilling sensitive cells and are subjected to considerable morphological and functional modifications after the cryopreservation process [78].

3.6.1 Sperm

The cryopreservation process determines osmotic damage (dehydration/rehydration) on the sperm cells reducing, after resumption, the viability, motility, membrane integrity, and fertilization capacity [75,79]. Usually, commercial freezing media for ram sperm cells contain permeable (glycerol 4%−6%) and nonpermeable (egg yolk 5%−20%) cryoprotectants (CPs) to reduce ice formation [80]). Presence of CPs provides a desirable postthaw motility recovery ranging between 44% and 85% [81]. The freezing media also contains sugars such as glucose, galactose, fructose, sucrose, and trehalose. The amount of these compounds can be variable and ram semen cryopreserved with media containing 50 or 100 mM trehalose, 5% glycerol, and 5% egg yolk [82] showed a good osmotic tolerance, cryosurvival rates, and ability to maintain acrosome integrity. Our results demonstrated that, during the nonbreeding season, the presence of trehalose in the extender enhances in mouflons the spermatozoa viability and fertilizability, [83]. Chilling sensitivity seems to depend also on semen recovery methods. Collection by electroejaculation yielded semen with reduced viability after cryopreservation compared to samples recovered with an artificial vagina [84].

Sperm cryopreservation leads to main modifications in membrane phospholipids [85], DNA structure [86], capacitation process [87], and protein composition in SP [88]. These modifications significantly reduce the success of cervical insemination in sheep, failing to achieve commercially viable pregnancy rates using cryopreserved semen [80]. The different ability of fresh and frozen spermatozoa to reach the fertilization site in the oviduct depends on the capacity to swim throughout the cervical mucus [89], which is mainly produced during estrus, and plays a physiological role as a lubricant and as a protective barrier to infection [90]. The ability of sperm cells to cross the cervical mucus is different among sheep breeds [91] and is determined not only by the grade of sperm motility, but also by the changes occurring in the SP after cryopreservation [92]. The addition of SP to the extender improved frozen/thawed ram sperm in terms of motility, viability, acrosome integrity, capacitation, and mitochondrial respiratory activity [93−96].

The SP represents a mixed secretion from several glands of the reproductive tract. Proteomic studies showed that SP contains different proteins, which could be involved in the protection of sperm cells exposed to hypothermic conditions [97]. Seminal plasma proteins (SPPs) are bound to the membrane of ejaculated ram spermatozoa that underwent cold shock [93,98]. These results were observed in the absence of egg yolk, suggesting direct protective SSPs effects on spermatozoa during cooling.

The SPPs, by stabilizing the sperm membrane, are also involved in the repair of cryopreservation damage, contrasting the effects on the capacitation and modulating sperm−zona pellucida (ZP) interaction of ram spermatozoa [79,93,97,99].

Ram sperm chilling sensitivity derives from the high levels of polyunsaturated fatty acids that, in the presence of reactive oxygen species, determine an intense lipid peroxidation [100]. However, SP contains a physiological antioxidant defense system that comprises superoxide dismutase (SOD), catalase (CAT), glutathione peroxidase (GPx) enzymes, and other scavengers such as amino acids [40]. This physiological antioxidant defense system, during sperm manipulation and cryopreservation, can be impaired with the generation of elevated amount of reactive oxygen species (ROS), although the addition of scavengers may reduce the negative effect of oxidative stress caused by ROS [101,102]. In fact, while the supplementation of melatonin in freezing media improves the fertilizing potential and embryo cleavage in a dose-dependent manner of cryopreserved ram sperm [36] by increasing antioxidant defense enzymes [39], amino acids [40], and vitamins [34], the addition of antioxidant enzymes did not enhance the fertilizing capacity of frozen/thawed ram sperm [103].

The supplementation of freezing extender with spices and herbs, containing high antioxidant molecules such as phenolic compounds (flavonoids), has been proposed for the composition of cryopreservation semen media. Good results on sperm quality have been obtained using a novel extender based on soybean lecithin (SL) [104].

The protein sources of freezing extender for ram semen should be taken into account in order to avoid the possible risk of disease transmission and microbial contamination derived by commercial egg yolk. Ram semen was successfully cryopreserved with an extender supplemented with 1.5% SL, to substitute animal protein sources [105]. Among the flavonoids with antioxidant activity, trans-resveratrol was found to improve sperm viability [106].

Sperm cryopreservation in liquid nitrogen is very expensive, presents transport difficulties, and has a high risk of disease transmission. To overcome these problems and create a less expensive storage system, freeze-drying was proposed as an alternative system to replace traditional cryopreservation [107].

Few studies on ram freeze-dried sperm have been performed until now. These studies have shown that ice crystallization/recrystallization during freezing and thawing procedures present great problems for freeze-drying sperm through a lyophilization process. The crystallization phases, combined with the rehydration process, are responsible for morphological damage of spermatozoa and the total absence of sperm motility [108,109]. For this reason, freezing dry sperm is only used for fertilization by intracytoplasmic sperm injection (ICSI) technique.

3.6.2 Oocytes

Mammalian oocytes are probably among the most difficult cells to be cryopreserved. The oocyte is chilling sensitive, has a low surface/volume ratio, and in sheep has a large amount of intracellular lipids. Among the different cryopreservation techniques employed for oocyte freezing, vitrification should be considered the first-choice method [110]. The vitrification procedure, consisting of the solidification of a solution at ultrarapid cooling without the formation of ice crystals, is the most suitable method for cryopreservation of oocytes and embryos in domestic species [111]. Different strategies have been proposed to improve the cryopreservation system, taking into account multifactorial aspects [112].

One of these aspects concerns the cryodevices used during vitrification of metaphase II (MII) oocytes. It has been demonstrated that cryotop is better than open pull straw [113], but less efficient than the cryoloop device [114]. Moawad et al. [115] obtained good blastocyst rates from ovine oocytes vitrified at the germinal vesicle (GV) stage using a cryoloop, but results were not compared with other cryodevices. It is possible that specific cryodevices are better suited to different developmental stages of the oocyte.

The optimal oocyte developmental stage for cryopreservation is still under debate among researchers. Sheep oocytes can be cryopreserved at GV or MII. In both stages, morphological and functional changes are induced by cryopreservation procedures. Modifications of α- and β-tubulin dimer microtubules polymerization and depolymerization, essential to build the correct meiotic spindle morphology, have been detected [116]. Similarly, using Raman microspectroscopy, biochemical and ultrastructural modification of the oocyte cortex, including F-actin cytoskeleton, have been reported in vitrified/warmed matured ovine oocytes [117,118]. Vitrification also affected in MII stage oocytes the amount and activity of cytoplasmic protein kinases involved in the cell cycle control, such as maturation-promoting factor (MPF) and mRNAs content of several functional important genes [116,119]. Furthermore, vitrification negatively affected the development competence of vitrified immature and matured oocytes acting on the expression pattern of certain genes involved in epigenetic modification and chromatin remodeling [120].

The addition to vitrification media of caffeine, a molecule capable of stabilizing and increasing MPF and mitogen-actived protein kinases activities, showed positive effects on sheep oocytes, reducing spontaneous parthenogenetic activation [121], while no improvement in oocyte quality and developmental competence of vitrified/warmed GV ovine oocytes was observed after in vitro maturation (IVM)/in vitro fertilization (IVF) [122]. The addition of antioxidants and glycine improved the developmental competence of vitrified/warmed ovine oocytes [123].

To reduce chilling injuries on the oocyte cytoskeleton structure during cryopreservation procedures, the addition of cytochalasin B before vitrification has been proposed in ovine GV and MII. Results showed an increased rate of IVF and subsequent improvement of embryonic developmental after IVF [124]. The same positive results have been obtained with the addition of cytochalasin B in vitrified/warmed ovine matured oocytes [125]. On the contrary, it has been reported that cytochalasin B added to cumulus-oocyte complexes (COCs) before vitrification reduced oocyte survival impairing MPF and mitogen-activated protein kinase activity, and resulted in heavy damage of functional coupling between the oocyte and cumulus cell through the transzonal projections [126]. No effects have been observed after pretreatment of cytochalasin before vitrification of immature oocytes of prepubertal sheep [127].

CPs used for oocyte vitrification induce a transient increase in intracellular calcium causing ZP hardening, probably through the triggering of cortical granule exocytosis, leading to parthenogenetic activation [116,117,128,129]. To avoid these problems, a decrease of calcium contents in the vitrification solutions has been adopted, with an improvement in the blastocyst rate of vitrified ovine oocytes, [130]. Sanaei et al. [131] confirmed these data by using a calcium chelator (Tetraacetic acid) in vitrification medium of ovine oocytes.

Low developmental rates of vitrified ovine oocytes can be also related to the toxicity of high concentration of penetrating CPs, such as ethylene glycol (EG) and dimethyl sulfoxide (DMSO), during vitrification and warming procedures [129].

Prepubertal-ovine oocytes exposed to nonpenetrating CPs such as trehalose during IVM showed a higher membrane integrity after vitrification/warming compared to control, but no difference was observed in fertilization and cleavage rates [132]. The addition to the IVM medium of delipidated estrous sheep serum decreased the presence of cytoplasmic lipid droplets in oocytes after maturation, although it did not increase oocyte capacity to respond to vitrification procedures [133].

Prepubertal oocytes at MII reacted with impaired ability to the insults provoked by vitrification showing more severe alterations of the meiotic spindle conformation, tubulin and a higher percentage of parthenogenetic activation compared to adult ones [134,135]. Furthermore, it has been shown that after warming, an additional in vitro culture (IVC) time is needed to resume all the metabolic activities and to repair the induced cryopreservation damages [136]. Under the best conditions, the actual low cleavage (47%) and blastocyst rates (17%) obtained after oocyte cryopreservation represent a great limit in the practical application of this technology and more studies need to be carried out to reach desirable results.

3.6.3 Embryos

The first cryopreserved ovine embryo was obtained by employing a slow freezing technique [137]. Currently, the vitrification method is gradually replacing the slow freezing approach, thanks to its more efficient results, particularly for in vitro—produced embryos. The effectiveness of vitrification depends on many factors such as the developmental stage, in vivo or in vitro production system, CP media, volume, and cooling rate [121]. Earlier embryonic stages in sheep, such as precompacted and compacted morulae, are more sensitive than blastocysts to vitrification procedures [138], treatment of donors [139,140], in vivo and in vitro embryo production system [141], CP media [142], volume, and cooling rate [143]. Vitrification methods provided good pregnancy rates following direct transfer of embryos collected on day 7 of the estrous cycle of FSH-stimulated sheep donors, when using 25% ethylene glycol (EG) and 25% glycerol vitrification media and loaded in straw [144]. The same results have been obtained in in vivo derived ovine embryo when using 20% EG and 20% DMSO vitrification media and loaded in open pulled straws (OPS) [145]. Several studies showed that vitrification methods using different cryodevices increased embryo survival rates, improving the number of lambs born in comparison with conventional slow freezing method, both in in vivo and in vitro—produced sheep embryos [143,146]. Several variations of vitrification media have been proposed. Among them serum have been replaced with defined macromolecules such as polyvinyl alcohol (PVA) to cryopreserve ovine blastocysts [147], but this led to a decrease in protein synthesis after warming and viability [148].

The ultrastructural evaluation of in vivo—produced ovine embryos cryopreserved in both OPS vitrification and slow freezing methods showed, regardless of the embryo stage, similar alterations such as disrupted cell membranes, severe mitochondrial injury, and cytoskeleton disorganization [149]. In vitro—produced embryos show a high level of lipids [150], and this aspect, during cryopreservation of the lipid phase transition, represents one of the major causes of cryodamage. Lipid reduction can be successfully applied to improve the cryotolerance, as observed by Romao et al. [151] in vitrified sheep embryos subjected to delipidation by centrifugation in absence or presence of cytochalasin D, a cytoskeleton modulator.

Cytoplasmic alterations after cryopreservation were observed in ovine morulae and blastocysts in terms of actin filaments disorganization [152]. Positive results to limit higher oxidative conditions, apoptotic index and lower ATP concentrations were obtained with the addition of Vitamin K2 to vitrification media of ovine embryos and melatonin in culture media after vitrification [153,154]. However, at high concentrations this hormone may exert some degree of toxic activity on preimplantation embryos; thus the dose to which the embryos are exposed is pivotal in obtaining positive effects. At the same time, supplementation of an antioxidant such as L-carnitine (3.72 mM) into vitrification or warming media for 6 and 7 days in vivo—produced ovine embryos did not improve quality parameters [155].

As observed in oocytes, it is possible that immediately after warming, vitrified embryos need a short time to restore cellular interrupted metabolic activity and to repair reversible damages [138]. The resumption of metabolic activity of vitrified blastocysts collected from superovulated ewes showed that vitrified/warmed embryos require 9—12 hours of culture to complete resumption of DNA synthesis and 29—35 hours to reacquire the full capacity of protein secretion, but not the qualitative secretion pattern [156].

3.7 In vitro embryo production

One of the solutions to overcome the low efficiency of MOET programs is to produce and transfer IVEP. The IVEP procedure does not necessitate superovulation of donors as oocytes can be recovered directly from unstimulated females. Oocytes for IVEP can be collected in large numbers in abattoirs, reducing costs. Moreover, oocytes for IVEP can be recovered from prepubertal females, reducing the generation interval and increasing the genetic gain in case of high-value animals. The ART technique of "in vitro embryo production (IVEP)" comprises different steps, which essentially include oocyte collection, IVM, fertilization and embryo culture up to the preimplantation stages, usually at the blastocyst stage. Each step of the complete process has been subjected to improvements over the last decades, reaching significant results. In sheep, the rates of IVM and cleavage are around 90% and 75%, respectively, and rates of 30%—50% blastocysts are achieved, depending on age, genetic background, nutritional management, and culture conditions.

Progress in sheep in vitro embryo production contributes extensively to the progress of ART in human and other animal breeding. For these reasons, the main steps of the process will be analyzed separately. In vitro embryo production involves three steps: IVM, IVF, and IVC.

3.7.1 In vitro maturation

IVM represented the base of ART technologies, particularly in large animals where the rate of IVM has resulted in large success rates compared to other domestic species and human-derived oocytes. Moreover, IVM is the crucial step and an integral part of IVEP because it influences oocyte quality, which subsequently affects embryonic development, fetal development, and even the health of the offspring (Fig. 3.3).

The first observations of in vitro meiotic resumption came from the studies of Pincus and Ezmann [157] where IVM oocyte was obtained in rabbit oocytes collected from donors stimulated with gonadotropin. The studies of Pincus and Ezman [157] indicated for the first time the need to remove the oocyte from the follicle structure to have the meiotic resumption in order to overcome the block exerted by follicular signals. Significant advances were introduced by Edwards et al. in the 1960s [158,159] which laid the foundation of modern IVM technology. Edwards' studies indicated that not only was the removal of the oocyte from the follicular context essential for meiotic maturation, but also was important to consider the time of in vitro incubation. Among the different species evaluated by these seminal researchers, sheep represented the first large animal model in which procedures were standardized and subsequently applied to

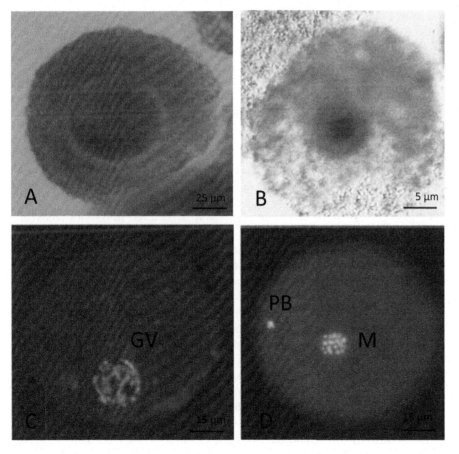

FIGURE 3.3 In vitro maturation of sheep oocytes (IVM) cultured for 24 hours: (A) cumulus-oocyte complex (COC); (B) cumulus expansion after 24 hours of incubation with gonadotropins; (C) germinal vesicle (GV) stage oocyte after Hoechst 33342 staining; (D) metaphase II oocyte after 24 hours of maturation stained with Hoechst 33342 which visualized the metaphase plate (M) and the polar body (PB).

other domestic animals. It is interesting to note that protocols (time and basic medium for maturation and supplements) established more than 40 years ago, are still being used without any significant modifications [160,161].

The establishment of IVM system in sheep depends on different factors that may influence its efficiency, such as the type of collection (in vivo and postmortem), ovarian status, the age of donors, and IVC conditions.

3.7.2 Oocyte collection

Essentially, IVEP in sheep are generated from oocytes collected from slaughtered animals or recovered in vivo by laparoscopic oocyte ovum pickup (LOPU) method. In the latter case, the collection of COCs from living donors requires surgical procedures. The surgical collection limits the reuse of the same donors to a few times [162]. Furthermore, the dimension of follicular size in nonstimulated ewes hinders the successful aspiration of all the follicle court. This limit can be overcome by hormonal treatments that may increase the follicular size, though they represent an extra cost that can restrict its application [163]. Due to this limitation, the number of studies is scarce and the LOPU technique is not much widespread and still requires some improvements.

Generally speaking, the cheapest way to harvest oocytes for IVM in sheep is to collect them from the ovaries of slaughtered animals. In this case, the ovaries and the subsequent collected oocytes are from ewes in the reproductive and nonreproductive seasons. Technically, the oocytes are recovered from the follicles by slicing the ovarian tissue or by follicular fluid aspiration. Both the techniques have been demonstrated to be efficient in sheep, providing an average of 5/6 oocytes from each ovary. In our experience, the oocyte-cumulus complex is better preserved when oocytes are retrieved by ovary slicing, while aspiration often causes the loss of cumulus cells due to mechanical forces exerted during aspiration through the syringe needle [164]. In very young females (1−3-month old), follicles are small and the smooth texture of the ovary makes it difficult to select oocytes by follicle aspiration. Consequently, oocytes from these ovaries are routinely obtained by means of slicing, from a heterogeneous degree of growth and atresia and from unknown stages of follicle development.

Previous studies [165], corroborated by our personal observation, have not revealed any significant difference in the number of oocytes collected between reproductive and nonreproductive seasons. Differences in oocyte recovery due to seasonality were reported in other studies, with a higher rate of developmental competent oocytes collected in the reproductive season compared to those recovered out of the reproductive season [165]. It is possible that breeds and geographic distribution may influence the different developmental competence of oocytes collected during and outside the reproductive season.

3.7.3 Ovarian status: follicular size, presence or absence of corpus luteum, and dominant follicle

Follicular size and the presence of corpus luteum are factors that are taken into account during oocyte ovarian collection in sheep, as they can influence their rate of maturation and developmental capacity.

Developmental competence of oocytes is acquired progressively, as females reach puberty and as follicular diameter increases to maximum size. It has been determined that follicular size does not influence meiotic progression rates though it has a significant effect on blastocyst development [166,167].

In sheep, it has been reported that a higher percentage of blastocysts [168,169] is achieved in oocytes recovered from follicles larger than 3.5 mm (40%) compared with those smaller than 3.5 mm (26%). In very young lambs (1−3 month old), the largest population of follicles are less than 3 mm, with IVM rates comparable to those observed from adult oocytes [170], although blastocyst development is significantly lower than those obtained from oocytes of adult donors [166,171]. For this reason, oocytes are mainly selected according to their diameter and morphologic appearance of cumulus cells and ooplasm [161,170].

It has been reported that the presence or absence of the corpus luteum could influence the developmental competence of harvested oocytes [172]. There are contradictory results and in some cases no difference in oocyte quality was observed, while in other studies it was reported that, in the presence of CL, the quality of oocytes and embryos was higher compared to those recovered in the absence of CL [173,174]. It would seem that the presence of corpora lutea and high levels of progesterone could modulate the gonadotropins. For instance, it has been suggested that progesterone is a key endocrine signal in sheep, governing periodic increases in both serum FSH concentrations and number of follicular waves per cycle. However, there is some debate on whether these effects of luteal progesterone on antral follicle lifespan are local, systemic (i.e., mediated by changes in FSH/LH secretion), or both. Hence, under the influence of luteal progesterone, the sensitivity of FSH-producing gonadotropins to GnRH may increase, resulting in a higher secretion of FSH from the pituitary gland. Secondly, circulating progesterone concentrations may dictate the clearance rate of circulating FSH [175].

3.7.4 Age of donors

IVM techniques in sheep have been established and improved mainly using oocytes collected from postpubertal animals as they reach the full reproductive maturity and ovarian dynamics are under the well-defined local and systemic endocrinological control [176]. It has been observed that optimal oocyte donors are 2/3-year-old animals, as a reduction of oocyte quality can be detected in older ewes, even if the reduced ability of oocytes harvested in old sheep is less consistent than what has been observed in humans or other large animal models [177–179]. However, collection of oocytes can be performed also in prepubertal animals and several studies reported the feasibility and efficiency of oocyte collection in donors in a wide age range (ranging from 30 days to 5/6 months). The main reasons for this interest are firstly, ovaries from prepubertal females have a high percentage of antral follicles from which high numbers of oocytes can be collected; secondly, the use of juvenile donors in ET programs offers considerable potential for accelerated genetic gain in domestic livestock through reduced generation interval.

Oocytes can be harvested in vivo from unstimulated or stimulated prepubertal donors or from slaughtered animals. In the first case, no difference could be detected between stimulated or unstimulated animals [166,180], although ovarian stimulation can increase the follicular size and allow a better follicle aspiration, improving the rate of oocyte recovery. The number of collected oocytes is highly variable and can range from 10 up to 200 oocytes from single donors. This large variation is related to the ovarian status as, particularly in the youngest ewes (30–40 days old), active waves of follicular growth and follicular atresia have been reported. The IVM rates of oocytes collected from stimulated and nonstimulated prepubertal donors are similar to those derived from adult ewes, but a slight delay in the meiotic progression was reported [181,182]. However, the feasibility and efficiency of oocyte collection from prepubertal animal still present some limits due to the high variability of oocytes that can be recovered and to the reduced developmental ability to reach the blastocyst stage after IVF and embryo culture compared to adult animals (15%/25% vs 30%/60%). The reduced viability seems to be related to morphological and metabolic differences observed in oocytes derived from prepubertal animals such as minor oocyte diameter and cytoplasm volume, reduced coupling between oocyte and cumulus cells, decreased stockpile of specific developmental transcripts and reduced content of calcium storage [183–185].

The evaluation of ovarian status of prepubertal-ovine donors can be taken into account to carry on an efficient IVEP program.

Dosage of AMH in lambs during the early weeks after parturition was found to be a good predictive marker of antral follicle population [64,186]. Lambs that at 3 weeks of age showed the highest level of AMH gave not only the largest number of oocytes, but also oocytes with a better in vitro developmental rate to reach the blastocyst stage [186].

3.7.5 Strategies to improve in vitro maturation in sheep

The first criteria used for oocyte collection is based on follicle size and generally oocytes contained in the small- and mid-sized antral follicles have inherently low developmental competence. These incompetent oocytes exhibit an impaired cytoplasmic maturation, even if they can resume and complete meiosis in relatively high percentages when incubated with gonadotropins in IVM systems. The presence of asynchronous nuclear and cytoplasmic maturation has suggested the application of different strategies to improve oocyte developmental competence in sheep. It has been suggested to inhibit, before processing the oocytes for IVM, the spontaneous oocyte meiotic maturation in vitro for 6–12 hours, in order to allow the oocyte to reach the full growth before being fertilized. This prematuration time can extend the oocyte-cumulus cells gap-junctional communication in vitro, determining a continued storage of mRNAs and proteins in the oocyte.

Among the different molecules that can reversibly inhibit the meiotic maturation, Roscovitine (Rosco) has been observed to efficiently inhibit the activity of MPF, maintaining the oocyte at GV stage. Adult sheep oocytes cultured with Rosco were in meiosis arrest and, after their release, they reached the rate of MII at a rate similar to not arrested ones. However, after IVF and IVC, no statistical differences were observed between Rosco-treated and control oocytes [187].

Alternatively, the use of exogenous oocyte-secreted growth factors (OSFs), that play a role in cell communication between oocyte and cumulus and granulosa cells, was suggested to improve cytoplasmic maturation. OSFs are released as soluble paracrine factors by fully grown oocytes and possibly exert their effects on granulosa cells by improving the functional coupling between oocytes and cumulus cells, as reviewed by Gilchrist and coworkers [188,189]. Recently, in order to increase the paracrine factors produced during IVM, a novel technique for IVM of sheep oocytes has been proposed [190] based on oocyte maturation in a small microbioreactor formed by the coating of small amount of IVM medium with a superhydrophobic powder.

More recently, another strategy to increase oocyte and embryo developmental competence has been advanced and is based on the culture with the simulated physiological oocyte maturation (SPOM) system. This IVM system is based on a sequential exposition of oocyte to different treatments: pre-IVM phase of $1-2$ hours, which increases oocyte cAMP levels; an IVM phase containing a type-3 phosphodiesterase inhibitor (cilostamide) with hormonal induced (FSH) maturation and an extended IVM interval [191]. Contradictory results have been obtained with this method and at the moment it appears that it does not significantly improve the developmental competence of oocytes matured, when compared to standard procedures.

Finally, it has been reported how oocytes with a different developmental competence can be selected in a noninvasive way before the maturation by staining procedures. Immature oocytes of adult and prepubertal oocytes can be stained by Brillant cresyl blue (BCB), which expresses the different amount of glucose-6-phosphate dehydrogenase (G6PDH) activity. The amount of G6PDH is higher in growing oocytes, while it is lower in fully grown oocytes. Oocytes selected through the BCB staining produced more blastocysts in vitro and of better quality. However, due to different staining and methodology procedures, this technique needs to be fully validated before it is considered a discriminating system for the evaluation of developmental capacity [192].

3.8   In vitro fertilization

IVF is a complex procedure requiring appropriate sequential actions such as sperm selection and sperm capacitation. It has also been observed that, even if motility parameters are equivalent between rams, their ability to fertilize in vitro can be significantly different, possibly as a consequence of the capacitating sensitivity during sperm preparation [193]. These differences can also occur in the same ram during the year. Before IVF, ejaculates need to be prepared prior to coincubation with oocytes. Therefore, after collection, semen is washed to remove the SP by centrifugation or by dilution with proper extenders. Subsequently, a selection of the sperm cells with the best motility is performed using, for ram semen, two main techniques: swim-up and discontinuous density gradient centrifugation [194,195]. Both systems can efficiently select the best motile population, but the swim-up technique is mainly used because ram semen is very sensitive to centrifugation. Once spermatozoa are selected, capacitation is carried out in vitro using several agents such as heparin and estrous sheep serum (ESS). Actually, ram semen is incubated with 50 mg/mL heparin for 45 minutes prior to IVF, whereas $2\%-20\%$ ESS is added to IVF in sheep. IVF is usually carried out in synthetic oviductal fluid (SOF) medium with a final sperm concentration of $1-4 \times 10^6$ spermatozoa/mL, and sperm and oocytes are coincubated between 16 and 24 hours at $38°C/39°C$. As far as the atmosphere used during the IVF period is concerned, both low oxygen tension (5% O_2) and atmosphere of 5% CO_2 in air resulted in similar fertilization rates [196]. High-quality ram semen, when regularly prepared for IVF, should guarantee IVF rates of around 75%.

3.9   Intracytoplasmic sperm injection

Another possibility to achieve IVF is through the use of ICSI technique, which consists of the injection of a single spermatozoon into the ooplasm of a matured oocyte. This technique offers the possibility to avert polyspermic fertilization and can permit the selection of the best male germ cells, reducing the abnormalities of parental origin. Moreover, ICSI enables the production of offspring from nonmotile spermatozoa and allows to predetermine the sex of the offspring by using sex-sorted spermatozoa. At present, ICSI is not very efficient in sheep and many studies have been carried out to discover the full potential of this technique [197]. Moreover, the ordinary IVF protocol is able to guarantee high-oocyte fertilization rates in this species. Another limiting factor in the use of this technique is represented by the need to induce a chemical activation of the oocyte as the sole injection is not sufficient to determine high rates of activated oocytes. Pretreatment of sperm cells with detergents before ICSI seems to improve the rate of spontaneous activation and blastocyst production [108].

3.10   Embryo culture in vitro

After fertilization, presumptive zygotes are usually cultured in vitro up to the blastocyst stage (Fig. 3.4). Over the last decades, different systems of IVC of sheep embryos have been experimented. Essentially, they can be summarized in three systems: (1) a culture of zygotes in temporary recipients [198]; (2) in vitro coculture with somatic cells, especially

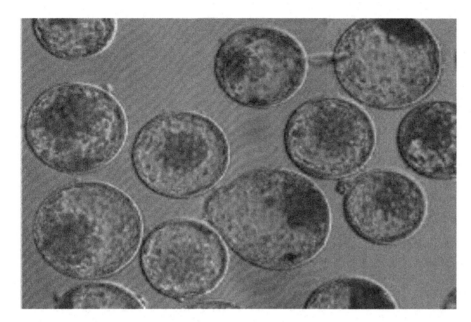

FIGURE 3.4 Expanded and hatching blastocysts obtained after 7 days of in vitro culture (IVP).

oviductal cells [195,199,200]; (3) culture in semi defined media, which is formulated with the composition of the oviductal secretion [195,200]. IVC culture media are formulated according to embryo metabolic requests and contain antioxidant molecules, growth factors, and protein sources, thanks to the addition of serum (5% or 10%) or bovine serum albumin (BSA). Currently, SOF with amino acids is the most widely used medium for IVC in sheep [195]. Culture conditions do not change during the 6/7 days of culture and are performed in 5% of O_2, 5% CO_2, and 90% N_2. The IVC of sheep embryos by coculture with somatic cells as oviductal cell monolayer requires a different gas atmosphere. In this case, a hyperbaric condition (20% of O_2) is necessary to maintain the viability of somatic cells. Comparison between 5% and 20% oxygen concentrations revealed a different ability to develop up to the blastocyst stage when embryos were cultured in these two different gas atmospheres [196]. Another approach to improve the system is the possibility to culture embryos in a biphasic medium, which should respond to the different metabolic requests of the embryos between the early and late preimplantation stages. However, the multiphase system did not determine a significant advantage compared to the system in which the embryo culture is performed for the whole time with the same medium [201]. It seems that, thanks to embryo plasticity, IVC embryo culture with no medium changes determines better developmental rates, avoiding manipulation stress. Moreover, this system reduces laboratory work for the embryologists. To evaluate the quality of IVC culture systems in sheep, several indicators have been proposed: percentage of cleavage embryos, rates of blastocysts, total cell blastocyst number, and in vitro hatching rates. Of course, pregnancy rates and number of offsprings after ET of IVEP into recipients still remain the best criteria. The speed of in vitro embryo development can be considered another possible criterion. This parameter is usually checked 24 hours after sperm coincubation with matured eggs and, according to different studies and from our personal observations, the presence of more than 50% of two-cell cleaved embryos is a good indicator of oocyte developmental competence. We have also observed that cleaved embryos in the first 24 hours reached in higher percentage the blastocyst stage and had a higher cell number at blastocyst stage compared to those with slower cleavage [202]. On the other hand, embryos with slow cleavage arrest their development at different preimplantation embryonic stages, usually before the embryonic genomic activation (8−16 cell stage) or at compacted morula stage. IVC is usually performed in groups in large culture system (350 μL) or small droplets volume (from 1−2 μL up to 50 μL per embryo) under paraffin oil at 38.5°C in 5% O_2, 5% CO_2, and 90% N_2 in humidified atmosphere.

Another criterion to evaluate the quality of IVC system in sheep is to quantify the in vitro hatching rate. High hatching rates are usually observed in those embryos that develop faster, and are correlated with the number of total cells [203].

Although significant progress in IVEP in sheep has been reported, different studies have shown that IVC culture conditions can be responsible for alteration detected during fetal development and in the derived offspring. It has been reported that IVEP, when transferred into recipient ewes, can generate larger size lambs with several abnormalities, usually termed "large offspring syndrome". Inappropriate IVC conditions may also induce defects in the placenta and absent or reduced signs of birth [204]. The highest incidence of these alterations has been reported in the culture system containing serum, while it is less frequent with the supplementation of BSA. It has been presumed that suboptimal embryo culture conditions and the presence of unknown substances can alter the gene expression of imprinted and non-imprinted genes functionally important during the early embryonic stages. In particular the insulin growth factor system is affected by the IVC conditions, with latent effects during fetal growth and early postnatal life [205].

3.11 Embryo manipulation and cloning

Willandsen [206] was the first to produce monozygotic sheep twins through the blastomere separation technique. This involves the positioning of individual blastomeres in empty ZP. The part of embryos was embedded in an agar cylinder and transferred into the oviduct of temporary recipients, or alternatively, they can be cultured in vitro [207].

When blastocysts from 1/2, 1/4, and 1/8 embryos are transferred, 80%, 50%, and 6%, respectively, come to term [208]. Therefore 1/4 embryos are the minimal size for achieving blastomere separation in sheep. Observations in sheep [208,209] showed that the time of blastulation is not affected by the division of embryos into halves or quarters. In fact, 1/2 and 1/4 embryos become blastocyst at the same time as normal embryos, but their cell number is proportionally reduced. A simple and more practical procedure, known as embryo splitting, can be used at both morula and blastocyst stage. Sheep embryos have been split with a fine glass needle [210,211] or with a blade [212].

The developmental stage affects the efficiency of this technique. Best results are obtained by bisecting blastocysts [213]. No differences are reported, if the split embryos are transplanted with or without ZP [212]. The developmental characteristics of preimplantation embryos limit the number of genetically identical offspring that can be produced using this technology.

3.12 Embryo biopsy

Embryo genomic selection by genetic screening of preimplantation embryos is increasingly being used to select the best embryos for modern breeding programs. The procedure starts with the collection of a few cells through micromanipulation biopsy from each embryo. Recovered cells are then genotyped and the genomic estimation is performed by molecular biology analysis to predict the possible improvements in animal production. The method can also be used to prevent the transfer of embryos that are a carrier of known recessive lethal genetic defects or other transmitted genetic diseases. Significant progress has been made in the bovine breed, while this methodology in sheep has been applied to a smaller extent. Most of the studies concerning sheep have been focused on the optimization of the biopsy techniques, estimating the effects of the biopsy methods on the subsequent viability and the stage of embryonic preimplantation development.

There are different biopsy methodologies available for cell collection from preimplantation embryos in sheep: the blade method in which a portion of compacted morulae or the polar trophectoderm from blastocyst is cut, and the needle biopsy method, which requires a micromanipulation procedure to aspirate cells from embryos at different cleavage stages. Both methods are designed to collect a sufficient number of cells for genetic analysis while avoiding the reduction of postmanipulation viability. For this reason, manipulation is usually performed only on high-quality embryos at blastocyst stage [76], as they most often contain a large number of cells, and is less suitable for poor-quality embryos, particularly those deriving from IVEP system. The collection of a small number of cells through embryo biopsy that can then be cultured in vitro to expand their number and facilitate subsequent molecular analysis was suggested as a viable alternative. Sex determination in sheep embryos has been performed after biopsy and subsequent in vitro cell culture of trophectoderm cells [214].

3.13 Cloning by nuclear transplantation

Initial nuclear transplantation experiments in Xenopus eggs provided the first evidence for the conservation of the genome after cellular differentiation. The successful nuclear transfer (NT) in amphibians and the establishment of totipotency of a differentiated cell paved the way for similar cloning experiments in other organisms.

The first real success in transplanting living nuclei from one cell to another was achieved by Briggs and King in 1952 [215]. They showed that the blastula nucleus of a Rana pipiens egg could be transplanted into the enucleated egg of the same species and, in a significant number of cases, they were able to obtain swimming tadpoles.

The nuclear transplantation techniques employed at this time, have been borrowed from those developed in mice by McGrath and Solter [216] and subsequently used in sheep [217]. A single blastomere from early-stage embryos, or a somatic cell, is transferred into the perivitelline space of an MII oocyte, previously enucleated by aspiration of the metaphase plate in a micropipette. While this can be carried out easily in laboratory animals [216], egg opacity in sheep imposes blind aspiration of the cytoplasm below the first polar body. Blastomere−oocyte pairs were fused by fusogenic agents such as the Sendai virus [216] or simply by electrofusion [217]. The fusion rate was around 90% [217], and after fusion the reconstituted embryos were placed in cytochalasin B to prevent haploidization. After 5 days of culture in temporary recipients, 48% of embryos developed to the blastocyst stage. Lambing rate of the transferred embryos to synchronized ewes resulted in around 18%. Even if the technique used is not actually so different from the previous procedures, significant progress has been made over the last decades.

Different areas of developmental biology have been investigated further to improve nuclear transplantation technology in sheep in order to obtain a full reprogramming of transferred nuclei: (1) improve enucleation procedures to obtain maximum enucleation rate and minimum cytoplasm removal, (2) use of in vitro−matured oocytes with good developmental potential instead of using limited sources of in vivo−matured oocytes, (3) use of nuclei donor of less differentiated cells such as pluripotent stem cells, and (4) improve cell cycle synchronization between the recipient cytoplast and the transplanted nucleus.

Currently, the possibility to stain the metaphase plate chromatin with fluorescent probes permits to have a 100% enucleation success rate and to remove a small portion of oocyte cytoplasm. However, it has been reported that ultraviolet (UV) exposure impairs oocyte development [218], reducing cleavage rate and blastocyst formation (cleavage, 59% for untreated oocytes vs 8% UV-exposed; blastocyst stage, 32% untreated vs 0% UV-exposed). Enucleation procedures can be facilitated by exposing the oocytes to demecolcine treatment and performing demecolcine-assisted enucleation, However, authors comparing these alternative approaches with the classical straight enucleation assessed that the latter remains the most reliable and least harmful protocol for somatic cell nuclear transfer (SCNT) [218].

IVM oocytes are currently used for NT in sheep with results that are reaching those obtained with in vivo−produced oocytes. IVM oocyte quality for NT appears to be improved when gametes are recovered from stimulated animals [219].

The source of the donor nucleus is a crucial aspect when the aim is to increase nuclear reprogramming efficiency. Most of the information available is based on mouse model, where different kinds of donor cells have been used. Cloned animals were obtained using less differentiated cells such as embryonic stem cells [220], primordial germ cells [221], and fetal fibroblasts [222], and from well-differentiated somatic cells such as cumulus cells [223], skin fibroblasts [224], and Sertoli cell [225]. In large animals, adult and fetal fibroblasts, cumulus cells, and mammary gland cells were used [226,227]. However, even though improved development to the blastocyst stage has been sometimes reported in NT embryos, especially in those reconstructed with embryonic cells, the percentage of clones developing to term is still disappointedly low. A new era of sheep cloning was heralded by the birth of Dolly, born in July 1996 [228], after transfer of the nucleus of a mammary cell of adult ewes to enucleated oocyte at MII. This extraordinary success was obtained by the observation that a greater proportion of reconstructed embryos developed to term if they were produced by transfer of nucleus in G0/G1 to oocyte at MII. G0 cell stage is a quiescent phase that can be easily obtained by serum deprivation without any chemical treatment [229,230]. This impressive success opened the way to cloning by nuclear transplantation in sheep with adult cells and underlined the importance of cell oocyte synchronization during embryo reconstruction [231].

More recently, following the idea that reprogramming is largely depending on the best synchronization between the recipient cytoplast and the nuclear donor cells, an interesting alternative for the cell donor preparation has been proposed by Iuso et al. [232] and Czernik et al. [233]. These authors demonstrated the possibility to replace histones with protamines, through the heterogenous expression of human protamine 1 (hPrm1) gene in sheep fibroblasts. They also reported that protaminization of somatic nucleus can be optimized by adjusting the best concentration and exposure time to trichostatin A (TSA) in serum-starved fibroblasts. Protaminized nuclei injected into enucleated oocytes efficiently underwent protamine to maternal histone TH2B exchange and developed into normal blastocyst-stage embryos in vitro. Altogether, these findings present a model to study male-specific chromatin remodeling, which can be exploited for the improvement of SCNT.

Other approaches also consider that cloned embryos display a wide array of epigenetic alterations, including DNA methylation, histone acetylation, methylation, and noncoding RNA transcripts expression [234]. Correcting these shortcomings will certainly improve mammalian cloning efficiency. Along this line, it has been reported that treatment with histone deacetylase inhibitors, or impeding Xist expression, and injection of histone lysine demethylase family member mRNA into embryos can facilitate the establishment of nuclear reprogramming [235−237].

3.14 Production of transgenic sheep

Transgenic sheep livestock was heralded in 1985 with the production of the first transgenic sheep [238], and Hammer's work was followed by numerous studies from several groups [239,240]. Transgenic sheep were produced to investigate the possibility that high growth hormone secretion would enhance growth characteristics and enhance meat production. Pronuclei of Merino sheep embryos were injected with the fusion gene construct ovine metallothionein ovine growth hormone. The generation of transgenic livestock holds considerable promise for the development of biomedical and agricultural systems. The main areas of application of transgenesis in ruminants include modification of protein production in the mammary gland, increase of disease resistance [241], enhanced wool production [242], and modification of biochemical pathways to increase the efficiency of the ruminant digestive system (Ward, 2000) as well as growth and development [240].

Great interest has also been addressed to the production of recombinant human proteins in the mammary gland of transgenic sheep. These human pharmaceutical proteins include α-antitrypsin, clotting factors VIII and IX, and fibrinogen [243,244].

The techniques to produce transgenic sheep changed consistently over the last decades. The early methods used to transfer new genes were based on the direct transfer of numerous copies of genes into the male pronuclear formation during the zygote phase. With this methodology, the level of efficiency was relatively low and significant progress was obtained when new approaches were used for the production of transgenic sheep, for instance, somatic cell cloning and sperm-mediated gene transfer.

Concomitant with the success of SCNT, the first cloning transgenic sheep was produced by NT with stable transgenic somatic cells [245]. The generation of transgenic livestock in sheep shows several significant shortcomings, despite the use of pronuclear microinjection or SCNT, such as low efficiency, high costs, random integration, and frequent incidence of mosaicism. Efficient generation of transgenic livestock at low costs remains to be developed in the transgenic animal field.

Further assisted tools have emerged to enhance the efficiency of genetic modification and to simplify the generation of genetically modified founders. These tools include viral vectors, recombinases, transposons, and endonucleases. The classes of site-specific endonucleases (meganucleases, zinc finger nucleases, transcription activator-like effector nucleases, and clustered regularly interspaced short palindromic repeats (CRISPRs) have attracted a great interest due to their DNA double-strand break-inducing role, which enables desired DNA modifications based on the stimulation of native cellular DNA repair mechanisms. Currently, gene editing in sheep using CRISPR/Cas9 systems have attracted a great deal of attention with the idea of producing a viable transgenic sheep model for investigating gene functions, improving animal breeding, producing pharmaceuticals in milk, improving animal disease resistance, and treating human diseases [4].

Conclusion

Over the last 50 years basic and applied research has provided the foundation for the development of ART, increasing significantly the efficiency of reproduction in sheep. The efficiency of ART techniques has produced fluctuating results due to the difficulties related to the structural and functional features of the genital tract, such as reproductive seasonality, typical of monotocous species such as sheep. Another aspect that objectively limits the development and application of ART can be ascribed to the low cost of small ruminants when compared to the value of the drugs employed and the costs involved in the applied procedures. Some practical protocols, such as induction and synchronization of the estrous cycle and AI, were successfully initiated in farms in developed countries. Other techniques related to the establishment of genetic cryobanking for the preservation of both gametes and embryos, MOET, the potential of IVEP derived from oocyte of prepubertal and adult, still require further development in order to avoid surgical approaches. More advanced technologies such as transgenic ICSI, cloning, and dry freezing remain at an experimental stage.

Yet the development and application of ART could bring benefits to those populations in rural areas of the world that have adopted sheep rearing in order to maintain biodiversity and improving production, providing thus real opportunities for social and economic progress.

References

[1] Amiridis GS, Cseh S. Assisted reproductive technologies in the reproductive management of small ruminants. Anim Reprod Sci 2012;130 (3-4):152−61.
[2] Ledda S, Gonzales-Bulnes A. ET-Technologies in small ruminants. In: Niemann H, Wrennzycki C, editors. Animal biotechnology, vol. 1. Cham: Springer; 2018.
[3] Cognie Y, Baril G, Poulin N, Mermillod P. Current status of embryo technologies in sheep and goat. Theriogenology 2003;59(1):171−88.
[4] Kalds P, Zhou S, Cai B, Liu J, Wang Y, Petersen B, et al. Sheep and goat genome engineering: from random transgenesis to the CRISPR era. Front Genet 2019;10:750.

[5] Schally AV, Arimura A, Baba Y, Nair RM, Matsuo H, Redding TW, et al. Isolation and properties of the FSH and LH-releasing hormone. Biochem Biophys Res Commun 1971;43(2):393−9.

[6] Tsutsui K, Ubuka T. How to contribute to the progress of neuroendocrinology: discovery of GnIH and progress of GnIH research. Front Endocrinol 2018;9:662.

[7] Clarke IJ, Sari IP, Qi Y, Smith JT, Parkington HC, Ubuka T, et al. Potent action of RFamide-related peptide-3 on pituitary gonadotropes indicative of a hypophysiotropic role in the negative regulation of gonadotropin secretion. Endocrinology 2008;149(11):5811−21.

[8] Lehman MN, Coolen LM, Goodman RL. Minireview: kisspeptin/neurokinin B/dynorphin (KNDy) cells of the arcuate nucleus: a central node in the control of gonadotropin-releasing hormone secretion. Endocrinology 2010;151(8):3479−89.

[9] Goodman RL, Hileman SM, Nestor CC, Porter KL, Connors JM, Hardy SL, et al. Kisspeptin, neurokinin B, and dynorphin act in the arcuate nucleus to control activity of the GnRH pulse generator in ewes. Endocrinology 2013;154(11):4259−69.

[10] Nestor CC, Bedenbaugh MN, Hileman SM, Coolen LM, Lehman MN, Goodman RL. Regulation of GnRH pulsatility in ewes. Reproduction 2018;156(3):R83−99.

[11] Malpaux B, Viguie C, Skinner DC, Thiery JC, Chemineau P. Control of the circannual rhythm of reproduction by melatonin in the ewe. Brain Res Bull 1997;44(4):431−8.

[12] Karsch FJ, Bittman EL, Foster DL, Goodman RL, Legan SJ, Robinson JE. Neuroendocrine basis of seasonal reproduction. Recent Prog Hormone Res 1984;40:185−232.

[13] de Reviers MM, Ravault JP, Tillet Y, Pelletier J. Melatonin binding sites in the sheep pars tuberalis. Neurosci Lett 1989;100(1-3):89−93.

[14] Li Q, Roa A, Clarke IJ, Smith JT. Seasonal variation in the gonadotropin-releasing hormone response to kisspeptin in sheep: possible kisspeptin regulation of the kisspeptin receptor. Neuroendocrinology 2012;96(3):212−21.

[15] Dardente H, Birnie M, Lincoln GA, Hazlerigg DG. RFamide-related peptide and its cognate receptor in the sheep: cDNA cloning, mRNA distribution in the hypothalamus and the effect of photoperiod. J Neuroendocrinol 2008;20(11):1252−9.

[16] Goodman RL, Gibson M, Skinner DC, Lehman MN. Neuroendocrine control of pulsatile GnRH secretion during the ovarian cycle: evidence from the ewe. Reprod Suppl 2002;59:41−56.

[17] Robinson TJ. Use of progestagen-impregnated sponges inserted intravaginally or subcutaneously for the control of the oestrous cycle in the sheep. Nat Biotechnol 1965;206:39−41.

[18] Hosseini Panah SM, Anvarian M, Mousavinia M, Alimardan M, Hamzei S, Zengir SBM. Effects of progesterone in synchronization of estrus and fertility in Shal ewes in nonproductive season. Eur J Exp Biol 2014;4:83−6.

[19] Martinez-Ros P, Astiz S, Garcia-Rosello E, Rios-Abellan A, Gonzalez-Bulnes A. Effects of of short-term intravaginal progestagens on the onset and features of estrus, preovulatory LH surge and ovulation in sheep. Anim Reprod Sci 2018;197:317−23.

[20] Martinez MF, McLeod B, Tattersfield G, Smaill B, Quirke LD, Juengel JL. Successful induction of oestrus, ovulation and pregnancy in adult ewes and ewe lambs out of the breeding season using a GnRH + progesterone oestrus synchronisation protocol. Anim Reprod Sci 2015;155:28−35.

[21] Swelum AAA, Alowaimer AN, Abouheif MA. Use of fluorogestone acetate sponges or controlled internal drug release for estrus synchronization in ewes: effects of hormonal profiles and reproductive performance. Theriogenology 2015;84(4):498−503.

[22] Piper PJ, Vane JR, Wyllie JH. Inactivation of prostaglandins by the lungs. Nature 1970;225(5233):600−4.

[23] Fierro S, Gil J, Vinoles C, Olivera-Muzante J. The use of prostaglandins in controlling estrous cycle of the ewe: a review. Theriogenology 2013;79(3):399−408.

[24] Martin GB, Milton JTB, Davidson RH, Hunzicker GEB, Lindsay DR, Blache D. Natural methods for increasing reproductive efficiency in small ruminants. Anim Reprod Sci 2004;82-3:231−46.

[25] Acritopoulou S, Haresign W. Response of ewes to a single injection of an analogue of PGF-2 alpha given at different stages of the oestrous cycle. J Reprod Fertil 1980;58(1):219−21.

[26] Nephew KP, Mcclure KE, Ott TL, Dubois DH, Bazer FW, Pope WF. Relationship between variation in conceptus development and differences in estrous-cycle duration in ewes. Biol Reprod 1991;44(3):536−9.

[27] Fierro S, Olivera-Muzante J, Gil J, Vinoles C. Effects of prostaglandin administration on ovarian follicular dynamics, conception, prolificacy, and fecundity in sheep. Theriogenology 2011;76(4):630−9.

[28] Bartlewski PM, Duggavathi R, Aravindakshan J, Barrett DMW, Cook SJ, Rawlings NC. Effects of a 6-day treatment with medroxyprogesterone acetate after prostaglandin F-2 alpha-induced luteolysis at midcycle on antral follicular development and ovulation rate in nonprolific western white-faced ewes. Biol Reprod 2003;68(4):1403−12.

[29] Vinoles C, Rubianes E. Origin of the preovulatory follicle after induced luteolysis during the early luteal phase in ewes. Can J Anim Sci 1998;78(3):429−31.

[30] Martemucci G, D'Alessandro AG. Estrous and fertility responses of dairy ewes synchronized with combined short term GnRH, PGF(2 alpha) and estradiol benzoate treatments. Small Rumin Res 2010;93(1):41−7.

[31] Fierro S, Olivera-Muzante J. Long interval prostaglandin as an alternative to progesterone-eCG based protocols for timed AI in sheep. Anim Reprod Sci 2017;180:78−84.

[32] Forcada F, Abecia JA, Zuniga O, Lozano JM. Variation in the ability of melatonin implants inserted at two different times after the winter solstice to restore reproductive activity in reduced seasonality ewes. Aust J Agr Res 2002;53(2):167−73.

[33] Abecia JA, Forcada F, Zuniga O. The effect of melatonin on the secretion of progesterone in sheep and on the development of ovine embryos in vitro. Veterinary Res Commun 2002;26(2):151−8.

[34] Manca ME, Manunta ML, Spezzigu A, Torres-Rovira L, Gonzalez-Bulnes A, Pasciu V, et al. Melatonin deprival modifies follicular and corpus luteal growth dynamics in a sheep model. Reproduction 2014;147(6):885—95.

[35] Abecia JA, Chemineau P, Gomez A, Keller M, Forcada F, Delgadillo JA. Presence of photoperiod-melatonin-induced, sexually-activated rams in spring advances puberty in autumn-born ewe lambs. Anim Reprod Sci 2016;170:114—20.

[36] Succu S, Berlinguer F, Pasciu V, Satta V, Leoni GG, Naitana S. Melatonin protects ram spermatozoa from cryopreservation injuries in a dose-dependent manner. J Pineal Res 2011;50(3):310—18.

[37] Casao A, Mendoza N, Perez-Pe R, Grasa P, Abecia JA, Forcada F, et al. Melatonin prevents capacitation and apoptotic-like changes of ram spermatozoa and increases fertility rate. J Pineal Res 2010;48(1):39—46.

[38] Mura MC, Luridiana S, Farci F, Di Stefano MV, Daga C, Pulinas L, et al. Melatonin treatment in winter and spring and reproductive recovery in Sarda breed sheep. Anim Reprod Sci 2017;185:104—8.

[39] Casao A, Perez-Pe R, Abecia JA, Forcada F, Muino-Blanco T, Cebrian-Perez JA. The effect of exogenous melatonin during the non-reproductive season on the seminal plasma hormonal profile and the antioxidant defence system of Rasa Aragonesa rams. Anim Reprod Sci 2013;138(3-4):168—74.

[40] Satta V, Manca ME, Torres-Rovira L, Succu S, Mereu P, Nehme M, et al. Effects of melatonin administration on seminal plasma metabolites and sperm fertilization competence during the non-reproductive season in ram. Theriogenology 2018;115:16—22.

[41] Kaya A, Aksoy M, Baspinar N, Yildiz C, Ataman MB. Effect of melatonin implantation to sperm donor rams on post-thaw viability and acrosomal integrity of sperm cells in the breeding and non-breeding season. Reprod Domest Anim 2001;36(3-4):211—15.

[42] Song Y, Wu H, Wang X, Haire A, Zhang X, Zhang J, et al. Melatonin improves the efficiency of super-ovulation and timed artificial insemination in sheep. PeerJ 2019;7:e6750.

[43] Forcada F, Abecia JA, Cebrian-Perez JA, Muino-Blanco T, Valares JA, Palacin I, et al. The effect of melatonin implants during the seasonal anestrus on embryo production after superovulation in aged high-prolificacy Rasa Aragonesa ewes. Theriogenology 2006;65(2):356—65.

[44] Kershaw CM, Khalid M, McGowan MR, Ingram K, Leethongdee S, Wax G, et al. The anatomy of the sheep cervix and its influence on the transcervical passage of an inseminating pipette into the uterine lumen. Theriogenology 2005;64(5):1225—35.

[45] Anel L, Kaabi M, Abroug B, Alvarez M, Anel E, Boixo JC, et al. Factors influencing the success of vaginal and laparoscopic artificial insemination in churra ewes: a field assay. Theriogenology 2005;63(4):1235—47.

[46] Buckrell BC, Buschbeck C, Gartley CJ, Kroetsch T, McCutcheon W, Martin J, et al. Further development of a transcervical technique for artificial insemination in sheep using previously frozen semen. Theriogenology 1994;42(4):601—11.

[47] Casali R, Pinczak A, Cuadro F, Guillen-Munoz JM, Mezzalira A, Menchaca A. Semen deposition by cervical, transcervical and intrauterine route for fixed-time artificial insemination (FTAI) in the ewe. Theriogenology 2017;103:30—5.

[48] Falchi L, Taema M, La Clanche S, Scaramuzzi RJ. The pattern of cervical penetration and the effect of topical treatment with prostaglandin and/or FSH and oxytocin on the depth of cervical penetration in the ewe during the peri-ovulatory period. Theriogenology 2012;78(2):376—84.

[49] Pau S, Falchi L, Ledda M, Bogliolo L, Ariu F, Zedda MT. Surgery on cervical folds for transcervical intrauterine artificial insemination with frozen-thawed semen enhances pregnancy rates in the sheep. Theriogenology 2019;126:28—35.

[50] Evans G, Maxwell WMC. Frozen storage of semen. In: Butterworths W, editor. Salamon's artificial insemination of sheep and goats. Sydney: Butterworths; 1987. p. 122—41.

[51] Salamon S, Maxwell WMC. Frozen storage of ram semen 2. Causes of low fertility after cervical insemination and methods of improvement. Anim Reprod Sci 1995;38(1-2):1—36.

[52] Vilarino M, Cuadro F, dos Santos-Neto PC, Garcia-Pintos C, Menchaca A. Time of ovulation and pregnancy outcomes obtained with the prostaglandin-based protocol Synchrovine for ETA! in sheep. Theriogenology 2017;90:163—8.

[53] Forcada F, Amer-Meziane MA, Abecia JA, Maurel MC, Cebrian-Perez JA, Muino-Blanco T, et al. Repeated superovulation using a simplified FSH/eCG treatment for in vivo embryo production in sheep. Theriogenology 2011;75(4):769—76.

[54] Bartlewski PM, Seaton P, Oliveira MEF, Kridli RT, Murawski M, Schwarz T. Intrinsic determinants and predictors of superovulatory yields in sheep: circulating concentrations of reproductive hormones, ovarian status, and antral follicular blood flow. Theriogenology 2016;86(1):130—43.

[55] Gonzalez-Bulnes A, Baird DT, Campbell BK, Cocero MJ, Garcia-Garcia RM, Inskeep EK, et al. Multiple factors affecting the efficiency of multiple ovulation and embryo transfer in sheep and goats. Reprod Fertil Dev 2004;16(4):421—35.

[56] de Bulnes AG, Moreno JS, Brunet AG, Sebastian AL. Relationship between ultrasonographic assessment of the corpus luteum and plasma progesterone concentration during the oestrous cycle in monovular ewes. Reprod Domest Anim 2000;35(2):65—8.

[57] Loi P, Ptak G, Dattena M, Ledda S, Naitana S, Cappai P. Embryo transfer and related technologies in sheep reproduction. Reprod Nutr Dev 1998;38(6):615—28.

[58] Rubianes E, Ibarra D, Ungerfeld R, Carbajal B, Decastro T. Superovulatory response in anestrous ewes is affected by the presence of a large follicle. Theriogenology 1995;43(2):465—72.

[59] Gonzalez-Bulnes A, Santiago-Moreno J, Cocero MJ, Lopez-Sebastian A. Effects of FSH commercial preparation and follicular status on follicular growth and superovulatory response in Spanish Merino ewes. Theriogenology 2000;54(7):1055—64.

[60] Berlinguer F, Gonzalez-Bulnes A, Succu S, Leoni G, Mossa F, Bebbere D, et al. Effects of progestagens on follicular growth and oocyte developmental competence in FSH-treated ewes. Domest Anim Endocrin 2007;32(4):303—14.

[61] Gonzalez-Bulnes A, Berlinguer F, Cocero MJ, Garcia-Garcia RM, Leoni G, Naitana S, et al. Induction of the presence of corpus luteum during superovulatory treatments enhances in vivo and in vitro blastocysts output in sheep. Theriogenology 2005;64(6):1392—403.

[62] Juengel JL, Quirke LD, Lun S, Heath DA, Johnstone PD, McNatty KP. Effects of immunizing ewes against bone morphogenetic protein 15 on their responses to exogenous gonadotrophins to induce multiple ovulations. Reproduction 2011;142(4):565—72.

[63] Mossa F, Duffy P, Naitana S, Lonergan P, Evans ACO. Association between numbers of ovarian follicles in the first follicle wave and superovulatory response in ewes. Anim Reprod Sci 2007;100(3-4):391−6.

[64] Torres-Rovira L, Gonzalez-Bulnes A, Succu S, Spezzigu A, Manca ME, Leoni GG, et al. Predictive value of antral follicle count and anti-Mullerian hormone for follicle and oocyte developmental competence during the early prepubertal period in a sheep model. Reprod Fertil Dev 2014;26(8):1094−106.

[65] Evans AC, Flynn JD, Duffy P, Knight PG, Boland MP. Effects of ovarian follicle ablation on FSH, oestradiol and inhibin A concentrations and growth of other follicles in sheep. Reproduction 2002;123(1):59−66.

[66] Rubianes E, Ungerfeld R, Vinoles C, Rivero A, Adams GP. Ovarian response to gonadotropin treatment initiated relative to wave emergence in ultrasonographically monitored ewes. Theriogenology 1997;47(8):1479−88.

[67] Torres-Rovira L, Succu S, Pasciu V, Manca ME, Gonzalez-Bulnes A, Leoni GG, et al. Postnatal pituitary and follicular activation: a revisited hypothesis in a sheep model. Reproduction 2016;151(3):215−25.

[68] Oliveira MEF, Ribeiro IF, Rodriguez MGK, Maciel GS, Fonseca JF, Brandao FZ, et al. Assessing the usefulness of B-mode and colour Doppler sonography, and measurements of circulating progesterone concentrations for determining ovarian responses in superovulated ewes. Reprod Domest Anim 2018;53(3):742−50.

[69] Bari F, Khalid M, Wolf B, Haresign W, Murray A, Merrell B. The repeatability of superovulatory response and embryo recovery in sheep. Theriogenology 2001;56(1):147−55.

[70] Ledda S, Naitana S, Loi P, Dattena M, Gallus M, Branca A, et al. Embryo recovery from superovulated mouflons (Ovis-gmelini-musimon) and viability after transfer into domestic sheep. Anim Reprod Sci 1995;39(2):109−17.

[71] Fonseca JF, Zambrini FN, Guimaraes JD, Silva MR, Oliveira MEF, Bartlewski PM, et al. Cervical penetration rates and efficiency of non-surgical embryo recovery in estrous-synchronized Santa Ines ewes after administration of estradiol ester (benzoate or cypionate) in combination with d-cloprostenol and oxytocin. Anim Reprod Sci 2019;203:25−32.

[72] Mckelvey WAC, Robinson JJ, Aitken RP, Robertson IS. Repeated recoveries of embryos from ewes by laparoscopy. Theriogenology 1986;25(6):855−65.

[73] Li QY, Guan H, Hou J, An XR, Chen YF. Technical note: transfer of ovine embryos through a simplified mini-laparotomy technique. J Anim Sci 2008;86(11):3224−7.

[74] Dixon AB, Knights M, Winkler JL, Marsh DJ, Pate JL, Wilson ME, et al. Patterns of late embryonic and fetal mortality and association with several factors in sheep. J Anim Sci 2007;85(5):1274−84.

[75] Armstrong DT, Evans G. Factors influencing success of embryo transfer in sheep and goats. Theriogenology 1983;19:31−42.

[76] Naitana S, Loi P, Ledda S, Cappai P, Dattena M, Bogliolo L, et al. Effect of biopsy and vitrification on in vitro survival of ovine embryos at different stages of development. Theriogenology 1996;46(5):813−24.

[77] Naitana S, Ledda S, Leoni G, Bogliolo L, Loi P, Cappai P. Membrane integrity and fertilizing potential of cryopreserved spermatozoa in European mouflon. Anim Reprod Sci 1998;52(2):105−12.

[78] Arav A, Yavin S, Zeron Y, Natan D, Dekel I, Gacitua H. New trends in gamete's cryopreservation. Mol Cell Endocrinol 2002;187(1-2):77−81.

[79] Pini T, Leahy T, de Graaf SP. Sublethal sperm freezing damage: manifestations and solutions. Theriogenology 2018;118:172−81.

[80] Salamon S, Maxwell WMC. Storage of ram semen. Anim Reprod Sci 2000;62(1-3):77−111.

[81] Anel L, de Paz P, Alvarez M, Chamoro CA, Boixo JC, Manso A, et al. Field and in vitro assay of three methods for freezing ram semen. Theriogenology 2003;60(7):1293−308.

[82] Ahmad E, Naseer Z, Aksoy M, Kucuk N, Ucan U, Serin I, et al. Trehalose enhances osmotic tolerance and suppresses lysophosphatidylcholine-induced acrosome reaction in ram spermatozoon. Andrologia 2015;47(7):786−92.

[83] Berlinguer F, Leoni GG, Succu S, Mossa F, Galioto M, Madeddu M, et al. Cryopreservation of European mouflon (Ovis gmelini musimon) semen during the non-breeding season is enhanced by the use of trehalose. Reprod Domest Anim 2007;42(2):202−7.

[84] Jimenez-Rabadan P, Soler AJ, Ramon M, Garcia-Alvarez O, Maroto-Morales A, Iniesta-Cuerda M, et al. Influence of semen collection method on sperm cryoresistance in small ruminants. Anim Reprod Sci 2016;167:103−8.

[85] Barrios B, Perez-Pe R, Gallego M, Tato A, Osada J, Muino-Blanco T, et al. Seminal plasma proteins revert the cold-shock damage on ram sperm membrane. Biol Reprod 2000;63(5):1531−7.

[86] Fraser L, Strzezek J, Kordan W. Effect of freezing on sperm nuclear DNA. Reprod Domest Anim 2011;46:14−17.

[87] Naresh S, Atreja SK. The protein tyrosine phosphorylation during in vitro capacitation and cryopreservation of mammalian spermatozoa. Cryobiology 2015;70(3):211−16.

[88] Soleilhavoup C, Tsikis G, Labas V, Harichaux G, Kohnke PL, Dacheux JL, et al. Ram seminal plasma proteome and its impact on liquid preservation of spermatozoa. J Proteom 2014;109:245−60.

[89] Lightfoot RJ, Salamon S. Fertility of ram spermatozoa frozen by the pellet method. I. Transport and viability of spermatozoa within the genital tract of the ewe. J Reprod Fertil 1970;22(3):385−98.

[90] Fair S, Meade KG, Reynaud K, Druart X, de Graaf SP. The biological mechanisms regulating sperm selection by the ovine cervix. Reproduction 2019;158(1):R1−13.

[91] Richardson L, Hanrahan JP, O'Hara L, Donovan A, Fair S, O'Sullivan M, et al. Ewe breed differences in fertility after cervical AI with frozen-thawed semen and associated differences in sperm penetration and physicochemical properties of cervical mucus. Anim Reprod Sci 2011;129(1-2):37−43.

[92] Rickard JP, Pini T, Soleilhavoup C, Cognie J, Bathgate R, Lynch GW, et al. Seminal plasma aids the survival and cervical transit of epididymal ram spermatozoa. Reproduction 2014;148(5):469−78.

[93] Bernardini A, Hozbor F, Sanchez E, Fornes MW, Alberio RH, Cesari A. Conserved ram seminal plasma proteins bind to the sperm membrane and repair cryopreservation damage. Theriogenology 2011;76(3):436−47.

[94] Dominguez MP, Falcinelli A, Hozbor F, Sanchez E, Cesari A, Alberio RH. Seasonal variations in the composition of ram seminal plasma and its effect on frozen-thawed ram sperm. Theriogenology 2008;69(5):564−73.

[95] Leahy T, Marti JI, Evans G, Maxwell WMC. Seasonal variation in the protective effect of seminal plasma on frozen-thawed ram spermatozoa. Anim Reprod Sci 2010;119(1-2):147−53.

[96] Rovegno M, Feitosa WB, Rocha AM, Mendes CM, Visintin JA, D'Avila Assumpcao ME. Assessment of post-thawed ram sperm viability after incubation with seminal plasma. Cell Tissue Bank 2013;14(2):333−9.

[97] Pinia T, Farmer K, Druart X, Teixeira-Gomes AP, Tsikis G, Labas V, et al. Binder of sperm proteins protect ram spermatozoa from freeze-thaw damage. Cryobiology 2018;82:78−87.

[98] Pini T, Leahy T, Soleilhavoup C, Tsikis G, Labas V, Combes-Soia L, et al. Proteomic investigation of ram spermatozoa and the proteins conferred by seminal plasma. J Proteome Res 2016;15(10):3700−11.

[99] Ledesma A, Fernandez-Alegre E, Cano A, Hozbor F, Martinez-Pastor F, Cesari A. Seminal plasma proteins interacting with sperm surface revert capacitation indicators in frozen-thawed ram sperm. Anim Reprod Sci 2016;173:35−41.

[100] Watson PF. The causes of reduced fertility with cryopreserved semen. Anim Reprod Sci 2000;60:481−92.

[101] Maxwell WMC, Stojanov T. Liquid storage of ram semen in the absence or presence of some antioxidants. Reprod Fertil Dev 1996;8 (6):1013−20.

[102] Coyan K, Baspinar N, Bucak MN, Akalin PP. Effects of cysteine and ergothioneine on post-thawed Merino ram sperm and biochemical parameters. Cryobiology 2011;63(1):1−6.

[103] Camara DR, Silva SV, Almeida FC, Nunes JF, Guerra MM. Effects of antioxidants and duration of pre-freezing equilibration on frozen-thawed ram semen. Theriogenology 2011;76(2):342−50.

[104] Masoudi R, Sharafi M, Zareh Shahneh A, Towhidi A, Kohram H, Esmaeili V, et al. Fertility and flow cytometry study of frozen-thawed sperm in cryopreservation medium supplemented with soybean lecithin. Cryobiology 2016;73(1):69−72.

[105] Sharafi M, Zhandi M, Akbari Sharif A. Supplementation of soybean lecithin-based semen extender by antioxidants: complementary flowcytometric study on post-thawed ram spermatozoa. Cell Tissue Bank 2015;16(2):261−9.

[106] Silva EC, Cajueiro JF, Silva SV, Soares PC, Guerra MM. Effect of antioxidants resveratrol and quercetin on in vitro evaluation of frozen ram sperm. Theriogenology 2012;77(8):1722−6.

[107] Kaneko T. Sperm freeze-drying and micro-insemination for biobanking and maintenance of genetic diversity in mammals. Reprod Fertil Dev 2016.

[108] Anzalone DA, Palazzese L, Iuso D, Martino G, Loi P. Freeze-dried spermatozoa: an alternative biobanking option for endangered species. Anim Reprod Sci 2018;190:85−93.

[109] Arav A, Idda A, Nieddu SM, Natan Y, Ledda S. High post-thaw survival of ram sperm after partial freeze-drying. J Assist Reprod Gen 2018;35(7):1149−55.

[110] Ledda S, Bogliolo L, Succu S, Ariu F, Bebbere D, Leoni GG, et al. Oocyte cryopreservation: oocyte assessment and strategies for improving survival. Reprod Fertil Dev 2007;19(1):13−23.

[111] Rall WF, Fahy GM. Ice-free cryopreservation of mouse embryos at -196-degrees-C by vitrification. Nature 1985;313(6003):573−5.

[112] Quan GB, Wu GQ, Hong QH. Oocyte cryopreservation based in sheep: the current status and future perspective. Biopreserv Biobank 2017;15 (6):535−47.

[113] Succu S, Leoni GG, Bebbere D, Berlinguer F, Mossa F, Bogliolo L, et al. Vitrification devices affect structural and molecular status of in vitro matured ovine oocytes. Mol Reprod Dev 2007;74(10):1337−44.

[114] Quan GB, Wu GQ, Wang YJ, Ma Y, Lv CR, Hong QH. Meiotic maturation and developmental capability of ovine oocytes at germinal vesicle stage following vitrification using different cryodevices. Cryobiology 2016;72(1):33−40.

[115] Moawad A, Zhu J, Choi I, Amarnath D, Chen W, Campbell K. Production of good-quality blastocyst embryos following IVF of ovine oocytes vitrified at the germinal vesicle stage using a cryoloop. Reprod Fertil Dev 2013;25(8):1204−15.

[116] Succu S, Leoni GG, Berlinguer F, Madeddu M, Bebbere D, Mossa F, et al. Effect of vitrification solutions and cooling upon in vitro matured prepubertal ovine oocytes. Theriogenology 2007;68(1):107−14.

[117] Bogliolo L, Ledda S, Innocenzi P, Ariu F, Bebbere D, Rosati I, et al. Raman microspectroscopy as a non-invasive tool to assess the vitrification-induced changes of ovine oocyte zona pellucida. Cryobiology 2012;64(3):267−72.

[118] Bogliolo L, Murrone O, Piccinini M, Ariu F, Ledda S, Tilocca S, et al. Evaluation of the impact of vitrification on the actin cytoskeleton of in vitro matured ovine oocytes by means of Raman microspectroscopy. J Assist Reprod Genet 2015;32(2):185−93.

[119] Succu S, Bebbere D, Bogliolo L, Ariu F, Fois S, Leoni GG, et al. Vitrification of in vitro matured ovine oocytes affects in vitro pre-implantation development and mRNA abundance. Mol Reprod Dev 2008;75(3):538−46.

[120] Shirazi A, Naderi MM, Hassanpour H, Heidari M, Borjian S, Sarvari A, et al. The effect of ovine oocyte vitrification on expression of subset of genes involved in epigenetic modifications during oocyte maturation and early embryo development. Theriogenology 2016;86(9):2136−46.

[121] Ariu F, Bogliolo L, Leoni G, Falchi L, Bebbere D, Nieddu SM, et al. Effect of caffeine treatment before vitrification on MPF and MAPK activity and spontaneous parthenogenetic activation of in vitro matured ovine oocytes. Cryoletters 2014;35(6):530−6.

[122] Moawad A, Choi I, Zhu J, El-Wishy A, Amarnath D, Chen W, et al. Caffeine and oocyte vitrification: sheep as an animal model. Int J Vet Sci Med 2018;6(Suppl):S41−8.

[123] Ahmadi E, Shirazi A, Shams-Esfandabadi N, Nazari H. Antioxidants and glycine can improve the developmental competence of vitrified/warmed ovine immature oocytes. Reprod Domest Anim 2019;54(3):595−603.

[124] Moawad A, Zhu J, Choi I, Amarnath D, Campbell K. Effect of Cytochalasin B pretreatment on developmental potential of ovine oocytes vitrified at the germinal vesicle stage. Cryo Lett 2013;34(6):634−44.

[125] Zhang J, Nedambale TL, Yang M, Li J. Improved development of ovine matured oocyte following solid surface vitrification (SSV): effect of cumulus cells and cytoskeleton stabilizer. Anim Reprod Sci 2009;110(1-2):46−55.

[126] Bogliolo L, Ariu F, Fois S, Rosati I, Zedda MT, Leoni G, et al. Morphological and biochemical analysis of immature ovine oocytes vitrified with or without cumulus cells. Theriogenology 2007;68(8):1138−49.

[127] Silvestre MA, Yaniz J, Salvador I, Santolaria P, Lopez-Gatius F. Vitrification of pre-pubertal ovine cumulus-oocyte complexes: effect of cytochalasin B pre-treatment. Anim Reprod Sci 2006;93(1-2):176−82.

[128] Mattioli M, Barboni B, Gioia L, Loi P. Cold-induced calcium elevation triggers DNA fragmentation in immature pig oocytes. Mol Reprod Dev 2003;65(3):289−97.

[129] Tian SJ, Yan CL, Yang HX, Zhou GB, Yang ZQ, Zhu SE. Vitrification solution containing DMSO and EG can induce parthenogenetic activation of in vitro matured ovine oocytes and decrease sperm penetration. Anim Reprod Sci 2007;101(3-4):365−71.

[130] Succu S, Berlinguer F, Leoni G, Bebbere D, Satta V, Marco-Jimenez F, et al. Calcium concentration in vitrification medium affects the developmental competence of in vitro matured ovine oocytes. Theriogenology 2011;75(4):715−21.

[131] Sanaei B, Movaghar B, Valojerdi MR, Ebrahimi B, Bazrgar M, Hajian M, et al. Developmental competence of in vitro matured ovine oocytes vitrified in solutions with different concentrations of trehalose. Reprod Domest Anim 2018;53(5):1159−67.

[132] Berlinguer F, Succu S, Mossa F, Madeddu M, Bebbere D, Leoni GG, et al. Effects of trehalose co-incubation on in vitro matured prepubertal ovine oocyte vitrification. Cryobiology 2007;55(1):27−34.

[133] Barrera N, Neto PCD, Cuadro F, Bosolasco D, Mulet AP, Crispo M, et al. Impact of delipidated estrous sheep serum supplementation on in vitro maturation, cryotolerance and endoplasmic reticulum stress gene expression of sheep oocytes. PLoS One 2018;13(6).

[134] Serra E, Gadau SD, Berlinguer F, Naitana S, Succu S. Morphological features and microtubular changes in vitrified ovine oocytes. Theriogenology 2019.

[135] Serra E, Succu S, Berlinguer F, Porcu C, Leoni GG, Naitana S, et al. Tubulin posttranslational modifications in in vitro matured prepubertal and adult ovine oocytes. Theriogenology 2018;114:237−43.

[136] Succu S, Gadau SD, Serra E, Zinellu A, Carru C, Porcu C, et al. A recovery time after warming restores mitochondrial function and improves developmental competence of vitrified ovine oocytes. Theriogenology 2018;110:18−26.

[137] Willadsen SM, Polge C, Rowson LE, Moor RM. Deep freezing of sheep embryos. J Reprod Fertil 1976;46(1):151−4.

[138] Naitana S, Dattena M, Gallus M, Loi P, Branca A, Ledda S, et al. Recipient synchronization affects viability of vitrified ovine blastocysts. Theriogenology 1995;43(8):1371−8.

[139] Berlinguer F, Leoni G, Bogliolo L, Pintus PP, Rosati I, Ledda S, et al. FSH different regimes affect the developmental capacity and cryotolerance of embryos derived from oocytes collected by ovum pick-up in donor sheep. Theriogenology 2004;61(7-8):1477−86.

[140] Leoni G, Bogliolo L, Pintus P, Ledda S, Naitana S. Sheep embryos derived from FSH/eCG treatment have a lower in vitro viability after vitrification than those derived from FSH treatment. Reprod Nutr Dev 2001;41(3):239−46.

[141] Papadopoulos S, Rizos D, Duffy P, Wade M, Quinn K, Boland MP, et al. Embryo survival and recipient pregnancy rates after transfer of fresh or vitrified, in vivo or in vitro produced ovine blastocysts. Anim Reprod Sci 2002;74(1-2):35−44.

[142] Bath ML. Optimized cryopreservation of mouse sperm based on fertilization rate. J Reprod Dev 2011;57(1):92−8.

[143] dos Santos-Neto PC, Cuadro F, Barrera N, Crispo M, Menchaca A. Embryo survival and birth rate after minimum volume vitrification or slow freezing of in vivo and in vitro produced ovine embryos. Cryobiology 2017;78:8−14.

[144] Baril G, Traldi AL, Cognie Y, Leboeuf B, Beckers JF, Mermillod P. Successful direct transfer of vitrified sheep embryos. Theriogenology 2001;56(2):299−305.

[145] Green RE, Santos BFS, Sicherle CC, Landim-Alvarenga FC, Bicudo SD. Viability of OPS vitrified sheep embryos after direct transfer. Reprod Domest Anim 2009;44(3):406−10.

[146] Dattena M, Ptak G, Loi P, Cappai P. Survival and viability of vitrified in vitro and in vivo produced ovine blastocysts. Theriogenology 2000;53(8):1511−19.

[147] Naitana S, Ledda S, Loi P, Leoni G, Bogliolo L, Dattena M, et al. Polyvinyl alcohol as a defined substitute for serum in vitrification and warming solutions to cryopreserve ovine embryos at different stages of development. Anim Reprod Sci 1997;48(2-4):247−56.

[148] Leoni G, Bogliolo L, Berlinguer F, Rosati I, Pintus PP, Ledda S, et al. Defined media for vitrification, warming, and rehydration: effects on post-thaw protein synthesis and viability of in vitro derived ovine embryos. Cryobiology 2002;45(3):204−12.

[149] Bettencourt EM, Bettencourt CM, Silva JN, Ferreira P, de Matos CP, Oliveira E, et al. Ultrastructural characterization of fresh and cryopreserved in vivo produced ovine embryos. Theriogenology 2009;71(6):947−58.

[150] Amstislavsky S, Mokrousova V, Brusentsev E, Okotrub K, Comizzoli P. Influence of cellular lipids on cryopreservation of mammalian oocytes and preimplantation embryos: a review. Biopreserv Biobank 2019;17(1):76−83.

[151] Romao R, Marques CC, Baptista MC, Barbas JP, Horta AE, Carolino N, et al. Cryopreservation of in vitro-produced sheep embryos: effects of different protocols of lipid reduction. Theriogenology 2015;84(1):118−26.

[152] Dalcin L, Silva RC, Paulini F, Silva BDM, Neves JP, Lucci CM. Cytoskeleton structure, pattern of mitochondrial activity and ultrastructure of frozen or vitrified sheep embryos. Cryobiology 2013;67(2):137−45.

[153] Sefid F, Ostadhosseini S, Hosseini SM, Ghazvini Zadegan F, Pezhman M, Nasr Esfahani MH. Vitamin K2 improves developmental competency and cryo-tolerance of in vitro derived ovine blastocyst. Cryobiology 2017;77:34—40.

[154] Succu S, Pasciu V, Manca ME, Chelucci S, Torres-Rovira L, Leoni GG, et al. Dose-dependent effect of melatonin on postwarming development of vitrified ovine embryos. Theriogenology 2014;81(8):1058—66.

[155] Saraiva HFRA, Batista RITP, Alfradique VAP, Pinto PHN, Ribeiro LS, Oliveira CS, et al. L-carnitine supplementation during vitrification or warming of in vivo-produced ovine embryos does not affect embryonic survival rates, but alters CrAT and PRDX1 expression. Theriogenology 2018;105:150—7.

[156] Leoni G, Berlinguer F, Rosati I, Bogliolo L, Ledda S, Naitana S. Resumption of metabolic activity of vitrified/warmed ovine embryos. Mol Reprod Dev 2003;64(2):207—13.

[157] Pincus G, Enzmann EV. The comparative behavior of mammalian eggs in vivo and in vitro: I. The activation of ovarian eggs. J Exp Med 1935;62(5):665—75.

[158] Edwards RG. Meiosis in ovarian oocytes of adult mammals. Nat Biotechnol 1962;196:446—50.

[159] Edwards RG. Maturation in vitro of mouse, sheep, cow, pig, rhesus monkey and human ovarian oocytes. Nature 1965;208(5008):349—51.

[160] de Souza-Fabjan JMG, Panneau B, Duffard N, Locatelli Y, De Figueiredo JR, Freitas VJDF, et al. In vitro production of small ruminant embryos: late improvements and further research. Theriogenology 2014;81(9):1149—62.

[161] Paramio MT, Izquierdo D. Recent advances in in vitro embryo production in small ruminants. Theriogenology 2016;86(1):152—9.

[162] Baldassarre H, Furnus CC, deMatos DG, Pessi H. In vitro production of sheep embryos using laparoscopic folliculocentesis: alternative gonadotrophin treatments for stimulation of oocyte donors. Theriogenology 1996;45(3):707—17.

[163] Alberio R, Olivera J, Roche A, Alabart J, Folch J. Performance of a modified ovum pick-up system using three different FSH stimulation protocols in ewes. Small Rumin Res 2002;46(2-3):81—7.

[164] Rodriguez C, Anel L, Alvarez M, Anel E, Boixo JC, Chamorro CA, et al. Ovum pick-up in sheep: a comparison between different aspiration devices for optimal oocyte retrieval. Reprod Domest Anim 2006;41(2):106—13.

[165] Mara L, Sanna D, Casu S, Dattena M, Munoz IMM. Blastocyst rate of in vitro embryo production in sheep is affected by season. Zygote 2014;22(3):366—71.

[166] Ledda S, Bogliolo L, Leoni G, Naitana S. Follicular size affects the meiotic competence of in vitro matured prepubertal and adult oocytes in sheep. Reprod Nutr Dev 1999;39(4):503—8.

[167] Moor RM, Trounson AO. Hormonal and follicular factors affecting maturation of sheep oocytes in vitro and their subsequent developmental capacity. J Reprod Fertil 1977;49(1):101—9.

[168] Cognie Y. State of the art in sheep-goat embryo transfer. Theriogenology 1999;51(1):105—16.

[169] Cognie Y, Poulin N, Locatelli Y, Mermillod P. State-of-the-art production, conservation and transfer of in-vitro-produced embryos in small ruminants. Reprod Fertil Dev 2004;16(4):437—45.

[170] Ledda S, Bogliolo L, Calvia P, Leoni G, Naitana S. Meiotic progression and developmental competence of oocytes collected from juvenile and adult ewes. J Reprod Fertil 1997;109(1):73—8.

[171] Ptak G, Dattena M, Loi P, Tischner M, Cappai P. Ovum pick-up in sheep: efficiency of in vitro embryo production, vitrification and birth of offspring. Theriogenology 1999;52(6):1105—14.

[172] Widyastuti R, Rizky M, Syamsunarno AA, Saili T, Boediono A. Oocyte quality and subsequent in vitro maturation of sheep oocyte-cumulus complex from ovary with presence and absence of corpus luteum. VMIC 2017. Available from: https://doi.org/10.18502/kls.v3i6.1125.

[173] Berlinguer F, Gonzalez-Bulnes A, Succu S, Leoni G, Mossa F, Bebbere D, et al. Effects of progestagens on follicular growth and oocyte developmental competence in FSH-treated ewes. Domest Anim Endocrinol 2007;32(4):303—14.

[174] Gonzalez-Bulnes A, Berlinguer F, Cocero MJ, Garcia-Garcia RM, Leoni G, Naitana S, et al. Induction of the presence of corpus luteum during superovulatory treatments enhances in vivo and in vitro blastocysts output in sheep. Theriogenology 2005;64(6):1392—403.

[175] Bartlewski PM, Baby TE, Giffin JL. Reproductive cycles in sheep. Anim Reprod Sci 2011;124(3-4):259—68.

[176] Padmanabhan V, Cardoso RC. Neuroendocrine, autocrine, and paracrine control of follicle-stimulating hormone secretion. Mol Cell Endocrinol 2020;500:110632.

[177] Berlinguer F, Gonzalez-Bulnes A, Spezzigu A, Contreras-Solis I, Succu S, McNeilly AS, et al. Effect of aging on follicular function may be relieved by exogenous gonadotropin treatment in a sheep model. Reproduction 2012;144(2):245—55.

[178] Cimadomo D, Fabozzi G, Vaiarelli A, Ubaldi N, Ubaldi FM, Rienzi L. Impact of maternal age on oocyte and embryo competence. Front Endocrinol 2018;9:327.

[179] Ruggeri E, Ruggeri E, DeLuca KF, Galli C, Lazzari G, DeLuca JG, et al. Cytoskeletal alterations associated with donor age and culture interval for equine oocytes and potential zygotes that failed to cleave after intracytoplasmic sperm injection. Reprod Fertil Dev 2015;27(6):944—56.

[180] O'Brien JK, Beck NFG, Maxwell WMC, Evans G. Effect of hormone pre-treatment of prepubertal sheep on the production and developmental capacity of oocytes in vitro and in vivo. Reprod Fertil Dev 1997;9(6):625—31.

[181] Leoni GG, Palmerini MG, Satta V, Succu S, Pasciu V, Zinellu A, et al. Differences in the kinetic of the first meiotic division and in active mitochondrial distribution between prepubertal and adult oocytes mirror differences in their developmental competence in a sheep model. PLoS One 2015;10(4):e0124911.

[182] Palmerini MG, Nottola SA, Leoni GG, Succu S, Borshi X, Berlinguer F, et al. In vitro maturation is slowed in prepubertal lamb oocytes: ultrastructural evidences. Reprod Biol Endocrin 2014;12:115.

[183]   Ledda S, Bogliolo L, Leoni G, Naitana S. Cell coupling and maturation-promoting factor activity in in vitro-matured prepubertal and adult sheep oocytes. Biol Reprod 2001;65(1):247—52.

[184]   Leoni GG, Bebbere D, Succu S, Berlinguer F, Mossa F, Galioto M, et al. Relations between relative mRNA abundance and developmental competence of ovine oocytes. Mol Reprod Dev 2007;74(2):249—57.

[185]   OBrien JK, Dwarte D, Ryan JP, Maxwell WMC, Evans G. Developmental capacity, energy metabolism and ultrastructure of mature oocytes from prepubertal and adult sheep. Reprod Fertil Dev 1996;8(7):1029—37.

[186]   McGrice H, Kelly JM, Kleemann DO, Kind KL, Hampton AJ, Hannemann P, et al. Plasma anti-Müllerian hormone concentration as a predictive endocrine marker for selection of donor lambs to improve success in juvenile in vitro embryo transfer programs. Reprod Fertil Dev. 2020;32(4):383—91. Available from: https://doi.org/10.1071/RD18498.

[187]   Crocomo LF, Ariu F, Bogliolo L, Bebbere D, Ledda S, Bicudo SD. In vitro developmental competence of adult sheep oocytes treated with roscovitine. Reprod Domest Anim 2016;51(2):276—81.

[188]   Gilchrist RB. Recent insights into oocyte-follicle cell interactions provide opportunities for the development of new approaches to in vitro maturation. Reprod Fertil Dev 2011;23(1):23—31.

[189]   Gilchrist RB, Lane M, Thompson JG. Oocyte-secreted factors: regulators of cumulus cell function and oocyte quality. Hum Reprod Update 2008;14(2):159—77.

[190]   Ledda S, Idda A, Kelly J, Ariu F, Bogliolo L, Bebbere D. A novel technique for in vitro maturation of sheep oocytes in a liquid marble microbioreactor. J Assist Reprod Gen 2016;33(4):513—18.

[191]   Rose RD, Gilchrist RB, Kelly JM, Thompson JG, Sutton-McDowall ML. Regulation of sheep oocyte maturation using cAMP modulators. Theriogenology 2013;79(1):142—8.

[192]   Opiela J, Katska-Ksiazkiewicz L. The utility of Brilliant Cresyl Blue (BCB) staining of mammalian oocytes used for in vitro embryo production (IVP). Reprod Biol 2013;13(3):177—83.

[193]   Fukui Y, Glew AM, Gandolfi F, Moor RM. Ram-specific effects on in-vitro fertilization and cleavage of sheep oocytes matured in vitro. J Reprod Fertil 1988;82(1):337—40.

[194]   Gomez MC, Catt JW, Gillan L, Evans G, Maxwell WM. Effect of culture, incubation and acrosome reaction of fresh and frozen-thawed ram spermatozoa for in vitro fertilization and intracytoplasmic sperm injection. Reprod Fertil Dev 1997;9(7):665—73.

[195]   Walker SK, Heard TM, Seamark RF. Invitro culture of sheep embryos without coculture — successes and perspectives. Theriogenology 1992;37(1):111—26.

[196]   Leoni GG, Rosati I, Succu S, Bogliolo L, Bebbere D, Berlinguer F, et al. A low oxygen atmosphere during IVF accelerates the kinetic of formation of in vitro produced ovine blastocysts. Reprod Domest Anim 2007;42(3):299—304.

[197]   Olaciregui M, Luno V, Domingo P, Gonzalez N, Gil L. In vitro developmental ability of ovine oocytes following intracytoplasmic injection with freeze-dried spermatozoa. Sci Rep 2017;7(1):1096.

[198]   Czlonkowska M, Eysymont U, Guszkiewicz A, Kossakowski M, Dziak J. Birth of lambs after in vitro maturation, fertilization, and coculture with oviductal cells. Mol Reprod Dev 1991;30(1):34—8.

[199]   Gandolfi F, Moor RM. Stimulation of early embryonic-development in the sheep by coculture with oviduct epithelial-cells. J Reprod Fertil 1987;81(1):23—8.

[200]   Gardner DK, Lane M, Spitzer A, Batt PA. Enhanced rates of cleavage and development for sheep zygotes cultured to the blastocyst stage invitro in the absence of serum and somatic-cells — amino-acids, vitamins, and culturing embryos in groups stimulate development. Biol Reprod 1994;50(2):390—400.

[201]   Ledda S, Bogliolo L, Leoni G, Loi P, Cappai P, Naitana S. Two culture systems showing a biphasic effect on ovine embryo development from the 1-2 cell stage to hatched blastocysts. Reprod Nutr Dev 1995;35(6):629—37.

[202]   Leoni GG, Succu S, Berlinguer F, Rosati I, Bebbere D, Bogliolo L, et al. Delay on the in vitro kinetic development of prepubertal ovine embryos. Anim Reprod Sci 2006;92(3-4):373—83.

[203]   Coello A, Meseguer M, Galan A, Alegre L, Remohi J, Cobo A. Analysis of the morphological dynamics of blastocysts after vitrification/warming: defining new predictive variables of implantation. Fertil Steril 2017;108(4):659—66.

[204]   Holm P, Walker SK, Seamark RF. Embryo viability, duration of gestation and birth weight in sheep after transfer of in vitro matured and in vitro fertilized zygotes cultured in vitro or in vivo. J Reprod Fertil 1996;107(2):175—81.

[205]   Bebbere D, Bauersachs S, Furst RW, Reichenbach HD, Reichenbach M, Medugorac I, et al. Tissue-specific and minor inter-individual variation in imprinting of IGF2R is a common feature of *Bos taurus* concepti and not correlated with fetal weight. PLoS One 2013;8(4):e59564.

[206]   Willadsen SM. A method for culture of micromanipulated sheep embryos and its use to produce monozygotic twins. Nature 1979;277 (5694):298—300.

[207]   Loi P, Ledda S, Filia F, Gallus M, Cappai P, Naitana S. [In vitro separation and development of sheep blastomeres]. Boll Soc Ital Biol Sper 1990;66(12):1173—9.

[208]   Willadsen SM. The development capacity of blastomeres from 4- and 8-cell sheep embryos. J Embryol Exp Morphol 1981;65:165—72.

[209]   Willadsen SM. The viability of early cleavage stages containing half the normal number of blastomeres in the sheep. J Reprod Fertil 1980;59 (2):357—62.

[210]   Gatica R, Boland MP, Crosby TF, Gordon I. Micromanipulation of sheep morulae to produce monozygotic twins. Theriogenology 1984;21 (4):555—60.

[211]   Willadsen SM, Godke RA. A simple procedure for the production of identical sheep twins. Vet Rec 1984;114(10):240—3.

[212] Szell A, Hudson RH. Factors affecting the survival of bisected sheep embryos in vivo. Theriogenology 1991;36(3):379−87.

[213] Filia F, Loi P, Ledda S, Branca A, Fadda P. [In vitro dissection and development of sheep embryos after preservation at very low temperature (−196 degrees C)]. Boll Soc Ital Biol Sper 1990;66(12):1165−71.

[214] Leoni G, Ledda S, Bogliolo L, Naitana S. Novel approach to cell sampling from preimplantation ovine embryos and its potential use in embryonic genome analysis. J Reprod Fertil 2000;119(2):309−14.

[215] Briggs R, King TJ. Transplantation of living nuclei from blastula cells into enucleated frogs' eggs. Proc Natl Acad Sci U S A 1952;38 (5):455−63.

[216] McGrath J, Solter D. Nuclear transplantation in mouse embryos. J Exp Zool 1983;228(2):355−62.

[217] Willadsen SM. Nuclear transplantation in sheep embryos. Nature 1986;320(6057):63−5.

[218] Iuso D, Czernik M, Zacchini F, Ptak G, Loi P. A simplified approach for oocyte enucleation in mammalian cloning. Cell Reprogram 2013;15 (6):490−4.

[219] Yuan Y, Liu R, Zhang X, Zhang J, Zheng Z, Huang C, et al. Effects of recipient oocyte source, number of transferred embryos and season on somatic cell nuclear transfer efficiency in sheep. Reprod Domest Anim 2019;54(11):1443−8.

[220] Wakayama T, Rodriguez I, Perry AC, Yanagimachi R, Mombaerts P. Mice cloned from embryonic stem cells. Proc Natl Acad Sci U S A 1999;96(26):14984−9.

[221] Miki H, Inoue K, Kohda T, Honda A, Ogonuki N, Yuzuriha M, et al. Birth of mice produced by germ cell nuclear transfer. Genesis 2005;41 (2):81−6.

[222] Wakayama T, Yanagimachi R. Mouse cloning with nucleus donor cells of different age and type. Mol Reprod Dev 2001;58(4):376−83.

[223] Wakayama T, Perry AC, Zuccotti M, Johnson KR, Yanagimachi R. Full-term development of mice from enucleated oocytes injected with cumulus cell nuclei. Nature 1998;394(6691):369−74.

[224] Wakayama T, Yanagimachi R. Cloning of male mice from adult tail-tip cells. Nat Genet 1999;22(2):127−8.

[225] Ogura A, Inoue K, Ogonuki N, Noguchi A, Takano K, Nagano R, et al. Production of male cloned mice from fresh, cultured, and cryopreserved immature Sertoli cells. Biol Reprod 2000;62(6):1579−84.

[226] Akagi S, Matsukawa K, Takahashi S. Factors affecting the development of somatic cell nuclear transfer embryos in cattle. J Reprod Dev 2014;60(5):329−35.

[227] Kato Y, Tsunoda Y. Role of the donor nuclei in cloning efficiency: can the ooplasm reprogram any nucleus? Int J Dev Biol 2010;54 (11-12):1623−9.

[228] Wilmut I, Schnieke AE, McWhir J, Kind AJ, Campbell KH. Viable offspring derived from fetal and adult mammalian cells. Nature 1997;385 (6619):810−13.

[229] Campbell KH, Loi P, Cappai P, Wilmut I. Improved development to blastocyst of ovine nuclear transfer embryos reconstructed during the presumptive S-phase of enucleated activated oocytes. Biol Reprod 1994;50(6):1385−93.

[230] Campbell KH, Loi P, Otaegui PJ, Wilmut I. Cell cycle co-ordination in embryo cloning by nuclear transfer. Rev Reprod 1996;1(1):40−6.

[231] Campbell KH, McWhir J, Ritchie WA, Wilmut I. Sheep cloned by nuclear transfer from a cultured cell line. Nature 1996;380(6569):64−6.

[232] Iuso D, Czernik M, Toschi P, Fidanza A, Zacchini F, Feil R, et al. Exogenous expression of human protamine 1 (hPrm1) remodels fibroblast nuclei into spermatid-like structures. Cell Rep 2015;13(9):1765−71.

[233] Czernik M, Iuso D, Toschi P, Khochbin S, Loi P. Remodeling somatic nuclei via exogenous expression of protamine 1 to create spermatid-like structures for somatic nuclear transfer. Nat Protoc 2016;11(11):2170−88.

[234] Matoba S, Zhang Y. Somatic cell nuclear transfer reprogramming: mechanisms and applications. Cell Stem Cell 2018;23(4):471−85.

[235] Cao H, Li J, Su WL, Li JJ, Wang ZG, Sun SC, et al. Zebularine significantly improves the preimplantation development of ovine somatic cell nuclear transfer embryos. Reprod Fertil Dev 2019;31(2):357−65.

[236] Czernik M, Anzalone DA, Palazzese L, Oikawa M, Loi P. Somatic cell nuclear transfer: failures, successes and the challenges ahead. Int J Dev Biol 2019;63(3-4-5):123−30.

[237] Wakayama S, Wakayama T. Improvement of mouse cloning using nuclear transfer-derived embryonic stem cells and/or histone deacetylase inhibitor. Int J Dev Biol 2010;54(11-12):1641−8.

[238] Hammer RE, Pursel VG, Rexroad CE, Wall RJ, Bolt DJ, Ebert KM, et al. Production of transgenic rabbits, sheep and pigs by microinjection. Nature 1985;315(6021):680−3.

[239] Rexroad Jr. CE. Production of sheep transgenic for growth hormone genes. Biotechnology 1991;16:259−63.

[240] Rexroad Jr. CE, Hammer RE, Bolt DJ, Mayo KE, Frohman LA, Palmiter RD, et al. Production of transgenic sheep with growth-regulating genes. Mol Reprod Dev 1989;1(3):164−9.

[241] Clements JE, Wall RJ, Narayan O, Hauer D, Schoborg R, Sheffer D, et al. Development of transgenic sheep that express the visna virus envelope gene. Virology 1994;200(2):370−80.

[242] Su HY, Jay NP, Gourley TS, Kay GW, Damak S. Wool production in transgenic sheep: results from first-generation adults and second-generation lambs. Anim Biotechnol 1998;9(2):135−47.

[243] Clark AJ, Bessos H, Bishop JO, Brown P, Harris S, Lathe R, et al. Expression of human anti-hemophilic factor-Ix in the milk of transgenic sheep. Nat Biotechnol 1989;7(5):487−92.

[244] Wright G, Carver A, Cottom D, Reeves D, Scott A, Simons P, et al. High level expression of active human alpha-1-antitrypsin in the milk of transgenic sheep. Biotechnology 1991;9(9):830−4.

[245] Campbell KH. Transgenic sheep from cultured cells. Methods Mol Biol 2002;180:289−301.

Chapter 4

Reproductive technologies in goats

Maria-Teresa Paramio, Sandra Soto-Heras and Dolors Izquierdo

Department of Animal and Food Sciences, Faculty of Veterinary, University Autonomus of Barcelona, Barcelona, Spain

4.1 Importance of artificial reproductive technologies in goats

Goat farming is an important source of milk and meat in countries where the exploitation of large ruminant species can be difficult due to local ecological constraints. From FAO statistics [1], in 2017 the number of goats was following a linear trend, although lower than sheep and cattle, being 1034, 1202, and 1491 M, respectively. Regardless of the high number of animals, goat breeds have not been genetically improved at the same level of sheep and cattle breeds. The use of assisted reproductive technologies (ARTs) such as artificial insemination (AI) has been a key tool for traditional genetic breeding programs. Nowadays, to improve the genetic background of livestock breeds, new ARTs have been developed such as in vitro embryo production (IVEP), juvenile in vitro embryo transfer, embryo sexing, and cryopreservation of embryos and gametes. Moreover, the in vitro culture of gametes opens new possibilities for improving genetic gains by using intensively selected animals. These ARTs, together with new achievements in transgenesis and cloning by somatic nuclear cell transfer, have changed traditional goat farming products in a new "pharming" activity, using the goat mammary gland as a bioreactor for therapeutic recombinant proteins. Several laboratories are in fact focusing on transgenesis and cloning technologies, whereas only few research groups aim at improving goat ARTs solely for enhancing reproductive efficiency and increasing food production. From recent data of the Association of Embryo Technology in Europe [2] in vitro embryo commercial activity in goats is significantly lower when compared to sheep and cattle. In fact, the magnitude of embryos produced is in the order of hundreds in goat (358 of which 85 transferred), thousands in sheep (12,239 of which 1282 transferred) and hundreds of thousands in cattle (148,851 of which 130,635 transferred). Similar trend is reported for in vivo−produced embryos from data of the International Embryo Technology Society [3].

In goats as in other domestic ruminants, in vivo−derived embryos are produced by multiple ovulation and embryo transfer technology, although this procedure suffers from limitations due to the ovulation variability in response to hormonal treatments and the traumatic surgical procedure for embryo recovery (reviewed by Paramio and Izquierdo [4]). IVEP overcomes these problems because oocytes are directly recovered from the follicles by laparoscopic ovum pickup without previous superovulation. Moreover, IVEP allows us to obtain embryos from nonfertile, pregnant, lactating, juvenile, and even dead females.

Due to the importance of milk production in the dairy industry, other ARTs are also relevant in goats. For instance, sex determination of embryos can add economic value to established pregnancies. Selection according to the embryo genotype is also of interest for breeding programs by identifying inherited genetic disorders or economically advantageous trait loci. The cryopreservation of gametes and embryos allows us to reduce the sanitary risks when livestock breeds must be moved between countries or even among farms. Finally, cloning and transgenic technologies allow us to use goats for more than just milk and meat production.

4.2 In vitro embryo production

4.2.1 Current status of in vitro embryo production in goats

IVEP in goats has been improved in the last years thanks to a constant and fruitful research. However, the most important problems related to IVEP remain unsolved, such as the high variability and unpredictability of the results obtained among laboratories and even among trials within the same laboratory. IVEP involves different procedures starting with

Reproductive Technologies in Animals. DOI: https://doi.org/10.1016/B978-0-12-817107-3.00004-7

TABLE 4.1 In vitro embryo production from oocytes of prepubertal and adult goats.

Goat age	Cleavage rate (%)	Blastocyst rate (%)	References
Adult	51	7	[5]
	83	13–23	[6]
	62	19	[7]
	63–82	16–31	[8]
	55	18	[9]
	53	21	[10]
	85	46	[11]
	39–68	28–47	[12]
	72	66	[13]
	71	10	[14]
Prepubertal	46	6	[5]
	28	7	[9]
	42	5–18	[10]
	75–85	11–23	[15]
	85	29	[16]

Cleavage rate: cleaved oocytes in relation to the number of total fertilized oocytes assessed at 48 h postfertilization. Blastocyst rate: total blastocysts related to the total number of fertilized oocytes assessed at 7–8 days postfertilization.

the recovery of immature oocytes up to the production of 8-day embryos that could be either cryopreserved or directly transferred to a recipient female. The main steps involved are: (1) in vitro maturation (IVM) of oocytes directly recovered from the ovarian follicles at the germinal vesicle (GV) stage until they reach the metaphase II (MII) nuclear stage, (2) in vitro fertilization (IVF) or coincubation of matured oocytes with previously capacitated spermatozoa, and (3) in vitro culture of presumptive zygotes up to the blastocyst stage. Table 4.1 shows a summary of blastocyst production in goats, using oocytes from adult and prepubertal females.

4.2.2 In vitro maturation

Oocyte IVM is a key and limiting step for the final efficiency of IVEP programs. Crozet et al. [17] showed that goat IVM oocytes develop to the blastocyst stage at a lower rate than ovulated oocytes after IVF and in vitro embryo culture. Moreover, the blastocyst rates obtained from IVM oocytes are positively and directly related with the follicle diameter from which the oocytes are recovered. Oocytes from small follicles (<3 mm) have lower competence to develop up to the blastocyst stage. Ovaries of adult females have a higher number of large follicles (>3 mm) than ovaries of prepubertal females. Thus oocytes recovered from prepubertal females are less competent and lead to poorer results in terms of blastocyst production than those recovered from adult females (Table 4.1).

The most conventional culture medium used for IVM in goats is tissue culture medium 199 (TCM199), bicarbonate-buffered with Earle's salts, containing minerals, energy sources (glucose, glutamine), vitamins and amino acids, hormones, antioxidants and growth factors, among other components. TCM199 is usually supplemented with serum from different sources, although it has an unknown composition that produces an unpredictable effect in embryo development. A summary of the IVM media most commonly used in goats is shown in Table 4.2.

4.2.3 In vitro fertilization

Following IVM, oocytes are in vitro fertilized with capacitated spermatozoa. Semen preparation for IVF involves first a protocol for selecting the best spermatozoa followed by a treatment to induce capacitation.

TABLE 4.2 Composition of in vitro maturation media for goat oocytes.

IVM medium	References
TCM199 + Cys + EGF + FBS	[6]
TCM199 + FSH[a] + E$_2$ + EGS[d]	[7]
TCM199 + Cys + LH[f] + FSH[b] + EGS[e]	[9]
TCM199 + Pyr[i] + Gln[d] + Cys + LH[f] + FSH[b] + E$_2$ + SS	[10]
TCM199 + Pyr[j] + Cys + EGF + LH[f] + FSH[b] + E$_2$ + Shh + FBS	[13]
TCM199 + Cys + Y27632 + LIF + LH[g] + FSH[c] + E$_2$ + FBS	[14]
TCM199 + Pyr[j] + Gln[k] + Cys + EGF + LH[h] + FSH[b] + E$_2$ + FBS	[15]
TCM199 + Pyr[j] + Gln[l] + melatonin + EGF + LH[h] + FSH[b] + E$_2$ + FBS	[16]

Cys, Cysteamine (100 μM); *EGF*, epidermal growth factor (10 ng/mL); *FSH*, follicle-stimulating hormone ([a]0.03 μg/mL, [b]5 μg/mL, [c]0.5 μg/mL); *E$_2$*, estradiol 17β (1 μg/mL); *EGS*, estrus goat serum ([d]5%, [e]10%); *LH*, luteinizing hormone ([f]10 μg/mL, [g]0.5 μg/mL, [h]5 μg/mL); *Pyr*, sodium pyruvate ([i]2.5 mM, [j]0.2 mM); *Gln*, glutamine ([k]2 mM, [l]1 mM); *SS*, steer serum (10%); *Shh*, sonic hedgehog protein (0.5 μg/mL); *FBS*, fetal bovine serum (10%); Y27632 (10 μM); *LIF*, leukemia inhibitory factor (100 IU/mL); melatonin (0.1 μM).

Two methods are currently implemented for selecting spermatozoa: swim-up and Percoll gradient. Swim-up is generally used for buck fresh semen [18]. In this procedure, the semen is placed at the bottom of a tube and layered with Tyrode's medium (TALP) medium. After 30−60 minutes at 38.5°C, the top layer is recovered, where highly motile spermatozoa are located. Percoll density gradient is also used for fresh semen but it is better used for frozen-thawed semen [19]. The gradient is formed with two density phases (45% and 90%) of colloidal silica particles. The semen is placed at the top of the 45% layer and centrifuged. The resulting pellet will contain the best spermatozoa. With regard to sperm capacitation, incubating sperm cells with heparin (50 mg/mL) for 15−60 minutes at 38.5°C has given good results in goat IVEP in our laboratory. Groups of 15−30 cumulus-oocyte complexes are usually cultured with $1-4 \times 10^6$ spermatozoa/mL in microdrops (50−100 μL) in modified TALP, covered with mineral oil for 17−24 hours at 38.5°C with humidified air and 5% CO_2. Afterward, the presumptive zygotes are pipetted to remove cumulus cells and attached spermatozoa and placed in the embryo-culture medium.

4.2.4 Embryo culture

Synthetic oviductal fluid (SOF), first described by Tervit et al. [20] and based on the composition of ovine oviductal fluid, is ordinarily used in goat embryo culture. SOF is usually supplemented with 10% fetal calf serum due to its stimulating effect on mitosis, although it can also induce chromosomal abnormalities [21]. In fact, Romaguera et al. [10] observed that 90% of in vitro−produced goat blastocysts presented mixoploidy. Another embryo-culture method is the coculture with oviductal epithelial cells. In goat, Rodríguez-Dorta et al. [6] showed that this system could lead to higher embryo survival rates after vitrification and transfer to recipient females, compared to embryos cultured with SOF. Coculture with somatic cells brings, with the risk of contamination, unpredictable results depending on the cell state and require a long preparation, hence a defined medium is preferred. Finally, sequential media can be used in order to support specific embryo stages. The sequential G1.2/G2.2 supplemented with bovine serum albumin (BSA) has been tested in goats with good results [22]. In our research group, goat embryos are routinely cultured in 10-μL drops (1 embryo/1−2 μL SOF) under paraffin oil at 38.5°C in humidified air with 5% CO_2, 5% O_2, and 90% N_2.

4.3 New strategies to improve in vitro production of goat embryos

During the first years of goat IVEP research, we have tested different protocols and culture media in order to standardize a replicable production system. However, variable oocyte quality (related to nutrition, health, age, follicle wave, apoptosis stage, and other factors) before the start of oocyte IVM causes unpredictable results following the completion of IVEP steps. Therefore, more recently, different strategies have been studied to improve oocyte quality in the course of IVM and thus enhance the overall IVEP efficiency.

4.3.1 Supplementation of culture medium with antioxidants

One of the main factors that can impair oocyte competence on in vitro conditions is the oxidative stress, induced by an imbalance between reactive oxygen species (ROS) production and elimination (reviewed by Tamura et al. [23]). It is well known that gametes and embryos handled and manipulated outside of controlled environments suffer from an excess of available oxygen. Conventional IVM atmospheric conditions present 3–4 times more O_2 than the oviduct [24]. Thus a number of factors such as the processes of oocyte recovery from the follicle, oocyte washing and selection, sperm cell selection and capacitation, and finally assessment of embryo development under the microscope all impair oocyte and embryo development, and this could be overcome by supplementing culture media with antioxidants. There are different antioxidants that have been tested for IVM in goats, but cysteamine is probably the most used, added at a dose of 100 μM. A previous study from our group in juvenile goats showed that the supplementation of the IVM medium with cysteamine, cysteine, cystine, and beta-mercaptoethanol increases oocyte intracytoplasmic glutathione (GSH), but only cysteamine enhances blastocyst yield [25]. Recently, other potentially more powerful antioxidants have been tested for goat IVM, such as melatonin and resveratrol.

Melatonin is a powerful antioxidant as it acts by different mechanisms of action including directly scavenging ROS and receptor-mediated effects [26]. Soto-Heras et al. [16] described that the supplementation of the IVM medium with 10^{-7} M melatonin increased blastocyst production compared to a control group (28.9% vs 11.7%, respectively) and blastocyst quality in juvenile goat. In a follow-up study [27], the positive effect of melatonin on oocyte competence was related to its antioxidant power, with the result of a lower cytoplasmic ROS content, and the increase of mitochondrial activity and ATP production. Regarding resveratrol, Piras et al. [28] found that the supplementation of the IVM medium with 1 μM resveratrol increased significantly blastocyst production compared to a control group in juvenile-goat oocytes (20.1% vs 6.8%, respectively).

The use of these antioxidants on other steps of goat IVEP is yet to be investigated, but the final positive effect could be maximized as oxidation is known also to impair fertilization [29] and embryo development [30].

4.3.2 Biphasic in vitro maturation system

Conventional IVM can impair embryo development by inducing precocious nuclear maturation, which interrupts the process of oocyte competence acquisition (reviewed by Gilchrist and Thompson [31]). During folliculogenesis, oocytes undergo changes at nuclear and cytoplasmic levels, essential for acquiring oocyte competence. Inside antral follicles, oocytes are arrested at GV stage, but spontaneously resume meiosis when released and placed in culture media in vitro [32]. In juvenile goats, Velilla et al. [33] observed that 30% of oocytes released from the follicle resume meiosis during the selection and washing processes before being placed in the IVM medium. Therefore new strategies for IVM are being developed based on the inhibition of spontaneous meiosis with cyclic AMP (cAMP) modulators, such as C-type natriuretic peptide (CNP). The cAMP modulators are used during a prematuration phase (pre-IVM), in a biphasic IVM system, followed by a conventional IVM phase (reviewed by Gilchrist et al. [34]). The pre-IVM culture prolongs cumulus-oocyte communication and increases mRNA and protein accumulation, enabling the oocyte to fully acquire developmental competence before IVM (reviewed by Gilchrist [35]).

In goats, a pre-IVM culture with 100 ng/mL CNP and 10 nM estradiol maintained oocytes in meiotic arrest for 6 hours [36]. This pre-IVM followed by 18 hours of conventional IVM improved blastocyst development after parthenogenetic activation compared to control 24-hour IVM (45% vs 30% blastocysts/cleaved oocytes, respectively). A similar biphasic IVM system has been developed in our research group to enhance IVEP in juvenile goats. In a recent study [37], juvenile-goat oocytes were maintained at meiotic arrest for 6 hours (75.4% vs 27.8% GV-oocytes, respectively) using a TCM199 medium supplemented with 100 nM CNP, 10 nM estradiol, 100 μM cysteamine, and 0.4% BSA. This pre-IVM protocol was applied to a biphasic IVM system (6 hours of pre-IVM culture followed by 24 hours of conventional IVM), which improved significantly blastocyst development after IVF compared to control IVM (30% and 17%, respectively). The pre-IVM positive effect was related in this case to an improvement of the cumulus-oocyte communication and the cytoplasmic GSH content. However, meiotic arrest can only be temporarily maintained with the current protocols; oocytes resume meiosis if pre-IVM is prolonged for 8 hours. This probably limits the benefits of the pre-IVM culture on the oocyte developmental competence, compared to results with mice oocytes, which can be maintained in meiotic arrest for 48 hours [38].

4.4 Cryopreservation of gametes and embryos

Cryopreservation technology allows us to preserve both gametes and embryos during a long period of time while maintaining their viability. All cryopreservation methods include temperature reduction, cellular dehydration, freezing, and thawing. Moreover, for avoiding ice crystal formation and shock effects during the freezing procedure, cryoprotective agents such as ethylene glycol (EG) or glycerol are used (reviewed by Mara et al. [39]).

Two methods of cryopreservation are employed: slow freezing and vitrification. Vitrification is faster and cheaper than slow freezing, and hence it is currently used for cryopreserving oocytes and embryos, whereas semen is usually frozen by ordinary slow-freezing protocols.

4.4.1   Oocyte cryopreservation

Oocytes are very sensitive to chilling and cryopreservation, and the protocols used for their cryopreservation are not as established and efficient as for semen or embryos (reviewed by Mara et al. [39]). The maturation stage of the oocyte, the composition and volume of the vitrification solution, and the size of the carrier all affect the cooling and warming rates and the cell viability. It has been reported that the tolerance to vitrification and thawing of matured goat oocytes seem to be higher than immature oocytes [40].

The combination of 20% EG, 20% dimethyl sulfoxide (DMSO), and 0.5−0.65 M sucrose has been used as a vitrification solution for cryopreserving both immature [41] and matured [42] goat oocytes. With regard to the effect caused by the carrier used [41], the best results following vitrification of immature oocytes were obtained when cryotop or hemistraw carriers were used instead of conventional straws, cryoloop, and open pulled straws (OPS). In matured oocytes, two cryopreservation methods, solid-surface vitrification and cryoloop vitrification, have been tested and both have produced acceptable levels of survival and cleavage [43].

4.4.2   Sperm cryopreservation

Fertility rates after AI with frozen-thawed goat semen varies from 40% to 65% (reviewed by Paramio and Izquierdo [4]). Heat transfer in sperm cells is too slow to allow vitrification, and therefore semen is always cryopreserved using slow-freezing protocols. Ram and buck semen freezing methods are similar when considering type of extenders, cryoprotectants, and cooling rates. The bulbo-urethral gland of the buck though secretes enzymes into the seminal plasma that interacts with some constituents of the egg yolk−based and milk-based diluents, generating a detrimental effect on spermatozoa. For this reason, seminal plasma from the buck is removed by washing the spermatozoa immediately after collection and before using egg yolk−based diluents, with the result of increasing the percentage of viable sperm cells and their motility [44]. It seems that low levels of egg yolk ($\sim 2.5\%$) may avoid toxic effects (reviewed by Cseh and Faigl [45]). However, skimmed milk [46] and soybean lecithin [47,48] can be considered as suitable alternatives to egg yolk. In fact, Chelucci et al. [47] obtained a better protection of sperm damage by cold shock and a higher fertilization potential, with soybean lecithin than egg yolk.

4.4.3   Embryo cryopreservation

In 1976, Bilton and Moore [49] reported the first kid born after transfer of a frozen-thawed embryo. Some years later, the first kid born from a vitrified goat embryo was reported [50].

Hong et al. [51] compared the effect of different vitrification solutions and obtained similar kidding rates when transferring OPS-vitrified embryos with 0.5 M sucrose, 15% EG, and 15% DMSO, and fresh ones (51% and 57%, respectively). The sucrose supplementation to the vitrification solution provides more repeatable embryo survival rates [52].

Different studies [51−53] have not found significant differences in kidding rates between slow-frozen (range from 31% to 47%) and OPS-vitrified (range from 35% to 52%) blastocysts. However, in other studies [54,55] significantly higher kidding rates have been achieved after transferring OPS-vitrified blastocysts (range from 82% to 93%), when compared to slow-frozen blastocysts (from 40% to 50%).

Regarding the cryotolerance of embryos according to their developmental stage, a lower embryo survival of goat morulae has been described both after slow freezing [56] and vitrification [55,57], compared with embryos cryopreserved at the blastocyst stage.

Finally, in our laboratory, Morató et al. [58] tested the effect of oocyte donor age (adult vs prepubertal goat) and the in vitro blastocyst stage (nonexpanded, expanded, hatching, and hatched) on the ability of the blastocyst to survive the vitrification/warming procedure using 15% DMSO, 15% EG, and 0.5 M sucrose as vitrification solution. The total blastocyst survival was not affected by the goat age (40.7% in adults and 51.8% in prepubertal animals). However, expanded, hatching, and hatched blastocysts showed the highest in vitro survival rates in embryos coming from adult-goat oocytes, while hatching blastocysts were the most cryotolerant among the embryos produced from prepubertal-goat oocytes.

4.5 In vitro culture of gametes

The high number of competent gametes in both ovaries and testes may give us the possibility to maximize the reproductive potential capacity of both males and females. In females, from all the oocytes available in the ovaries at birth (more than 500,000 per ovary), only a few of them will be fertilized and give a newborn. In males, the loss of reproductive cells is even higher. Thus the in vitro culture of gonadal tissue could allow us to improve the genetic diffusion of selected animals.

4.5.1 In vitro culture of ovarian follicles

The in vitro culture of ovarian follicles is a promising ART that consists in the culture of preantral (primordial, primary, and secondary) and antral follicles until their oocytes reach the competent stage to be in vitro matured, fertilized, and developed up to a viable embryo. The culture of immature oocytes inside follicles could increase in a drastic way the number of embryos and offspring produced by a female (reviewed by Gupta and Nandi [59]). This technique is also useful for understanding the processes of folliculogenesis and oocyte maturation. As reviewed by Figueiredo et al. [60], caprine follicles are cultured either enclosed in ovarian cortical slices (in situ) or isolated (after mechanical or enzymatic follicle removal from the ovary).

Silva et al. [61] established a culture medium that promotes in vitro growth and survival of caprine primordial follicles inside ovarian cortical slices. The slices are cultured in minimum essential medium (MEM) supplemented with insulin-transferrin-selenium (ITS), pyruvate, glutamine, hypoxanthine, BSA, and antibiotics at 39°C with 5% CO_2 for up to 5 days. This medium was further improved by the addition of 100 ng/mL epidermal growth factor (EGF) and 100 ng/mL follicle-stimulating hormone (FSH) [62]. Other supplements, such as GDF9 [63], also increase the survival rate of preantral follicles and promote the activation of primordial follicles. On the other hand, Huanmin and Yong [64] showed for the first time that isolated caprine preantral follicles grow and develop in vitro when cultured in agar gel with fetal calf serum, hypoxanthine, dbcAMP, FSH, ITS, IGF-I, hydrocortisone, and antibiotics at 39°C with 5% CO_2. Culture of individual preantral follicles in microdrops covered with mineral oil is also possible. For instance, Rajarajan et al. [65] cultured caprine preantral follicles in microdrops of TCM199 supplemented with TGF-α, IGF-II, EGF, and FSH.

Despite the amount of studies carried out in the past 20 years on culture of isolated follicles, meiotic resumption and nuclear maturation after IVM of these oocytes is still low, and few studies have reported blastocyst production after IVF. In a study by Ferreira et al. [66], 5.4% of the oocytes from preantral follicles and 37.5% from antral follicles reached MII stage after being cultured for 18 days in microdrops of α-MEM supplemented with BSA, ITS, glutamine, hypoxanthine, and ascorbic acid (known as α-MEM +) plus 50 mIU/mL FSH. Saraiva et al. [67] were able to produce two blastocysts after culture of preantral follicles with a sequential medium for 18 days. The sequential medium consisted in α-MEM + , supplemented with 100 ng/mL FSH from day 0 to 6, 500 ng/mL FSH from day 6 to 12, and 1000 ng/mL FSH plus 100 ng/mL LH and 100 ng/mL EGF from day 12 to 18. Therefore recent studies are focusing on improving the viability and meiotic resumption of oocytes after preantral and antral culture [66], and this would lead to higher rates of embryo production, especially from follicles of early stages.

4.5.2 In vitro culture of spermatogonia

Spermatogonial stem cells (SSCs), also termed male germline stem cells (mGSCs), can either divide and maintain their own population or differentiate into mature spermatids (reviewed by de Rooij and Grootegored [68]). The in vitro culture of SSCs can be used as a model for the study of spermatogenesis and cellular reprogramming. Moreover, SSCs can be genetically manipulated and used for transfer of genetic information. Long-term culture systems for mGSCs could reduce the time and costs for producing transgenic animals (reviewed by Sahare et al. [69]). However, maintenance and proliferation of SSCs on an in vitro culture system are difficult [70].

Conventionally, SSCs are cultured on cell feeder layers (somatic testicular cells) with serum supplementation. However, culture with feeder cells and serum make difficult the SSC self-renewal analyses, and hence a defined culture system would be preferred [71]. A 1% fetal bovine serum supplementation of an enriched Dulbecco's modified Eagle medium maintains goat SSCs and enables them to proliferate, although higher concentrations are detrimental [72]. In this culture system, SSCs are maintained at 37°C with 5% CO_2 for 7 days. Wang et al. [73] further improved this culture system with the addition of vitamin C, which maintains a certain physiological level of ROS. Some growth factors (glial cell line—derived neurotrophic factor, leukemia inhibitory factor, fibroblast growth factor, EGF) have been also tested in goat SSCs culture, which can maintain the SCCs in a more undifferentiated state [74]. On the other hand, Pramod and Mitra [75] developed a 2-month caprine SSCs culture system by isolating the cells with an enzymatic digestion method, enriching the cells with a Percoll density gradient and culturing with a feeder layer of Sertoli cells. More recently, Deng et al. [76] were able to obtain functional spermatids through SSCs in vitro culture in Saanen dairy goats. The supplementation of the culture medium with testosterone increased the efficiency of the SSC differentiation and the rate of embryo development after intracytoplasmic sperm injection.

4.6 Embryo micromanipulation

The manipulation of embryos through blastomere biopsy enables the selection of the embryo according to sex and specific genes related to productive characters, as well as the production of clones, transgenic animals, and stem cell lines.

4.6.1 Embryo sexing and genotyping

Leoni et al. [77] described for the first time a method for sexing goat embryos using polymerase chain reaction (PCR) and the restriction fragment length polymorphism (RFLP) analysis. They amplified a DNA fragment derived from blastocyst biopsied cells. However, the RFLP entangles risk of contamination and misdiagnosis due to the low amount of DNA in embryo biopsies and requires extra time for specific endonuclease digestion [78]. Chen et al. [79] developed a more efficient and faster PCR sexing protocol for goat embryos based on both the X- and Y-amelogenin genes (*AMELX* and *AMELY*). This system was improved by Tsai et al. [80] who used a triple primer set for amplifying a fragment of the X female chromosome and two fragments from the X and Y male chromosomes. They reached 100% accuracy with the use of a single biopsied blastomere in different goat breeds (Alpine, Saanen, Nubian, and Taiwan goats).

Guignot et al. [81] developed an accurate genotyping diagnosis in goat embryos: the whole genome amplification using multiple displacement amplification. The genotyping was performed in biopsied cells from morulae and blastocysts and did not affect the embryo viability after vitrification-warming and transfer. The genotyping was focused on assessing sex (through the amplification of ZFX/ZFY and Y chromosome−specific sequences) and prion protein (PRNP) in order to select scrapie-resistant genotypes. This procedure seems to be efficient for embryo selection based on production traits and diseases and a useful tool for enhancing genetic gain in breeding programs. However, no further studies in goat embryo genotyping have been carried out. Moreover, genes controlling desirable and undesirable traits still need to be characterized in goats.

4.6.2 Cloning and transgenesis

The methodology of cloning animals using somatic cell nuclear transfer (SCNT) follows the protocol used by Wilmut et al. [82] to produce Dolly, the first sheep cloned from an adult animal. In 2001, the first cloned goat using cumulus cells as somatic nuclei was reported [83]. The utilization of SCNT as a reproductive technology is still inefficient. In goats, SCNT presents a high embryo loss, as in calves. However, no signs of placental abnormalities, large fetal syndrome, respiratory or cardiovascular dysfunction, organ dysplasia, high perinatal mortality, or abnormal postnatal development have been reported in goats as have been observed in calves [84].

The intense effort placed on SCNT is due to the importance of transgenic research. Dairy goats have an exceptional potential to serve as bioreactors by producing milk containing therapeutic recombinant proteins. Goats have a shorter generation interval compared to cattle, and a lower incidence of scrapie, and from two to threefold greater volume of milk production compared to sheep. Moreover, the high concentration of recombinant protein in milk (1−5 g/L) allows for herds of transgenic goats of manageable size that could easily yield 1−300 kg of purified product per year. A special dwarf goat breed (BELE: breed early, lactate early) was used to produce transgenic goats with nuclear cloning [85].

Ebert et al. [86] produced the first transgenic goat, which expressed in milk the human longer acting tissue plasminogen activator at a concentration of 1−3 mg/mL, by using direct microinjection of DNA constructs into the zygote pronuclei. This technique has been used in goats to produce several proteins, such as human factor IX [87], human antithrombin III [88], lysozyme [89], butyrylcholinesterase [90], lactoferrin [91], and a novel human plasminogen activator [92]. The overall efficiency of gene microinjection is poor, as only around 1% of the injected zygotes give birth to a transgenic kid compared with 0.1% in cow and 5%−10% in mice (see review [93]).

More recently transgenic goats have been produced by SCNT. Somatic cells are genetically modified before transfer to enucleated oocytes. The advantages of cloning via nuclear transfer are that all produced animals are transgenic, the creation of transgenic animals can be shortened by one generation, and the cultured cells can be stored almost indefinitely. The overall efficiency of nuclear transfer in goats averages 2.6% (total live kids/embryos transferred), making this technology an attractive alternative to microinjection. Using this method, transgenic goats have been obtained, which produce malaria antigen [94], human acid beta-glucosidase [95], human lactoferrin [96], human granulocyte-colony stimulating factor (hG-CSF) [97], caprine growth hormone [98], human α-lactalbumin [99], and a novel human plasminogen activator [92]. Baguisi et al. [100] reported three cloned transgenic goats producing high-level of human antithrombin III, similar to the parental transgenic line. In March 2012, Blash et al. [101] reported that the three cloned transgenic goats born on October 14−15th, 1998, were alive and in good health. These goats did not show any mastitis, and both growth and reproductive parameters were normal. Recently, He et al. [92] reported two live goat kids from 256 reconstructed oocytes that expressed the recombinant human plasminogen activator (rhPA) in the mammary glands. The rhPA concentration in milk of the F0 and F1 was 78.32 μg/mL.

So far, the only recombinant protein produced by farm animals approved by the European Agency for the Evaluation of Medicinal Products in 2006 and the Food and Drug Administration in 2009 is recombinant human antithrombin III (rHAT) produced in goats (ATryn by GTC Biotherapeutics, United States). Apart from gene pharming, production traits have been enhanced by producing transgenic goats, such as growth rate, milk quantity and composition, fiber production, and resistance to diseases. However, to date, transgenic goats generated for the benefit of an increased productivity in milk and meat have not been reported.

References

[1] FAOSTAT. Retrieved from: <http://www.fao.org/faostat/en/#data/QA>; 2019.

[2] Mikkola M. Commercial embryo transfer activity in Europe 2016. In: Proceedings of the 33rd annual meeting of European Embryo Transfer Association (AETE). Bath, 8th and 9th September 2017. <https://www.aete.eu/previous-meetings/>; 2017. p. 23−29.

[3] Viana J. 2017 statistics of embryo production and transfer in domestic farm animals. Embryo Technol Newsl 2018;36(4):8−25 <https://www.iets.org/pdf/Newsletter/Dec18_IETS_Newsletter.pdf>.

[4] Paramio MT, Izquierdo D. Current status of in vitro embryo production in sheep and goats. Reprod Domest Anim 2014;49(Suppl. 4):37−48. Available from: https://doi.org/10.1111/rda.12334.

[5] Koeman J, Keefer CL, Baldassarre H, Downey BR. Developmental competence of prepubertal and adult goat oocytes cultured in semi-defined media following laparoscopic recovery. Theriogenology 2003;60(5):879−89. Available from: https://doi.org/10.1016/S0093-691X(03)00090-6.

[6] Rodríguez-Dorta N, Cognié Y, González F, Poulin N, Guignot F, Touzé J-L, et al. Effect of coculture with oviduct epithelial cells on viability after transfer of vitrified in vitro produced goat embryos. Theriogenology 2007;68(6):908−13. Available from: https://doi.org/10.1016/j.theriogenology.2007.07.004.

[7] Kątska-Książkiewicz L, Opiela J, Ryńska B. Effects of oocyte quality, semen donor and embryo co-culture system on the efficiency of blastocyst production in goats. Theriogenology 2007;68(5):736−44. Available from: https://doi.org/10.1016/j.theriogenology.2007.06.016.

[8] Berlinguer F, Leoni GG, Succu S, Spezzigu A, Madeddu M, Satta V, et al. Exogenous melatonin positively influences follicular dynamics, oocyte developmental competence and blastocyst output in a goat model. J Pineal Res 2009;46(4):383−91. Available from: https://doi.org/10.1111/j.1600-079X.2009.00674.x.

[9] Leoni GG, Succu S, Satta V, Paolo M, Bogliolo L, Bebbere D, et al. In vitro production and cryotolerance of prepubertal and adult goat blastocysts obtained from oocytes collected by laparoscopic oocyte-pick-up (LOPU) after FSH treatment. Reprod Fertil Dev 2009;21(7):901−8. Available from: https://doi.org/10.1071/RD09015.

[10] Romaguera R, Moll X, Morató R, Roura M, Palomo MJ, Catalá MG, et al. Prepubertal goat oocytes from large follicles result in similar blastocyst production and embryo ploidy than those from adult goats. Theriogenology 2011;76(1):1−11. Available from: https://doi.org/10.1016/j.theriogenology.2010.12.014.

[11] De AK, Malakar D, Akshey YS, Jena MK, Garg S, Dutta R, et al. In vitro development of goat (*Capra hircus*) embryos following cysteamine supplementation of the in vitro maturation and in vitro culture media. Small Rumin Res 2011;96(2):185−90. Available from: https://doi.org/10.1016/j.smallrumres.2011.01.001.

[12] Souza-Fabjan JMG, Locatelli Y, Duffard N, Corbin E, Touzé J-L, Perreau C, et al. In vitro embryo production in goats: slaughterhouse and laparoscopic ovum pick up−derived oocytes have different kinetics and requirements regarding maturation media. Theriogenology 2014;81(8):1021−31. Available from: https://doi.org/10.1016/j.theriogenology.2014.01.023.

[13] Wang DC, Huang JC, Lo NW, Chen LR, Mermillod P, Ma WL, et al. Sonic Hedgehog promotes in vitro oocyte maturation and term development of embryos in Taiwan native goats. Theriogenology 2017;103:52−8. Available from: https://doi.org/10.1016/j.theriogenology.2017.07.029.

[14] An L, Liu J, Du Y, Liu Z, Zhang F, Liu Y, et al. Synergistic effect of cysteamine, leukemia inhibitory factor, and Y27632 on goat oocyte maturation and embryo development in vitro. Theriogenology 2018;108:56−62. Available from: https://doi.org/10.1016/j.theriogenology.2017.11.028.

[15] Catala MG, Roura M, Soto-Heras S, Menéndez I, Contreras-Solis I, Paramio MT, et al. Effect of season on intrafollicular fatty acid concentrations and embryo production after in vitro fertilization and parthenogenic activation of prepubertal goat oocytes. Small Rumin Res 2018;168:82−6. Available from: https://doi.org/10.1016/j.smallrumres.2018.10.003.

[16] Soto-Heras S, Roura M, Catalá MG, Menéndez-Blanco I, Izquierdo D, Fouladi-Nashta AA, et al. Beneficial effects of melatonin on in vitro embryo production from juvenile goat oocytes. Reprod Fertil Dev 2018;30(2):253−61. Available from: https://doi.org/10.1071/RD17170.

[17]   Crozet N, Ahmed-Ali M, Dubos MP. Developmental competence of goat oocytes from follicles of different size categories following maturation, fertilization and culture in vitro. J Reprod Fertil 1995;103(2):293−8. Available from: https://doi.org/10.1530/jrf.0.1030293.

[18]   Palomo M, Izquierdo D, Mogas T, Paramio MT. Effect of semen preparation on IVF of prepubertal goat oocytes. Theriogenology 1999;51 (5):927−40. Available from: https://doi.org/10.1016/S0093-691X(99)00039-4.

[19]   Rho GJ, Hahnel AC, Betteridge KJ. Comparisons of oocyte maturation times and of three methods of sperm preparation for their effects on the production of goat embryos in vitro. Theriogenology 2001;56(3):503−16. Available from: https://doi.org/10.1016/S0093-691X(01)00581-7.

[20]   Tervit HR, Whittingham DG, Rowson LE. Successful culture in vitro of sheep and cattle ova. J Reprod Fertil 1972;30(3):493−7. Available from: https://doi.org/10.1530/jrf.0.0300493.

[21]   Lonergan P, Pedersen HG, Rizos D, Greve T, Thomsen PD, Fair T, et al. Effect of the post-fertilization culture environment on the incidence of chromosome aberrations in bovine blastocysts. Biol Reprod 2004;71(4):1096−100. Available from: https://doi.org/10.1095/biolreprod.104.030635.

[22]   Ongeri EM, Bormann CL, Butler RE, Melican D, Gavin WG, Echelard Y, et al. Development of goat embryos after in vitro fertilization and parthenogenetic activation by different methods. Theriogenology 2001;55(9):1933−45. Available from: https://doi.org/10.1016/S0093-691X(01) 00534-9.

[23]   Tamura H, Takasaki A, Taketani T, Tanabe M, Kizuka F, Lee L, et al. Melatonin as a free radical scavenger in the ovarian follicle. Endocr J 2013;60(1):1−13. Available from: https://doi.org/10.1507/endocrj.EJ12-0263.

[24]   Mastroianni L, Jones R. Oxygen tension within the rabbit fallopian tube. J Reprod Fertil 1965;9:99−102 <http://www.ncbi.nlm.nih.gov/pubmed/14257722>.

[25]   Rodríguez-González E, López-Bejar M, Mertens MJ, Paramio MT. Effects on in vitro embryo development and intracellular glutathione content of the presence of thiol compounds during maturation of prepubertal goat oocytes. Mol Reprod Dev 2003;65(4):446−53. Available from: https://doi.org/10.1002/mrd.10316.

[26]   Reiter RJ, Rosales-Corral SA, Manchester LC, Tan DX. Peripheral reproductive organ health and melatonin: ready for prime time. Int J Mol Sci 2013;14(4):7231−72. Available from: https://doi.org/10.3390/ijms14047231.

[27]   Soto-Heras S, Catalá MG, Roura M, Menéndez-Blanco I, Piras AR, Izquierdo D, et al. Effects of melatonin on oocyte developmental competence and the role of melatonin receptor 1 in juvenile goats. Reprod Domest Anim 2019;54:381−90. Available from: https://doi.org/10.1111/rda.1337.

[28]   Piras AR, Menéndez-Blanco I, Soto-Heras S, Catalá MG, Izquierdo D, Bogliolo L, et al. Resveratrol supplementation during in vitro maturation improves embryo development of prepubertal goat oocytes selected by brilliant cresyl blue staining. J Reprod Dev 2019;65(2):113−20. Available from: https://doi.org/10.1262/jrd.2018-077.

[29]   Takahashi T, Takahashi E, Igarashi H, Tezuka N, Kurachi H. Impact of oxidative stress in aged mouse oocytes on calcium oscillations at fertilization. Mol Reprod Dev 2003;66(2):143−52. Available from: https://doi.org/10.1002/mrd.10341.

[30]   Guérin P, El Mouatassim S, Ménézo Y. Oxidative stress and protection against reactive oxygen species in the pre-implantation embryo and its surroundings. Hum Reprod Update 2001;7(2):175−89. Available from: https://doi.org/10.1093/humupd/7.2.175.

[31]   Gilchrist RB, Thompson JG. Oocyte maturation: emerging concepts and technologies to improve developmental potential in vitro. Theriogenology 2007;67(1):6−15. Available from: https://doi.org/10.1016/j.theriogenology.2006.09.027.

[32]   Edwards RG. Maturation in vitro of mouse, sheep, cow, pig, rhesus monkey and human ovarian oocytes. Nature 1965;208:349−51. Available from: https://doi.org/10.1038/208349a0.

[33]   Velilla E, Rodríguez-Gonzalez E, Vidal F, Paramio T. Microtubule and microfilament organization in immature, in vitro matured and in vitro fertilized prepubertal goat oocytes. Zygote 2005;13(2):155−65. Available from: https://doi.org/10.1017/S0967199405003229.

[34]   Gilchrist RB, Luciano AM, Richani D, Zeng HT, Wang X, De Vos M, et al. Oocyte maturation and quality: role of cyclic nucleotides. Reproduction 2016;152(5):143−57. Available from: https://doi.org/10.1530/REP-15-0606.

[35]   Gilchrist RB. Recent insights into oocyte−follicle cell interactions provide opportunities for the development of new approaches to in vitro maturation. Reprod Fertil Dev 2010;23(1):23−31. Available from: https://doi.org/10.1071/RD10225.

[36]   Zhang J, Wei Q, Cai J, Zhao X, Ma B. Effect of C-type natriuretic peptide on maturation and developmental competence of goat oocytes matured in vitro. PLoS One 2015;10(7):e0132318. Available from: https://doi.org/10.1371/journal.pone.0132318.

[37]   Soto-Heras S, Menéndez-Blanco I, Catalá MG, Izquierdo D, Thompson JG, Paramio MT. Biphasic in vitro maturation with C-type natriuretic peptide enhances the developmental competence of juvenile-goat oocytes. PLoS One 2019;14(8):e0221663. Available from: https://doi.org/10.1371/journal.pone.0221663.

[38]   Romero S, Sánchez F, Lolicato F, Van Ranst H, Smitz J. Immature oocytes from unprimed juvenile mice become a valuable source for embryo production when using C-type natriuretic peptide as essential component of culture medium. Biol Reprod 2016;95(3):1−10. Available from: https://doi.org/10.1095/biolreprod.116.139808.

[39]   Mara L, Casu S, Carta A, Dattena M. Cryobanking of farm animal gametes and embryos as a means of conserving livestock genetics. Anim Reprod Sci 2013;138:25−38. Available from: https://doi.org/10.1016/j.anireprosci.2013.02.006.

[40]   Quan GB, Li WJ, Lan ZG, Wu SS, Shao QY, Hong QH. The effects of meiotic stage on viability and developmental capability of goat oocytes vitrified by the cryoloop method. Small Rumin Res 2014;116(1):32−6. Available from: https://doi.org/10.1016/j.smallrumres.2013.10.005.

[41]   Rao BS, Mahesh YU, Charan KV, Suman K, Sekhar N, Shivaji S. Effect of vitrification on meiotic maturation and expression of genes in immature goat cumulus oocyte complexes. Cryobiology 2012;64(3):176−84. Available from: https://doi.org/10.1016/j.cryobiol.2012.01.005.

[42]   Srirattana K, Sripunya N, Sangmalee A, Imsoonthornruksa S, Liang Y, Ketudat-Cairns M, et al. Developmental potential of vitrified goat oocytes following somatic cell nuclear transfer and parthenogenetic activation. Small Rumin Res 2013;112(1−3):141−6. Available from: https://doi.org/10.1016/j.smallrumres.2012.10.011.

[43] Begin I, Bhatia B, Baldassarre H, Dinnyes A, Keefer CL. Cryopreservation of goat oocytes and in vivo derived 2- to 4-cell embryos using the cryoloop (CLV) and solid-surface vitrification (SSV) methods. Theriogenology 2003;59(8):1839−50. Available from: https://doi.org/10.1016/S0093-691X(02)01257-8.

[44] Leboeuf B, Restall B, Salamon S. Production and storage of goat semen for artificial insemination. Prod Anim. 2003;16(2):91−9 <https://www6.inra.fr/productions-animales_eng/2003-Volume-16/Issue-2-2003/Production-and-storage-of-goat-semen-for-artificial-insemination>.

[45] Cseh S, Faigl V. Semen processing and artificial insemination in health management of small ruminants. Anim Reprod Sci 2012;130:187−92. Available from: https://doi.org/10.1016/j.anireprosci.2012.01.014.

[46] Küçük N, Aksoy M, Uçan U, Ahmad E, Naseer Z, Ceylan A, et al. Comparison of two different cryopreservation protocols for freezing goat semen. Cryobiology 2014;68(3):327−31. Available from: https://doi.org/10.1016/j.cryobiol.2014.04.009.

[47] Chelucci S, Pasciu V, Succu S, Addis D, Leoni GG, Manca ME, et al. Soybean lecithin-based extender preserves spermatozoa membrane integrity and fertilizing potential during goat semen cryopreservation. Theriogenology 2015;83(6):1064−74. Available from: https://doi.org/10.1016/j.theriogenology.2014.12.012.

[48] Salmani H, Towhidi A, Zhandi M, Bahreini M, Sharafi M. In vitro assessment of soybean lecithin and egg yolk based diluents for cryopreservation of goat semen. Cryobiology 2014;68(2):276−80. Available from: https://doi.org/10.1016/j.cryobiol.2014.02.008.

[49] Bilton RJ, Moore NW. In vitro culture, storage and transfer of goat embryos. Aust J Biol Sci 1976;29(2):125−30. Available from: https://doi.org/10.1071/BI9760125.

[50] Yuswiati E, Holtz W. Work in progress: successful transfer of vitrified goat embryos. Theriogenology 1990;34(4):629−32. Available from: https://doi.org/10.1016/0093-691X(90)90018-O.

[51] Hong QH, Tian SJ, Zhu SE, Feng JZ, Yan CL, Zhao XM, et al. Vitrification of boer goat morulae and early blastocysts by straw and open-pulled straw method. Reprod Domest Anim 2007;42(1):34−8. Available from: https://doi.org/10.1111/j.1439-0531.2006.00720.x.

[52] Guignot F, Bouttier A, Baril G, Salvetti P, Pignon P, Beckers JF, et al. Improved vitrification method allowing direct transfer of goat embryos. Theriogenology 2006;66(4):1004−11. Available from: https://doi.org/10.1016/j.theriogenology.2006.02.040.

[53] Ferreira-Silva JC, Moura MT, Silva TD, Oliveira LRS, Chiamenti A, Figueirêdo Freitas VJ, et al. Full-term potential of goat in vitro produced embryos after different cryopreservation methods. Cryobiology 2017;75:75−9. Available from: https://doi.org/10.1016/j.cryobiol.2017.01.009.

[54] El-Gayar M, Holtz W. Technical note: vitrification of goat embryos by the open pulled-straw method. J Anim Sci 2001;79(9):2436−8. Available from: https://doi.org/10.2527/2001.7992436x.

[55] Yacoub ANA, Gauly M, Holtz W. Open pulled straw vitrification of goat embryos at various stages of development. Theriogenology 2010;73(8):1018−23. Available from: https://doi.org/10.1016/j.theriogenology.2009.11.028.

[56] Li R, Cameron AWN, Batt PA, Trounson AO. Maximum survival of frozen goat embryos is attained at the expanded, hatching and hatched blastocyst stages of development. Reprod Fertil Dev 1990;2(4):345−50. Available from: https://doi.org/10.1071/RD9900345.

[57] Gibbons A, Cueto MI, Pereyra Bonnet F. A simple vitrification technique for sheep and goat embryo cryopreservation. Small Rumin Res 2011;95(1):61−4. Available from: https://doi.org/10.1016/j.smallrumres.2010.08.007.

[58] Morató R, Romaguera R, Izquierdo D, Paramio MT, Mogas T. Vitrification of in vitro produced goat blastocysts: effects of oocyte donor age and development stage. Cryobiology 2011;63(3):240−4. Available from: https://doi.org/10.1016/j.cryobiol.2011.09.002.

[59] Gupta PS, Nandi S. Isolation and culture of preantral follicles for retrieving oocytes for the embryo production: present status in domestic animals. Reprod Domest Anim 2012;47:513−19. Available from: https://doi.org/10.1111/j.1439-0531.2011.01904.x.

[60] Figueiredo JR, Celestino JJH, Faustino LR, Rodrigues APR. In vitro culture of caprine preantral follicles: advances, limitations and prospects. Small Rumin Res 2011;98:192−5. Available from: https://doi.org/10.1016/j.smallrumres.2011.03.039.

[61] Silva JR, van den Hurk R, Costa SH, Andrade ER, Nunes AP, Ferreira FV, et al. Survival and growth of goat primordial follicles after in vitro culture of ovarian cortical slices in media containing coconut water. Anim Reprod Sci 2004;81(3−4):273−86. Available from: https://doi.org/10.1016/j.anireprosci.2003.09.006.

[62] Silva J, van den Hurk R, de Matos M, dos Santos R, Pessoa C, de Moraes M, et al. Influences of FSH and EGF on primordial follicles during in vitro culture of caprine ovarian cortical tissue. Theriogenology 2004;61(9):1691−704. Available from: https://doi.org/10.1016/j.theriogenology.2003.09.014.

[63] Martins FS, Celestino JJH, Saraiva MVA, Matos MHT, Bruno JB, Rocha-Junior CMC, et al. Growth and differentiation factor-9 stimulates activation of goat primordial follicles in vitro and their progression to secondary follicles. Reprod Fertil Dev 2008;20(8):916−24. Available from: https://doi.org/10.1071/RD08108.

[64] Huanmin Z, Yong Z. In vitro development of caprine ovarian preantral follicles. Theriogenology 2000;54(4):641−50. Available from: https://doi.org/10.1016/S0093-691X(00)00379-4.

[65] Rajarajan K, Rao BS, Vagdevi R, Tamilmani G, Arunakumari G, Sreenu M, et al. Effect of various growth factors on the in vitro development of goat preantral follicles. Small Rumin Res 2006;63(1−2):204−12. Available from: https://doi.org/10.1016/j.smallrumres.2005.02.013.

[66] Ferreira ACA, Cadenas J, Sá NAR, Correia HHV, Guerreiro DD, Lobo CH, et al. In vitro culture of isolated preantral and antral follicles of goats using human recombinant FSH: concentration-dependent and stage-specific effect. Anim Reprod Sci 2018;196:120−9. Available from: https://doi.org/10.1016/j.anireprosci.2018.07.004.

[67] Saraiva MVA, Rossetto R, Brito IR, Celestino JJH, Silva CMG, Faustino LR, et al. Dynamic medium produces caprine embryo from preantral follicles grown in vitro. Reprod Sci 2010;17(12):1135−43. Available from: https://doi.org/10.1177/1933719110379269.

[68] de Rooij DG, Grootegored JA. Spermatogonial stem cells. Curr Opin Cell Biol 1998;10:694−701. Available from: https://doi.org/10.1016/S0955-0674(98)80109-9.

[69] Sahare MG, Suyatno, Imai H. Recent advances of in vitro culture systems for spermatogonial stem cells in mammals. Reprod Med Biol 2018;17(2):134−42. Available from: https://doi.org/10.1002/rmb2.12087.

[70] Kanatsu-Shinohara M, Ogonuki N, Inoue K, Miki H, Ogura A, Toyokuni S, et al. Long-term proliferation in culture and germline transmission of mouse male germline stem cells. Biol Reprod 2003;69(2):612−16. Available from: https://doi.org/10.1095/biolreprod.103.017012.

[71] Kanatsu-Shinohara M, Inoue K, Ogonuki N, Morimoto H, Ogura A, Shinohara T. Serum- and feeder-free culture of mouse germline stem cells. Biol Reprod 2011;84(1):97−105. Available from: https://doi.org/10.1095/biolreprod.110.086462.

[72] Bahadorani M, Hosseini SM, Abedi P, Hajian M, Hosseini SE, Vahdati A, et al. Short-term in-vitro culture of goat enriched spermatogonial stem cells using different serum concentrations. J Assist Reprod Genet 2012;29(1):39−46. Available from: https://doi.org/10.1007/s10815-011-9687-5.

[73] Wang J, Cao H, Xue X, Fan C, Fang F, Zhou J, et al. Effect of vitamin C on growth of caprine spermatogonial stem cells in vitro. Theriogenology 2014;81(4):545−55. Available from: https://doi.org/10.1016/j.theriogenology.2013.11.007.

[74] Heidari B, Rahmati-Ahmadabadi M, Akhondi MM, Zarnani AH, Jeddi-Tehrani M, Shirazi A, et al. Isolation, identification, and culture of goat spermatogonial stem cells using c-kit and PGP9.5 markers. J Assist Reprod Genet 2012;29(10):1029−38. Available from: https://doi.org/10.1007/s10815-012-9828-5.

[75] Pramod RK, Mitra A. In vitro culture and characterization of spermatogonial stem cells on Sertoli cell feeder layer in goat (*Capra hircus*). J Assist Reprod Genet 2014;31(8):993−1001. Available from: https://doi.org/10.1007/s10815-014-0277-1.

[76] Deng S, Wang X, Wang Z, Chen S, Wang Y, Hao X, et al. In vitro production of functional haploid sperm cells from male germ cells of Saanen dairy goat. Theriogenology 2017;90:120−8. Available from: https://doi.org/10.1016/j.theriogenology.2016.12.002.

[77] Leoni GG, Schmoll F, Ledda S, Bogliolo L, Marogna G, Calvia P, et al. Sex determination in goat embryos. In: Proc Conf Adv Biotechnol Agriculture, Nutrition and Environment, S5-3-7(Abstr), Ferrara, Italy; 1996. p. 8−11.

[78] Aasen E, Medrano JF. Amplification of the ZFY andZFX genes for sex identification in humans, cattle, sheep and goats. Bio/Technology 1990;8(12):1279−81. Available from: https://doi.org/10.1038/nbt1290-1279.

[79] Chen AQ, Xu ZR, Yu SD. Sexing goat embryos by PCR amplification of X- and Y- chromosome specific sequence of the amelogenin gene. Asian-Aust J Anim Sci 2007;20(11):1689−93. Available from: https://doi.org/10.5713/ajas.2007.1689.

[80] Tsai TC, Wu SH, Chen HL, Tung YT, Cheng WTK, Huang JC, et al. Identification of sex-specific polymorphic sequences in the goat amelogenin gene for embryo sexing. J Anim Sci 2011;89(8):2407−14. Available from: https://doi.org/10.2527/jas.2010-3698.

[81] Guignot F, Perreau C, Cavarroc C, Touzé JL, Pougnard JL, Dupont F, et al. Sex and PRNP genotype determination in preimplantation caprine embryos. Reprod Domest Anim 2011;46(4):656−63. Available from: https://doi.org/10.1111/j.1439-0531.2010.01724.x.

[82] Wilmut I, Schnieke AE, McWhir J, Kind AJ, Campbell KHS. Viable offspring derived from fetal and adult mammalian cells. Nature 1997;385:810−13. Available from: https://doi.org/10.1038/385810a0.

[83] Zou X, Chen Y, Wang Y, Luo J, Zhang Q, Zhang X, et al. Production of cloned goats from enucleated oocytes injected with cumulus cell nuclei or fused with cumulus cells. Cloning 2001;3(1):31−7. Available from: https://doi.org/10.1089/152045501300189312.

[84] Wilmut I, Beaujean N, de Sousa PA, Dinnyes A, King TJ, Paterson LA, et al. Somatic cell nuclear transfer. Nature 2002;419:583−6. Available from: https://doi.org/10.1038/nature01079.

[85] Keefer CL, Keyston R, Lazaris A, Bhatia B, Begin I, Bilodeau AS, et al. Production of cloned goats after nuclear transfer using adult somatic cells. Biol Reprod 2002;66(1):199−203. Available from: https://doi.org/10.1095/biolreprod66.1.199.

[86] Ebert KM, Selgrath JP, DiTullio P, Denman J, Smith TE, Memon MA, et al. Transgenic production of a variant of human tissue-type plasminogen activator in goat milk: generation of transgenic goats and analysis of expression. Bio/Technology 1991;9(9):835−8. Available from: https://doi.org/10.1038/nbt0991-835.

[87] Zhang K, Lu D, Xue J, Huang Y, Huang S. Construction of mammary gland-specific expression vectors for human clotting factor IX and its secretory expression in goat milk. Chin J Biotechnol 1997;13(4):271−6.

[88] Edmunds T, Van Patten SM, Pollock J, Hanson E, Bernasconi R, Higgins E, et al. Transgenically produced human antithrombin: structural and functional comparison to human plasma-derived antithrombin. Blood 1998;91:4561−71. Available from: https://doi.org/10.1182/blood.V91.12.4561.

[89] Maga EA, Shoemaker CF, Rowe JD, Bondurant RH, Anderson GB, Murray JD. Production and processing of milk from transgenic goats expressing human lysozyme in the mammary gland. J Dairy Sci 2006;89(2):518−24. Available from: https://doi.org/10.3168/jds.S0022-0302(06)72114-2.

[90] Huang YJ, Huang Y, Baldassarre H, Wang B, Lazaris A, Leduc M, et al. Recombinant human butyrylcholinesterase from milk of transgenic animals to protect against organophosphate poisoning. Proc Natl Acad Sci U S A 2007;104(34):13603−8. Available from: https://doi.org/10.1073/pnas.0702756104.

[91] Zhang J, Li L, Cai Y, Xu X, Chen J, Wu Y, et al. Expression of active recombinant human lactoferrin in the milk of transgenic goats. Protein Expr Purif 2008;57:127−35. Available from: https://doi.org/10.1016/j.pep.2007.10.015.

[92] He Z, Lu R, Zhang T, Jiang L, Zhou M, Wu D, et al. A novel recombinant human plasminogen activator: efficient expression and hereditary stability in transgenic goats and in vitro thrombolytic bioactivity in the milk of transgenic goats. PLoS One 2018;13(8):e0201788. Available from: https://doi.org/10.1371/journal.pone.0201788.

[93] Boulanger L, Passet B, Pailhoux E, Vilotte JL. Transgenesis applied to goat: current applications and ongoing research. Transgenic Res 2012;2012(21):1183−90. Available from: https://doi.org/10.1007/s11248-012-9618-y.

[94] Behboodi E, Ayres SL, Memili E, O'Coin M, Chen LH, Reggio BC, et al. Health and reproductive profiles of malaria antigen-producing transgenic goats derived by somatic cell nuclear transfer. Cloning Stem Cell 2005;7:107−18. Available from: https://doi.org/10.1089/clo.2005.7.107.

[95]   Zhang YL, Wan YJ, Wang ZY, Xu D, Pang XS, Meng L, et al. Production of dairy goat embryos, by nuclear transfer, transgenic for human acid beta-glucosidase. Theriogenology 2010;73:681–90. Available from: https://doi.org/10.1016/j.theriogenology.2009.11.008.

[96]   An LY, Yuan YG, Yu BL, Yang TJ, Cheng Y. Generation of human lactoferrin transgenic cloned goats using donor cells with dual markers and a modified selection procedure. Theriogenology 2012;78(6):1303–11. Available from: https://doi.org/10.1016/j.theriogenology.2012.05.027.

[97]   Freitas VJF, Melo LM, Teixeira DIA, Andreeva LE, Serova IA, Serov OL. Goats as bioreactors for the production of human granulocyte colony stimulating factor (hG-CSF). Cloning Transgenesis 2014;3:3. Available from: https://doi.org/10.4172/2168-9849.1000I101.

[98]   Zhang Y, Zhu Z, Xu Q, Chen G. Association of polymorphisms of exon 2 of the growth hormone gene with production performance in Huoyan goose. Int J Mol Sci 2014;15(1):670–83. Available from: https://doi.org/10.3390/ijms15010670.

[99]   Feng X, Cao S, Wang H, Meng C, Li J, Jiang J, et al. Production of transgenic dairy goat expressing human α-lactalbumin by somatic cell nuclear transfer. Transgenic Res 2015;24(1):73–85. Available from: https://doi.org/10.1007/s11248-014-9818-8.

[100] Baguisi A, Behboodi E, Melican DT, Pollock JS, Destrempes MM, Cammuso C, et al. Production of goats by somatic cell nuclear transfer. Nat Biotechnol 1999;17:456–61. Available from: https://doi.org/10.1038/8632.

[101] Blash S, Schofield M, Echelard Y, Gavin W. Update on the first cloned goats. Nat Biotechnol 2012;30:229–30. Available from: https://doi.org/10.1038/nbt.2140.

Chapter 5

Reproductive technologies in swine

Joaquín Gadea, Pilar Coy, Carmen Matás, Raquel Romar and Sebastián Cánovas
Department of Physiology, University of Murcia, Murcia, Spain

5.1 General aspects of oocyte collection, evaluation, and in vitro maturation

Porcine cumulus-oocyte complexes (COCs) are commonly collected from ovaries of prepubertal gilts or cycling sows from crossbred commercial breeds, and a higher oocyte developmental competence has been reported in COCs from adult animals [1−3]. Nevertheless, COCs obtained from prepubertal animals can be successfully matured and fertilized in vitro, producing alive offspring after embryo transfer of 2−4-cell embryos [4]. Nowadays, the majority of in vitro studies use female gametes derived from prepubertal gilts since they represent the most common class of female pigs slaughtered in developed countries.

Ovaries are collected at the slaughterhouse immediately after sacrifice and transported to specialized laboratories in a thermoflask containing simple solutions (usually phosphate-buffered saline (PBS) or saline containing antibiotics at 38.5°C) within a short time from death. The transport of ovaries has received little attention and scarce research has been done in pigs on the effect of temperature, media, and time of ovaries transport on the progression of embryo development [5,6]. The use of organ preservation solutions used in human organ transplantation should be explored since the specific formulation is intended to prevent cell swelling, changes in pH and free radical formation, among others [7]. These emerging solutions [i.e., Solution de Conservation des Organes et des Tissus 15 (SCOT 15) [8]] might be of interest and effective in maintaining metabolic rate and functional ovarian cells, reducing the injury on COCs brought by ischemia, and thus improving further development. The temperature and the duration of the ovarian tissue transport to the laboratory are also important components in maintaining further oocyte viability, with temperatures and times ranging from 4°C to 39°C and 1 to 24 hours, respectively. As a general rule, as the storage temperature for the ovaries increases, the transportation time should be decreased (reviewed by Barberino et al. [9]). The use of new preservation solutions and standardized temperature/time ratio between labs for pig ovaries transport might help in implementing more efficient protocols.

Once in the lab, ovaries are subjected to several washes prior to recovering COCs, task commonly done by aspiration or slicing of ovarian preantral follicles 3−6 mm in diameter. The selection of suitable COCs for in vitro maturation (IVM) is done under stereomicroscope following morphological criteria such as presence of several compact layers of cumulus cells and appearance of dark and granulated cytoplasm. COCs with greater developmental competence can be selected with brilliant cresyl blue (BCB) staining, a simple test based on glucose-6-phosphate dehydrogenase (G6PDH) activity [10]. The use of this test yields controversial results since BCB-selected oocytes are more competent for nuclear and cytoplasmic maturation but did not improve cleavage and blastocyst formation [11]. Moreover, in porcine oocytes exposed to a BCB solution a specific pZP3 translocation from the zona pellucida to the cytoplasm has been reported [12], with the consequence of affecting further embryo development [10].

Once selected, COCs are subjected to IVM, which in porcine species lasts for 40−44 hours, with the objective of obtaining a female gamete competently matured both at the nuclear and cytoplasmic levels. Traditionally, the most commonly used maturation medium is tissue culture medium (TCM)-199 but other media have been already employed such as Waymouth MB 752/1, North Carolina State University (NCSU37 and NCSU23), modified Tyrode's solution and, more recently, porcine oocyte medium (reviewed by Redel et al. [13]). The incidence of in vitro matured COCs that develop to the blastocyst stage after in vitro fertilization (IVF) is higher when COCs are matured in a two-step system, where they are exposed to dibutyryl cyclic adenosine monophosphate for the first 20 hours of IVM [14].

Reproductive Technologies in Animals. DOI: https://doi.org/10.1016/B978-0-12-817107-3.00005-9

In pigs the low efficiency in cytoplasmic maturation has been related to low rates of male pronuclear formation after fertilization, high rates of polyspermy, and low development to the blastocyst stage (reviewed by Romar et al. [15]). Over the years, these problems have been partially overcome by replacing fetal calf serum with porcine follicular fluid [16] and supplementing the IVM medium with cysteine [17,18]. Other additives with beneficial effects on oocyte's maturation are epidermal growth factor [19], beta-mercaptoethanol [20], and FLI [21], a combination of fibroblast growth factor 2, leukemia inhibitory factor, and insulin-like growth factor 1, which was recently reported to improve maturation of immature pig oocytes and increasing the efficiency of blastocyst production by twofold after IVF.

In the current porcine IVM systems, rate of metaphase II (nuclear maturation) and male pronuclear formation after IVF (one of the parameters directly correlated to oocyte's cytoplasmic maturation) are both over 90% [2,14,16,17,21]. However, the incidence of polyspermic penetration in pig IVF is greater than 40% of inseminated oocytes (reviewed by Romar et al. [22]).

More studies and improvements are still necessary in the IVM system in pigs to produce competent matured oocytes that will further develop to high-quality blastocysts capable of becoming live and healthy offspring.

5.2 General aspects of sperm assessment and preparation for assisted reproductive technology

Advanced sperm selection techniques are thought to improve the chance that structurally intact and mature sperm cells with high DNA integrity are selected for fertilization. Basically, the methods used for sperm preparation prior to assisted reproductive technology (ART) have been designed, on the one hand, to remove diluent media, cryoprotectants, or decapacitating factors (from the seminal plasma or epididymal fluid), and on the other hand, to select motile and mature sperm cells. However, depending on the method of sperm preparation, the outcome of ART is very different [23−25].

Basically four methods for sperm selection can be considered prior to ART in the porcine species: (1) dilution of semen and washing, (2) sperm migration: known as the direct swim-up method, (3) selective washing of semen (density gradient method: single or multiple layers), and (4) adherence method (glass wool, glass beads, or Sephadex columns) [26−28]. However, the latter are not practically used in pigs.

The first success in porcine IVF was achieved by Cheng [29] who used ejaculated spermatozoa that were washed and preincubated for 4−5 hours at 37°C or stored at 20°C for 16 hours, washed, and preincubated only at 37°C for 40 minutes. Today, washing procedures consist of two or three dilutions and washing in a saline solution supplemented with albumin (from 0.1% to 1%). Albumin promotes sperm capacitation acting as acceptor of cholesterol by removing it from the plasma membrane. A decreased cholesterol/phospholipid ratio consequently contributes to an increased membrane fluidity, which in turn promotes an increase of ion permeability (revised by López-Úbeda and Matás [30]).

The conventional swim-up procedure was originally described in human by Mahadevan and Baker in 1984 [31]. The methodology is based on the active movement of spermatozoa from the pellet generated after a centrifugation into an overlaying medium. This technique serves to select a very high population (>90%) of motile sperm cells that are morphologically normal, with low DNA fragmentation and free of debris and other cells [32,33]. The swim-up method has also been modified by the use of a strainer as permeable barrier [34], using specific designed medium supplemented with natural additives such as oviductal fluid [35−37].

Increased quality sperm cells may also be selected based on their density differential. For this, they are centrifuged through a continuous or discontinuous density gradient of either colloidal or silanized colloidal silica [4,38]. Several studies have shown better performance of IVF using sperm cells selected with discontinuous density gradients than with washing procedures [23−25,39,40]. However, when spermatozoa came from the epididymis, the outcome was not affected by the method of sperm selection [24]. A *caveat* of this procedure, though, is that seminal plasma proteins and cholesterol can be removed from the sperm cell membrane [41], although this aspect does not seem to negatively affect IVF performance, as similar results were obtained when spermatozoa were centrifugated without a continuous gradient [42].

The preparation and selection of spermatozoa for intracytoplasmic sperm injection (ICSI) require procedures beyond those described thus far (revised by Garcia-Rosello et al. [43]). For example, several approaches have been explored with the use of Percoll for selection of ejaculated and epididymal spermatozoa [44,45]. To disrupt their membranes, mechanical processes such as repeated freeze-thaw cycles or the use of compounds as Triton X-100 [46−49] as well as pretreatment with compounds such as lipase or hyaluronic acid have been employed [48,50]. Lastly, microfluidic systems have also been used for sperm selection [46].

5.3   In vitro fertilization and embryo culture

Cheng et al. (1986) successfully fertilized in vitro—ovulated oocytes and cultured the resulting embryos to the two- to four-cell stage. A total of 19 piglets were born after transfer of these embryos to recipient gilts [51]. The IVF system they described transformed the research conducted in this area, which had been plagued by a lack of success up until that time.

At the beginning of the porcine IVF development, there were several problems to solve. On the one hand, an insufficient maturation of the oocytes was achieved, and on the other hand the sperm preparation prior to fertilization was not perfected. Today, the main problem that remains unsolved is the high degree of polyspermy (reviewed by Romar et al. [15,22], Coy and Romar [52], and Dang-Nguyen et al. [53]). To solve this problem, a number of investigations have been performed with the goal of evaluating all the factors involved and needed in the fertilization process, leading to the establishment of different methods and protocols to improve the final outcome. Nevertheless, to date not a single solution to this problem is available nor a generalized protocol that guarantees a successful porcine IVF. Therefore nowadays the success rate (ratio of monospermic zygotes/number of inseminated oocytes) usually does not exceed 45% (reviewed by Romar et al. [15]).

The most frequently used source of oocytes for IVF comes from ovaries obtained from slaughterhouses. Subsequently, these are matured in vitro (described in previous paragraphs). However, in vivo—matured oocytes from gilts/sows collected by oviduct washing are also used [54,55]. COCs secrete various molecules, including progesterone, hyaluronic acid, and NO [56—58], and these molecules favor sperm capacitation and therefore fertilization [59,60]. However, the presence/absence of cumulus cells may have positive and/or negative effects on IVF parameters [61,62]. In this regard, some authors have shown that the effect of some molecules added to IVF media is opposite, depending on whether cumulus-enclosed or cumulus-free oocytes are inseminated [63].

With regard to sperm cells, epididymal and ejaculated spermatozoa have been compared in IVF system [24,64], together with fresh versus frozen/thawed samples, with no conclusive results [64,65]. This could be due to the high influence on IVF outcomes of other factors such as sperm concentration, sperm selection and preparation methodology, and IVF medium and additives [23,62,66].

IVF culture media share a common formulation containing various components including inorganic salts, nutrients, vitamins, and growth factors for gamete coculture (revised by Romar et al. [22]). The classical media used for porcine IVF include modified Tyrode's albumin lactate pyruvate, modified Tris-buffered medium, modified TCM-199 and porcine gamete medium (reviewed by Romar et al. [15,22]).

Under in vivo conditions, millions of spermatozoa are deposited in the female genital tract, but only a small subpopulation will reach the oviduct and site of fertilization. In in vitro systems, various methods have been devised to restrict the number of spermatozoa that will reach the oocyte within the insemination device, such as the climbing-over-a-wall method, biomimetic microchannel IVF system, straw IVF, modified swim-up method, and the microfluidic sperm sorter (revised by Grupen [67]). Each of these systems attempts to mimic the in vivo selection so that sperm cells have to overcome an obstacle while acquiring hypermotility.

IVF involves fertilization in an artificial environment. In this regard, temperature, pH, gases, and humidity must be maintained closer to what physiologically typical of the oviductal environment. Consequently, performing these procedures using precise temperatures and gas compositions that best mimic those found in reproductive organs would be a major step toward improving in vitro conditions for IVF. The pH in the porcine oviductal ampulla is close to 8.0 during the periovulatory phase [68]; however, IVF is usually done at a pH closer to neutrality (7.4) [69]. Modification in the percent of CO_2 in the incubator so that the pH was 8 induced a marked improvement in IVF results [60]. With regard to the atmosphere surrounding the gametes, the conventional gas composition is $\sim 20\%$ O_2 (i.e., air) and 5% CO_2. It has been shown though that by lowering the oxygen tension (7%), measured in vivo in the pig reproductive organs, the IVF output is increased [36]. A similar approach was followed with the temperature, and the use of 37°C for IVF, as measured in the oviduct, demonstrated to produce a higher efficiency than the pig body temperature (38.5°C) conventionally used, increasing monospermy rates, and blastocyst yield [70].

The in vitro culture of porcine zygotes to the blastocyst stage still remains very challenging since the proportion of success, after years of research, barely reaches 30%—40% today [62]. A comprehensive and thorough account of the different culture media on porcine preimplantation embryo has been recently published [13], and it is strongly suggested to read that review for further information pertaining to this topic. Briefly, the most common embryo culture media used in the porcine species are the North Carolina State University (NCSU-23 medium [71]), the porcine zygote medium [72], and the Beltsville Embryo Culture Medium 73], with their different modifications. As not all of them are commercially available, they need to be prepared in the laboratories. This fact may create variations in the batches if quality controls are not employed; hence the claims about the consistently high quality of the chemically defined media may be called into question. There is a tendency to use these kinds of media as it permits to pinpoint the impact of specific components on embryo development. However, these modifications to culture media have turned them "antinatural" as they no longer mimic the in vivo situation. Indeed, the reproductive fluids soaking the lumen of the oviduct and uterus have a chemical composition extremely complex, and the molecules included in the chemical definition of the media prepared in the lab represent a poor facsimile of a few components found in these environments. Artificial media are in fact made of electrolytes, amino acids, and some carbohydrates, together with antibiotics and, occasionally a protein source such as serum albumin. Even this last molecule is often being replaced by polyvinyl alcohol (PVA) arguing that chemical variations in serum albumin can have an impact on embryo development [74]. The net result in the use of defined media is to force the embryos to grow in a considerably stressful environment because, in nature, they on the contrary would develop in a milieu containing, to say the least, a considerable proportion of glycosaminoglycans, several hundreds of different proteins, and hormones [75–77].

The lipidomic variations in the endometrium can also affect the success of implantation after embryo transfer, making the picture even more complicated [78]. The increasing evidence supporting the idea that oviduct and uterine fluids play a key role in the "developmental origins of health and disease" concept in different species represents the best wake-up call for researchers to start re-thinking about the future of embryo culture systems. It is time to decide if the final goal, in the pig embryo production systems, consists of getting the maximum number of animals or the healthiest individuals [35,79–81].

5.4 Tissue culture and isolation of germline

Germinal cells are in charge of transmitting, mainly, genetic information from one generation to the next to ensure species conservation. Germline specification begins in the preimplantation embryo with the formation of the bipotential primordial germ cells (PGCs), which will result, after a highly orchestrated regulatory process, in male or female gametes. The final maturation stage will give haploid gametes, and by fusion of the female and the male counterparts, a new fertilization occurs and with it the development of a new organism.

PGC isolation is challenging in all mammalian species due to the low number of cells from which they originate and the position within the embryo. Isolation of PGC has been unsuccessful for decades in pigs, due to a lack of knowledge about the germline specification pathways and specific germline markers. Recently, Kobayashi et al. (2017) reported that PGC specification in pigs [82] starts from the posterior preprimitive-streak competent epiblast. Sequential upregulation of SOX17 and BLIMP1 seems critical in this process, similar to what has been observed in human PGC specification.

Identification and isolation of spermatogonial stem cells (SSCs) could be useful for achieving genetic gain or restore fertility, and genetic modification of SSCs combined with germ cell transplantation provides the possibility to produce transgenic animals. In the last years, the concept of surrogate sires has emerged as an alternative for breeding in livestock production, including the swine industry. This is based on the regenerative capacity of SSCs, which are isolated from selected boars, and transferred into male recipients with germline ablation to produce mature spermatozoa with the donor haplotype [83]. Reestablishment of fertility using this technology was demonstrated more than 25 years ago in mice. In that experiment, offspring were produced from infertile males with transplanted SSCs [84]. However, the scarcity of SSCs in the testes requires in vitro expansion of SSCs to obtain enough material for cell transplantation, which has limited progress in pigs. More promising results in swine have been recently reported by Zhang et al. (2017) who showed long-term maintenance of porcine neonatal testis cells with similar morphology to cattle and mouse spermatogonia and expression of conserved, but no specific, markers of undifferentiated spermatogonia (GFRA1, LIN28, and PLZF) [85]. However, they did not show SSC regenerative capacity after cell transplantation. Few studies have reported SSC transplantation in domestic pigs [86–88] and unequivocal evidence for SSC regeneration in recipient males have not been provided. Even with these advances, several challenges remain unsolved, such as identification of specific markers to distinguish SSCs from progenitor spermatogonia that lack regenerative capacity.

Another challenge already surpassed was the development of effective strategies for complete germline ablation in recipient males. Knockout (KO) of NANOS2, a male germline specific factor, in pigs has been reported recently, resulting in male specific germline ablation with normal physiology and histology of the seminiferous tubules [89]. Considering that an increased efficiency in livestock animal production is required to cover the unprecedented demand that is predicted for next decades, the use of surrogate sires is proposed as a valuable breeding tool in swine production, with some utility even in natural breeding.

Despite the fact that mammals are programmed to require male and female gametes to generate an embryo, because both genomes are not equivalent, recent technology has allowed to obtain new individuals without the necessity of both gametes.

5.5 Nuclear transfer, micromanipulation, and embryo biopsy

Somatic cell nuclear transfer (SCNT) provides the possibility to obtain embryos using enucleated oocytes and a somatic cell [90]. It is almost two decades since the first reports on cloned pigs using SCNT and adult cells were reported [91,92]. SCNT has been used to produce transgenic pigs for generation of humanized organs for xenotransplantation [93] or the development of models to investigate human diseases [94,95]. More recently, the interest of using pigs for xenotransplants has been rekindled after the publication showing genome-wide inactivation of porcine endogenous retroviruses by the use of CRISPR/Cas9 technology [96].

Unfortunately, the efficiency of SCNT in pigs remains poor, with only 1%−3% of transferred embryos developing to term [97,98]. Among the biggest challenges for cloning pigs are: (1) the ineffective in vitro embryo culture system, which is required to culture reconstructed embryos before transfer to surrogate sows and (2) the necessity for at least four good embryos for maternal recognition of pregnancy in sows. During the past few years, porcine embryo culture systems have been significantly improved by us and others (reviewed by Romar et al. [22]). This progress, in combination with different strategies to promote cell donor reprogramming [99] or XIST nullification in the donor cells [97], could result in enhanced cloning efficiency in the near future.

5.6 Molecular application to embryo identification, selection, and sex determination

In pigs embryo selection is critical in order to increase embryo production efficiency, especially when using a nonoptimized in vitro culture system. Even though noninvasive embryonic selection criteria based on morphokinetics still remain prevalent over other markers, a wide range of transcriptomic, epigenetic, and metabolomic markers have been developed during the last decade. Gene expression analysis of a few pluripotency, embryonic, and/or cell cycle regulation factors has been extensively used for embryo quality assessment with limited utility. Recently, transcriptomic and epigenetic profiling of single embryos has become affordable cutting-edge technologies. This fact has allowed comparison between in vivo and in vitro−produced embryos and permitted the identification of differentially expressed genes that could be used as quality markers to help in improving embryo culture and embryo selection. For example, our group [35] recently reported whole-genome DNA methylation and transcriptome analyses in single pig blastocyst, findings that will undoubtedly shed light on the mechanisms underlying the abnormalities observed in ART-derived offspring.

Imprinted genes, which show parental-specific methylation, play a key role in fetal and placental development and they have been associated to ART-derived defects (review by Canovas et al. [100]). Expression and DNA methylation of imprinted genes (such as H19, IGF2R, XIST, IGF2, NNAT, PEG10, or MEST) are also molecular markers of embryo viability. Similarly, expression of DNA methyltransferases, involved in de novo (DNMT3a/b) and maintenance (DNMT1) DNA methylation, together with TET enzymes responsible for the modification of methyl DNA, are also useful molecular marker. However, the inclusion of these markers in reliable and easy-to-use kits for embryo selection before transfer remains challenging.

On the other hand, metabolomic markers provide the possibility to select embryos based on the change in the levels of specific nutrients and metabolites and the consumption of key nutrients from the culture media [101,102]. In humans, early studies reported correlation between metabolomic profiles and implantation and pregnancy rates. However, randomized controlled trials using the same metabolomics-based viability index did not result in increased pregnancy rates compared with the routine morphology selection (reviewed by Krisher et al. [102]). In pigs, Krisher et al. (2015) reported some metabolic characteristics of the in vitro−derived porcine embryos, such as the dependency on glutamine, but there is still a lack of metabolic markers for embryo quality in pigs [101]. Moreover, embryo selection using metabolic analysis requires single-embryo culture, which is not a routine practice in porcine in vitro embryo culture.

In addition to embryo quality, preselection by sex is desired to manipulate the expected male and female ratio (50/50) in the offspring. In pigs, it could be useful to reduce the male ratio and avoid the unpleasant specific male boar taint, the peculiar male taste given to the muscles by the male hormones, which is rejected by consumers [103]. It would entail economic benefit by a faster female growth and the avoidance of male castration. Nonetheless, even though boar sexed semen is available through fluorescence-activated sperm sorting, its use is not a routine practice in commercial farms, even after more than 15 years from the first successful report using porcine sexed semen and nonsurgical insemination [104]. The susceptibility of sperm cells to the sorting procedure, the high number of sperm required for insemination in pigs (even with the development of an intrauterine insemination device), and the slow speed of the sex-sorting process (which affects the costs and availability of samples) are limitations to the commercial application of sexed semen in the swine industry [105]. Another alternative is the selection of the embryo by karyotyping or gene expression analysis to identify chromosome Y specific factors. However, the use of embryo transfer in pigs remains dramatically low, which seriously limits its use.

5.7 Transgenesis/genetically modified animals

Genome engineering in animals has become a reality for the scientific community since 1980, when transgenic mice were obtained by recombinant plasmid pronuclear microinjection [106]. A few years later, in 1985 the first transgenic pigs were produced by two different research groups by microinjection into the pronuclei with a random insertion of foreign DNA [107,108]. In the next two decades, other techniques were implemented to generate transgenic pigs with the use of retrovirus [109] and lentivirus as vectors [110], sperm-mediated gene transfer by artificial insemination [111,112], and ICSI [113,114]. All these techniques had limitations and produced only insertion of new genes in the pig genome [only knock in (KI)] in a random and not targeted manner and with low efficiency. A new situation was opened with the use of somatic nuclear transfer (NT) technology [91,92,115], giving thus the possibility to first make the modification in the somatic cells and then use the modified cells as donors for NT. SCNT was used to produce KI [116,117] and KO pigs [118,119]. One decade later, the use of transcription activator-like effector nuclease and zinc finger nuclease genome editing [120,121] opened the possibility of genome edition. Later, CRISPR-Cas9 gene editing made the rapid and economic generation of KO/KI pigs feasible [122].

In 2014, it was reported that the CRISPR/Cas9 system could be used to produce genetically engineered pigs by intracytoplasmic injection of zygotes [123,124] and SCNT [124,125]. Electroporation has also been used to lead the entry of CRISPR/Cas9 into oocytes and early embryos [126]. This last technique requires less sophisticated equipment and highly trained specialized personnel than microinjection or NT methodology, making more feasible its implementation. In the past few years an exponential number of studies have tried to optimize the factors that affect the efficiency of this methodology in pigs and in diverse areas and applications [127–133]. A new area of gene editing is dawning with the development of new methodologies associated with the use of base editors for multiple genes [134] or conditional and tissue-specific expression of mutations using Cre/loxP site-specific recombination systems [135,136].

Applications of transgenic and gene editing in pig animal production are associated with improvement in the carcass composition [137] and meat quality [138,139], lactational performance [140,141] and reduction of manure phosphorus contamination [142,143]. However, the most interesting application in the pig industry is related to the increase in the resistance to important diseases such as porcine reproductive and respiratory syndrome [144,145], coronavirus [146], classical swine fever virus [147], and foot and mouth disease [148].

In addition, there are applications of these technologies in biomedicine. Generation of KO and KI mice by targeting genes in embryonic stem (ES) cells revolutionized biomedicine by making it possible to study gene function in a living mammal [149,150]. However, mice are far from an ideal model for many human pathophysiological processes. Pigs are much more like humans in size, diet, physiology, and drug uptake/metabolism [151,152]. However, their usefulness as a biomedical model has been limited by a failure to isolate ES cells for generating KO/knocking versions of this species. The application of CRISPR/Cas technology opened a new era for the generation of pig models of human diseases [95,133,153,154]. Furthermore, gene-edited pigs hold promise for the development of humanized organs that may be used for xenotransplantation [155–158] as well as for the production of recombinant proteins in the modified animals for biopharmaceutical and nutrition industries [159,160].

5.8 Sperm, oocyte, and embryo cryopreservation

Semen cryopreservation greatly facilitates the distribution of male desirable genes over distant places and without time restraints, and this aspect alone could determine a rapid increase in swine industry productivity [161]. However, the application of frozen semen in commercial farms is limited to genetically selected animals, for long-distance transport or to preserve specific valuable individuals or breed/lines [162,163].

The fertilizing ability of frozen-thawed samples has increased in recent years, with farrowing rates well over 80% in experimental field trials [164–166]. The improvement in the results has been related to the selection of good freezing boars [167], changes in freezing procedures, and use of additives in the extenders [164,168], ovulatory synchrony with insemination [169], and by the use of intrauterine and postcervical insemination (revised by Knox [162], Yeste et al. [163], and Roca et al. [170]).

One other alternative to preserve material from a male could be the cryopreservation of testicular tissue that later could be xenografted into immunodeficient nude mice [171] or transferred to "recipient testis" by germ cell transplantation [86]. The implantation of testicular tissue in the nude mouse model leads to spermatogenesis and the recovery of normal spermatozoa with fertilizing capacity by ICSI, and to the production of live piglets after embryo transfer into recipients [172,173].

Porcine oocytes and embryos are very sensitive to cold shock, mainly related to the high content of lipids in the cytoplasm, and for this reason making it difficult to cryopreserve by using conventional freezing or vitrification protocols (reviewed by Zhang et al. [174], Men et al. [175], and Kikuchi et al. [176]). Nevertheless, some studies have reported embryo development and birth of piglets derived from cryopreserved oocytes [50,177] and embryos [178,179], indicating that it is feasible to use this technology.

Whole ovarian tissue [180] or fragments of ovarian cortex [181] could be cryopreserved and later implanted by xenografting in an immunodeficient nude mice [182], although the efficiency is limited [183]. Another alternative explored for fertility preservation is the xenotransplantation of the whole ovary into irradiated ovariectomized nude rats [184]. Ovarian autotransplantation in pigs is under study as a clinical method for the preservation of fertility and hormonal function [185].

References

[1] Grupen CG, McIlfatrick SM, Ashman RJ, Boquest AC, Armstrong DT, Nottle MB. Relationship between donor animal age, follicular fluid steroid content and oocyte developmental competence in the pig. Reprod Fertil Dev 2003;15(1-2):81−7.

[2] Ikeda K, Takahashi Y. Comparison of maturational and developmental parameters of oocytes recovered from prepubertal and adult pigs. Reprod Fertil Dev 2003;15(4):215−21.

[3] Bagg MA, Nottle MB, Armstrong DT, Grupen CG. Relationship between follicle size and oocyte developmental competence in prepubertal and adult pigs. Reprod Fertil Dev 2007;19(7):797−803.

[4] Mattioli M, Bacci ML, Galeati G, Seren E. Developmental competence of pig oocytes matured and fertilized in vitro. Theriogenology. 1989;31(6):1201−7.

[5] Lucci CM, Schreier LL, Machado GM, Amorim CA, Bao SN, Dobrinsky JR. Effects of storing pig ovaries at 4 or 20 degrees C for different periods of time on the morphology and viability of pre-antral follicles. Reprod Domest Anim 2007;42(1):76−82.

[6] Kim HJ, Choi SH, Son DS, Cho SR, Choe CY, Kim YK, et al. Effect of exposure duration of ovaries and oocytes at ambient temperature on parthenogenetic development of porcine follicular oocytes. J Reprod Dev 2006;52(5):633−8.

[7] Petrenko A, Carnevale M, Somov A, Osorio J, Rodriguez J, Guibert E, et al. Organ preservation into the 2020s: the era of dynamic intervention. Transfus Med Hemother 2019;46(3):151−72.

[8] Thuillier R, Renard C, Rogel-Gaillard C, Demars J, Milan D, Forestier L, et al. Effect of polyethylene glycol-based preservation solutions on graft injury in experimental kidney transplantation. Br J Surg 2011;98(3):368−78.

[9] Barberino RS, Silva JRV, Figueiredo JR, Matos MHT. Transport of domestic and wild animal ovaries: a review of the effects of medium, temperature, and periods of storage on follicular viability. Biopreserv Biobank 2019;17(1):84−90.

[10] Santos EC, Pradiee J, Madeira EM, Pereira MM, Mion B, Mondadori RG, et al. Selection of porcine oocytes in vitro through brilliant cresyl blue staining in distinct incubation media. Zygote. 2017;25(1):49−55.

[11] Ishizaki C, Watanabe H, Bhuiyan MM, Fukui Y. Developmental competence of porcine oocytes selected by brilliant cresyl blue and matured individually in a chemically defined culture medium. Theriogenology. 2009;72(1):72−80.

[12] Kempisty B, Jackowska M, Piotrowska H, Antosik P, Wozna M, Bukowska D, et al. Zona pellucida glycoprotein 3 (pZP3) and integrin beta2 (ITGB2) mRNA and protein expression in porcine oocytes after single and double exposure to brilliant cresyl blue test. Theriogenology. 2011;75(8):1525−35.

[13] Redel BK, Spate LD, Prather RS. In vitro maturation, fertilization, and culture of pig oocytes and embryos. Methods Mol Biol 2019;2006:93−103.

[14] Funahashi H, Cantley TC, Day BN. Synchronization of meiosis in porcine oocytes by exposure to dibutyryl cyclic adenosine monophosphate improves developmental competence following in vitro fertilization. Biol Reprod 1997;57(1):49−53.

[15] Romar R, Funahashi H, Coy P. In vitro fertilization in pigs: new molecules and protocols to consider in the forthcoming years. Theriogenology. 2016;85(1):125−34.

[16] Naito K, Fukuda Y, Toyoda Y. Effects of porcine follicular fluid on male pronucleus formation in porcine oocytes matured in vitro. Gamete Res 1988;21(3):289−95.

[17] Yoshida M, Ishigaki K, Pursel VG. Effect of maturation media on male pronucleus formation in pig oocytes matured in vitro. Mol Reprod Dev 1992;31(1):68−71.

[18] Sawai K, Funahashi H, Niwa K. Stage-specific requirement of cysteine during in vitro maturation of porcine oocytes for glutathione synthesis associated with male pronuclear formation. Biol Reprod 1997;57(1):1−6.

[19] Abeydeera LR, Wang WH, Cantley TC, Rieke A, Prather RS, Day BN. Presence of epidermal growth factor during in vitro maturation of pig oocytes and embryo culture can modulate blastocyst development after in vitro fertilization. Mol Reprod Dev 1998;51(4):395–401.

[20] Abeydeera LR, Wang WH, Cantley TC, Prather RS, Day BN. Presence of beta-mercaptoethanol can increase the glutathione content of pig oocytes matured in vitro and the rate of blastocyst development after in vitro fertilization. Theriogenology. 1998;50(5):747–56.

[21] Yuan Y, Spate LD, Redel BK, Tian Y, Zhou J, Prather RS, et al. Quadrupling efficiency in production of genetically modified pigs through improved oocyte maturation. Proc Natl Acad Sci U S A 2017;114(29) E5796-e804.

[22] Romar R, Canovas S, Matas C, Gadea J, Coy P. Pig in vitro fertilization: where are we and where do we go? Theriogenology. 2019;137:113–21.

[23] Matas C, Coy P, Romar R, Marco M, Gadea J, Ruiz S. Effect of sperm preparation method on in vitro fertilization in pigs. Reproduction. 2003;125(1):133–41.

[24] Matas C, Sansegundo M, Ruiz S, Garcia-Vazquez FA, Gadea J, Romar R, et al. Sperm treatment affects capacitation parameters and penetration ability of ejaculated and epididymal boar spermatozoa. Theriogenology. 2010;74(8):1327–40.

[25] Matas C, Vieira L, Garcia-Vazquez FA, Aviles-Lopez K, Lopez-Ubeda R, Carvajal JA, et al. Effects of centrifugation through three different discontinuous Percoll gradients on boar sperm function. Anim Reprod Sci 2011;127(1-2):62–72.

[26] Bussalleu E, Pinart E, Rivera MM, Briz M, Sancho S, Yeste M, et al. Effects of matrix filtration of low-quality boar semen doses on sperm quality. Reprod Domest Anim 2009;44(3):499–503.

[27] Ramio-Lluch L, Balasch S, Bonet S, Briz M, Pinart E, Rodriguez-Gil JE. Effects of filtration through Sephadex columns improve overall quality parameters and "in vivo" fertility of subfertile refrigerated boar-semen. Anim Reprod Sci 2009;115(1-4):189–200.

[28] Bussalleu E, Pinart E, Rivera MM, Arias X, Briz M, Sancho S, et al. Effects of filtration of semen doses from subfertile boars through neuter Sephadex columns. Reprod Domest Anim 2008;43(1):48–52.

[29] Cheng WTK. In vitro fertilization of farm animal oocytes. Ph.D. thesis, Council for National Academic Awards, Cambridge, UK; 1985.

[30] López-Úbeda R, Matás C. An approach to the factors related to sperm capacitation process. Androl Open Acces 2015;4(1):1000128.

[31] Mahadevan M, Baker G. Assessment and preparation of semen for in vitro fertilization. In: Wood C, Trounson A, editors. Clinical in vitro fertilization. London: Springer; 1984. p. 83–97.

[32] Holt WV, Hernandez M, Warrell L, Satake N. The long and the short of sperm selection in vitro and in vivo: swim-up techniques select for the longer and faster swimming mammalian sperm. J Evol Biol 2010;23(3):598–608.

[33] Morency E, Anguish L, Coonrod S. Subcellular localization of cytoplasmic lattice-associated proteins is dependent upon fixation and processing procedures. PLoS One 2011;6(2):e17226.

[34] Park CH, Lee SG, Choi DH, Lee CK. A modified swim-up method reduces polyspermy during in vitro fertilization of porcine oocytes. Anim Reprod Sci 2009;115(1-4):169–81.

[35] Canovas S, Ivanova E, Romar R, García-Martínez S, Soriano-Úbeda C, García-Vázquez FA, et al. DNA methylation and gene expression changes derived from assisted reproductive technologies can be decreased by reproductive fluids. eLife. 2017;6:e23670.

[36] Garcia-Martinez S, Sanchez-Hurtado MA, Gutierrez H, Sanchez-Margallo FM, Romar R, Latorre R, et al. Mimicking physiological O_2 tension in the female reproductive tract improves assisted reproduction outcomes in pig. Mol Hum Reprod 2018;24(5):260–70.

[37] Navarro-Serna S, Garcia-Martinez S, Paris-Oller E, Calderon PO, Gadea J. The addition of porcine oviductal fluid (OF) in swim-up media improves the selection and modifies motility patterns and capacitation potential of boar spermatozoa. Abstract of the 34th Annual Meeting of the ESHRE, Barcelona, Spain 1 to 4 July 2018 (P-099). Hum Reprod 2018;33:i186.

[38] Berger T, Horton MB. Evaluation of assay conditions for the zona-free hamster ova bioassay of boar sperm fertility. Gamete Res 1988;19(1):101–11.

[39] Grant SA, Long SE, Parkinson TJ. Fertilizability and structural properties of boar spermatozoa prepared by Percoll gradient centrifugation. J Reprod Fertil 1994;100(2):477–83.

[40] Jeong BS, Yang X. Cysteine, glutathione, and Percoll treatments improve porcine oocyte maturation and fertilization in vitro. Mol Reprod Dev 2001;59(3):330–5.

[41] Kruse R, Dutta PC, Morrell JM. Colloid centrifugation removes seminal plasma and cholesterol from boar spermatozoa. Reprod Fertil Dev 2011;23(7):858–65.

[42] Sjunnesson YC, Morrell JM, Gonzalez R. Single layer centrifugation-selected boar spermatozoa are capable of fertilization in vitro. Acta Vet Scand 2013;55:20.

[43] Garcia-Rosello E, Garcia-Mengual E, Coy P, Alfonso J, Silvestre MA. Intracytoplasmic sperm injection in livestock species: an update. Reprod Domest Anim 2009;44(1):143–51.

[44] Garcia-Rosello E, Matas C, Canovas S, Moreira PN, Gadea J, Coy P. Influence of sperm pretreatment on the efficiency of intracytoplasmic sperm injection in pigs. J Androl 2006;27(2):268–75.

[45] Nakai M, Suzuki SI, Ito J, Fuchimoto DI, Sembon S, Noguchi J, et al. Efficient pig ICSI using Percoll-selected spermatozoa; evidence for the essential role of phospholipase C-zeta in ICSI success. J Reprod Dev 2016;62(6):639–43.

[46] Matsuura K, Uozumi T, Furuichi T, Sugimoto I, Kodama M, Funahashi H. A microfluidic device to reduce treatment time of intracytoplasmic sperm injection. Fertil Steril 2013;99(2):400–7.

[47] Tian JH, Wu ZH, Liu L, Cai Y, Zeng SM, Zhu SE, et al. Effects of oocyte activation and sperm preparation on the development of porcine embryos derived from in vitro-matured oocytes and intracytoplasmic sperm injection. Theriogenology. 2006;66(2):439–48.

[48] Wei Y, Fan J, Li L, Liu Z, Li K. Pretreating porcine sperm with lipase enhances developmental competence of embryos produced by intracytoplasmic sperm injection. Zygote. 2016;24(4):594–602.

[49] Yong HY, Pyo BS, Hong JY, Kang SK, Lee BC, Lee ES, et al. A modified method for ICSI in the pig: injection of head membrane-damaged sperm using a 3−4 micro m diameter injection pipette. Hum Reprod 2003;18(11):2390−6.

[50] Casillas F, Betancourt M, Cuello C, Ducolomb Y, Lopez A, Juarez-Rojas L, et al. An efficiency comparison of different in vitro fertilization methods: IVF, ICSI, and PICSI for embryo development to the blastocyst stage from vitrified porcine immature oocytes. Porcine Health Manag 2018;4:16.

[51] Cheng WTK, Moor RM, Polge C. In vitro fertilization of pig and sheep oocytes matured in vivo and in vitro. Theriogenology. 1986;25(1):146.

[52] Coy P, Romar R. In vitro production of pig embryos: a point of view. Reprod Fertil Dev 2002;14(5-6):275−86.

[53] Dang-Nguyen TQ, Somfai T, Haraguchi S, Kikuchi K, Tajima A, Kanai Y, et al. In vitro production of porcine embryos: current status, future perspectives and alternative applications. Anim Sci J 2011;82(3):374−82.

[54] Coy P, Martinez E, Ruiz S, Vazquez JM, Roca J, Gadea J. Environment and medium volume influence in vitro fertilisation of pig oocytes. Zygote (Cambridge, Engl) 1993;1(3):209−13.

[55] Nakamura Y, Tajima S, Kikuchi K. The quality after culture in vitro or in vivo of porcine oocytes matured and fertilized in vitro and their ability to develop to term. Anim Sci J 2017;88(12):1916−24.

[56] Nagyova E. Regulation of cumulus expansion and hyaluronan synthesis in porcine oocyte-cumulus complexes during in vitro maturation. Endocr Regul 2012;46(4):225−35.

[57] Nagyova E. The biological role of hyaluronan-rich oocyte-cumulus extracellular matrix in female reproduction. Int J Mol Sci 2018;19(1).

[58] Chmelikova E, Jeseta M, Sedmikova M, Petr J, Tumova L, Kott T, et al. Nitric oxide synthase isoforms and the effect of their inhibition on meiotic maturation of porcine oocytes. Zygote. 2010;18(3):235−44.

[59] Suzuki K, Asano A, Eriksson B, Niwa K, Nagai T, Rodriguez-Martinez H. Capacitation status and in vitro fertility of boar spermatozoa: effects of seminal plasma, cumulus-oocyte-complexes-conditioned medium and hyaluronan. Int J Androl 2002;25(2):84−93.

[60] Soriano-Úbeda C, García-Vázquez FA, Romero-Aguirregomezcorta J, Matás C. Improving porcine in vitro fertilization output by simulating the oviductal environment. Sci Rep 2017;7:43616. Available from: https://doi.org/10.1038/srep43616.

[61] Romar R, Coy P, Ruiz S, Gadea J, Rath D. Effects of oviductal and cumulus cells on in vitro fertilization and embryo development of porcine oocytes fertilized with epididymal spermatozoa. Theriogenology. 2003;59(3-4):975−86.

[62] Li R, Liu Y, Pedersen HS, Callesen H. Effect of cumulus cells and sperm concentration on fertilization and development of pig oocytes. Reprod Domest Anim 2018;53(4):1009−12.

[63] Tatemoto H, Muto N, Yim SD, Nakada T. Anti-hyaluronidase oligosaccharide derived from chondroitin sulfate a effectively reduces polyspermy during in vitro fertilization of porcine oocytes. Biol Reprod 2005;72(1):127−34.

[64] Rath D, Niemann H. In vitro fertilization of porcine oocytes with fresh and frozen-thawed ejaculated or frozen-thawed epididymal semen obtained from identical boars. Theriogenology. 1997;47(4):785−93.

[65] Suzuki C, Yoshioka K, Itoh S, Kawarasaki T, Kikuchi K. In vitro fertilization and subsequent development of porcine oocytes using cryopreserved and liquid-stored spermatozoa from various boars. Theriogenology. 2005;64(6):1287−96.

[66] Rath D. Experiments to improve in vitro fertilization techniques for in vivo-matured porcine oocytes. Theriogenology. 1992;37(4):885−96.

[67] Grupen CG. The evolution of porcine embryo in vitro production. Theriogenology. 2014;81(1):24−37.

[68] Rodriguez-Martinez H. Role of the oviduct in sperm capacitation. Theriogenology. 2007;68(Suppl 1):S138−46.

[69] Soriano-Úbeda C, Romero-Aguirregomezcorta J, Matás C, Visconti PE, García-Vázquez FA. Manipulation of bicarbonate concentration in sperm capacitation media improvesin vitro fertilisation output in porcine species. J Anim Sci Biotechnol 2019;10(1):19.

[70] Garcia-Martinez S, Lopez Albors O, Latorre R, Romar R, Coy P, editors. Culture under the physiological temperature registered along the reproductive tract of female pigs improves the blastocysts yield in vitro. Abstract of the 35rd Scientific meeting Association of Embryo Technology in Europe (AETE); 2019. Murcia (Spain) Anim Reprod 2019;16(3):726.

[71] Petters RM, Wells KD. Culture of pig embryos. J Reprod Fertil Suppl 1993;48:61−73.

[72] Yoshioka K, Suzuki C, Itoh S, Kikuchi K, Iwamura S, Rodriguez-Martinez H. Production of piglets derived from in vitro-produced blastocysts fertilized and cultured in chemically defined media: effects of theophylline, adenosine, and cysteine during in vitro fertilization. Biol Reprod 2003;69(6):2092−9.

[73] Dobrinsky JR, Johnson LA, Rath D. Development of a culture medium (BECM-3) for porcine embryos: effects of bovine serum albumin and fetal bovine serum on embryo development. Biol Reprod 1996;55(5):1069−74.

[74] Fowler KE, Mandawala AA, Griffin DK, Walling GA, Harvey SC. The production of pig preimplantation embryos in vitro: current progress and future prospects. Reprod Biol 2018;18(3):203−11.

[75] Coy P, Yanagimachi R. The common and species-specific roles of oviductal proteins in mammalian fertilization and embryo development. Bioscience. 2015;65(10):973−84.

[76] Smolinska N, Szeszko K, Dobrzyn K, Kiezun M, Rytelewska E, Kisielewska K, et al. Transcriptomic analysis of porcine endometrium during implantation after in vitro stimulation by adiponectin. Int J Mol Sci 2019;20(6):1335.

[77] Leese HJ, Hugentobler SA, Gray SM, Morris DG, Sturmey RG, Whitear SL, et al. Female reproductive tract fluids: composition, mechanism of formation and potential role in the developmental origins of health and disease. Reprod Fertil Dev 2008;20(1):1−8.

[78] Li J, Gao Y, Guan L, Zhang H, Chen P, Gong X, et al. Lipid profiling of peri-implantation endometrium in patients with premature progesterone rise in the late follicular phase. J Clin Endocrinol Metab 2019;104(11):5555−65.

[79] Khosla S, Dean W, Brown D, Reik W, Feil R. Culture of preimplantation mouse embryos affects fetal development and the expression of imprinted genes. Biol Reprod 2001;64(3):918−26.

[80] Calle A, Fernandez-Gonzalez R, Ramos-Ibeas P, Laguna-Barraza R, Perez-Cerezales S, Bermejo-Alvarez P, et al. Long-term and transgenerational effects of in vitro culture on mouse embryos. Theriogenology. 2012;77(4):785−93.

[81] Fernandez-Gonzalez R, Ramirez MA, Bilbao A, De Fonseca FR, Gutierrez-Adan A. Suboptimal in vitro culture conditions: an epigenetic origin of long-term health effects. Mol Reprod Dev 2007;74(9):1149−56.

[82] Kobayashi T, Zhang H, Tang WWC, Irie N, Withey S, Klisch D, et al. Principles of early human development and germ cell program from conserved model systems. Nature. 2017;546(7658):416−20.

[83] Giassetti MI, Ciccarelli M, Oatley JM. Spermatogonial stem cell transplantation: insights and outlook for domestic animals. Annu Rev Anim Biosci 2019;7:385−401.

[84] Brinster RL, Avarbock MR. Germline transmission of donor haplotype following spermatogonial transplantation. Proc Natl Acad Sci U S A 1994;91(24):11303−7.

[85] Zhang P, Chen X, Zheng Y, Zhu J, Qin Y, Lv Y, et al. Long-term propagation of porcine undifferentiated spermatogonia. Stem Cell Dev 2017;26(15):1121−31.

[86] Honaramooz A, Megee SO, Dobrinski I. Germ cell transplantation in pigs. Biol Reprod 2002;66(1):21−8.

[87] Mikkola M, Sironen A, Kopp C, Taponen J, Sukura A, Vilkki J, et al. Transplantation of normal boar testicular cells resulted in complete focal spermatogenesis in a boar affected by the immotile short-tail sperm defect. Reprod Domest Anim 2006;41(2):124−8.

[88] Zeng W, Tang L, Bondareva A, Honaramooz A, Tanco V, Dores C, et al. Viral transduction of male germline stem cells results in transgene transmission after germ cell transplantation in pigs. Biol Reprod 2013;88(1):27.

[89] Park KE, Kaucher AV, Powell A, Waqas MS, Sandmaier SE, Oatley MJ, et al. Generation of germline ablated male pigs by CRISPR/Cas9 editing of the NANOS2 gene. Sci Rep 2017;7:40176. Available from: https://doi.org/10.1038/srep40176.

[90] Gurdon JB. The transplantation of nuclei between two species of Xenopus. Dev Biol 1962;5:68−83.

[91] Onishi A, Iwamoto M, Akita T, Mikawa S, Takeda K, Awata T, et al. Pig cloning by microinjection of fetal fibroblast nuclei. Science. 2000;289(5482):1188−90.

[92] Polejaeva IA, Chen SH, Vaught TD, Page RL, Mullins J, Ball S, et al. Cloned pigs produced by nuclear transfer from adult somatic cells. Nature. 2000;407(6800):86−90.

[93] Kurome M, Kessler B, Wuensch A, Nagashima H, Wolf E. Nuclear transfer and transgenesis in the pig. In: Beaujean N, Jammes H, Jouneau A, editors. Nuclear reprogramming. Methods in molecular biology (methods and protocols), vol 1222. New York: Humana Press; 2015. p. 37−59. Available from: http://dx.doi.org/10.1007/978-1-4939-1594-1_4.

[94] Keefer CL. Artificial cloning of domestic animals. Proc Natl Acad Sci U S A 2015;112(29):8874−8.

[95] Dorado B, Pløen GG, Barettino A, Macías A, Gonzalo P, Andrés-Manzano MJ, et al. Generation and characterization of a novel knockin minipig model of Hutchinson-Gilford progeria syndrome. Cell Discov 2019;5(1):16.

[96] Yang L, Guell M, Niu D, George H, Lesha E, Grishin D, et al. Genome-wide inactivation of porcine endogenous retroviruses (PERVs). Science. 2015;350(6264):1101−4.

[97] Ruan D, Peng J, Wang X, Ouyang Z, Zou Q, Yang Y, et al. XIST Derepression in active X chromosome hinders pig somatic cell nuclear transfer. Stem Cell Rep 2018;10(2):494−508.

[98] Lee K, Prather R. Cloning pigs by somatic cell nuclear transfer. In: Cibelli J, Gurdon J, Wilmut I, Jaenisch R, Lanza R, West MD, et al., editors. Principles of cloning. 2nd ed. San Diego: Academic Press; 2014. p. 245−54.

[99] Miyoshi K, Kawaguchi H, Maeda K, Sato M, Akioka K, Noguchi M, et al. Birth of cloned microminipigs derived from somatic cell nuclear transfer embryos that have been transiently treated with valproic acid. Cell Reprogram 2016;18(6):390−400.

[100] Canovas S, Ross PJ, Kelsey G, Coy P. DNA methylation in embryo development: epigenetic impact of ART (assisted reproductive technologies). Bioessays. 2017;39:11.

[101] Krisher RL, Heuberger AL, Paczkowski M, Stevens J, Pospisil C, Prather RS, et al. Applying metabolomic analyses to the practice of embryology: physiology, development and assisted reproductive technology. Reprod Fertil Dev 2015;27(4):602−20.

[102] Krisher RL, Schoolcraft WB, Katz-Jaffe MG. Omics as a window to view embryo viability. Fertil Steril 2015;103(2):333−41.

[103] Borrisser-Pairo F, Panella-Riera N, Zammerini D, Olivares A, Garrido MD, Martinez B, et al. Prevalence of boar taint in commercial pigs from Spanish farms. Meat Sci 2016;111:177−82.

[104] Rath D, Ruiz S, Sieg B. Birth of female piglets following intrauterine insemination of a sow using flow cytometrically sexed boar semen. Vet Rec 2003;152(13):400−1.

[105] Spinaci M, Perteghella S, Chlapanidas T, Galeati G, Vigo D, Tamanini C, et al. Storage of sexed boar spermatozoa: limits and perspectives. Theriogenology. 2016;85(1):65−73.

[106] Gordon JW, Scangos GA, Plotkin DJ, Barbosa JA, Ruddle FH. Genetic transformation of mouse embryos by microinjection of purified DNA. Proc Natl Acad Sci U S A 1980;77(12):7380−4.

[107] Brem G, Brenig B, Goodman HM, Selden RC, Graf F, Kruff B, et al. Production of transgenic mice, rabbits and pigs by microinjection into pronuclei. Zuchthygiene Reprod Domest Anim 1985;20(5):251−2.

[108] Hammer RE, Pursel VG, Rexroad Jr. CE, Wall RJ, Bolt DJ, Ebert KM, et al. Production of transgenic rabbits, sheep and pigs by microinjection. Nature. 1985;315(6021):680−3.

[109] Petters RM, Shuman RM, Johnson BH, Mettus RV. Gene transfer in swine embryos by injection of cells infected with retrovirus vectors. J Exp Zool 1987;242(1):85−8.

[110] Hofmann A, Kessler B, Ewerling S, Weppert M, Vogg B, Ludwig H, et al. Efficient transgenesis in farm animals by lentiviral vectors. EMBO Rep 2003;4(11):1054–60.

[111] Sperandio S, Lulli V, Bacci ML, Forni M, Maione B, Spadafora C, et al. Sperm mediated gene transfer in bovine and swine species. Anim Biotech 1996;7:59–77.

[112] Lavitrano M, Bacci ML, Forni M, Lazzereschi D, Di Stefano C, Fioretti D, et al. Efficient production by sperm-mediated gene transfer of human decay accelerating factor (hDAF) transgenic pigs for xenotransplantation. Proc Natl Acad Sci U S A 2002;99(22):14230–5.

[113] Garcia-Vazquez FA, Ruiz S, Matas C, Izquierdo-Rico MJ, Grullon LA, De Ondiz A, et al. Production of transgenic piglets using ICSI-sperm-mediated gene transfer in combination with recombinase RecA. Reproduction. 2010;140(2):259–72.

[114] Gadea J, Garcia-Vazquez FA, Canovas S, Parrington J. Sperm-mediated gene transfer in agricultural species. In: Smith K, editor. Sperm-mediated gene transfer: concepts and controversies. Oak Park. Bentham Books; 2012. p. 76–91.

[115] Prather RS, Sims MM, First NL. Nuclear transplantation in early pig embryos. Biol Reprod 1989;41(3):414–18.

[116] Park KW, Cheong HT, Lai LX, Im GS, Kuhholzer B, Bonk A, et al. Production of nuclear transfer-derived swine that express the enhanced green fluorescent protein. Anim Biotechnol 2001;12(2):173–81.

[117] Fujimura T, Kurome M, Murakami H, Takahagi Y, Matsunami K, Shimanuki S, et al. Cloning of the transgenic pigs expressing human decay accelerating factor and N-acetylglucosaminyltransferase III. Cloning Stem Cell 2004;6(3):294–301.

[118] Lai L, Kolber-Simonds D, Park KW, Cheong HT, Greenstein JL, Im GS, et al. Production of alpha-1,3-galactosyltransferase knockout pigs by nuclear transfer cloning. Science. 2002;295(5557):1089–92.

[119] Harrison S, Boquest A, Grupen C, Faast R, Guildolin A, Giannakis C, et al. An efficient method for producing alpha(1,3)-galactosyltransferase gene knockout pigs. Cloning Stem Cell 2004;6(4):327–31.

[120] Carlson DF, Tan W, Lillico SG, Stverakova D, Proudfoot C, Christian M, et al. Efficient TALEN-mediated gene knockout in livestock. Proc Natl Acad Sci U S A 2012;109(43):17382–7.

[121] Lillico SG, Proudfoot C, Carlson DF, Stverakova D, Neil C, Blain C, et al. Live pigs produced from genome edited zygotes. Sci Rep 2013;3:2847.

[122] Redel BK, Prather RS. Meganucleases revolutionize the production of genetically engineered pigs for the study of human diseases. Toxicol Pathol 2016;44(3):428–33.

[123] Hai T, Teng F, Guo R, Li W, Zhou Q. One-step generation of knockout pigs by zygote injection of CRISPR/Cas system. Cell Res 2014;24(3):372–5.

[124] Whitworth KM, Lee K, Benne JA, Beaton BP, Spate LD, Murphy SL, et al. Use of the CRISPR/Cas9 system to produce genetically engineered pigs from in vitro-derived oocytes and embryos. Biol Reprod 2014;91(3):78.

[125] Li P, Estrada JL, Burlak C, Montgomery J, Butler JR, Santos RM, et al. Efficient generation of genetically distinct pigs in a single pregnancy using multiplexed single-guide RNA and carbohydrate selection. Xenotransplantation. 2015;22(1):20–31.

[126] Tanihara F, Takemoto T, Kitagawa E, Rao S, Do LTK, Onishi A, et al. Somatic cell reprogramming-free generation of genetically modified pigs. Sci Adv 2016;2(9):e1600803.

[127] Yao J, Huang J, Zhao J. Genome editing revolutionize the creation of genetically modified pigs for modeling human diseases. Hum Genet 2016;135(9):1093–105.

[128] Petersen B. Basics of genome editing technology and its application in livestock species. Reprod Domest Anim 2017;52(Suppl 3):4–13.

[129] Ruan J, Xu J, Chen-Tsai RY, Li K. Genome editing in livestock: are we ready for a revolution in animal breeding industry? Transgenic Res 2017;26(6):715–26.

[130] Watanabe M, Nagashima H. Genome editing of pig. Methods Mol Biol 2017;1630:121–39.

[131] Wells KD, Prather RS. Genome-editing technologies to improve research, reproduction, and production in pigs. Mol Reprod Dev 2017;84(9):1012–17.

[132] Ryu J, Prather RS, Lee K. Use of gene-editing technology to introduce targeted modifications in pigs. J Anim Sci Biotechnol 2018;9(1):5.

[133] Yang H, Wu Z. Genome editing of pigs for agriculture and biomedicine. Front Genet 2018;9:360.

[134] Xie J, Ge W, Li N, Liu Q, Chen F, Yang X, et al. Efficient base editing for multiple genes and loci in pigs using base editors. Nat Commun 2019;10(1):2852.

[135] Li S, Flisikowska T, Kurome M, Zakhartchenko V, Kessler B, Saur D, et al. Dual fluorescent reporter pig for Cre recombination: transgene placement at the ROSA26 locus. PLoS One 2014;9(7):e102455.

[136] Song Y, Lai L, Li L, Huang Y, Wang A, Tang X, et al. Germ cell-specific expression of Cre recombinase using the VASA promoter in the pig. FEBS open bio 2016;6(1):50–5.

[137] Pursel VG, Mitchell AD, Bee G, Elsasser TH, McMurtry JP, Wall RJ, et al. Growth and tissue accretion rates of swine expressing an insulin-like growth factor I transgene. Anim Biotechnol 2004;15(1):33–45.

[138] Lai L, Kang JX, Li R, Wang J, Witt WT, Yong HY, et al. Generation of cloned transgenic pigs rich in omega-3 fatty acids. Nat Biotechnol 2006;24(4):435–6.

[139] Saeki K, Matsumoto K, Kinoshita M, Suzuki I, Tasaka Y, Kano K, et al. Functional expression of a Delta12 fatty acid desaturase gene from spinach in transgenic pigs. Proc Natl Acad Sci U S A 2004;101(17):6361–6.

[140] Donovan SM, Monaco MH, Bleck GT, Cook JB, Noble MS, Hurley WL, et al. Transgenic over-expression of bovine α-lactalbumin and human insulin-like growth factor-I in porcine mammary gland. J Dairy Sci 2001;84:E216–22.

[141] Noble MS, Rodriguez-Zas S, Cook JB, Bleck GT, Hurley WL, Wheeler MB. Lactational performance of first-parity transgenic gilts expressing bovine alpha-lactalbumin in their milk. J Anim Sci 2002;80(4):1090−6.

[142] Golovan SP, Meidinger RG, Ajakaiye A, Cottrill M, Wiederkehr MZ, Barney DJ, et al. Pigs expressing salivary phytase produce low-phosphorus manure. Nat Biotechnol 2001;19(8):741−5.

[143] Forsberg CW, Meidinger RG, Murray D, Keirstead ND, Hayes MA, Fan MZ, et al. Phytase properties and locations in tissues of transgenic pigs secreting phytase in the saliva. J Anim Sci 2014;92(8):3375−87.

[144] Chen J, Wang H, Bai J, Liu W, Liu X, Yu D, et al. Generation of pigs resistant to highly pathogenic-porcine reproductive and respiratory syndrome virus through gene editing of CD163. Int J Biol Sci 2019;15(2):481−92.

[145] Whitworth KM, Rowland RR, Ewen CL, Trible BR, Kerrigan MA, Cino-Ozuna AG, et al. Gene-edited pigs are protected from porcine reproductive and respiratory syndrome virus. Nat Biotechnol 2016;34(1):20−2.

[146] Whitworth KM, Rowland RRR, Petrovan V, Sheahan M, Cino-Ozuna AG, Fang Y, et al. Resistance to coronavirus infection in amino peptidase N-deficient pigs. Transgenic Res 2019;28(1):21−32.

[147] Xie Z, Pang D, Yuan H, Jiao H, Lu C, Wang K, et al. Genetically modified pigs are protected from classical swine fever virus. PLoS Pathog 2018;14(12):e1007193.

[148] Hu S, Qiao J, Fu Q, Chen C, Ni W, Wujiafu S, et al. Transgenic shRNA pigs reduce susceptibility to foot and mouth disease virus infection. eLife. 2015;4:e06951.

[149] Evans MJ, Kaufman MH. Establishment in culture of pluripotential cells from mouse embryos. Nature. 1981;292(5819):154−6.

[150] Capecchi MR. Generating mice with targeted mutations. Nat Med 2001;7(10):1086−90.

[151] Gun G, Kues WA. Current progress of genetically engineered pig models for biomedical research. BioResearch open access 2014;3 (6):255−64.

[152] Klymiuk N, Seeliger F, Bohlooly YM, Blutke A, Rudmann DG, Wolf E. Tailored pig models for preclinical efficacy and safety testing of targeted therapies. Toxicol Pathol 2016;44(3):346−57.

[153] Navarro-Serna S, Romar R, Dehesa-Etxebeste M, Lopes JS, Lopez de Munain A, Gadea J, editors. First production of Calpain3 KO pig embryo by CRISPR/Cas9 technology for human disease modelling: efficiency comparison between electroporation and intracytoplasmic microinjection. Abstract of the 35rd Scientific meeting Association of Embryo Technology in Europe (AETE); 2019. Murcia (Spain) Anim Reprod 2019;16(3):770.

[154] Gadea J, Garcia-Vazquez FA, Hachem A, Bassett A, Romero-Aguirregomezcorta J, Canovas S, et al. Generation of TPC2 knock out pig embryos by CRISPR-Cas technology. Reprod Domest Anim 2018;53:87−8.

[155] Niemann H, Petersen B. The production of multi-transgenic pigs: update and perspectives for xenotransplantation. Transgenic Res 2016;25 (3):361−74.

[156] Iwase H, Liu H, Wijkstrom M, Zhou H, Singh J, Hara H, et al. Pig kidney graft survival in a baboon for 136 days: longest life-supporting organ graft survival to date. Xenotransplantation. 2015;22(4):302−9.

[157] Naeimi Kararoudi M, Hejazi SS, Elmas E, Hellstrom M, Naeimi Kararoudi M, Padma AM, et al. Clustered regularly interspaced short palindromic repeats/Cas9 gene editing technique in xenotransplantation. Front Immunol 2018;9:1711.

[158] Fischer K, Kind A, Schnieke A. Assembling multiple xenoprotective transgenes in pigs. Xenotransplantation. 2018;25(6):e12431.

[159] Park JK, Lee YK, Lee P, Chung HJ, Kim S, Lee HG, et al. Recombinant human erythropoietin produced in milk of transgenic pigs. J Biotechnol 2006;122(3):362−71.

[160] Bertolini LR, Meade H, Lazzarotto CR, Martins LT, Tavares KC, Bertolini M, et al. The transgenic animal platform for biopharmaceutical production. Transgenic Res 2016;25(3):329−43.

[161] Bailey JL, Lessard C, Jacques J, Breque C, Dobrinski I, Zeng W, et al. Cryopreservation of boar semen and its future importance to the industry. Theriogenology. 2008;70(8):1251−9.

[162] Knox RV. The fertility of frozen boar sperm when used for artificial insemination. Reprod Domes Anim 2015;50(Suppl 2):90−7.

[163] Yeste M, Rodriguez-Gil JE, Bonet S. Artificial insemination with frozen-thawed boar sperm. Mol Reprod Dev 2017;84(9):802−13.

[164] Estrada E, Rodriguez-Gil JE, Rocha LG, Balasch S, Bonet S, Yeste M. Supplementing cryopreservation media with reduced glutathione increases fertility and prolificacy of sows inseminated with frozen-thawed boar semen. Andrology 2014;2(1):88−99.

[165] Didion BA, Braun GD, Duggan MV. Field fertility of frozen boar semen: a retrospective report comprising over 2600 AI services spanning a four year period. Anim Reprod Sci 2013;137(3-4):189−96.

[166] Roca J, Parrilla I, Rodriguez-Martinez H, Gil MA, Cuello C, Vazquez JM, et al. Approaches towards efficient use of boar semen in the pig industry. Reprod Domest Anim 2011;46(Suppl 2):79−83.

[167] Thurston LM, Siggins K, Mileham AJ, Watson PF, Holt WV. Identification of amplified restriction fragment length polymorphism markers linked to genes controlling boar sperm viability following cryopreservation. Biol Reprod 2002;66(3):545−54.

[168] Gadea J, Selles E, Marco MA, Coy P, Matas C, Romar R, et al. Decrease in glutathione content in boar sperm after cryopreservation. Effect of the addition of reduced glutathione to the freezing and thawing extenders. Theriogenology 2004;62(3-4):690−701.

[169] Bolarin A, Roca J, Rodriguez-Martinez H, Hernandez M, Vazquez JM, Martinez EA. Dissimilarities in sows' ovarian status at the insemination time could explain differences in fertility between farms when frozen-thawed semen is used. Theriogenology. 2006;65(3):669−80.

[170] Roca J, Rodriguez-Martinez H, Vazquez JM, Bolarin A, Hernandez M, Saravia F, et al. Strategies to improve the fertility of frozen-thawed boar semen for artificial insemination. Soc Reprod Fertil Suppl 2006;62:261−75.

[171] Honaramooz A, Snedaker A, Boiani M, Scholer H, Dobrinski I, Schlatt S. Sperm from neonatal mammalian testes grafted in mice. Nature. 2002;418(6899):778−81.

[172] Nakai M, Kaneko H, Somfai T, Maedomari N, Ozawa M, Noguchi J, et al. Production of viable piglets for the first time using sperm derived from ectopic testicular xenografts. Reproduction. 2010;139(2):331−5.

[173] Kaneko H, Kikuchi K, Nakai M, Somfai T, Noguchi J, Tanihara F, et al. Generation of live piglets for the first time using sperm retrieved from immature testicular tissue cryopreserved and grafted into nude mice. PLoS One 2013;8(7):e70989.

[174] Zhang W, Yi K, Yan H, Zhou X. Advances on in vitro production and cryopreservation of porcine embryos. Anim Reprod Sci 2012;132(3-4):115−22.

[175] Men H, Walters EM, Nagashima H, Prather RS. Emerging applications of sperm, embryo and somatic cell cryopreservation in maintenance, relocation and rederivation of swine genetics. Theriogenology. 2012;78(8):1720−9.

[176] Kikuchi K, Kaneko H, Nakai M, Somfai T, Kashiwazaki N, Nagai T. Contribution of in vitro systems to preservation and utilization of porcine genetic resources. Theriogenology. 2016;86(1):170−5.

[177] Somfai T, Yoshioka K, Tanihara F, Kaneko H, Noguchi J, Kashiwazaki N, et al. Generation of live piglets from cryopreserved oocytes for the first time using a defined system for in vitro embryo production. PLoS One 2014;9(5):e97731.

[178] Nagashima H, Hiruma K, Saito H, Tomii R, Ueno S, Nakayama N, et al. Production of live piglets following cryopreservation of embryos derived from in vitro-matured oocytes. Biol Reprod 2007;76(5):900−5.

[179] Cuello C, Martinez CA, Nohalez A, Parrilla I, Roca J, Gil MA, et al. Effective vitrification and warming of porcine embryos using a pH-stable, chemically defined medium. Sci Rep 2016;6:33915. Available from: https://doi.org/10.1038/srep33915.

[180] Lotz L, Hauenstein T, Nichols-Burns SM, Strissel P, Hoffmann I, Findeklee S, et al. Comparison of whole ovary cryotreatments for fertility preservation. Reprod Domest Anim 2015;50(6):958−64.

[181] Gabriel PR, Torres P, Fratto MC, Cisale H, Claver JA, Lombardo DM, et al. Effects of different sucrose concentrations on vitrified porcine preantral follicles: qualitative and quantitative analysis. Cryobiology. 2017;76:1−7.

[182] Kaneko H, Kikuchi K, Noguchi J, Hosoe M, Akita T. Maturation and fertilization of porcine oocytes from primordial follicles by a combination of xenografting and in vitro culture. Biol Reprod 2003;69(5):1488−93.

[183] Kikuchi K, Kaneko H, Nakai M, Noguchi J, Ozawa M, Ohnuma K, et al. In vitro and in vivo developmental ability of oocytes derived from porcine primordial follicles xenografted into nude mice. J Reprod Dev 2006;52(1):51−7.

[184] Nichols-Burns SM, Lotz L, Schneider H, Adamek E, Daniel C, Stief A, et al. Preliminary observations on whole-ovary xenotransplantation as an experimental model for fertility preservation. Reprod Biomed Online 2014;29(5):621−6.

[185] Damasio LC, Soares-Junior JM, Iavelberg J, Maciel GA, de Jesus Simoes M, Dos Santos, et al. Heterotopic ovarian transplantation results in less apoptosis than orthotopic transplantation in a minipig model. J Ovarian Res 2016;9:14.

Chapter 6

Reproductive technologies in the buffalo (*Bubalus bubalis*)

Giorgio A. Presicce[1], Bianca Gasparrini[2], Angela Salzano[2], Gianluca Neglia[2], Giuseppe Campanile[2] and Luigi Zicarelli[2]

[1]*ARSIAL, Rome, Italy,* [2]*Department of Veterinary Medicine and Animal Production, University of Naples "Federico II,"Napoli, Italy*

6.1 General introduction to the buffalo

Around 199 million heads around the world and in all continents account for the totality of the two subspecies, breeds, and varieties of domestic buffaloes that can be found. According to some authors, buffaloes perform better than cattle and predominate over the latter species under tropical conditions also because, among other variables, they are considered to be better converter of poor-quality feed resources into milk and meat [1]. These assumptions though are not shared by other authors, as shown in some nutritional studies [2,3]. Actually, it is likely that as milk production increases and high nutritive value diets with low crude fiber content are used, the difference between buffalo and other ruminants regarding gross food digestibility decreases [4]. More realistically, the numeric prevalence of buffaloes in tropical territories is very likely due to the inability of dairy cattle to express its productive potential when reared in humid and hot climate. Conditions pertaining to other aspects such as commercial and political facets make the buffalo farming, like in Italy, more profitable than dairy cattle as buffalo milk is paid to the farmers three to four times higher, and therefore a significant increase in the number of heads has been witnessed around the country over the last decades. It is true that the domestic buffalo can be an ideal transformer of poor-quality forage within marginal areas into animal products to be locally stored and commercialized. Nevertheless, whenever higher quality forage is available, the buffalo can be more selective than cattle under intensive production systems [5]. In the last decades, a varied pattern of development in buffalo both in terms of number of heads and production strategies has been witnessed around the world. Countries characterized mostly by the presence of swamp buffaloes mainly used for draft purposes, like the ones in the Far East such as the Philippines, Vietnam, Bangladesh, Sri Lanka, Thailand, and Indonesia, have seen their heads and production either stabilized or even reduced due to the advancement of mechanization in agriculture [6–9]. On the contrary, in countries where the dominant buffalo is given by the river subspecies such as India, Pakistan, and Italy, an improvement of those features has been reported. The countries where most of the advancement in reproduction and production strategies has been carried out are Brazil in South America, India and Pakistan in the East, and Italy in the Mediterranean area. This is due mainly to the presence of the most productive breeds in terms of milk yield, like Murrah, Nili-Ravi, Surti, and Jaffarabadi in India and the Mediterranean Buffalo in Italy. In addition, these are the countries where most of the efforts in terms of adopted genetic strategies have been and still are implemented, in order to improve both quality and quantity of production traits. In China, where the majority of available buffaloes belong to the less productive swamp type, an attempt at improving milk and meat production was made with the introduction of the river type Murrah and Nili-Ravi in the 1950s–1970s, and more recently by crossbreeding with the local swamp buffaloes [10]. It is noteworthy to mention that in the same 1950s, in Italy the domestic buffalo was on the verge of extinction, due to swamp reclamation and changed economic conditions that would not support buffalo breeding. Despite such economical constraints, from the end of World War II, on the contrary buffalo heads have increased steadily and by the year 2000 the Mediterranean Italian Buffalo breed has been officially recognized, while selection has continued throughout the years thanks especially to the natural isolation and the consequent lack of introduction of other breeds in the country.

Reproductive Technologies in Animals. DOI: https://doi.org/10.1016/B978-0-12-817107-3.00006-0

Different production strategies characterize those countries where the least productive swamp subspecies are utilized for both milk and meat production, and in some cases only for meat, as milk is enough only for sustaining calf growth up to weaning. It has to be underlined though that in all Asia, both river and swamp subspecies have always been reared as dual or three purpose animals. On the contrary in other European countries and especially in Italy, there has always been a strong association between the buffalo and its milk production and derived cheese products, of which the mozzarella cheese is certainly the most known worldwide. Recently though, a strong interest in buffalo meat, due to its qualitative characteristics and health benefits, has lifted breeder attention into this parallel production strategy by creating specific protocols for bringing animals to slaughter age and weight [11]. An incredible body of knowledge has been built around buffalo meat and its benefits derived from its consumption in the human species, as opposed to cattle meat, especially when fatty acid composition and cholesterol content are taken into consideration [12]. Although buffalo meat covers only a very small fraction of meat demand in the entire world, in countries where this species is prevalent such as in the Far East, or in countries where for religious beliefs cattle cannot be used for human consumption like India, the availability of buffalo meat has greatly increased for both national and international markets [13].

6.2 The domestic buffalo and its genetic merit

Two are the buffalo species in the world: the African buffalo (*Syncerus caffer*) and the water buffalo (*Bubalus bubalis*), and it is the domestic water buffalo the object of our interest in this chapter. The name buffalo may generate some confusion, as in some parts of the world, like in the United States, often the bison (*Bison bison*, different genus and species) is referred to as buffalo. Of the water buffalo species then, although several haplotypes both at genomic and mitochondrial level have been found to be shared, two distinct subspecies can be described, namely the river and the swamp buffalo, with 50 and 48 diploid number of chromosomes, respectively [14], and their phylogenetic separation predates domestication. As previously synthetically reported, the two subspecies characterize the buffalo species in different countries of the world, and their individual prevalence in each country is mainly dictated by the specific management system that can be found in place, for the milk and meat production system adopted and the recorded yield, and for the interest in enhancing the genetic potential through the implementation of old or newly developed reproductive technologies. Finally, the reason why such varied management system is in place for buffaloes has to be found in the intermingling of a number of different factors such as climate, local geography and economical wealth or constraints, cropping systems, size of farms, and finally the primary purpose for which buffaloes are reared in each different country, among milk, meat, and draught, singularly taken or in combination.

Cytogenetic studies in buffaloes have revealed that the karyotype characterizing the swamp type has derived from the tandem fusion translocation of chromosome 4 (BBU4) and 9 (BBU9) of the river type, giving rise to the large characteristic chromosome number 1 in swamp buffaloes [15]. Their genetic similarity can be inferred by G- and R- banding homologies, in addition to the hypothesis of a common genetic ancestor among all bovids (*urus−Bos primigenius taurus*) as evidenced by centric fusion translocation between two of the ten homologous cattle autosomes, that gives rise to each of the five river buffalo biarmed pairs [16].

The knowledge of a complete genome sequence in farm animals allows a technical and conceptual step to be taken, namely the prediction of phenotypes from available genotypes, reducing thus the time constraints posed by older acquired protocols such as progeny testing. Therefore bulls acquire a genomic breeding value, through the use of genome DNA markers, and semen doses can be sold solely on this basis. In cattle, genome-wide association studies have been widely applied, and therefore either marker-assisted selection or gene-assisted selection constitutes today a strong basis for the application of breeding selection schemes. The buffalo genome has been fully sequenced, although only a very small percentage of genes with known function have been reported so far [17]. In turn, this limited knowledge is reflected into little information for the genetic variability of the species, which prevents a fast and correct genetic improvement. Nevertheless, following comparison with the bovine genome, some knowledge of single-nucleotide polymorphism (SNP) sites in buffaloes has been achieved, bringing to the availability of SNP chips and making it possible to delve into buffalo genetics among all the available breeds and to spot those markers responsible for exerting a greater effect not only on production of the traits of interest, but also on health and efficiency [18]. It has to be underlined though, that still missing the complete notation of the entire buffalo genome, the utilization of the available existing tools for the genetic enhancement of the buffalo species is not yet satisfying. At present, only few genomic studies have been performed in buffaloes. A genome-wide association study (GWAS) was carried out in Italian Mediterranean Buffaloes, in which four candidate SNPs were found in two genomic regions associated with buffalo milk production traits [19]. One region is linked to milk fat and protein percentage and was located on the equivalent of Bos taurus autosome (BTA3), while the other region, linked to total milk yield, fat yield, and protein yield, was located on the equivalent of BTA14. Interestingly, both of the regions were reported to have quantitative trait loci affecting milk performance also in dairy cattle. Although usually milk production in ruminants is not associated to prolactin, in another study, high resolution melting techniques were developed for genotyping Italian Mediterranean Buffaloes and showed that polymorphisms of the buffalo prolactin gene (PRL) were associated with milk production traits [19]. Hence, PRL could be used as a candidate gene for marker-assisted selection in Italian Mediterranean river buffalo breeding. Regarding reproductive activity, another GWAS study performed in buffaloes showed that a total of 40 suggestive loci (related to 28 genes) were identified to be associated with six reproductive traits (first, second, and third calving age, calving interval, the number of services per conception and open days) [20]. Nevertheless, marker-assisted selection for the traits of interest in the buffalo, with forthcoming increased knowledge and improved efficient technology, can still be seen today for the near future as the best choice for the identification of animals with high breeding value and for planning appropriate breeding schemes [21].

6.3 Reproductive physiology

Buffaloes adapt well to a broad variety of climatic conditions and, like other production species such as horses, small ruminants and some cattle breeds kept under extensive pasture, are tendentially seasonal animals although they can breed all year round. This is an evolutionary strategy adopted in the wild in order to match birth and weaning under the best environmental conditions, to better satisfy nutritional and thermal requirements typical of the species. In fact, being an animal of tropical origin where most of the green pasture is available from the end of the rainy season, coincident with the end of summer, the peak of reproductive activity in natural condition will be therefore witnessed in the semester going from the end of summer into the first months of winter. This will ensure that birth, after around 310–320 days of pregnancy, will coincide with the highest forage availability. The nocturnal release of melatonin from the epiphysis will be triggered by the coincident decrease of daylight hours, and thanks to a domino effect, melatonin in turn will exert an effect on the hypothalamus hypophysis gonadal axis, heading thus toward the beginning of estrous cycles. We can say then that, under such conditions, the buffalo increases its reproductive activity in the period of the year characterized by reduced daylight hours or, even if light hours increase, night hours are still prevalent within the day. Of course, whenever buffaloes are reared in environments where daylight and night light hours are equivalent, reproductive activity is similar throughout the year, and possibly other variables, such as nutritional requirements will represent the key element for reproductive success [22].

Traditionally, the domestic buffalo under farm conditions and in comparison to cattle, has always been considered to be characterized by a reduced reproductive efficiency due to late sexual maturity, seasonality, long time interval calving to first estrus, reduced signs of estrous behavior and conception rate. Up to a very large extent this is true, although recently the efficiency in conception rate following artificial insemination (AI) in conjunction to the newly developed synchronization protocols has approached similar results with cattle, even when sexed semen is utilized. Differently, if we consider the parameters concerning reproductive efficiency in other contexts, where the buffalo is naturally mated without human intervention and/or constraints posed by local governing laws of milk and cheese production and commercialization, then this species can be regarded as highly efficient and productive. As an example of how buffalo reproductive efficiency is put into play, let us consider what happens in Italy, where most of the milk and cheese produced is traditionally required by the market in the time of the year where, by physiologic limits of the species, the least number of calvings are recorded. To offset for such intrinsic species-specific seasonality, a tendency for selecting animals less sensitive to photoperiod is in place in farms, in addition to the practice of skewing the time of the year when most of the natural mating and AI is practiced. Obviously, this contrived protocol may inevitably bring some strong financial loss to the breeders, as a consequence of the attempt at increasing the number of births in the period of the year where naturally a physiological anestrus, typical of this species, will occur. In fact, at these latitudes during increasing day length months, buffaloes sensibly reduce their capacity to reboot an ovarian cycle heading toward a new pregnancy. In this period, if pregnancy is not established, the animal may delay a new fertile follicular development within the ovaries and will possibly restore fertility into the successive autumn to winter months. This will determine a significant and deleterious lengthening of the intercalving period and a strong reduction of milk production. This challenge to buffalo reproductive efficiency is still active in most farms, even though more and more cheese makers are devoting part of their production to being exported in other countries, such as United States and Japan, a context in which the time of the year in which milk and cheese are consumed can also be important for the trade. The strong influence of the environment over reproductive function and success can also be seen, in addition to higher rates of reduced or failed ovarian function as above described, in the case of reported higher embryo mortality. In fact, this is typically witnessed in the same countries such as Italy where animals are challenged with AI in the period of winter-spring transition, characterized by increasing day length [23].

Despite such species-specific limitation, the buffalo is characterized by a reproductive and productive longevity by far not comparable to cattle. In fact, dairy cattle are ordinarily culled following only few lactations, whereas in contrast usually most buffaloes remain in production for many years and for many lactations mainly due to an intrinsic lower milk production and to a more physiological feed rationing. This is surely due to its more rustic characteristics and shorter time available in the hands of breeders for adopting selection strategies for the characters of interest [24].

Again, an additional element that contributes to lower the overall reproductive efficiency, whenever manipulative technologies are applied to this species, is given by the 10-fold lower content of ovarian primordial follicles when compared to cattle. This accounts for a significant lower number of available antral follicles even following hormonal stimulation within adopted protocols for multiple ovulation and embryo transfer (MOET), and ovum pickup (OPU) in conjunction to in vitro embryo production (IVEP) procedures [25,26]. It is noteworthy to mention though, that prepuberal buffalo calves display the same high number of antral follicles as recorded in cattle calves. In addition, and similarly to cattle, buffaloes are both qualitatively and quantitatively characterized by two to three waves of follicular development within the ovaries in puberal and pluriparous animals. As expected, the pattern of hypothalamic-hypophyseal-gonadal hormonal release is in fact superimposable to other ruminant species too [22]. All the above aspects have to be aligned and put in the correct perspective whenever any of the newly developed reproductive technologies are going to be implemented in this species.

6.4 The role of artificial insemination

The first documented attempt at AI goes back to 1783, when Lazzaro Spallanzani achieved the first pregnancy in a bitch, although it was only at the beginning of the 20th century that the Russian I.I. Ivanov devised the first artificial vaginas for equine and ruminants. Only after the Second World War though, the technology spread over the most important species of zootechnical interest with dairy cattle benefitting the most. Among ruminants, buffaloes have been subjected to AI practice much later than cattle for a number of reasons. Firstly, in most parts of the world the least productive buffaloes have always been considered labor animals, and only residual milk production other than the milk used by the calves is processed for family and local cheese production, and in some countries such as Nepal and India, milk used by families for direct consumption is high, from roughly 50%−80% of the whole production, respectively. Meat from buffalo is made available to families and for local trade only at the end of the animal career within small holders. The small number of animals per owner and the poor management conditions typical of farmers at the village level, especially in Southeast Asia has been a substantial halt to the application of AI. On the contrary, in countries characterized by the availability of the most productive subspecies and breeds, and where intensive management is run, animals have been selected for their production by massal selection, prior to the introduction of reproductive technologies. In these last countries a proper buffalo genetic selection has started few decades ago, and the earlier studies on male and female reproductive characteristics have been performed. Nevertheless, a very slow beginning has characterized the implementation of AI in buffaloes, mostly due to the resistance and doubts posed by the same breeders, skeptical over the years that this technique may prove beneficial for buffalo genetic enhancement.

Therefore, and always in comparison with the cattle counterpart, buffaloes have had less time to be selected for the traits of breeders interest. This can be seen in both males and females. Buffalo bulls are in fact characterized by a poorer sperm quality and larger individual variability in fertility, which is exacerbated in the season of the year of reduced reproductive performance, where also libido is usually affected. The reduced fertility may be attributed to a higher sensitivity to oxidative stress due to an increased lipid peroxidation, in conjunction to a reduced activity of antioxidant enzymes [27]. Attempts have been made in order to overcome the natural sensitivity of buffalo semen to cryopreservation damage, by including in semen extenders proteins derived from the isthmic portion of oviducts [28,29]. Similarly to other ruminant species and mammals in general, nutrition, thermal stress, and other environmental variables may also be responsible for higher or lower semen quality, depending on the intermingling of the variables involved [30].

On the female side, under intensive management and correct and balanced feeding regimen, photoperiod, as previously stated, is known to affect reproductive function. This sensitivity though is different between heifers and pluriparous cows, the first being less sensitive to the light stimulus. Therefore whenever AI protocols are applied in the period of the year characterized by increasing daylight hours, a higher conception rate should be expected from heifers [31]. Along this line, in countries where photoperiod is a critical factor in reproductive efficiency for buffaloes and where higher production is differentially required among seasons, breeders tend to select animals less sensitive to the light stimulus. On a different note, and likewise cattle, additional nutritional and metabolic factors may negatively affect pregnancy rates following the implementation of AI programs [32].

The very first attempts at synchronization protocols in buffalo heifers and cows have been made by using double prostaglandin administration with variable and inconsistent results. Typically, such well-established protocols would be taken from cattle and strictly applied to buffaloes with the assumption that similar synchronization and conception rate would be achieved. Differently, poor results have been reported initially, discouraging breeders from accepting the introduction of a new concept of genetic improvement in the buffalo species. One of the most important reasons behind such initial failure in AI has to be searched in the physiological difference between cattle and buffaloes control of reproductive function, following exogenous administration of hormones in the course of synchronization protocols. It may be hard to spot the best moment to perform AI in individual buffaloes as, under natural conditions and differently from cattle, they do not frequently express homosexual behavior and are characterized by a reduced frequency of mucus discharge, together with a similar coat color. A vasectomized bull within the herd though [33] usually is effective in significantly increasing sexual behavior among females. The reduction in sexual behavior and signs is even more pronounced when exogenous hormones are applied to synchronize animals, making the heat detection and hence the identification of the best time for AI even more difficult. Following an estrous cycle, which is much more variable in length than cattle, a similar higher variable timing of inner hormonal discharge progression, leading finally to ovulation, is recorded in buffaloes. In early seminal studies by Seren [34], the reproductive hormones (luteinizing hormone (LH), follicle-stimulating hormone (FSH), estradiol, progesterone, prostaglandin, and prolactin) involved and needed to a fertile ovulation were evaluated during the estrous cycle successive to the one induced by double prostaglandin administration. From that study, it emerged that the interval between the beginning of behavioral estrus and ovulation lies between 50 and 55 hours in the case of single ovulation. As previously reported [35], a high frequency of double ovulations, up to more than 30%, was in fact recorded. In that case the timing to the first and second ovulation would be of 40 versus 110 hours, respectively. In case of a single ovulation the interval between LH peak and ovulation would be of 35 hours, whereas in the event of double ovulations, the interval would be of 30 and 65 hours to the first and second ovulation, respectively. A more recent study following the implementation of two different estrus synchronization protocols has revealed a similar time interval between LH peak to ovulation [36]. Of course, such unexpected variability in hormonal discharge and progression followed by ovulation did explain part of the reasons of unsuccessful attempts of AI at that time.

Far better results have been obtained more recently, through the combined implementation of protocols for synchronization of ovulation and ultrasound monitoring of the reproductive tract and ovarian follicular dynamics. Synchronization protocols have been largely taken from those in use in cattle, especially following the widespread efficiency of the Ovsynch with its various modifications when applied to both heifers and cows [37]. The availability of protocols for the synchronization of ovulation has allowed the possibility to reduce the incidence of double ovulations and to more precisely target the timing of ovulation with a timed fixed AI (TAI). In fact, De Araujo Berber et al. [38] have shown that following the second gonadotropin-releasing hormone (GnRH) administration, most of the ovulations lie between 26 and 38 hours, justifying therefore the insemination of animals up to 20 hours after the second GnRH. The highest efficiency has been reported on cyclic animals with an average, although variable, conception rate of 50%−60% when ultrasound diagnosis is made up to the first 30 days following AI. This protocol, cheap and easy to be performed, is nowadays largely applied to buffaloes. Protocols have been subjected to modifications on acyclic animals by adding progesterone as a removable implant, with a resulting efficiency variable in terms of restoring cyclicity and conception rates [39,40].

The efficiency of AI in conjunction to protocols for synchronization of ovulation has in recent years greatly benefitted by the use of ultrasound technology. A first and important improvement has derived from the significant reduction of mistakes incurred during manual exploration and diagnosis within the reproductive tract. In comparison to cattle, smaller size of the ovaries and corpora lutea (CLs) at times deeply embedded into the ovarian stroma is responsible for misjudgment and clinical false interpretation. In addition, ultrasound technology enables a reliable and correct diagnosis of early pregnancy, as well as detecting embryonic mortality, anticipating though the new course of action to be followed on every single animal. An additional significant aid to reproductive management is given by the possibility to follow the events occurring at the ovarian level with great accuracy. In fact, available ultrasound machines allow easy interpretation of ovarian structures and follicular dynamics, with the result of a better-timed implementation of protocols for synchronization of ovulation and a more suitable selection of animals to be enrolled into AI programs. De Rensis et al. [40] have in fact reported a higher conception rate in the course of Ovsynch programs, whenever a large dominant follicle responsive to the first GnRH administration is found. In the same line, Neglia et al. [41] have found that the ovulation after the first GnRH administration within an Ovsynch protocol results into a larger CL area, a higher

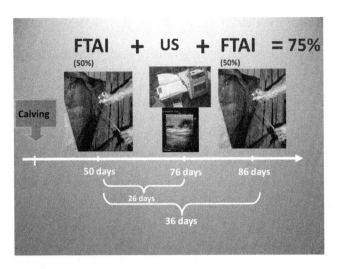

FIGURE 6.1 Pregnancy rates following two consecutive TAIs.

response to the following prostaglandin administration, and finally into a higher response to the second GnRH adminis-tration leading to the ovulation of a larger dominant follicle. As expected, the corresponding CLs show a greater proges-terone concentration paralleled by significant higher pregnancy rates. Higher efficiency and conception rates are then usually accomplished when such treatments run in parallel to an ideal reproductive management within the herd. In recent years, modifications of TAI programs have mushroomed and tried repeatedly in many farms, usually with such satisfying results, to convince breeders of the usefulness of this technology and make them even more confident as to include synchronization programs into their farm reproductive management. Under this light Rossi et al. [42] have shown that buffaloes, whenever found nonpregnant, can be submitted into a continuously running TAI program without prejudice to their health or future reproductive function, and that the first three consecutive TAI programs hold for the majority of pregnancies established. The same study demonstrated that the intercalving period between animals under-going TAI programs in comparison to animals following natural mating within the herd is shorter of a little less than 2 months. This translates into an optimization of reproductive efficiency in the herd, higher conception rates, and calving, heading toward a higher economic efficiency of the farm.

The ultrasound technology comes in help when proper assessment of CL functionality is needed. This can be of course accomplished by the ultrasound image referring to size, texture, and shades of gray on the screen. However, an additional important help can be received by the understanding of blood flow through the CL itself, by the adoption of the color Doppler technique. In fact, a reduced CL vascularization on day 5 following ovulation is inductive of signifi-cant lower conception rates on day 45, as substantiated by a differential progesterone value made on day 5 [43]. In con-clusion, in our experience with the use of ultrasound technology as ancillary aid in well-managed buffalo farms, fixed TAI programs can result in approximately 75% of the animals being pregnant from calving onward up to less than 3 months if a 50% conception rate is assumed (Fig. 6.1).

From all the above, and comparisons over the years run among synchronization protocols, including also the ones incorporating progesterone devices, it can be stated that an ideal synchronization treatment does not exist. In fact, in order to achieve the highest results from AI practice, the following four most important variables have to be taken into account: season of the year, category of animals, metabolic status of the animals and of course, and the protocol for synchronization of ovulation that would fit best the first three variables.

In conclusion, farm and environmental conditions, feeding regimen and availability, season, age and parity of the animals, and implemented protocols are in various degrees responsible for the high variability registered across the years in terms of conception rates following AI.

6.5 Sexed semen

The implementation of sexed semen technology in buffaloes, as for other domestic species, is especially helpful among others for optimizing female replacement and rapidly proliferating the genotypes of interest within the herd. This is especially true also when, due to health problems, there is a need to have an internal turnover of replacement heifers, or when it is mandatory to increase the number of heads in a fast-growing buffalo farm. The technology behind this new tool for genetic improvement relies on the possibility of cell sorting cells bearing differential amount of DNA. In both cattle and buffaloes this difference in sperm cells bearing either X or Y chromosome is large enough to allow an efficient separation of the two cell populations to obtain a roughly 90% purity within samples. The use of sexed semen has become a potentially powerful tool for the genetic enhancement of *B. bubalis* since the first successful application of the technology in Mediterranean Italian buffaloes [44]. In the intervening years, with a similar number of preselected sperm cells, successful attempts have been carried out in Murrah and Nili-Ravi breeds following either IVEP procedures or AI [45–47]. Finally, the same reduced dose of 2 million live preselected sperm cells has been used in Mediterranean Italian buffalo heifers, showing a similar pregnancy rate when compared to nonsexed conventional frozen-thawed semen. That same study showed that insemination into the body of the uterus ensures higher rates of pregnancies when compared to deposition of preselected semen into the horn ipsilateral to the ovary where ovulation had occurred, possibly as a consequence of more severe trauma received by the horn endometrium by the AI gun [48]. The chronology of studies in buffaloes on preselecting sperm cell populations and first attempts via AI or IVEP procedures can be seen in Fig. 6.2. More recently, in line with a number of ideas tested over the years to select sperm cells to alter sex ratio, a commercial based group in the United States has brought into the market a product that would both increase fertility rate and skew the sex ratio in favor of born female calves in cattle, by adding to conventional frozen/thawed semen a preprepared solution containing molecules enhancing X-bearing sperm cells motility and allowing them to reach the fertilization site earlier than Y-bearing sperm cells, in conjunction to a delayed insemination time as otherwise required by the Ovsynch protocol. Despite the success that has been publicized in cattle, the first attempt in buffaloes has not been equally successful [49], and a more recent trial in buffaloes, in the period of the year characterized by decreasing light hours, has revealed similar results, confirming that even if AI is delayed as requested by the new protocol, both herd fertility and born female calves rates are not altered (Presicce, 2019, unpublished results).

In buffaloes, a similar pregnancy rate with preselected frozen-thawed spermatozoa and conventional nonpreselected cryopreserved semen was recorded, in contrast with reports in cattle, where usually rates are lower down to 60%–80% when compared to controls [50,51]. Although speculative at this point, it could be possible that buffalo semen can better survive to physical and chemical stress during the sorting process. Therefore sexed sperm cells may thus maintain the same fertilizing capacity as controls, as it has also been demonstrated in sheep, by assuming the hypothesis that the sex-sorting process may select a superior population of sperm cells. In fact, it may be possible that cell sorting will select sperm cells against a 15 kDa protein (SLLP1—acrosome indicator protein), which is as a result abundant in control sperm cell samples but absent in sorted population of spermatozoa [52].

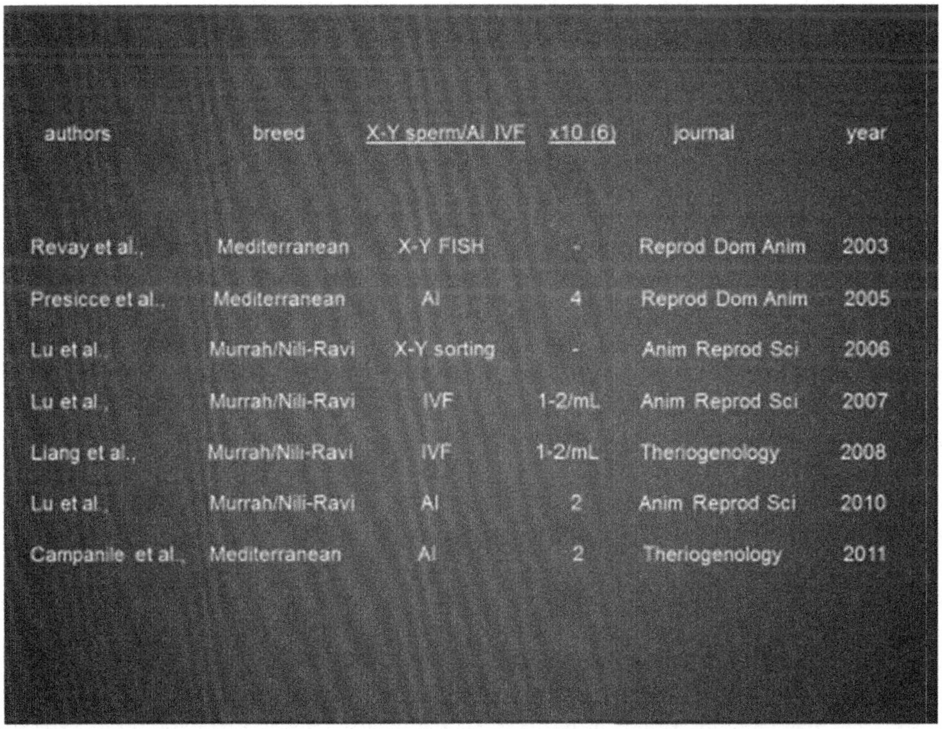

authors	breed	X-Y sperm/AI IVF	x10 (6)	journal	year
Revay et al.,	Mediterranean	X-Y FISH	-	Reprod Dom Anim	2003
Presicce et al.,	Mediterranean	AI	4	Reprod Dom Anim	2005
Lu et al.,	Murrah/Nili-Ravi	X-Y sorting	-	Anim Reprod Sci	2006
Lu et al.,	Murrah/Nili-Ravi	IVF	1-2/mL	Anim Reprod Sci	2007
Liang et al.,	Murrah/Nili-Ravi	IVF	1-2/mL	Theriogenology	2008
Lu et al.,	Murrah/Nili-Ravi	AI	2	Anim Reprod Sci	2010
Campanile et al.,	Mediterranean	AI	2	Theriogenology	2011

FIGURE 6.2 Early chronological events related to sperm cell technology and its implementation in the domestic buffalo (*B. bubalis*).

TABLE 6.1 Projected benefits following use of sexed semen in an intensive buffalo farming system.

Farm turnover 20%; bull effect = 0; mortality = 0; mean farm milk yield = 2550 kg					
	Semen				
	Sexed	Nonsexed			
To get	20♀	20♀			
% conception	40	45			
% top buffaloes to AI	50	90[a]			
Mean milk (kg) of inseminated buffaloes	2800	2550			
Selective differential (kg)	250	50			
Hereditability (30%)	75	15			
Difference from starting mean milk production of 2550 kg					
	Semen				
After years	Sexed	Nonsexed	Milk difference	× head	× 10 heads
	(kg)	(kg)	(kg)	(€)	(€)
4	175	163	12	15	150
8	579	436	143	178	1782

[a]*Half of them will be ♂.*

The use of sexed semen, considering its final cost to the farmer, type of farming, geographical location, and its socioeconomical features, may have different impact worldwide. As an example (Table 6.1), it is interesting to highlight what the benefits may be within an intensive buffalo farming system such as the one in place in Italy where the implementation of sexed semen can be an integral part of the reproductive management of the herd. If we assume some fixed starting conditions such as a farm turnover of 20%, a nonexisting bull effect, and mortality, a heritability for milk production of 30% and a mean farm milk yield of 2550 kg, then a reduced number and more productive buffaloes could be used to obtain the same number of replacement heifers. Over the years, the use of the top production animals and sexed semen would ensure a significant milk difference and a corresponding higher cash gain (Zicarelli, unpublished results).

6.6 In vivo embryo production (multiple ovulation and embryo transfer)

In line with a delayed attention to the implementation of reproductive technologies compared to cattle, the first attempts on in vivo embryo production in the buffalo date back to the beginning of 1980s, with the first report by Drost et al. [53]. As expected, procedures for MOET did rely on already existing protocols of hormone administration in cattle and have been since then improved or modified, in order to enhance the success in terms of ovulating follicles and recovered embryos.

Hormones available such as pituitary extracts usually from porcine origin, equine chorionic gonadotropin, and human menopausal gonadotropin [54–56] are administered in mid-cycle in conjunction to the controlled emergence of a new follicular wave, by hormonal or, more recently, mechanical means [57,58]. Pitfalls in the use of hormonal treatment for control of multiple follicle development are always possible, although a number of strategies, depending on the type of used hormone, have been devised. In buffaloes too, the use of equine chrorionic gonadotropin (eCG) may possibly lead to persistent stimulation due to the prolonged half-life of the molecule, and in order to reduce the incidence of such occurrences, monoclonal or polyclonal antibodies have been administered, although no improvement in terms of produced and recovered embryos have been described when compared to conventional protocols [59,60]. Indeed, nowadays the most used hormones for superovulatory treatment in buffaloes are pituitary extracts of FSH with variable levels of LH content, which may additionally account for the large variability in the final response. Such variability could be overcome by using bovine FSH produced by recombinant DNA technology, although such option has not been employed in buffaloes so far. Overall, the use of both pituitary extracts, eCG or human Menopausal Gonadotropin (hMG), has given over the years similar values in terms of ovulating follicles and recovered embryos in the two buffalo subspecies, river and swamp [61–63].

Nevertheless, despite many efforts carried out over the years, the efficiency of MOET in buffalo is still poor. In addition to the inconsistent response to hormonal stimulation in this species, the major limitation is unquestionably the low number of transferable embryos flushed per donor. Several factors may play a role in the efficiency of MOET programs in buffaloes. Age and parity do not seem to significantly alter the rate of embryo recovery, although in heifers a higher number of ovulations are recorded [56,64]. Likewise cattle, the time of initiation of hormonal administration has to respect the physiological command of follicle development within waves, and therefore it is key to start the protocol at the beginning of the new follicular wave by hormonal or mechanical means, as previously highlighted. With regard to the hormonal status of the animal, higher levels of hematic progesterone are inductive of a more satisfactory follicle development following hormonal administration, and this is confirmed by a higher rate of available CLs and recovered embryos [65,66]. When considering environmental variables, in an earlier study the effects of photoperiod on MOET programs were not in line with the seasonality pattern of the species [67,68]. This unexpected result may be accounted for by the fact that during the unfavorable season only the most fertile cows are cycling and hence can be recruited for MOET, leading to a sort of selection of the most responsive donors. The effect of nutrition on reproductive function has been amply demonstrated in ruminants, and this is substantiated in buffaloes by a higher and more efficient response to reproductive technologies such as AI and MOET programs when body condition score lies between 2.5 and 4 on a scale from 1 to 5 [32,69]. The time to start MOET programs in relation to calving has also been considered and a higher response has been reported when protocols are implemented between 60 and 220 days postpartum [68]. In addition, according to Di Palo et al. [70], better results when implementing MOET programs in buffaloes are achieved when richer diets containing forages are given to the animals, as opposed to concentrates, being the former more physiologically fit to ruminants.

Other attempts at improving the efficiency of MOET programs in buffaloes include the possibility to increase the number of developing ovarian follicles by recombinant-Bovine Somatotropin (rBST) priming on a dosage effect manner [71,72]. Similar improvement can be obtained also by immunoneutralization of buffaloes against inhibin, reflected also in the consequent rate of recovered transferable embryos [73]. An additional improvement has been reported by the combined exogenous administration of a GnRH agonist and LH, restricting thus the timing of ovulation in a shorter period of time [63]. The administration of prostaglandin $F_{2\alpha}$ at the end of the superovulatory treatment has also been implicated in some improvement in the rate of recovered ova and embryos, due to effects on endothelin-1 and angiotensin-II gene expression inhibition within the CL, then by increasing ampulla and fimbriae contractility of smooth muscles and finally by favoring follicle rupture [74]. Nevertheless, there is a clear evidence that the low efficiency of MOET, in terms of embryo yields, in buffalo is not simply due to the reduced follicular reservoir at birth, as hormonal stimulation is effective in inducing the growth of multiple follicles up to ovulation. In contrast, a poor embryo/CL rate is commonly observed (7–9), suggesting an impairment of ovum capture by the oviduct at ovulation. Further studies have confirmed the difficulty in improving the efficiency of MOET programs in buffaloes [75].

It has been previously hypothesized that estradiol plasma level, which under natural estrous cycles is lower in comparison to cattle, is particularly elevated following hormonal treatment resulting in abnormal estradiol to progesterone ratio, affecting the capacity of fimbriae to capture ovulated oocytes, or worse push them back into the peritoneal cavity by reversed peristalsis [76]. Similar assumption has been suggested by Baruselli et al. [77], invoking a lower quality of the ovulated oocytes characterized by a reduced number of surrounding granulosa cells and preventing them to be rightly captured by the fimbriae resulting in a failure entry into the oviduct. We have recently demonstrated that both inappropriate cumulus expansion and disruption of contraction-relaxation of the oviduct contribute to OPU failure in superovulated buffaloes, as a result of impaired steroid synthesis [78]. Indeed, in superovulated buffaloes granulosa cell function was altered, as indicated by the overexpression of *FSHR* and *CYP19A1* and the decreased expression of the STAR protein, leading to lower estradiol synthesis. The altered intrafollicular steroid profile was associated to compromised cumulus expansion in most of the oocytes recovered from periovulatory follicles. Finally, the reduced expression of steroid receptors and vascular endothelial growth factor in the oviduct indicated that the contraction-relaxation of the oviduct in the periovulatory period may also play a role.

6.7 In vitro embryo production

As previously reported, the implementation of MOET programs in buffaloes has brought limited success, despite the attempts over the years at improving the recovery of high-quality embryos. Such lack of efficiency has paved the road to the introduction of another younger strategy, again borrowed from cattle, dealing with the possibility to produce embryos in vitro following procedures devised to mature oocytes recovered from antral follicles and to fertilize them with capacitated frozen-thawed spermatozoa [79]. The mass production of embryos derived from the availability of ovaries from slaughtered buffaloes though would have only a minor impact on the genetic improvement of the species, unless this strategy could target the best available animals. This coupling has been brought on the scene by the possibility to recover oocytes from the ovaries in live animals, by ultrasound puncture of antral follicles [80]. The combined IVEP/OPU technology has been proved feasible in buffaloes since more than a couple of decades and has resulted, over the world, in live offspring from the transfer of either fresh or cryopreserved embryos [81−85]. In line with the promising results obtained with the use of sexed semen for AI, embryos have also been produced in vitro with sex-sorted sperm [47]. The possibility to link the OPU technology to the IVEP procedures has been proved feasible even in prepuberal buffaloes in which immature and mature oocytes from antral follicles were recovered, with and without hormonal treatment. In that study, the high number of recorded antral follicles within prepuberal ovaries, similar to cattle calves, highlighted the possibility to bypass the intrinsic genetic paucity of antral follicles recorded in adult females, although no attempts were made at fertilizing the recovered mature oocytes [86]. Recently, embryos were in vitro produced from oocytes recovered by laparoscopy from buffalo calves and pregnancies were obtained after embryo transfer [87]. This approach, commonly known as juvenile OPU, would dramatically increase the genetic progress by shortening generation interval. The successes reported in a quarter of century in buffaloes have been and still are hindered though by a series of: (1) in vivo aspects intrinsic to the species, such as lower follicle count and oocyte recovery, partly surpassed by hormonal treatment, high individual variability [88], lower quality of developing CL responsible for inadequate progesterone production to sustain pregnancy, and poor oocyte competence during the long day length season, leading to early and late embryo mortality [89−91]; and (2) in vitro aspects such as reduced embryo quality according to international evaluation standards and corresponding viability, very likely as a consequence of suboptimal culture conditions, together to a reduced capacity and resilience to withstand cryopreservation procedures, possibly due to a higher lipid content in comparison to cattle [92].

Following is a synthetic series of up-to-date findings and recent improvements related to the protocols implemented in buffaloes in order to obtain in vitro embryos: (1) oocytes: in adult animals lower numbers of antral follicles and corresponding retrieved oocytes, the latter usually of lower quality, are recorded. Across the available literature, differences within this variable are reported according to the subspecies or breeds investigated, age, nutrition, and health status. A number of other variables such as time interval between oocyte retrieval and laboratory processing, oocyte holding temperature, time of the year (short or long day length), and environmental temperature, may overall affect the efficiency of the procedure in buffaloes [93]. To enhance quality and number of punctured follicles and recovered oocytes, especially when OPU is integral part of the in vitro procedure, hormonal treatment with FSH and rBST has been evaluated earlier, resulted in limited success [94,95]. More recently, we demonstrated the efficacy of an OPU priming, with FSH treatment in the presence of progesterone, at increasing the number of follicles, oocytes, and, more importantly, transferable embryos. It consists in the ablation of dominant follicle and insertion of a progesterone-releasing device on day 0, followed by 3 days of FSH given twice daily and a coasting period of around 40−44 hours, with OPU carried out on day 6. A remarkable effect was observed in both seasons, with a fourfold increase of blastocysts number. This certainly ensures a better exploitation of the germinal material especially when programming OPU trials; (2) in vitro maturation: nuclear and cytoplasmic maturation in buffalo oocytes is usually completed between 20 and 24 hours after immature oocytes are placed into maturation media. It is worth noting that buffalo oocytes undergo earlier aging, and hence the in vitro fertilization (IVF) should be carried out as soon as possible since 18 hours post-IVM and not later than 24 hours [96]. Indeed, reduced efficiency in terms of both cleavage and blastocyst rates together with higher proportion of degenerated oocytes has been reported at increasing maturation times. Improvements in oocyte maturation have been reported following inclusion into maturation media of growth factors such as insulin-like growth factor (IGF)-1, IGF-2, epidermal growth factor (EGF), and vasoactive intestinal peptide [97−99]. It has furthermore been reported that antioxidants such as thiol compounds promoting glutathione (GSH) synthesis [100−102], melatonin or taurine [103], are fundamental to improve maturational processes and subsequent embryo development. Finally, lectin was reported to ameliorate oocyte maturation and upregulate the expression of genes involved in gap junction, cell communication and cell cycle proteins (Cx43 and growth differentiation factor-9), cell membrane protein (fibronectin), and basic growth factor (fibroblast growth factor-4) [104]; (3) IVF: the fertilization of mature oocytes in vitro is another fundamental and multifactorial milestone whose efficiency depends in turn on a number of finely tuned steps. The most important and limiting factor is the so-called "bull effect," as only roughly 10% of all the bulls screened for IVF will satisfyingly respond with an adequate fertilizing efficiency, despite the fact that the quality of cryopreserved semen has definitely been improved when compared to the past [105]. The most reliable method to screen bulls for IVF is inevitably to test semen against batches of in vitro matured oocytes and assess the rate of late embryo development. Lately, a simple double staining Trypan Blue/Giemsa technique was used, giving evidence of a correlation between acrosome intact sperm cells and blastocyst development rate. Therefore, this staining technique can be considered predictive of the in vitro fertilizing capability of sperm cells [106]. Both Percoll density gradient and swim-up are efficiently used to select motile sperm cells prior to fertilization, although even with this procedure, the one or the other may perform better depending on the individual bull [107]. While in coincubation with matured oocytes, sperm cells need the support of capacitating agents of which heparin is mostly used also in buffaloes. Other molecules have been tested and proved successful in promoting the capacitation process in buffalo sperm cells, such as progesterone, sodium nitroprusside, and melatonin [108−110]. Finally, factors available in both follicular fluid and serum or present in the oviduct such as osteopontin, also detected in semen in high concentration [111], when included in fertilization media significantly, increase the capacitation process and embryo development [84,112]. The "bull effect" should also be considered when dealing with the time needed for sperm/oocytes coincubation. Although such time has been set at 16 hours according to the results available [96], individual bulls may display a different timing in the kinetics of sperm penetration [113]. Therefore the importance of testing each bull for this parameter prior to entering into fertilizing trials has to be taken into consideration; (4) in vitro embryo culture: differently from cattle, buffalo in vivo and in vitro embryos are known to be characterized by an earlier minor genomic activation at the two- to four-cell stage and a more important one at the four- to eight-cell stage development [114], suggesting from the start of possible differential in vitro culture requirements and explaining a faster development similarly to other domestic ruminants [115]. Initial attempts at culturing buffalo embryos in vitro have relied on the combination of complex media and cell coculture [116]. The need to understand the real requirements of embryos in the course of their early in vitro preimplantation development has brought the attention to the implementation of defined media, which are still today the elective choice [117]. Additionally, since the oviduct represents the only natural milieu where the initial development of the embryos occurs, an understanding of the molecular environment present at the time of embryo presence in the oviduct may well give the hint for the correct adjustments to make in the media used for embryo culture. With this aim, some differences between buffaloes and cattle have been found, following analysis of the oviductal environment in the two species. In fact, the total amount of proteins and their concentration was reported to be lower in buffaloes when compared to cattle, suggesting thus to reduce their level in the formulation for media culture [118]. One of the strategies adopted in cattle in order to reduce ammonium concentration, toxic catabolites, and free radicals in medium for embryo culture, known to be potentially involved in the large offspring syndrome, is to periodically change the same medium during in vitro culture. In contrast, not only this methodology does not improve buffalo embryo development, but on the contrary in our experience it seems that better results are achieved if a minimum amount of manipulation is made on embryos prior to the final evaluation. This is possibly due to the involvement of other variables more likely to affect embryo development during manipulation such as temperature and pH fluctuation. Another significant difference to other domestic ruminants for embryo energy requirement during culture is the need for higher amount of glucose especially in the course of buffalo early development [119,120]. The addition into culture media of cysteamine, taurine, melatonin, or by lowering O_2 concentration, has increased the quality of late-stage embryo development in buffaloes. Such supplementation into culture media has in fact the merit to overcome the oxidative stress in the course of embryo culture as evidenced by a higher expression of antiapoptotic genes in the embryos [121]. Higher quality of embryo development together with increased cryotolerance has been proved successful by inclusion into media of L-carnitine, hyaluronic acid, and leukemia inhibitory factor [122−124].

6.8 Oocyte and embryo cryopreservation

The possibility to store oocytes, similarly to sperm cells, through cryopreservation procedures will ensure the availability of both germ cells whenever wanted and needed for the implementation of in vitro reproductive technologies. The feasibility of cryopreserving buffalo oocytes is of particular interest, as they could be more efficiently harvested and made available in a more suitable manner whenever needed, considering their intrinsic reduced number at birth when compared to cattle [26]. The technology is still far from being commercially viable in all species where it has been tested, and despite the challenges faced due to the injuries received by the oocytes during the cryopreservation procedure, live offspring have been reported in a number of species including humans [107]. Both conventional slow freezing and vitrification procedures on immature oocytes have been characterized by low efficiency in terms of embryo production, due possibly to low permeability of plasma membrane and hence higher sensitivity of chilling injuries. The attention has then shifted to the use of matured oocytes in parallel to an improvement of vitrification technology, characterized by the adoption of very small volume of media and direct contact with liquid nitrogen, such as Cryoloop, Cryotop vitrification, and solid-state vitrification, characterized by a very fast cooling and warming rate, allowing to reduce the concentration of cryoprotectants. Among these vitrification methods, Cryotop proved to be probably the most efficient for in vitro matured bovine and human oocytes [125] and later was successfully used for Swamp and River buffaloes oocytes in IVF [126,127] and nuclear transfer trials [117]. Cleavage rate can be significantly increased by removing granulosa cells from matured oocytes prior to vitrification, followed by coculturing rewarmed oocytes with intact COCs, although late-stage embryo development still remains low [128].

Vitrification has also been the method of choice for cryopreservation of buffalo embryos, and live offspring have been reported from such technology in this species [81,82,85,129]. The importance and the interplay of many variables on the viability of cryopreserved embryos has been amply demonstrated in all animal species. In the buffalo species, cryotolerance is also affected by in vitro maturation and culture conditions, as well as the source of oocytes [130]. It has also been documented in buffaloes that faster developing embryos are more viable and more adequately face cryopreservation procedures as indicated by increased blastocoele re-expansion and hatching [115], and pregnancy rate to term [81]. An additional and possibly a more critical component in buffaloes is represented by the degree of synchrony to be reached between stage of embryo development and recipients, which according to Kasiraj et al. [131] should not exceed 12 hours in the case of in vivo produced embryos cryopreserved by conventional slow freezing, suggesting therefore a need for a more stringent synchrony in case of in vitro produced and vitrified embryos. In addition, it has recently been reported that during the nonbreeding season the cryotolerance of in vitro produced buffalo embryos is negatively affected, possibly due to a reduced oocyte developmental competence [132]. The relatively paucity of information, in comparison to cattle, and the difference among protocols followed by researchers make more difficult to lucidly delineate a clear indication toward the most appropriate and efficient protocol to be used for embryo cryopreservation. In recent years though two different trials have given far better results than ever before by the combination of IVEP, vitrification, and fixed-time embryo transfer into synchronized recipients [85,133], introducing thus a realistic possibility to put this technology into practice for the benefit of buffalo production and genetic improvement.

6.9 More recently developed technologies

The reproductive technologies described so far are instrumental for the development of newly strategic technological steps, for both dissemination of genetic superior animals and/or the introduction of genetic modifications leading to a faster genetic progression or the creation of new genetic lines of animals, for the betterment of animal production and human welfare. Cloning and transgenesis are the relatively new methodologies developed to the scope and applied with varying degrees of efficiency and success across animal species. Somatic cell nuclear transfer (SCNT) is the laboratory procedure meant to clone an individual animal by de-differentiating a somatic cell through nuclear reprogramming. Through the somatic cell methodology though, the first report came from the use of fetal fibroblasts to produce cloned swamp buffalo embryos [134]. Later studies have shown that the expression level of some genes and the epigenetic status can be affected by the sex of the cloned embryos [135]. Within the needed laboratory passages to create a clone, both fertilized zygotes and two-cell embryos can be used as recipient cells [136], and in buffaloes oocytes at metaphase II are ordinarily employed in the process of SCNT as cytoplasmic substrate for cloned embryos production and successful birth of calves [137]. Following preparation of donor and recipient cells, fusion in buffaloes is usually achieved by one or two DC electric pulses [134], and similarly to cattle, a number of successful activation protocols are employed, especially if performed few hours after fusion [138]. Variations of the well-established SCNT cloning protocols, such as the hand-made cloning have also been attempted in buffaloes and pregnancies have been reported [139], together with the birth of calves following vitrification and rewarming of embryos resulting from adult, newborn, and fetal fibroblast donor cells [140]. Lately, nonviable fibroblast from buffalo skin was used for SCNT and embryo production, showing the possibility to use altered nonviable cells for cloning as a possible strategy for rescuing and preserving endangered species [141]. The viability of recently produced cloned buffaloes has also been linked to the DNA methylation pattern and expression of imprinted genes [142]. Adult somatic cells, to be used as donor cells for nuclear transfer, can be induced to be reprogrammed into pluripotent stem cells (iPS) and therefore being characterized by an embryonic stem cell-like state. In buffaloes, recently some attempts have been made in this direction in improving the reprogramming efficiency and the production of iPS [143].

The use of SCNT into the production of transgenic animals has significant higher potential advantages when compared to other approaches. Although transgenic animals have been produced in a number of species, in buffaloes to date, only transgenic embryos have been produced by (1) transfecting buffalo oviductal epithelial cells with enhanced green fluorescent protein followed by transfer into enucleated oocytes [144] and (2) by transfer of embryonic germ-like cells expressing the green fluorescent protein into in vitro−derived buffalo blastocysts [145]. The efficiency of transgenic integration into cloned embryos is being currently pursued by, for example, optimizing electroporation conditions for integration into fetal fibroblast to be used as donor cells [146].

6.10 Conclusion

In the last few decades, the domestic buffalo has greatly benefitted by the implementation of reproductive technologies toward the enhancement of genetic improvement and production. It all started by overcoming the initial difficulties in applying AI to the animals, and the common prejudice of most farmers against the introduction of emerging technologies in a species characterized by evident rustic behavior and a seasonal reproductive performance. Significant progress has been made over the intervening years as clearly highlighted in this review, and we have now a stronger springboard toward higher achievements in buffalo production.

References

[1] Thanh VTK, Orskov ER. Protein digestion and metabolism in buffalo ARSIAL — Regione Lazio, Rome, Italy In: Presicce GA, editor. The buffalo (*Bubalus bubalis*): production and research. Bentham Science Publishers Ltd; 2017. p. 180−95.

[2] Abdullah N, Ho YW, Mahayuddin M, Jalaludin S. Comparative studies of fibre digestion in cattle and buffaloes. In: Proceedings of domestic buffalo production in Asia. Rockhampton, Australia: IAEA; 1989. p. 75−87.

[3] Raghavan GV. La produzione della carne nella specie bufalina. In: Ferrara B, et al., editors. Proceedings II Conv. Internaz. "Allevamento bufalino nel mondo". 1982. p. 620−34.

[4] Bartocci S, Amici A, Verna M, Terramoccia S, Martillotti F. Solid and fluid passage rate in buffalo, cattle and sheep feds diets with different forage to concentrate ratios. Livest Prod Sci 1997;52:201−8.

[5] Zicarelli L. Nutrition in dairy buffaloes. Bubalus bubalis 2001;1−66.

[6] Zicarelli L. Influence of seasonality on buffalo production ARSIAL — Regione Lazio, Rome, Italy In: Presicce GA, editor. The buffalo (*Bubalus bubalis*): production and research. Bentham Science Publishers Ltd; 2017. p. 196−224.

[7] Mingala CN, Villanueva MA, Cruz L. River and swamp buffaloes: history, distribution and their characteristics ARSIAL — Regione Lazio, Rome, Italy In: Presicce GA, editor. The buffalo (*Bubalus bubalis*): production and research. Bentham Science Publishers Ltd; 2017. p. 3−31.

[8] Zicarelli L. Enhancing reproductive performance in domestic dairy water buffalo (*Bubalus bubalis*). Soc Reprod Fertil Suppl 2010;67:443−55.

[9] Zicarelli L. Management of buffaloes. In: Proceedings of world Buiatrics congress, Santiago, Chile; 2010, p. 326−42.

[10] Borghese A, Mazzi M. Buffalo population and strategies in the world. In: Buffalo production and strategies. FAO Regional Office for Europe, REU Technical Series 67; 2005. p. 1−39.

[11] Calabrò S, Cutrignelli MI, Gonzalez OJ, Chiofalo B, Grossi M, Tudisco R, et al. Meat quality of buffalo young bulls fed baba bean as protein source. Meat Sci 2014;96:591−6.

[12] Giordano G, Guarini P, Ferrari P, Biondi-Zoccai G, Schiavone B, Giordano A. Beneficial impact on cardiovascular risk profile of water buffalo meat consumption. Eur J Clin Nutr 2010;64:1000−6.

[13] Murthy TRK, Prince Devadason I. Buffalo meat and meat products — an overview. In: Proceedings of the fourth Asian buffalo congress, New Delhi, India, vol. 28; 2003. p. 194−99.

[14] Di Berardino D, Di Meo GP, Gallagher DS, Hayes H, Iannuzzi L. ISCNDB2000—international system for chromosome nomenclature of domestic bovids. Cytogenet Cell Genet 2001;92:283−99.

[15] Di Berardino D, Iannuzzi L. Chromosome banding homologies in Swamp and Murrah buffalo. J Hered 1981;72(3):183−8 [PMID: 6168678].

[16] Wurster DH, Benirschke K. Chromosome studies in the superfamily Bovoidea. Chromosoma 1968;25(2):152−71 [PMID: 5709393]. <https://doi.org/10.1007/BF00327175>.

[17] Michelizzi VN, Dodson MV, Pan Z, et al. Water buffalo genome science comes of age. Int J Biol Sci 2010;6(4):333−49 [PMID: 20582226]. <https://doi.org/10.7150/ijbs.6.333>.

[18] Iamartino D, Williams JL, Sonstegard T, Reecy J, Van Tassell C, Nicolazzi AL, et al. The buffalo genome and the application of genomics in animal management and improvement. In: The 10th world buffalo congress and the 7th Asian buffalo congress, Phuket, Thailand; 2013. p. 151−58.

[19] Li J, Liang A, Li Z, Du C, Hua G, Salzano A, et al. An association analysis between PRL genotype and milk production traits in Italian Mediterranean river buffalo. J Dairy Res 2017;84(4):430−3.

[20] Li J, Liu J, Liu S, Plastow G, Zhang C, Wang Z, et al. Integrating RNA-seq and GWAS reveals novel genetic mutations for buffalo reproductive traits. Anim Reprod Sci 2018;197:290−5.

[21] Pauciullo A, Iannuzzi L. Molecular genetics and selection in dairy buffaloes: the Italian situation ARSIAL — Regione Lazio, Rome, Italy In: Presicce GA, editor. The buffalo (*Bubalus bubalis*): production and research. Bentham Science Publishers Ltd; 2017. p. 50−68.

[22] Presicce GA. Reproduction in the water buffalo. Reprod Domest Anim 2007;42(Suppl. 2):24−32.

[23] Campanile G, Neglia G, D'Occhio MJ. Embryonic and fetal mortality in river buffalo (*Bubalus bubalis*). Theriogenology 2016;86:207−13.

[24] Peeva R, Ilieva Y. Longevity of buffalo cows and reasons for their culling. Ital J Anim Sci 2007;2007(6):378−80. Available from: https://doi.org/10.4081/ijas.2007.s2.378 (sup2).

[25] Manik RS, Palta P, Singla SK, Sharma V. Folliculogenesis in buffalo (*Bubalus bubalis*): a review. Reprod Fertil Dev 2002;14(5−6):315−25.

[26] Baruselli PS, Presicce GA. Folliculogenesis and ovarian physiology applied to reproductive biotechnologies in buffaloes ARSIAL — Regione Lazio, Rome, Italy In: Presicce GA, editor. The buffalo (*Bubalus bubalis*): production and research. Bentham Science Publishers Ltd; 2017. p. 313−39.

[27] Nair SJ, Brar AS, Ahuja CS, Sangha SP, Chaudhary KC. A comparative study on lipid peroxidation, activities of antioxidant enzymes and viability of cattle and buffalo bull spermatozoa during storage at refrigeration temperature. Anim Reprod Sci 2006;96:21−9.

[28] Kumaresan A, Ansari MR, Garg A. Modulation of post-thaw sperm functions with oviductal proteins in buffaloes. Anim Reprod Sci 2005;90:73−84.

[29] Kumaresan A, Ansari MR, Garg A, Kataria M. Effect of oviductal proteins on sperm functions and lipid peroxidation levels during cryopreservation in buffaloes. Anim Reprod Sci 2006;93:246−57.

[30] Sansone G, Nastri MJF, Fabbrocini A. Storage of buffalo (*Bubalus bubalis*) semen. Anim Reprod Sci 2000;62:55−76.

[31] de Carvalho NA, Soares JG, Baruselli PS. Strategies to overcome seasonal anestrus in water buffalo. Theriogenology 2016;86(1):200−6. Available from: https://doi.org/10.1016/j.theriogenology.2016.04.032.

[32] Campanile G, Neglia G, Di Palo R, Gasparrini B, Pacelli C, D'Occhio MJ, et al. Relationship of body condition score and blood urea and ammonia to pregnancy in Italian Mediterranean buffaloes. Reprod Nutr Dev 2006;46:57−62.

[33] Zicarelli L, Esposito L, Campanile G, Di Palo R, Armstrong AR. Effect of using vasectomized bulls in A.I. practice on the reproductive efficiency of Italian buffalo cows. Anim Reprod Sci. 1997;47:171−80.

[34] Seren E. Periestrous endocrine changes in Italian buffaloes. Agricoltura e Ricerca 1994;17−24.

[35] Zicarelli L, Campanile G, Infascelli F, Esposito L, Ferrari G Incidence and fertility of heats with double ovulations in the buffalo cows. In: Proceedings of the II world buffalo congress, New Delhi, India, vol. 3; 1988. p. 57−62.

[36] Barile VL, Terzano GM, Pacelli C, Todini L, Malfatti A, Barbato O. LH peak and ovulation after two different estrus synchronization treatments in buffalo cows in the daylight-lengthening period. Theriogenology 2015;84:286−93. Available from: https://doi.org/10.1016/j.theriogenology.2015.03.019 Mar. pii: S0093-691X(15)00146-6.

[37] Pursley JR, Mee MO, Wiltbank MC. Synchronization of ovulation in dairy cows using PGF2a and GnRH. Theriogenology 1995;44:915−23.

[38] De Araujo Berber RC, Madureira EH, Baruselli PS. Comparison of two Ovsynch protocols (GnRH versus LH) for fixed timed insemination in buffalo (*Bubalus bubalis*). Theriogenology 2002;57:1421−30.

[39] Neglia G, Gasparrini B, Di Palo R, De Rosa C, Zicarelli L, Campanile G. Comparison of pregnancy rates with two estrus synchronization protocols in Italian Mediterranean buffalo cows. Theriogenology 2003;60:125−33.

[40] De Rensis F, Ronci G, Guarneri P, Nguyen BX, Presicce GA, Huszenicza G, et al. Conception rate after fixed time insemination following Ovsynch protocol with and without progesterone supplementation in cyclic and noncyclic Mediterranean Italian buffaloes (*Bubalus bubalis*). Theriogenology 2005;63:1824−31.

[41] Neglia G, Gasparrini B, Salzano A, Vecchio D, De Carlo E, Cimmino R, et al. Relationship between the ovarian follicular response at the start of an Ovsynch-TAI program and pregnancy outcome in the Mediterranean river buffalo. Theriogenology 2016;86(9):2328−33. Available from: https://doi.org/10.1016/j.theriogenology.2016.07.027.

[42] Rossi P, Vecchio D, Neglia G, Di Palo R, Gasparrini B, D'Occhio MJ, et al. Seasonal fluctuations in the response of Italian Mediterranean buffaloes to synchronization of ovulation and timed artificial insemination. Theriogenology 2014;82(1):132−7. Available from: https://doi.org/10.1016/j.theriogenology.2014.03.005.

[43] Neglia G, Restucci B, Russo M, Vecchio D, Gasparrini B, Prandi A, et al. Early development and function of the corpus luteum and relationship to pregnancy in the buffalo. Theriogenology 2015;83(6):959−67. Available from: https://doi.org/10.1016/j.theriogenology.2014.11.035.

[44] Presicce GA, Verberckmoes S, Senatore EM, Rath D. First established pregnancies in Mediterranean Italian buffaloes (*Bubalus bubalis*) following deposition of sexed spermatozoa near the utero tubal junction. Reprod Domest Anim 2005;73−5.

[45] Lu YQ, Liang XW, Zhang M, Wang WL, Kitiyanant Y, Lu SS, et al. Birth of twins after in vitro fertilization with flow-cytometric sorted buffalo (*Bubalus bubalis*) sperm. Anim Reprod Sci 2007;100(1−2):192−6.

[46] Lu Y, Zhang M, Lu S, Xu D, Huang W, Meng B, et al. Sex-preselected buffalo (*Bubalus bubalis*) calves derived from artificial insemination with sexed sperm. Anim Reprod Sci 2010;119(3−4):169−71. Available from: https://doi.org/10.1016/j.anireprosci.2010.01.001.

[47] Liang XW, Lu YQ, Chen MT, Zhang XF, Lu SS, Zhang M, et al. In vitro embryo production in buffalo (*Bubalus bubalis*) using sexed sperm and oocytes from ovum pick up. Theriogenology 2008;69(7):822−6. Available from: https://doi.org/10.1016/j.theriogenology.2007.11.021.

[48] Campanile G, Gasparrini B, Vecchio D, Neglia G, Senatore EM, Bella A, et al. Pregnancy rates following AI with sexed semen in Mediterranean Italian buffalo heifers (*Bubalus bubalis*). Theriogenology 2011;76(3):500−6. Available from: https://doi.org/10.1016/j.theriogenology.2011.02.029.

[49] Barile VL, Pacelli C, De Santis G, Baldassi D, Mazzi M, Terzano GM. Use of commercial available bovine semen sexing agent in buffalo: preliminary report of the effect on the conception rate. In: The 10th world buffalo congress and the 7th Asian buffalo congress, May 6−8, 2013; p. 52.

[50] Underwood SL, Bathgate R, Maxwell WM, Evans G. Birth of offspring after artificial insemination of heifers with frozen-thawed, sex-sorted, re-frozen-thawed bull sperm. Anim Reprod Sci 2010;118(2−4):171−5. Available from: https://doi.org/10.1016/j.anireprosci.2009.08.007.

[51] Seidel Jr1 GE. Update on sexed semen technology in cattle. Animal 2014;8(Suppl. 1):160−4. Available from: https://doi.org/10.1017/S1751731114000202.

[52] Leahy T, Marti JI, Crossett B, Evan G, Maxwell WM. Two-dimensional polyacrylamide gel electrophoresis of membrane proteins from flow cytometrically sorted ram sperm. Theriogenology 2011;75(5):962−71. Available from: https://doi.org/10.1016/j.theriogenology.2010.11.003.

[53] Drost M, Wright Jr JM, Cripe WS, Richter AR. Embryo transfer in water buffalo (*Bubalus bubalis*). Theriogenology 1983;20:579−84.

[54] Karaivanov C. Comparative studies on the superovulatory effect of PMSG and FSH in water buffalo (*Bubalus bubalis*). Theriogenology 1986;26:51−9.

[55] Singla SK, Madan ML. Response of superovulation in buffaloes (*Bubalus bubalis*) with Super-OV and FSH-P. Theriogenology 1990;33:327.

[56] Misra AK, Tyagi S. *In vivo* embryo production in buffalo: present and perspectives. Ital J Anim Sci 2007;6(Suppl. 2):74–91 <https://doi.org/10.4081/ijas.2007.s2.74>.

[57] Bó GA, Adams GP, Pierson RA, Mapletoft RJ. Exogenous control of follicular wave emergence in cattle. Theriogenology 1995;43:31–40 <https://doi.org/10.1016/0093-691X(94)00010-R>.

[58] Zicarelli L, Boni R, Campanile G, et al. Effect of OPU priming on superovulatory response in Italian buffalo cows. In: 5th world buffalo conference, Royal Palace, Caserta, Italy; October 1997. p. 13–6.

[59] Palta P, Kumar M, Jailkhani S, Manik RS, Madan ML. Changes in peripheral inhibin levels and follicular development following treatment of buffalo with PMSG and neutra-PMSG for superovulation. Theriogenology 1997;48(2):233–40 [PMID: 16728122]. <https://doi.org/10.1016/S0093-691X(97)84070-8>.

[60] Matharoo JS, Cheema S, Ranjna SJ, Mehar HMC, Tiwana MS. Use of hetero and homologous anti-PMSG with PMSG for superovulation and embryo recovery in buffaloes. In: Proceedings of the 5th world buffalo congress, Caserta; 13–16 October, 1997. p. 754–57.

[61] Situmorang P. Superovulation in different buffalo genotypes. Indones J Anim Vet Sci 2003;8:40–5.

[62] Baruselli PS, Mucciolo RG, Arruda R, Madureira RH, Amaral R, Assumpcao ME. Embryo recovery rate in superovulated buffaloes. Theriogenology 1999;51:401 <https://doi.org/10.1016/S0093-691X(99)91960-X>.

[63] Carvalho NA, Baruselli PS, Zicarelli L, Madureira EH, Visintin JA, DOcchio MJ. Control of ovulation with a GnRH agonist after superstimulation of follicular growth in buffalo: fertilization and embryo recovery. Theriogenology 2002;58(9):1641–50 [PMID: 12472135]. <https://doi.org/10.1016/S0093-691X(02)01057-9>.

[64] Boni R, Roviello S, Gasparrini B, Zicarelli L. Pregnancies established after transferring embryos yielded by ovum pick-up and *in vitro* embryo production in Italian buffalo cows. In: Proceedings of V world buffalo congress, Caserta, Italia; 1997. p. 787–92.

[65] Madan ML, Singla SK, Singh C, Prakash BS, Jailkhani S. Embryo transfer technology in buffaloes: endocrine responses and limitations. In: Proceeding of II world buffalo congress, New Delhi, vol. III; 1988. p. 195–211.

[66] Neglia G, Gasparrini B, Vecchio D, et al. Progesterone supplementation during multiple ovulation treatment in buffalo species (*Bubalus bubalis*). Trop Anim Health Prod 2010;42(6):1243–7 [PMID: 20411328]. <https://doi.org/10.1007/s11250-010-9556-8>.

[67] Zicarelli L, Di Palo R, Palladino M, Campanile G, Esposito L. Embryo transfer in Mediterranean *Bubalus bubalis*. In: Proceeding of the international symposium "prospects of buffalo production in the Mediterranean and the Middle East," Cairo, Egypt; 1993. p. 73–5.

[68] Zicarelli L, Boni R, Campanile G, Roviello S, Gasparrini B, Di Palo R. Response to superovulation according to parity, OPU priming, presence/absence of dominant follicle and season. In: V world buffalo congress, Caserta, Italy; 1997. p. 748–53.

[69] Baruselli PS, Madureira EH, Visintin JA, Barnabe VH, Barnabe RC, Amaral R. Inseminação artificial em tempo fixo com sincronização da ovulação em bubalinos. Rev Bras Reprod Anim 1999;23:360–2.

[70] Di Palo R, Campanile G, Esposito L, Barbieri V, Zicarelli L. Characteristics of diet and FSH-p superovulatory response in Italian Mediterranean buffaloes. In: Proceedings of IV world buffalo congress, Sao Paolo; 1994. p. 474–76.

[71] Zicarelli L, Di Palo R, Campanile G, Boni, R, Langella M. rBST + FSH-P in superovulatory treatment of Italian Mediterranean buffaloes. In: Proceedings of IV world buffalo congress, Sao Paolo; 1994. p. 459–461.

[72] Songsasen N, Yiengvisavakul V, Buntaracha B, Pharee S, Apimeteetumrong M, Sukwongs Y. Effect of treatment with recombinant bovine somatotropin on responses to superovulatory treatment in swamp buffalo (*Bubalus bubalis*). Theriogenology 1999;52(3):377–84 [PMID: 10734373]. <https://doi.org/10.1016/S0093-691X(99)00136-3>.

[73] Li DR, Qin GS, Wei YM, et al. Immunisation against inhibin enhances follicular development, oocyte maturation and superovulatory response in water buffaloes. Reprod Fertil Dev 2011;23(6):788–97 [PMID: 21791180]. <https://doi.org/10.1071/RD10279>.

[74] Neglia G, Natale A, Esposito G, et al. Effect of prostaglandin F2α at the time of AI on progesterone levels and pregnancy rate in synchronized Italian Mediterranean buffaloes. Theriogenology 2008;69(8):953–60 [PMID: 18346780]. <https://doi.org/10.1016/j.theriogenology.2008.01.008>.

[75] Zicarelli L. Advanced reproductive technologies for improving buffalo production. In: Proceedings of the first buffalo symposium of Americas, Estacao das Docas, Belèm, Parà, Brazil; 2002. 186–97.

[76] Misra AK, Kasiraj R, Mutha Rao M, Rangareddy NS, Jaiswal RS, Pant HC. Rate of transport and development of preimplantation embryo in the superovulated buffalo (*Bubalus bubalis*. Theriogenology 1998;50:637–49.

[77] Baruselli PS, Madureira EH, Visintin JA, Porto-Filho R, Carvalho NAT, Campanile G, et al. Failure of oocyte entry into oviduct in superovulated buffalo. Theriogenology 2000;53:491.

[78] Salzano A, De Canditiis C, Della Ragione F, Prandi A, Zullo G, Neglia G, et al. Evaluation of factors involved in the failure of ovum capture in superovulated buffaloes. Theriogenology 2018;122:102–8.

[79] Fukuda Y, Ichikawa M, Naito K, Toyoda Y. Birth of normal calves resulting from bovine oocytes, matured, fertilized, and cultured with cumulus cells in vitro up to the blastocyst stage. Biol Reprod 1990;42:114–19.

[80] Boni R, Roviello S, Zicarelli L. Repeated ovum pick-up in Italian Mediterranean buffalo cows. Theriogenology 1996;46(5):899–909 [PMID: 16727954]. <https://doi.org/10.1016/S0093-691X(96)00248-8>.

[81] Neglia G, Gasparrini B, Caracciolo di Brienza V, Di Palo R, Zicarelli L. First pregnancies to term after transfer of buffalo vitrified embryos entirely produced *in vitro*. Vet Res Commun 2004;28:233–326 [PMID: 15372965]. <https://doi.org/10.1023/B:VERC.0000045414.65331.6a>.

[82] Hufana-Duran D, Pedro PB, Venturina HV, Hufana RD, Salazar AR, Duran PG, et al. Post-warming hatching and birth of live calves following transfer of *in vitro*-derived vitrified water buffalo (*Bubalus bubalis*) embryos. Theriogenology 2004;61(7–8):1429–39 [PMID: 15036974]. <https://doi.org/10.1016/j.theriogenology.2003.08.011>.

[83] Huang Y, Zhuang X, Gasparrini B, Presicce GA. Oocyte recovery by ovum pick-up and embryo production in Murrah and Nili-Ravi buffaloes (*Bubalus bubalis*) imported in China. Reprod Fertil Dev 2005;17(2):273.

[84] Boccia L, Di Francesco S, Neglia G, De Blasi M, Longobardi V, Campanile G, et al. Osteopontin improves sperm capacitation and in vitro fertilization efficiency in buffalo (*Bubalus bubalis*). Theriogenology 2013;80(3):212−17.

[85] Saliba W, Gimenes L, Drumond R, Bayao H, Alvim M, Baruselli P, et al. Efficiency of OPU-IVEP-ET of fresh and vitrified embryos in buffaloes. 10th World Buffalo Congress, the 7th Asian Buffalo congress, May 6−8, Phuket, Thailand: 10. Buffalo Bull, 2013;32(Special issue 2):385−88.

[86] Presicce GA, Senatore EM, De Santis G, Stecco R, Terzano GM, Borghese A, et al. Hormonal stimulation and oocyte maturational competence in prepuberal Mediterranean Italian buffaloes (*Bubalus bubalis*). Theriogenology 2002;57:1877−84.

[87] Baldassarre H, Bordignon V. Laparoscopic ovum pick-up for in vitro embryo production from dairy bovine and buffalo calves. In: Proceedings of the 32nd annual meeting of the Brazilian Embryo Technology Society (SBTE), Florianópolis, SC, Brazil; August 16th−18th, 2018. doi: 10.21451/1984−3143-AR2018-0057.

[88] Gasparrini B, Neglia G, Di Palo R, Vecchio D, Albero G, Esposito L, et al. Influence of oocyte donor on in vitro embryo production in buffalo. Anim Reprod Sci 2014;144(3−4):95−101. Available from: https://doi.org/10.1016/j.anireprosci.2013.11.010.

[89] Campanile G, Baruselli PS, Neglia G, et al. Ovarian function in the buffalo and implications for embryo development and assisted reproduction. Anim Reprod Sci 2010;121(1−2):1−11 [PMID: 20430540]. <https://doi.org/10.1016/j.anireprosci.2010.03.012>.

[90] Di Francesco S, Boccia L, Campanile G, Di Palo R, Vecchio D, Neglia G, et al. The effect of season on oocyte quality and developmental competence in Italian Mediterranean buffaloes (*Bubalus bubalis*). Anim Reprod Sci 2011;123(1−2):48−53 PMID: 21168984.

[91] Di Francesco S, Novoa MV, Vecchio D, et al. Ovum pick-up and *in vitro* embryo production (OPUIVEP) in Mediterranean Italian buffalo performed in different seasons. Theriogenology 2012;77(1):148−54 [PMID: 21872310]. <https://doi.org/10.1016/j.theriogenology.2011.07.028>.

[92] Gasparrini B. In vitro embryo production in buffalo species: state of the art. Theriogenology 2002;57:237−56.

[93] Misra V, Misra AK, Sharma R. Effect of ambient temperature on *in vitro* fertilization of bubaline oocyte. Anim Reprod Sci 2007;100 (3−4):379−84 [PMID: 17125942]. <https://doi.org/10.1016/j.anireprosci.2006.10.020>.

[94] Boni R, Di Palo R, Barbieri V, Zicarelli L. Ovum pick-up in deep anestrus buffaloes. In: Proceedings of the 1994 IV world buffalo congress, Sao Paolo, Brazil; 1994. p. 480−82.

[95] Sá Filho MF, Carvalho NA, Gimenes LU, et al. Effect of recombinant bovine somatotropin (bST) on follicular population and on *in vitro* buffalo embryo production. Anim Reprod Sci 2009;113(1−4):51−9 [PMID: 18691835]. <https://doi.org/10.1016/j.anireprosci.2008.06.008>.

[96] Gasparrini B, De Rosa A, Attanasio L, et al. Influence of the duration of *in vitro* maturation and gamete co-incubation on the efficiency of *in vitro* embryo development in Italian Mediterranean buffalo (*Bubalus bubalis*). Anim Reprod Sci 2008;105(3−4):354−64 [PMID: 17481834]. <https://doi.org/10.1016/j.anireprosci.2007.03.022>.

[97] Pawshe CH, Appa Rao KBC, Totey SM. Effect of insulin-like growth factor I and its interaction with gonadotrophins on in vitro maturation and embryonic development, cell proliferation, and biosynthetic activity of cumulus-oocytes complexes and granulosa cells in buffalo. Mol Reprod Dev 1998;49:277−85.

[98] Chauhan MS, Singla SK, Palta P, Manik RS, Madan ML. Effect of epidermal growth factor on the cumulus expansion, meiotic maturation and development of buffalo oocytes in vitro. Vet Ret 1999;144:266−7.

[99] Nandi S, Ravindranatha BM, Gupta PS, Raghu HM, Sarma PV. Developmental competence and post-thaw survivability of buffalo embryos produced in vitro: effect of growth factors in oocyte maturation medium and of embryo culture system. Theriogenology 2003;60(9):1621−31.

[100] Gasparrini B, Neglia G, Di Palo R, Campanile G, Zicarelli L. Effect of cysteamine during in vitro maturation on buffalo embryo development. Theriogenology 2000;54:1537−42.

[101] Gasparrini B, Sayoud H, Neglia G, Matos DG, Donnay I, Zicarelli L. Glutathione synthesis during *in vitro* maturation of buffalo (*Bubalus bubalis*) oocytes: effects of cysteamine on embryo development. Theriogenology 2003;60(5):943−52 [PMID: 12935871]. <https://doi.org/10.1016/S0093-691X(03)00098-0>.

[102] Gasparrini B, Boccia L, Marchandise J, Di Palo R, George F, Donnay I, et al. Enrichment of in vitro maturation medium for buffalo (*Bubalus bubalis*) oocytes with thiol compounds: effects of cystine on glutathione synthesis and embryo development. Theriogenology 2006;65 (2):275−87.

[103] Manjunatha BM, Devaraj M, Gupta PSP, Ravindra JP, Nandi S. Effect of taurine and melatonin in the culture medium on buffalo in vitro embryo development. Reprod Domest Anim 2009;44(1):12−16.

[104] Pandey A, Gupta N, Gupta SC. Improvement of *in vitro* oocyte maturation with lectin supplementation and expression analysis of Cx43, GDF-9, FGF-4 and fibronectin mRNA transcripts in buffalo (*Bubalus bubalis*). J Assist Reprod Genet 2009;26(6):365−71 [PMID: 19629675]. <https://doi.org/10.1007/s10815-009-9314-x>.

[105] Wilding M, Gasparrini B, Neglia G, Dale B, Zicarelli L. Mitochondrial activity and fertilization potential of fresh and cryopreserved buffalo sperm. Theriogenology 2003;59:466.

[106] Boccia L, Di Palo R, De Rosa A, Attanasio L, Mariotti E, Gasparrini B. Evaluation of buffalo semen by Trypan blue/Giemsa staining and related fertility *in vitro*. 8th World Buffalo Congress, Caserta, Italy October 19−22, 2007. Ital J Anim Sci 2007;6(2):739−42 <https://doi.org/10.4081/ijas.2007.s2.739>.

[107] Gasparrini B. Applied reproductive technologies in the buffalo species ARSIAL − Regione Lazio, Rome, Italy In: Presicce GA, editor. The buffalo (*Bubalus bubalis*): production and research. Bentham Science Publishers Ltd; 2017. p. 374−433.

[108] Boccia L, Attanasio L, Monaco E, De Rosa A, Di Palo R, Gasparrini B. Effect of progesterone on capacitation of buffalo (*Bubalus bubalis*) spermatozoa *in vitro*. Reprod Domest Anim 2006;41(4):311.

[109] Boccia L, Attanasio L, De Rosa A, Pellerano G, Di Palo R, Gasparrini B. Effect of sodium nitroprusside on buffalo sperm capacitation *in vitro*. Reprod Fertil Dev 2007;19:276 <https://doi.org/10.1071/RDv19n1Ab322>.

[110] Di Francesco S, Mariotti E, Tsantarliotou M, et al. Melatonin promotes *in vitro* sperm capacitation in buffalo (*Bubalus bubalis*). Proceedings of 36th annual conference of the International Embryo Transfer Society, 9–12 January, Cordoba, Argentina. Reprod Fertil Dev 2010;22 (1):311–12. <https://doi.org/10.1071/RDv22n1Ab311>.

[111] Pero ME, Killian GJ, Lombardi P, Zicarelli L, Avallone L, Gasparrini B. Identification of osteopontin in water buffalo semen. Reprod Fertil Dev 2007;19:279.

[112] Siniscalchi C, Mariotti E, Boccia L, et al. Co-culture with oviduct epithelial cells promotes capacitation of buffalo (*Bubalus bubalis*) sperm. Proceedings of 36th Annual Conference of the International Embryo Transfer Society, 9–12 January, Cordoba, Argentina. J Reprod Fertil Dev 2010;22(1):316. <https://doi.org/10.1071/RDv22n1Ab321>.

[113] Rubessa M, Di Fenza M, Mariotti E, Di Francesco S, De Dilectis C, Di Palo R, et al. Kinetics of sperm penetration is correlated with in vitro fertility of buffalo (*Bubalus bubalis*) bulls. Reprod Fertil Dev 2009;21(I):206–7.

[114] Verma A, Kumar P, Rajput S, Roy B, De S, Datta TK. Embryonic genome activation events in buffalo (*Bubalus bubalis*) preimplantation embryos. Mol Reprod Dev 2012;79(5):321–8 [PMID: 22461405]. <https://doi.org/10.1002/mrd.22027>.

[115] Gasparrini B, Neglia G, Caracciolo, di Brienza V, Campanile G, Di Palo R, et al. Preliminary analysis of vitrified *in vitro* produced embryos. Theriogenology 2001;55:307.

[116] Boni R, Roviello S, Gasparrini B, Langella M, Zicarelli L. *In vitro* production of buffalo embryos in chemically defined medium. Buffalo J 1999;15:115–20.

[117] Caracciolo di Brienza V, Neglia G, Masola N, Gasparrini B, Di Palo R, Campanile G. In vitro embryo production in chemically defined media. In: Proceedings of 1° Congresso Nazionale sull'Allevamento del Bufalo; 2001. p. 341–44.

[118] Vecchio D, Neglia G, Di Palo R, et al. Ion, protein, phospholipid and energy substrate content of oviduct fluid during the oestrous cycle of buffalo (*Bubalus bubalis*). Reprod Domest Anim 2010;45(5):e32–9 [PMID: 19761531].

[119] Monaco E, De Rosa A, Attanasio L, Boccia L, Zicarelli L, Gasparrini B. *In vitro* culture of buffalo (*Bubalus bubalis*) embryos in the presence or absence of glucose. Reprod Domest Anim 2006;41(4):332.

[120] Suárez Novoa MV, Di Francesco S, Rubessa M, Boccia L, Longobardi V, De Blasi M, et al. Effect of reducing glucose concentration during in vitro embryo culture in buffalo (*Bubalus bubalis*). Reprod Fertil Dev 2011;23(1):168.

[121] Elamaran G, Singh KP, Singh MK, et al. Oxygen concentration and cysteamine supplementation during in vitro production of buffalo (*Bubalus bubalis*) embryos affect mRNA expression of BCL-2, BCL-XL, MCL-1, BAX and BID. Reprod Domest Anim 2012;47(6):1027–36 [PMID: 22452597]. <https://doi.org/10.1111/j.1439-0531.2012.02009.x>.

[122] Gasparrini B, Longobardi V, Zullo G, et al. Effect of L-carnitine on buffalo *in vitro* embryo development. In: Proceedings of 20th ASPA congress, Bologna, Italy; 2013.

[123] Boccia L, Rubessa M, De Blasi M, et al. Hyaluronic acid improve cryotolerance of buffalo (*Bubalus bubalis*) in vitro derived embryos. Proceeding of 38th annual conference of the International Embryo Transfer Society, 7–10 January 2012, Phoenix, Arizona. Reprod Fertil Dev 2012;24(I):138. <https://doi.org/10.1071/RDv24n1Ab52>.

[124] Eswari S, Sai Kumar G, Sharma GT. Expression of mRNA encoding leukaemia inhibitory factor (LIF) and its receptor (LIFRβ) in buffalo pre-implantation embryos produced *in vitro*: markers of successful embryo implantation. Zygote 2013;21(2):203–13 [PMID: 22892066]. <https://doi.org/10.1017/S0967199412000172>.

[125] Kuwayama M, Vajta G, Kato O, Leibo SP. Highly efficient vitrification method for cryopreservation of human oocytes. Reprod Biomed Online 2005;11(3):300–8 [PMID: 16176668]. <https://doi.org/10.1016/S1472-6483(10)60837-1>.

[126] Attanasio L, De Rosa A, De Blasi M, Neglia G, Zicarelli L, Campanile G, et al. The influence of cumulus cells during *in vitro* fertilization of buffalo (*Bubalus bubalis*) denuded oocytes that have undergone vitrification. Theriogenology 2010;74(8):1504–8 [PMID: 20615538]. <https://doi.org/10.1016/j.theriogenology.2010.05.014>.

[127] Attanasio L, Boccia L, Vajta G, Kuwayama M, Campanile G, Zicarelli L, et al. Cryotop vitrification of buffalo (*Bubalus bubalis*) in vitro matured oocytes: effects of cryoprotectant concentrations and warming procedures. Reprod Domest Anim 2010;45(6):997–1002. Available from: https://doi.org/10.1111/j.1439-0531.2009.01475.x.

[128] Muenthaisong S, Laowtammathron C, Ketudat-Cairns M, Parnpai R, Hochi S. Quality analysis of buffalo blastocysts derived from oocytes vitrified before or after enucleation and reconstructed with somatic cell nuclei. Theriogenology 2007;67(4):893–900 [PMID: 17161454]. <https://doi.org/10.1016/j.theriogenology.2006.11.005>.

[129] Boccia L, De Rosa A, Attanasio L, Neglia G, Vecchio D, Campanile G, et al. Developmental speed affects the cryotolerance of in vitro produced buffalo (*Bubalus bubalis*) embryos. Ital J Anim Sci 2013;12:e80.

[130] Manjunatha BM, Gupta PS, Ravindra JP, Devaraj M, Nandi S. *In vitro* embryo development and blastocyst hatching rates following vitrification of river buffalo embryos produced from oocytes recovered from slaughterhouse ovaries or live animals by ovum pick-up. Anim Reprod Sci 2008;104(2–4):419–26 [PMID: 17689038]. <https://doi.org/10.1016/j.anireprosci.2007.06.030>.

[131] Kasiraj R, Misra AK, Mutha Rao M, Jaiswal RS, Rangareddi NS. Successful culmination of pregnancy and live birth following the transfer of frozen-thawed buffalo embryos. Theriogenology 1993;39(5):1187–92 [PMID: 16727286]. <https://doi.org/10.1016/0093-691X(93)90016-X>.

[132] Kosior MA, Parente E, Salerno F, Annes K, Annunziata R, Albero G, et al. Season affects cryotolerance of *in vitro*-produced buffalo embryos. Reprod Fertil Dev 2019;31(1):139.

[133] Gasparrini B. *In vitro* embryo production in buffalo: Yesterday, today and tomorrow. Buffalo Bull 2013;32(1):188−95.

[134] Parnpai R, Tasripoo K, Kamonpatana M. Development of cloned swamp buffalo embryos derived from fetal fibroblasts: comparison in vitro cultured with or without buffalo and cattle epithelial cells. Buffalo J 1999;15:371−84.

[135] Sandhu A, Mohapatra SK, Agrawal H, Singh MK, Palta P, Singla SK, et al. Effect of sex of embryo on developmental competence, epigenetic status, and gene expression in buffalo (*Bubalus bubalis*) embryos produced by hand-made cloning. Cell Reprogram. 2016;18(5):356−65.

[136] Campbell KH, Fisher P, Chen WC, et al. Somatic cell nuclear transfer: past, present and future perspectives. Theriogenology 2007;68(Suppl. 1):S214−31 [PMID: 17610946]. <https://doi.org/10.1016/j.theriogenology.2007.05.059>.

[137] Shi D, Lu F, Wei Y, et al. Buffalos (*Bubalus bubalis*) cloned by nuclear transfer of somatic cells. Biol Reprod 2007;77(2):285−91 [PMID: 17475931]. <https://doi.org/10.1095/biolreprod.107.060210>.

[138] Simon L, Veerapandian C, Balasubramanian S, Subramanian A. Somatic cell nuclear transfer in buffalos: effect of the fusion and activation protocols and embryo culture system on preimplantation embryo development. Reprod Fertil Dev 2006;18(4):439−45 [PMID: 16737637]. <https://doi.org/10.1071/RD05079>.

[139] Shah RA, George A, Singh MK, et al. Pregnancies established from handmade cloned blastocysts reconstructed using skin fibroblasts in buffalo (*Bubalus bubalis*). Theriogenology 2009;71(8):1215−19 [PMID: 19168209]. <https://doi.org/10.1016/j.theriogenology.2008.10.004>.

[140] Saha A, Panda SK, Chauhan MS, Manik RS, Palta P, Singla SK. Birth of cloned calves from vitrified-warmed zona-free buffalo (*Bubalus bubalis*) embryos produced by hand-made cloning. Reprod Fertil Dev 2013;25(6):860−5. Available from: https://doi.org/10.1071/RD12061.

[141] Duah EK, Mohapatra SK, Sood TJ, Sandhu A, Singla SK, Chauhan MS, et al. Production of hand-made cloned buffalo (*Bubalus bubalis*) embryos from non-viable somatic cells. In Vitro Cell Dev Biol Anim 2016;52(10):983−8.

[142] Ruan Z, Zhao X, Qin X, Luo C, Liu X, Deng Y, et al. DNA methylation and expression of imprinted genes are associated with the viability of different sexual cloned buffaloes. Reprod Domest Anim 2018;53(1):203−12. Available from: https://doi.org/10.1111/rda.13093.

[143] Mahapatra PS, Singh R, Kumar K, Sahoo NR, Agarwal P, Mili B, et al. Valproic acid assisted reprogramming of fibroblasts for generation of pluripotent stem cells inbuffalo (*Bubalus bubalis*). Int J Dev Biol 2017;61(1−2):81−8. Available from: https://doi.org/10.1387/ijdb.160006sb.

[144] Wadhwa N, Kunj N, Tiwari S, Saraiya M, Majumdar SS. Optimization of embryo culture conditions for increasing efficiency of cloning in buffalo (*Bubalus bubalis*) and generation of transgenic embryos via cloning. Cloning Stem Cells 2009;11(3):387−95 [PMID: 19594388]. <https://doi.org/10.1089/clo.2009.0003>.

[145] Huang B, Li T, Wang XL, et al. Generation and characterization of embryonic stem-like cell lines derived from in vitro fertilization buffalo (*Bubalus bubalis*) embryos. Reprod Domest Anim 2010;45(1):122−8 [PMID: 19144015]. <https://doi.org/10.1111/j.1439-0531.2008.01268.x>.

[146] Kumar D, Sharma P, Vijayalakshmy K, Selokar NL, Kumar P, Rajendran R, et al. Generation of venus fluorochrome expressing transgenic handmade cloned buffalo embryos using sleeping beauty transposon. Tissue Cell 2018;51:49−55. Available from: https://doi.org/10.1016/j.tice.2018.02.005.

Chapter 7

Reproductive technologies for the conservation of wildlife and endangered species

Gabriela F. Mastromonaco[1] and Nucharin Songsasen[2]

[1]*Reproductive Sciences, Toronto Zoo, Toronto, ON, Canada,* [2]*Center for Species Survival, Smithsonian Conservation Biology Institute, Front Royal, VA, United States*

7.1 Introduction

Reproductive technologies have been implemented for more than 100 years in food and laboratory animals. The ability to propagate desired genetic traits has been instrumental for increasing production yield in farm animals [1] and maintaining specialized biomedical research models in laboratory species [2]. In both these cases, the outcomes primarily benefit humans, including providing nutritional resources and medical therapies to support the health and well-being of an increasing global population. However, this unparalleled growth in human population, and the concomitant demand for natural and man-made resources, has resulted in a rapid and ongoing decline in global biodiversity [3]. As more vertebrate species face extinction, conservation program managers are recognizing the importance of incorporating assisted reproductive technologies (ARTs) to enhance reproductive performance and ensure long-term preservation of valuable genetic material. However, routine implementation of these technologies has been difficult, and the number of live offspring born from ARTs does not reflect the considerable efforts invested to date. This chapter focuses on the successes and challenges of developing ARTs for species with significant diversity in reproductive anatomy, mating strategy, and natural life history.

7.2 Assisted reproductive technologies for species conservation

7.2.1 Advantages of assisted reproductive technologies

The highest priority of conservation breeding programs, that is, ex situ propagation of threatened and endangered animals, is the long-term preservation of valuable and rare alleles. The current model works to maintain 90% genetic diversity over 100 years in wildlife populations maintained in zoos and breeding centers [4]. There are significant challenges to this plan, which ideally requires equal representation of all founders in the captive population, and maintenance of a minimum number of animals to avoid inbreeding and loss of distinct alleles [5]. However, preserving heterozygosity through natural breeding is not always successful, often due to sexual incompatibility within the chosen breeding pair or difficulty in transporting animals between geographically isolated locations [6,7]. In recent years, concerns over demographic stability and inbreeding in these small, widely dispersed populations have led to some consideration in shifting from regional to global population management plans to form larger, interconnected populations [8]. For these reasons, incorporating reproductive technologies originally established in livestock and humans into conservation breeding programs [9,10] is necessary to ensure the contribution of founders and subsequent generations whenever or wherever their genetics are required. Unlike its primary use in farm animals, which follows a herd medicine approach (value of overall output efficiency), the genetic importance of a specific animal from a threatened species warrants an individualized medicine approach (value of individual output efficiency). To this end, an animal is assessed for its genetic background and relatedness, age, gender, and health to determine a treatment plan, and evaluate the effort and cost that will be committed to the individual (Fig. 7.1).

Reproductive Technologies in Animals. DOI: https://doi.org/10.1016/B978-0-12-817107-3.00007-2

In most circumstances, the need to obtain offspring from unrepresented individuals results in attempts to develop treatment options for aged, subfertile animals. Thus wildlife ART has a tendency to more closely resemble the application in humans rather than livestock.

In the 1980s, it became widely recognized that ARTs could become valuable tools to support captive breeding programs in meeting their targets for maintaining genetically and demographically stable populations with limited numbers of animals [11]. Genome resource banks containing primarily cryopreserved semen samples were initiated in a number of zoological and academic institutions around the world. This coincided with a sudden increase in research and application of the complete spectrum of reproductive techniques available at the time: semen evaluation and cryopreservation, ovarian stimulation, artificial insemination (AI), in vitro fertilization (IVF), and intra- and interspecific embryo transfer (ET) (reviewed by Herrick [9] and Andrabi and Maxwell [12]). As ARTs continued to advance, such as intracytoplasmic sperm injection (ICSI) and somatic cell nuclear transfer (SCNT), so did the attempts in wildlife species. Despite the extensive commitment of time and funds, application of ARTs for species conservation remains challenging, and pregnancy and birth rates are significantly lower than standard applications in livestock. The primary limiting steps are: (1) lack of knowledge of the basic reproductive biology of each given species, including ovarian function and semen characteristics, the building blocks of all ARTs, and (2) extensive diversity in reproductive mechanisms among species even within the same taxon [9,13]. Furthermore, limited holding space within zoological institutions dictates that most wildlife species are kept in small numbers with few individuals available for research, thereby restricting access to animals and samples for experimental studies. As a result, although live offspring have been produced by ARTs in multiple taxa, assisted breeding (AI, IVF, and ET) has been incorporated into the ex situ management of only a handful of species. To date, success continues to be associated with reproductive knowledge of the species, level of invasiveness of the technique, and technical difficulty and cost (Fig. 7.2).

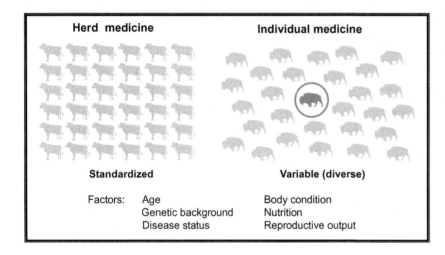

FIGURE 7.1 Standard approach for the implementation of ARTs for conservation purposes: individualized medicine.

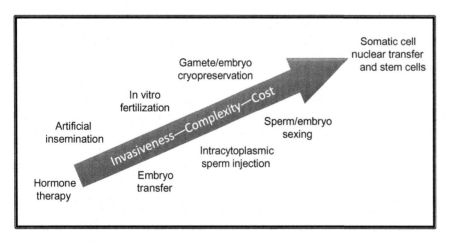

FIGURE 7.2 Full range of ARTs applied in wildlife species as related to invasiveness, complexity, and cost.

7.2.2 Diversity in natural reproductive biology

Farm and laboratory animals have undergone many years of domestication and selective breeding that have augmented their reproductive efficiency by advancing the age of sexual maturity, increasing litter size or eliminating reproductive seasonality [14]. In contrast, wildlife species each have their own natural life histories that have allowed them to evolve to be highly adapted to their specific environment. This has led to variation in reproductive strategies (e.g., female mate choice, sperm competition, and monogamy vs polygyny/polyandry), reproductive seasonality (e.g., nonseasonal polyestrus, seasonal polyestrus, and monoestrus), and ovarian function (e.g., spontaneous vs induced ovulation) [15,16]. Also, many species exhibit some key differences in reproductive anatomy that have resulted from coevolution of penile and cervical structures (e.g., porcine corkscrew shape) [17], and also sperm morphology (Fig. 7.3).

This extensive diversity in natural reproductive biology adds a layer of complexity for developing ARTs in novel species and impacts the ability to easily extrapolate techniques and protocols from one species to another. For example, most wild felid species are induced ovulators; however, some felid species, including the lion (*Panthera leo*), clouded leopard (*Neofelis nebulosa*), and margay cat (*Leopardus wiedii*), spontaneously ovulate [18]. Additionally, felid response to ovarian stimulation protocols does not typically follow a pattern based on family or size: for example, follicle growth and ovulation is induced with 500 IU equine chorionic gonadotropin (eCG) + 225 IU human chorionic gonadotropin (hCG) in an ocelot (small cat), 200 IU eCG + 100 IU hCG in a cheetah (mid-sized cat), and 600 IU eCG + 300 IU hCG in a snow leopard (also mid-sized cat) [19]. These inherent differences demand species-specific protocols to be developed for each phase of the reproductive technology of interest, from hormone therapy onwards. Such challenges put constraints on the progress of ARTs in wildlife species and have significantly impacted their widespread application.

7.3 Assisted reproductive technologies for ex situ population management

There have been numerous reviews on the use of ARTs in wildlife species that have covered detailed aspects of the techniques involved, including ovarian hormone monitoring and manipulation and gamete/embryo cryopreservation (such as Refs. [6,12]). For this review, we are focusing on the application of the technologies rather than the technical components necessary for carrying them out successfully. In this section, we highlight specific problems in captive animal management that have been overcome with the use of ARTs.

7.3.1 Successful incorporation of assisted reproductive technologies—artificial insemination

AI is the most applicable ART in wildlife species [9,20]. Due to the technical simplicity and minimal level of invasiveness (for most species), it is to be expected that AI (with or without hormone induction) has been the most successful technique for achieving consistent pregnancy and live birth rates. Owing to the advancements in livestock, AI was initially applied to wild ungulates; the first offspring produced from AI was reported in 1973 in reindeer (*Rangifer tarandus*) [21]. Since then, offspring have been produced in six additional cervid species and across other wildlife taxa (summarized in Table 7.1).

Despite successful outcomes in more than 50 species, routine application of this assisted breeding technique is limited to wildlife species that are managed for commercial purposes, including red deer, fallow deer, white-tailed deer, and reindeer. Whereas in conservation species, ARTs have been incorporated into the breeding management plans of only a handful of species as discussed below.

Dedication of significant time and resources has led to some key successes in a variety of species, each presenting unique challenges for their long-term sustainability under human care.

Giant panda (*Ailuropoda melanoleuca*): AI has been used to overcome behavioral incompatibility within giant panda breeding pairs in facilities around the world, resulting in a large number of cubs being born. Despite its expected implementation when breeding introductions fail, successful outcomes continue to be difficult to obtain. A recent analysis of 304 insemination attempts over 21 years showed that birth rates following insemination of natural cycles are approximately one-third of the rates obtained with natural mating (18.5% vs 60.7%) [35]. The authors suggest that the effects of anesthesia and exact timing of ovulation may be impacting AI success in this species, an issue that may be overcome with the use of ovulation induction.

Black-footed ferret (*Mustela nigripes*): AI has played a critical role in the recovery of this species, which experienced a severe genetic bottleneck when the remaining wild individuals were brought into captivity. Between 1996 and

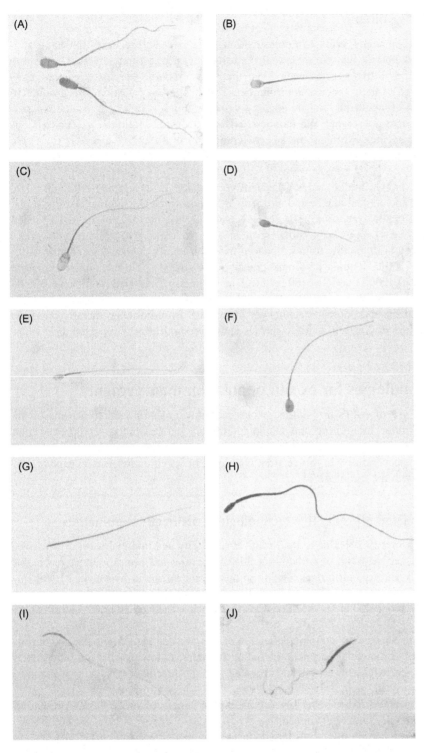

FIGURE 7.3 Sperm morphology of taxonomically diverse vertebrate species. (A) wood bison (*Bison bison athabascae*), (B) Bactrian camel (*Camelus bactrianus*), (C) mouflon (*Ovis aries musimon*), (D) hippopotamus (*Hippopotamus amphibius*), (E) spotted hyena (*Crocuta crocuta*), (F) polar bear (*Ursus maritimus*), (G) brush-tailed bettong (*Bettongia penicillata*), (H) feathertail glider (*Acrobates pygmaeus*), (I) Komodo dragon (*Varanus komodoensis*), and (J) Puerto Rican crested toad (*Peltophryne lemur*). All images were taken at 100× magnification under oil except for (G) (40× magnification).

2008, close to 140 ferret kits were produced using laparoscopic insemination of fresh or cryopreserved sperm following ovulation induction [81]. Unlike the giant panda, deposition of the sperm in utero resulted in pregnancy rates of 67%–75% [81]. Most importantly, the birth of eight ferret kits from inseminations with sperm frozen for 20 years resulted in a genetic revival of the captive population with an increase in gene diversity by 0.2%, and reduction in inbreeding by 5.8% when compared to natural mating alone [82]. This is an important example of the ability to exceed generation intervals and input genetics across generations. The availability of a model species [domestic ferret (*Mustela putorius furo*) and Siberian

TABLE 7.1 List of wildlife species producing live offspring following artificial insemination.

Order	Common name	Scientific name	Sperm type	Results (live offspring/AI attempt)	References
Columbiformes					
	Blue rock pigeon	Columba livia	Frozen	1/11[a]	[22]
Galliformes					
	Mikado pheasant	Syrmaticus mikado	Fresh	2/14[a]	[23]
	Swinhoe's pheasant	Lophura swinhoii	Fresh	4/12[a]	[23]
	Temminck's tragopan	Tragopan temminckii	Fresh	1/2[a]	[23]
	Blyth's tragopan	Tragopan blythii	Fresh	3/4[a]	[23]
	Cabot's tragopan	Tragopan caboti	Fresh	11/34[a]	[23]
			Frozen	1/2[a]	[23]
	Himalayan monal	Lophophorus impeyanus	Fresh	2/10[a]	[23]
	Lady Amherst's pheasant	Chrysolophus amherstiae	Frozen	5/6[a]	[23]
	Berlioz's silver pheasant	Lophura berliozi	Frozen	3/22[a]	[23]
Gruiformes					
	Sandhill crane	Grus canadensis	Frozen	43/83[a]	[24]
	Houbara bustard	Chlamydotis undulata undulata	Frozen	3/6[a]	[25]
Sphenisciformes					
	Magellanic penguin	Spheniscus magellanicus	Frozen	4/15[a]	[26]
Artiodactyla					
	Fallow deer	Dama dama	Fresh	23/55	[27]
			Fresh	2/3	[28]
			Frozen	38/83	[27]
			Frozen	1/3	[28]
	Axis deer	Axis Axis	Fresh	4/9	(reviewed in Ref. [29])
			Fresh	2/10	[30]
			Frozen	8/41	[29]
	Eld's deer	Cervus eldi thamin	Frozen	7/20	[31]
	Red deer[d]	Cervus elaphus	Frozen	267/347	(reviewed in Ref. [29])
	White-tailed deer	Odocoileus virginianus	Chilled	4/2 (twin)	[32]
			Frozen	57/53 (twins)	[33]
			Frozen	11/9 (twin)	[34]
	Reindeer	Rangifer tarandus	Fresh	2/10	[35]

(Continued)

TABLE 7.1 (Continued)

Order	Common name	Scientific name	Sperm type	Results (live offspring/AI attempt)	References
	Sambar deer	*Cervus unicolor*	Frozen	5/10	[36]
	Eland	*Taurotragus oryx*	Frozen	1/4	(reviewed in Ref. [29])
	Banteng	*Bos javanicus*	Frozen	3/21	(reviewed in Ref. [29])
	Gaur	*Bos gaurus*	Frozen	1/6	[37]
	Addax	*Addax nasomaculatus*	Frozen	1/2	[38]
	Scimitar-horned oryx	*Oryx dammah*	Frozen	14/64	(reviewed in Ref. [29])
	Blackbuck	*Antilope cervicapra*	Fresh	6/8	[39]
			Frozen	1/3	[39]
	Speke's gazelle	*Gazella spekei*	Fresh	1/2	(reviewed in Ref. [29])
	Mhorr gazelle	*Gazella dama mhorr*	Frozen	1/ND	[40]
	Spanish ibex	*Capra pyrenaica*	Frozen[b]	1/6	[41]
	Barbary sheep	*Ammotragus lervia*	Fresh	2/4	[42]
Cetacea					
	Killer whale	*Orcinus orca*	Chilled	2/3	[43]
			Frozen	1/5	[43]
	Bottlenose dolphin	*Tursiops truncatus*	Frozen	5/8	[44]
	White-sided dolphin	*Lagenorhynchus obliquidens*	Frozen[c]	5/10	[45]
Proboscidea					
	African elephant	*Loxodonta africana*	Frozen	1/1	[46]
	Asian elephant	*Elephas maximus*	Chilled	1/2	[47]
			Fresh	1 calf from 6 attempts in 1 female	[47]
			Frozen	1 calf from 55 attempts in 6 females	[48]
Perissodactyla					
	Persian onager	*Equus hemionus onager*	Chilled	1/1	[49]
			Frozen	1/2	[49]
	Przewalski's horse	*Equus Ferus przewalskii*	Fresh	1/1	(B. Pukazhenthi, pers. comm.)
	Southern white rhinoceros	*Ceratotherium simum simum*	Fresh	1 calf from 2 AI attempts in 1 female	[50]
			Frozen	1 calf from 2 AI attempts in 1 female	[51]
	Greater one-horned rhinoceros	*Rhinoceros unicornis*	Frozen	3 calves from 34 AI attempts in 3 females	[52]

(Continued)

TABLE 7.1 (Continued)

Order	Common name	Scientific name	Sperm type	Results (live offspring/AI attempt)	References
Carnivora					
	Gray wolf	*Canis lupus*	Fresh	1/3	[53]
			Frozen	6 pups from 4 AI attempts in 1 female	[54]
	Mexican gray wolf	*Canis lupus baileyi*	Fresh	3 litters/3	[55]
	Red wolf	*Canis rufus*	Fresh	2 litters/ND	[56]
	Blue fox/red fox	*Alopex lagopus/ Vulpes vulpes* L.	Fresh[e]	80% (over million females)	[57]
			Frozen[e]	30%–50%	[57]
	Clouded leopard	*Neofelis nebulosa*	Chilled	2/1 (twin)	[58]
	Amur leopard	*Pionailurus bengalensis eutilura*	Fresh	1/2	[59]
	Golden cat	*Catopuma temminckii*	Fresh	2/1 (twin)	[60]
	Cheetah	*Acinonyx jubatus*	Fresh	7/13	[61]
	Amur tiger	*Panthera tigris altaica*	Fresh	3/1 (triplet)	[62]
			Fresh	1/1	[63]
			Fresh	1/11	[64]
	Jaguar	*Panthera onca*	Fresh	1/ND	[65]
	Fishing cat	*Prionailurus viverrinus*	Fresh	1/1	[66]
	Ocelot	*Leopardus pardalis*	Fresh	3/ND	[67]
	Pallas's cat	*Otocolobus manul*	Fresh	3/1 (triplet)	[68]
	Lion	*Panthera leo*	Fresh	2/1 (twin)	[69]
	Snow leopard	*Uncia uncia*	Fresh	1/15	[70]
	Giant panda	*Ailuropoda melanoleuca*	Fresh	1/2	[71]
			Frozen	5/14	[72]
	European ferret	*Mustela putorius*	Fresh	17/24	[73]
			Frozen	7/10	[74]
	Black-footed ferret	*Mustela nigripes*	Frozen	8/18	[75]
Pangidae					
	Chimpanzee	*Pan troglodytes*	Fresh	1/1	[76]
			Fresh	6/29	[77]
	Gorilla	*Gorilla gorilla*	Fresh	1/1	[78]
	Orangutan	*Pongo* sp.	Fresh	1/1	[79]
Marsupialia					
	Koala	*Phascolarctos cinereus*	ND	32/ND	[80]

ND, No Data.
[a]*Offspring per total number of eggs laid.*
[b]*Ongoing pregnancy at time of report.*
[c]*One ongoing pregnancy at time of report.*
[d]*Pregnancies have been reported (29%–70%) in 5083 additional attempts, but no data on live births.*
[e]*Pregnancy rate.*

polecat (*Mustela eversmanii*)] that permitted studies aimed at understanding the fundamental reproductive biology and factors affecting sperm cryosurvival contributed to the successful application of AI in the black-foot ferret breeding program [75,82].

Asian (*Elephas maximus*) and African elephant (*Loxodonta africana*): The need to support genetic management while reducing breeding-related transfers of large mega-vertebrates between institutions has led to significant progress in the insemination of elephants. With their unique reproductive hormone trait that includes a double luteinizing hormone surge [83], insemination of natural cycles with fresh and chilled sperm has resulted in the birth of more than 40 calves worldwide [46]. In 2012, an African elephant calf was born following insemination of a captive female with frozen-thawed semen from a wild bull [46]. This success is indicative of future possibilities, not only for global population management, but also for the transfer of genetic material between wild and captive populations.

Koala (*Phascolarctos cinereus*): Although the diversity among placental mammals is significant, marsupials present another challenge altogether, particularly due to their complex reproductive anatomy and physiology. However, extensive research in the koala resulted in the birth of 31 pouch young following inseminations with both fresh and chilled sperm with conception rates being similar between AI and natural breeding attempts (44% vs 43%−57%, respectively) [84]. This progress led to the development of a frozen sperm biobank in the hopes of mitigating the effects of chlamydia infection among free-ranging koalas [85]. As with food animals, control and eradication of disease are an important benefit of applying ARTs.

Bottlenose dolphin (*Tursiops truncatus*): One of the concerns of captive breeding programs is the production of male offspring, particularly male-dominant species exhibiting aggressive behaviors, which can be difficult to manage under human care, such as gorillas (*Gorilla gorilla*). Sperm sex selection prior to insemination has shown the most advancement in the bottlenose dolphin with more than 10 female calves born from chilled and frozen-thawed semen [86]. Interestingly, post-AI pregnancy rates were similar between nonsexed frozen-thawed sperm (control) and chilled sexed sperm (75% vs 71%), with a decrease being observed with frozen-thawed sexed sperm (50%) [86]. The ability to adjust the demographics of a small population can be advantageous for many species since space constraints make it difficult to continue housing groups of nonbreeding animals.

Compared to the minimal progress in mammalian species, there have been some significant successes in nonmammalian species. AI has been well documented in wild birds since the 1970s and has been instrumental in the captive breeding of species such as cranes (sandhill crane, *Grus canadensis*; whooping crane, *Grus americana*) and raptors (golden eagle, *Aquila chrysaetos*; peregrine falcon, *Falco peregrinus*; American kestrel, *Falco sparverius*) [87−89]. In most of these cases, AI was implemented to enhance the fertility and genetic management of the small captive populations. The first application of AI was in sandhill cranes in 1969. The practice was later adapted for other cranes, including the whooping crane [87], another species on the brink of extinction with only 16 birds remaining in the wild in the 1960s. Since then, AI has been used extensively in all whooping crane breeding centers for reproductive management, which has resulted in approximately 450 birds now living in captivity and the wild. Unlike the black-footed ferret, however, sperm cryopreservation has not been successful in cranes. Aside from the birth of one sandhill crane chick from frozen-thawed sperm, overall postthawed motility of crane sperm is only 15% [90], thereby limiting the ability to fully utilize AI in crane genetic management.

7.3.2 Challenges with embryo-based assisted reproductive technologies—in vitro fertilization

Although AI demonstrates the significant benefits of incorporating ARTs into genetic management programs for wildlife species, it focuses primarily on male genetics since it only permits the widespread distribution and long-term preservation of sperm. The importance of representing female genetics in current and future populations highlights the need for developing embryo-based techniques, such as IVF along with ET and cryopreservation. However, these techniques are typically more invasive (oocyte or embryo retrieval), complex (handling and culture media requirements), and costly (equipment and expertise) (Fig. 7.2). Furthermore, as previously mentioned, access to females for research is limited, which influences the development of reproducible and consistent protocols. As a result, the number of live offspring produced following IVF and ET has been minimal compared to AI (summarized in Table 7.2).

As with AI, advancements in livestock led to the application of IVF-ET in wild bovids with one of the first living offspring being reported in the gaur (*Bos gaurus*) in 1994 [92]. Once again, commercial interest led to extensive efforts in red deer (*Cervus elaphus*) that resulted in the births of multiple live offspring [96−99]. Similarly, the growing needs for ARTs in humans led to a rapid expansion of the techniques in nonhuman primates [e.g., rhesus macaque (*Macaca mulatta*), common marmoset (*Callithrix jacchus*) as models for human ART (reviewed by Bavister and Boatman [109])]; however, as they are considered laboratory animals, the outcomes will not be discussed here. Since the announcement of the

TABLE 7.2 List of wildlife species producing live offspring following in vitro fertilization and embryo transfer.

Order	Common name	Scientific name	Embryo type	Results (offspring/ET attempt)	References
Artiodactyla					
	Cape buffalo	*Syncerus caffer*	Fresh	1/ND	[91]
	Gaur	*Bos gaurus*	Fresh	1/2	[92]
	Wood bison	*Bison bison athabascae*	Fresh	3/10	[93]
			Frozen	1/4	(G. Mastromonaco, pers. comm.)
	Eld's deer	*Rucervus eldii*	Fresh	1/11	[94]
			Fresh	1/8	[95]
	Red deer	*C. elaphus*	Fresh	4/7	[96]
			Fresh	5/29	[97]
			Frozen	7/18	[98]
			Frozen	3/5	[99]
	Sika deer	*Cervus nippon*	Frozen	1/ND	[100]
	European mouflon	*Ovis orientalis musimon*	Frozen	4/10 (1 twin)	[101]
	Red sheep	*Ovis orientalis gmelini*	Fresh	3/8	[102]
Carnivora					
	Bengal tiger	*Panthera tigris*	Fresh	3/6 (triplet)	[103]
	Caracal	*Caracal caracal*	Fresh/ frozen	4/15 (1 twin)	[104]
	Fishing cat	*Prionailurus viverrinus*	Fresh	1/12	[104]
	Indian desert cat	*Felis silvestris ornata*	Fresh	2/9 (twin)	[105]
	Ocelot	*Leopardus pardalis*	Frozen	1/1	[106]
	Sand cat	*Felis margarita*	Fresh/ frozen	2/9 (twin)	[107]
	Black-footed cat	*Felis nigripes*	Fresh/ frozen	2/4 (twin)	[108]

births of three tiger (*Panthera tigris*) cubs by IVF-ET in 1990 [103], several research groups have dedicated themselves to improving this technique in small wild felids with some success. Live births have been documented in species including the fishing cat (*Prionailurus viverrinus*) [104], sand cat (*Felis margarita*) [107], and black-footed cat (*Felis nigripes*) [108]. Most notably, in vitro embryo production in the ocelot (*Leopardus pardalis*) is being developed as a strategy for long-term genetic management of captive ocelot populations [106]. With the knowledge gained from related domesticated species, there have been numerous attempts to produce mammalian embryos in vitro, either by IVF or ICSI, in a wide variety of wildlife species primarily using culture systems developed for a related model species with varying results. Unfortunately, in most cases, ETs were not attempted, and if so, live births were rare (Table 7.2). To date, IVF-ET is considered experimental and has not been incorporated into breeding management plans based on the substantial technical and financial challenges, but primarily due to inefficiency.

Although they undergo external fertilization, there have been some noteworthy successes in amphibian species (reviewed by Kouba and Vance [110]). Hormonal induction of both spermiation and spawning in breeding males and females followed by fertilization and embryo development under controlled conditions in vitro has resulted in the production of thousands of tadpoles that have successfully metamorphosed into adults, including the endangered Wyoming toads (*Bufo baxteri*) [111] and Mississippi gopher frogs (*Rana sevosa*) [112], to name a few. In both these cases, tadpoles produced by IVF have been released into the wild to help re-populate threatened areas, an important example of the value of implementing this assisted breeding technique.

7.4 Emerging technologies

As threats to natural spaces remain and global biodiversity continues to decline, many species are becoming extinct before any attempts to preserve their genetic material can be successful. The lack of widespread application of ARTs across all taxa after more than 30 years of efforts highlights the need for alternative approaches for genome resource banking that are feasible and viable, even if offspring production lies many years into the future. In this section, we focus on innovative technologies that utilize cryopreserved somatic and stem cells for the production of gametes and embryos.

7.4.1 Cell transplantation technologies

Precursor cells, such as germ cells or stem cells, provide novel sources of genetic material for breeding programs and thus, unique strategies for the long-term preservation of valuable animals. With their ability to establish themselves and subsequently regenerate and differentiate into functional end products (e.g., sperm), they offer a renewable source of genetic material compared to the finite possibilities of cryopreserved sperm and oocytes. Therefore although there are significant challenges in utilizing precursor cells, including isolating and transplanting them effectively, these technologies warrant further investigation.

1. Primordial germ cell (PGC) transplantation:

    PGCs, the precursors of functional gametes (i.e., sperm or eggs), can potentially alter our ability to indefinitely preserve genetic material from threatened and endangered wildlife species. For commercial purposes, PGC transplantation has been broadly investigated in poultry, especially chickens [113]. In chicken, PGCs circulate in the extraembryonic blood vessels (Day 2.2 of embryo development) until they reach the germinal ridges on Day 6 [114]. During the time they are circulating in blood vessels to gonadal differentiation, PGCs can be isolated and placed in culture to establish embryonic germ cell lines. Throughout the past decade, researchers have explored the potential of interspecific germline transplantation technology, originally developed in domestic fowl, to enhance the reproductive capacity of endangered bird species [115–117]. This technology involves injecting isolated PGCs of a rare species into an embryo of a more common counterpart to create germ line chimeras that can ultimately produce a functional donor gamete, and transmit the complete genetic information to the next generation [114]. To date, interspecific germline transmission (to produce germline chimeras) has been demonstrated in several species, including chicken × duck (donor cells × recipient egg) [118], pheasant (*Phasianus colchicus*) × chicken [116], chicken × guinea fowl (*Numida meleagris*) [115], and Houbara bustard (*Chlamydotis undulata*) × chicken [117]; however, the efficiency of offspring production is still very low. In the case of the Houbara bustard, PGCs were isolated from the gonads of Day 8 male embryos and injected into chicken embryos. Of 35 sexually mature chimeric roosters, eight were confirmed to be germline chimeras. AI with sperm collected from germline chimera males resulted in one live Houbara offspring [117]. In addition to the use of freshly collected PGCs, offspring have been produced following transplantation of cryopreserved PGCs in chicken [119]. To date, the application of cryopreserved PGCs has not been demonstrated in endangered birds. Although this approach offers interesting potential in preserving rare genetics, isolation of PGCs from early embryos presents a major drawback as these resources are highly valuable in themselves, and difficult to obtain.

2. Spermatogonial stem cell (SSC) transplantation:

    SSCs, which originate from gonocytes within the postnatal testis, are the foundation of spermatogenesis. SSCs have self-renewal and differentiation ability and give rise to the haploid spermatozoa. Since the first report in the mouse by Brinster and Avarbock [120], SSC transplantation has been applied to several mammalian species, including the human [121]. To date, it has been primarily used to understand the mechanism of genetic defect in spermatogenesis and for fertility preservation [121]. This technology involves the injection of germ cells (SSCs) from donor testes into the seminiferous tubules of recipients that either have undergone germ cell depletion or are naturally sterile. In a suitable environment, the SSCs then colonize in the recipient's seminiferous tubules and undergo proliferation and differentiation into spermatozoa [121]. Fertile offspring have been produced in the mouse following the transfer of embryos produced by microinjection of sperm derived from SSCs that had been frozen for 14 years [122]. To date, SSC transplantation has been explored in mammalian [123], avian [124] and aquatic species [125−127] with some promising results. For example, elongated spermatogonia and spermatozoa have been found in the testis and epididymis, respectively, at 13 weeks posttransplantation of ocelot (*L. pardalis*) germ cells into the testes of domestic cats [123]. In other attempts, SSCs have been characterized in Indian blackbuck (*Antilope cervicapra* L.) and transplanted into recipient mice [128]. Four weeks after transplantation, blackbuck spermatogonia were detected in the basement membrane of the seminiferous tubules of all recipient mice testes, although complete spermatogenesis was not observed [128]. Because of their self-renewal property, SSCs are crucial sources of valuable genetics, and when combined with cryopreservation and (syngeneic or xeno-) transplantation, they represent a valuable tool for species conservation [122]. Notably, compared to PGCs that are isolated from early embryos, SSCs can be obtained from living or recently deceased males.

7.4.2 Somatic cell technologies

Advancements in cell-based technologies during the past two decades, specifically the ability to reprogram differentiated cell nuclei into embryonic or germinal cell lineages, also present the potential for developing alternative methods for preserving and disseminating valuable genetics of rare and endangered wildlife species [129,130]. The use of somatic cells precludes the need for viable gametes (sperm or eggs), thereby allowing the contribution of individuals that: (1) experience reproductive dysfunction, (2) die unexpectedly during the nonbreeding season or prior to reaching sexual maturity, (3) lack the opportunity to breed due to geographic isolation, (4) have been castrated, or (5) whose maturation status is unknown [129].

To date, collection, establishment, and cryopreservation of fibroblast cell lines have been the primary focus since fibroblasts can be recovered fairly easily from living animals by biopsy dart, or opportunistically during routine health examinations or medical procedures, and from deceased animals up to several days postmortem. Furthermore, there are no or minimal requirements for species-specific protocols for the culture and cryopreservation of fibroblast cells. For these reasons, primary fibroblast cultures have been established and cryopreserved for several hundred mammalian and close to 100 aquatic wildlife species (reviewed by New York Post [65]). In addition to fibroblasts, studies have been conducted to explore the usefulness of somatic stem cells derived from adipose tissue biopsies. However, unlike fibroblasts, adult stem cell culture and cryopreservation are more complex and require fundamental knowledge in stem cell biology. Nevertheless, somatic stem cells have been recovered, cultured, and induced to differentiate, such as the recovered adipose-derived stem cells from brown bears (*Ursus arctos*), which undergo chondrogenic and osteogenic differentiation following culture for 3 weeks in osteogenic induction medium [131].

1. Induced pluripotent stem cells (iPSCs)

   Pluripotent stem cells are characterized by their robust self-renewal ability and potential to differentiate into multiple cell types [132], making them an attractive resource for long-term storage. Embryonic stem (ES) cells, which are derived from an embryo, can grow indefinitely while maintaining pluripotency and the ability to differentiate into cells of all three germ layers. However, the use of ES cells in the propagation of wildlife species is limited to the availability of gametes and embryos, which are not feasible in most cases [130,133]. Breakthroughs in the ability to induce pluripotency in somatic cells in the mouse [134], and then human [135], have stimulated significant interest in exploring the value of iPSC technology for wildlife conservation. As mentioned previously, this is due to the practicality of collecting and using somatic cells compared to embryo tissue. The first report of fertile animals from iPSC technology was mice produced from a reprogrammed embryonic fibroblast [136]. Although this initial success involved the use of an embryonic cell line, iPSCs have since been derived from adult skin cells from diverse wildlife taxa, including drill (*Mandrillus leucophaeus*) [137], snow leopard (*Panthera uncia*) [138], tiger (*P. tigris*) [139], serval (*Leptailurus serval*) [139], jaguar (*Panthera onca*) [139], orangutan (*Pongo abelii*) [140], and white rhinoceros (*Ceratotherium simum cottoni*) [137]. In all these cases, iPSC lines were created by transduction with retroviruses containing the pluripotent genes, *OCT4, SOX2, KLF4*, and *c-MYC*. However, in studies on wild felids, *Nanog* was found to be an essential factor in the induction of pluripotency, and in the absence of this factor, iPSC lines could not be maintained beyond passage 7 [139]. Notably, in addition to iPSC lines, ES cell lines have recently been established from Southern white rhinoceros blastocysts produced by IVF [141].

Because they are continuously self-renewing and pluripotent, iPSC technology holds great promise for endangered species conservation, including therapeutic applications for animals that suffer degenerative diseases, or rescue and preservation of valuable genomes through the production of iPSC-derived germ cells, which could then be used in conjunction with other ARTs [142]. Finally, iPSCs can be used as somatic donors to enhance embryo production following nuclear transfer [130]. Nevertheless, for the routine application of iPSCs and other stem cell technologies in endangered species conservation, a deeper understanding of the basic reproductive biology and mechanisms regulating pluripotency for each given wildlife species is still required.

2. Reproductive cloning

Reprogramming somatic cells using the ooplasmic machinery, as occurs with reproductive cloning or nuclear transfer (i.e., transplantation of a donor nucleus into an enucleated recipient oocyte), was first reported in frogs in the 1950s when an embryonic cell was injected into an unfertilized egg; the resulting embryos were able to develop to metamorphosed frogs [143]. However, it was not until 1996 when the birth of "Dolly," derived from an adult somatic cell, stimulated interest in the application of nuclear transfer for the propagation of genetically valuable livestock, laboratory animals, and even endangered wildlife [129]. During the past two decades, reproductive cloning has been attempted among diverse wildlife taxa, including bovids, felids, canids, ursids, cetaceans, and primates (summarized in Ref. [65]). Despite the number of cases where embryos developed to the blastocyst stage following nuclear transfer, there have been only minimal attempts to transfer the embryos and produce offspring. In fact, live births that went on to survive occurred in only a handful of species, including the European mouflon (*Ovis orientalis musimon*) [144], African wild cat (*Felis silvestris libica*) [145], and gray wolf (*Canis lupus*) [146]. Despite the low efficiency in producing live offspring, the technique not only provides an alternative method for offspring production, but also valuable knowledge on embryo development in species where embryos are difficult or impossible to obtain.

7.4.3 Gonadal tissue cryopreservation

Along with the production of functional gametes, gonadal tissues are also the source of gamete precursor cells in various stages of development. These valuable cells are maintained within the gonads throughout an animal's life despite the stage of sexual maturity or time of year. The ability to store the entire hierarchy of cells in the spermatogenic and oogenic pathway opens up possibilities for generating functional sperm and eggs when their production may not have been naturally possible.

1. Ovarian tissue

Each ovary contains large numbers of immature follicles enclosing oocytes that are never ovulated and thus, never contribute to reproduction [143]. Ovarian tissue cryopreservation is highly relevant to preserving fertility potential in wildlife, particularly from neonatal and prepubertal individuals that might die unexpectedly, or adult females spayed during the nonbreeding season [147,148]. Primordial follicles enclosed within the ovarian cortex are more resistant to cryoinjuries because their oocytes have a relatively inactive metabolism, lack of a metaphase spindle, and contain low amounts of lipids [147]. To date, ovarian tissue (or whole ovary) cryopreservation has been reported in various wildlife species, including the wombat (*Vombatus ursinus*) [149–151], African elephant (*L. africana*) [152], tammar wallaby (*Macropus eugenii*) [153], lion (*P. leo*) [154], Amur leopard (*Panthera pardus orientalis*) [155], black-footed cat (*F. nigripes*) [155], Geoffroy's cat (*Leopardus geoffroyi*) [155], Northern Chinese leopard (*Panthera pardus japonensis*) [155], oncilla (*Leopardus tigrinus*) [155], rusty-spotted cat (*Prionailurus rubiginosus*) [155], serval (*L. serval*) [155], and Sumatran tiger (*Panthera tigris sumatrae*) [155]. In another group of threatened species, including the black-footed ferret (*M. nigripes*), cheetah (*Acinonyx jubatus*), clouded leopard (*N. nebulosa*), scimitar-horned oryx (*Oryx dammah*), and Eld's deer (*Rucervus eldii*), routine cryopreservation of ovarian tissue also has been reported [10]. Of these attempts, cryopreserved ovarian tissues from wombats, elephants, wallabies, and lions have been transplanted into immunodeficient mice, and in all cases, it was possible to observe morphologically normal secondary or antral follicles in the grafted tissue [10]. Although significant research is still required to obtain functional oocytes that are successful during in vitro maturation and IVF, this technique offers a much-needed tool for the long-term preservation of female genetics since, compared to spermatozoa, immature and mature oocytes continue to present numerous challenges for cryopreservation.

2. Testicular tissue

Testicular tissue contains germ cells that can give rise to spermatozoa through the spermatogenic process. The ability to preserve testis tissue allows the rescue of valuable genomes from neonatal and prepubertal individuals as mentioned above for ovarian tissue [156]. Cryopreservation of testicular tissue can be used in conjunction with SSC transplantation or grafting. To date, live offspring have been produced from microinjection of sperm recovered from grafted cryopreserved tissue in the mouse [157], rabbit [157], pig [158], and rhesus macaque (*M. mulatta*) [159]. This technology has also been explored in avian species and fishes [160] and has resulted in the production of off-spring in chicken [161] and Japanese quails (*Coturnix japonica*) [162], as well as hatched larvae in the critically endangered cyprinid honmoroko (*Gnathopogon caerulescens*) [163]. Although live offspring have not been produced from cryopreserved testicular tissue in mammalian wildlife, there is growing interest in this technology, and testicular tissue cryopreservation has been reported in a number of wildlife species, including the jungle cat (*Felis chaus*) [164], lion (*P. leo*) [164], leopard (*Panthera pardus*) [164], Rusa deer (*Cervus timorensis*) [164], Fea's muntjac (*Muntiacus feae*) [164], Sumatran serow (*Capricornis sumatraensis*) [164], Eld's deer (*R. eldii*) [165], barking deer (*Muntiacus muntjac*) [166], sambar deer (*Rusa unicolor*) [166], hog deer (*Hyelaphus porcinus*) [166], Indian mouse deer (*Moschiola indica*) [167], and gray wolf (*C. lupus*) (J. Nagashima, pers. comm.) with varying success in terms of postthawed viability. In the case of the Indian mouse deer, pachytene spermatocytes have been observed in cryopreserved testicular tissue 24 weeks after grafting [167]. Similar to ovarian tissue, much progress is needed in testicular tissue cryopreservation and grafting; however, its value for male genome preservation warrants further efforts.

7.4.4 Oocyte cryopreservation

Unlike sperm, which can be handled and cryopreserved adequately in numerous species, the oocyte has several unique features (e.g., large size and amount of intracellular lipid) that contribute to its extreme susceptibility to damage during the cryopreservation procedure [168,169]. The banking of oocytes to preserve female genetic material has lagged significantly behind other cell types. However, the development of minimum volume vitrification (MVV) methods, such as open pulled straw and cryotop, that permit cooling rates exceeding −100,000°C/minute, has significantly improved the survival and function of frozen-thawed gametes. Since then, this technique has been widely adopted to preserve oocytes in several species, including the human. However, studies of oocyte cryopreservation in wildlife species have been limited due to the availability of good quality gametes [156]. To date, oocyte cryopreservation has been reported in serval (*L. serval*) [156], Pallas's cat (*Otocolobus manul*) [156], chousingha (*Tetracerus quadricornis*) [170], Tasmanian devil (*Sarcophilus harrisii*) [171], and Mexican gray wolf (*Canis lupus baileyi*) [172], all of which have employed an MVV procedure. With the exception of the chousingha study where meiotic maturation of cryopreserved oocytes was assessed, all studies only assessed oocyte viability using either SYBR/propidium iodine or fluorescein diacetate/ethidium bromide staining, and developmental competence of cryopreserved oocytes remains to be tested. As our understanding of oocyte cryobiology grows, implementation of oocyte cryopreservation protocols will change the contribution of wildlife biomaterials banks around the world.

7.5 Conclusion

Breeding management goals for wildlife species share some similarities with those of other industries, such as food and laboratory animals. The need to propagate specific individuals to ensure the retention of their alleles in current and future populations puts significant pressure on population managers to obtain breeding success, highlighting the importance of implementing ARTs to support immediate and long-term genetic goals. After more than 30 years of effort by numerous institutions working on close to 100 species, the number of live births, and more importantly, the routine application of reproductive techniques are minimal. Notably, we are not aware of the full extent of failures in applying ARTs due to the lack of documentation of negative attempts. Among the many challenges that exist for the widespread application of ARTs in wildlife species, including issues such as funding and regulatory restrictions, time and expertise to decipher and overcome species-specific parameters of reproductive biology and natural life history are not sufficient. Currently, application of ARTs in most wildlife species can be considered experimental, with low sample sizes and lack of access to animals, resulting in the slow progress toward the development of consistent and reproducible protocols. Due to the complexity of establishing certain ARTs (e.g., developing an effective in vitro embryo production system when gametes are difficult to obtain), somatic and stem cell−based technologies remain attractive options as they offer alternative approaches to the demand for viable and functional gametes, albeit they have their own set of challenges.

The future for wildlife species continues to be challenged by ongoing anthropogenic pressures. It is time to increase collaboration among professionals from zoological, academic, governmental, and other institutions to focus on a common goal: the long-term preservation of the remaining genetic diversity of wildlife species across the globe. It is important to continue supporting the incorporation of ARTs as routine tools in animal management programs to assist with the challenges of meeting biodiversity targets for the future.

References

[1]   Hansen PJ. Current and future assisted reproductive technologies for mammalian farm animals. In: Lamb GC, DiLorenzo N, editors. Current and future reproductive technologies and world food production, Advances in experimental medicine and biology. New York: Springer Science + Business Media; 2014. p. 1−22.

[2]   Kaneko T. Reproductive technologies for the generation and maintenance of valuable animal strains. J Reprod Fertil 2018;64(3):209−15.

[3]   WWF. Living Planet Report − 2018: aiming higherIn: Grooten M, Almond REA, editors. Gland: WWF; 2018. p. 1−146.

[4]   Foose TJ, de Boer L, Seal US, Lande R. Conservation management strategies based on viable populations. In: Ballou J, Gilpin M, Foose TJ, editors. Population management for survival and recovery. New York: Columbia University Press; 1995. p. 273−94.

[5]   Ralls K, Ballou J. Captive breeding programs for populations with a small number of founders. Trends Ecol Evol. 1986;1(1):19−22.

[6]   Pukazhenthi BS, Wildt DE. Which reproductive technologies are most relevant to studying, managing and conserving wildlife? Reprod Fertil Dev 2004;16(1-2):33−46.

[7]   Wielebnowski N. Contributions of behavioral studies to captive management and breeding of rare and endangered mammals. In: Caro T, editor. Behavioral ecology and conservation biology. New York: Oxford University Press; 1998. p. 130−62.

[8]   Lees CM, Wilken J. Global programmes for sustainability. WAZA Mag 2011;12:2−5.

[9]   Herrick JR. Assisted reproductive technologies for endangered species conservation: developing sophisticated protocols with limited access to animals with unique reproductive mechanisms. Biol Reprod 2019;100(5):1158−70.

[10]   Comizzoli P, Songsasen N, Wildt DE. Protecting and extending fertility options for females of wild and endangered species. Cancer Treat Res 2010;156:87−100.

[11]   Ballou J. Strategies for maintaining genetic diversity in captive populations through reproductive technology. Zoo Biol 1984;3(4):311−23.

[12]   Andrabi SMH, Maxwell WMC. A review on reproductive biotechnologies for conservation of endangered mammalian species. Anim Reprod Sci 2007;99(3-4):223−43.

[13]   Wildt DE, Comizzoli P, Pukazhenthi B, Songsasen N. Lessons from biodiversity—the value of nontraditional species to advance reproductive sciences, conservation, and human health. Mol Reprod Dev 2010;77(5):397−409.

[14]   Setchell BP. Domestication and reproduction. Anim Reprod Sci 1992;28(1-4):195−202.

[15]   Klug H. Animal mating systems. Chichester: John Wiley & Sons, Ltd; 2011. p. 1−7. Available from: https://doi.org/10.1002/9780470015902.a0022553.

[16]   Brown JL. Comparative ovarian function and reproductive monitoring of endangered mammals. Theriogenology 2018;109:2−13.

[17]   Brennan PLR, Prum RO. Mechanisms and evidence of genital coevolution: the roles of natural selection, mate choice, and sexual conflict. Cold Spring Harb Perspect Biol 2015;7:a017749.

[18]   Brown JL, Graham LH, Wielebnowski N, Swanson WF, Wildt DE, Howard JG. Understanding basic reproductive biology of wild felids by monitoring of faecal steroids. J Reprod Fertil 2001;57(Suppl.):71−82.

[19]   Pelican KM, Wildt DE, Pukazhenthi B, Howard JG. Ovarian control for assisted reproduction in the domestic cat and wild felids. Theriogenology 2006;66(1):37−48.

[20]   Comizzoli P. Biotechnologies for wildlife fertility preservation. Anim Front 2015;5(1):73−8.

[21]   Dott HM, Utsi MNP. Artificial insemination of reindeer (Rangifer tarandus). J Zool 1973;170(4):505−8.

[22]   Sontakke SD, Umapathy G, Sivaram V, Kholkute SD, Shivaji S. Semen characteristics, cryopreservation, and successful artificial insemination in the blue rock pigeon (Columba livia). Theriogenology 2004;62(1-2):139−53.

[23]   Saint Jalme M, Lecoq R, Seigneurin F, Blesbois E, Plouzeau E. Cryopreservation of semen from endangered pheasants: the first step towards a cryobank for endangered avian species. Theriogenology 2003;59(3-4):875−88.

[24]   Gee GF, Baskt MR, Sexton SF. Cryogenic preservation of semen from the greater sandhill crane. J Wildl Manag 1985;49(2):480−4.

[25]   Hartley PS, Dawson B, Lindsay C, McCormick P, Wishart G. Cryopreservation of Houbara semen: a pilot study. Zoo Biol 1999;18(2):147−52.

[26]   O'Brien JK, Steinman KJ, Montano GA, Dubach JM, Robeck TR. Chicks produced in the Magellanic penguin (Spheniscus magellanicus) after cloacal insemination of frozen-thawed semen. Zoo Biol 2016;35(4):326−38.

[27]   Asher GW, Adam JL, James RW, Barnes D. Artificial insemination of farmed fallow deer (Dama dama): fixed-time insemination at a synchronized oestrus. Anim Sci 1988;47(3):487−92.

[28]   Mulley RC, Moore NW, English AW. Successful uterine insemination of fallow deer with fresh and frozen semen. Theriogenology 1988;29(5):1149−53.

[29]   Morrow CJ, Penfold LM, Wolfe BA. Artificial insemination in deer and non-domestic bovids. Theriogenology 2009;71(1):149−65.

[30]   Umapathy G, Sontakke SD, Reddy A, Shivaji S. Seasonal variations in semen characteristics, semen cryopreservation, estrus synchronization, and successful artificial insemination in the spotted deer (Axis axis). Theriogenology 2007;67(8):1371−8.

[31]   Monfort SL, Asher GW, Wildt DE, Wood TC, Schiewe MC, Williamson LR, et al. Successful intrauterine insemination of Eld's deer (Cervus eldi thamin) with frozen-thawed spermatozoa. J Reprod Fertil 1993;99(2):459−65.

[32]   Haigh JC. Artificial insemination of two white-tailed deer. J Am Vet Med Assoc 1984;185(11):1446−7.

[33]   Jacobson HA, Bearden HJ, Whitehouse DB. Artificial insemination trials with white-tailed deer. J Wildl Manage 1989;53(1):224−7.

[34]   Magyar SJ, Biediger T, Hodges C, Kraemer DC, Seager SW. A method of artificial insemination in captive white-tailed deer (Odocoileus virginianus). Theriogenology 1989;31(5):1075−9.

[35]   Li D, Wintle NJP, Zhang G, Wang C, Luo B, Martin-Wintleb MS, et al. Analyzing the past to understand the future: natural mating yields better reproductive rates than artificial insemination in the giant panda. Biol Cons 2017;216:10−17.

[36]   Vongpralub T, Chinchiyanond W, Hongkuntod P, Sanchaisuriya P, Liangpaiboon S, Thongprayoon A, et al. Cryopreservation of Sambar deer semen in Thailand. Zoo Biol 2015;34(4):335–44.

[37]   Godfrey RW, Lunstra DD, French JA, Schwartz J, Armstrong DL, Simmons LG, et al. Estrous synchronization in the gaur (*Bos gaurus*): behavior and fertility to artificial insemination after prostaglandin treatment. Zoo Biol 1991;10(1):35–41.

[38]   Densmore MA, Bowen MJ, Magyar SJ, Amoss MS, Robinson RM, Harms PG, et al. Artificial insemination with frozen, thawed semen and pregnancy diagnosis in addax (*Addax nasomaculatus*). Zoo Biol 1987;6(1):21–9.

[39]   Holt WV, Moore HD, North RD, Hartman TD, Hodges JK. Hormonal and behavioural detection of oestrus in blackbuck, *Antilope cervicapra*, and successful artificial insemination with fresh and frozen semen. J Reprod Fertil 1988;82(2):717–25.

[40]   Roldan ER, Gomendio M, Garde JJ, Espeso G, Ledda S, Berlinguer F, et al. Inbreeding and reproduction in endangered ungulates: preservation of genetic variation through the organization of genetic resource banks. Reprod Domest Anim 2006;41(Suppl. 2):82–92.

[41]   Santiago-Moreno J, Toledano-Díaz A, Pulido-Pastor A, Gómez-Brunet A, López-Sebastián A. Birth of live Spanish ibex (*Capra pyrenaica hispanica*) derived from artificial insemination with epididymal spermatozoa retrieved after death. Theriogenology 2006;66(2):283–91.

[42]   Johnston SD, Blyde D, Pedrana R, Gibbs A. Laparoscopic intrauterine insemination in Barbary sheep (*Ammotragus lervia*). Aust Vet J 2000;78 (10):714–17.

[43]   Robeck TR, Steinman KJ, Gearhart S, Reidarson TR, McBain JF, Monfort SL. Reproductive physiology and development of artificial insemination technology in killer whales (*Orcinus orca*). Biol Reprod 2004;71(2):650–60.

[44]   Robeck TR, Steinman KJ, Yoshioka M, Jensen E, O'Brien JK, Katsumata E, et al. Estrous cycle characterisation and artificial insemination using frozen-thawed spermatozoa in the bottlenose dolphin (*Tursiops truncatus*). Reproduction 2005;129(5):659–74.

[45]   Robeck TR, Steinman KJ, Greenwell M, Ramirez K, Van Bonn W, Yoshioka M, et al. Seasonality, estrous cycle characterization, estrus synchronization, semen cryopreservation, and artificial insemination in the Pacific white-sided dolphin (*Lagenorhynchus obliquidens*). Reproduction 2009;138(2):391–405.

[46]   Hildebrandt TB, Hermes R, Saragusty J, Potier R, Schwammer HM, Balfanz F, et al. Enriching the captive elephant population genetic pool through artificial insemination with frozen-thawed semen collected in the wild. Theriogenology 2012;78(6):1398–404.

[47]   Thongtip N, Mahasawangkul S, Thitaram C, Pongsopavijitr P, Kornkaewrat K, Pinyopummin A, et al. Successful artificial insemination in the Asian elephant (*Elephas maximus*) using chilled and frozen-thawed semen. Reprod Biol Endocrin 2009;7:8.

[48]   Brown JL, Göritz F, Pratt-Hawkes N, Hermes R, Galloway M, Graham LH, et al. Successful artificial insemination of an Asian elephant at the National Zoological Park. Zoo Biol 2004;23(1):45–63.

[49]   Schook MW, Wildt DE, Weiss RB, Wolfe BA, Archibald KE, Pukazhenthi BS, et al. Fundamental studies of the reproductive biology of the endangered persian onager (*Equus hemionus onager*) result in first wild equid offspring from artificial insemination. Biol Reprod 2013;89(2):41.

[50]   Hildebrandt TB, Hermes R, Walzer C, Sós E, Molnar V, Mezösi L, et al. Artificial insemination in the anoestrous and the postpartum white rhinoceros using GnRH analogue to induce ovulation. Theriogenology 2007;67(9):1473–84.

[51]   Hermes R, Göritz F, Saragusty J, Sós E, Molnar V, Reid CE, et al. First successful artificial insemination with frozen-thawed semen in rhinoceros. Theriogenology 2009;71(3):393–9.

[52]   Stoops MA, Campbell MK, DeChant CJ, Hauser J, Kottwitz J, Pairan RD, et al. Enhancing captive Indian rhinoceros genetics via artificial insemination of cryopreserved sperm. Anim Reprod Sci 2016;172:60–75.

[53]   Asa CS, Bauman K, Callahan P, Bauman J, Volkmann DH, Jöchle W. Induction of fertile estrus with either natural mating or artificial insemination followed by birth of pups in gray wolves (*Canis lupus*). Theriogenology 2006;66(6-7):1778–82.

[54]   Seager SWJ, Platz Jr. CC, Hodge W. Successful pregnancy using frozen semen in the wolf. Anim Reprod Sci 1975;15(1):140–3.

[55]   Thomassen R, Farstad W. Artificial insemination in canids: a useful tool in breeding and conservation. Theriogenology 2009;71(1):190–9.

[56]   Goodrowe KL, Hay MA, Platz CC, Behrns SK, Jones MH, Waddell WT. Characteristics of fresh and frozen-thawed red wolf (*Canis rufus*) spermatozoa. Anim Reprod Sci 1998;53(1-4):299–308.

[57]   Farstad W. Reproduction in foxes: current research and future challenges. Anim Reprod Sci 1998;53(1-4):35–42.

[58]   Tipkantha W, Thuwanut P, Maikeaw U, Thongphakdee A, Yapila S, Kamolnorranath S, et al. Successful laparoscopic oviductal artificial insemination in the clouded leopard (*Neofelis nebulosa*) in Thailand. J Zoo Wildl Med 2017;48(3):804–12.

[59]   Tajima H, Yoshizawa M, Sasaki S, Yamamoto F, Narushima E, Tsutsui T, et al. Intrauterine insemination with fresh semen in Amur leopard cat (*Pionailurus bengalensis eutilura*) during non-breeding season. J Vet Med Sci 2017;79(1):92–9.

[60]   Lueders I, Ludwig C, Schroeder M, Mueller K, Zahmel J, Dehnhard M. Successful nonsurgical artificial insemination and hormonal monitoring in an Asiatic golden cat (*Catopuma temmincki*). J Zoo Wildl Med 2014;45(2):372–9.

[61]   Howard JG, Roth TL, Byers AP, Swanson WF, Wildt DE. Sensitivity to exogenous gonadotropins for ovulation induction and laparoscopic artificial insemination in the cheetah and clouded leopard. Biol Reprod 1997;56(4):1059–68.

[62]   Chagas de Silva JN, Leitao RM, Lapao NE, da Cunha MB, da Cunha TP, da Silva JP, et al. Birth of a Siberian tiger (*Panthera tigris altaica*) cubs after transvaginal artificial insemination. J Zoo Wildl Med 2000;31(4):566–9.

[63]   Donoghue AM, Johnston LA, Armstrong DL, Simmons LG, Wildt DE. Birth of a Siberian tiger cub (*Panthera tigris altaica*) following laparoscopic intrauterine artificial insemination. J Zoo Wildl Med 1993;24(2):185–9.

[64]   Minnesota Zoo. Amur tiger artificial insemination, <http://mnzoo.org/conservation/around-world/ulysses-s-seal-conservation-grant-program/amur-tiger-artificial-insemination-usa-zoos/>; 2019 [accessed 06.09.19].

[65]   New York Post. World's first jaguar born by artificial insemination eaten by mom, <https://nypost.com/2019/04/03/worlds-first-jaguar-born-by-artificial-insemination-is-eaten-by-mom/>; 2019 [accessed 06.09.19].

[66] Cincinnati Zoo Blog. Successful fixed time artificial insemination in the fishing cat, <http://blog.cincinnatizoo.org/2014/04/11/successful-fixed-time-artificial-insemination-in-the-fishing-cat/>; 2014 [accessed 06.09.19].

[67] Cincinnati Zoo & Botanical Garden. Brazilian ocelot births validate importance of artificial insemination for species conservation, <http://cincinnatizoo.org/news-releases/brazilian-ocelot-births-validate-importance-of-artificial-insemination-for-species-conservation/>; 2016 [accessed 06.09.19].

[68] Swanson WF. Laparoscopic oviductal embryo transfer and artificial insemination in felids — challenges, strategies and successes. Reprod Dom Anim 2012;47(Suppl. 6):136—40.

[69] CTV News. First lion cubs conceived through artificial insemination born, <https://www.ctvnews.ca/sci-tech/first-lion-cubs-conceived-through-artificial-insemination-born-1.4115402>; 2018 [accessed 06.09.19].

[70] Roth TL, Armstrong DL, Barrie MT, Wildt DE. Seasonal effects on ovarian responsiveness to exogenous gonadotrophins and successful artificial insemination in the snow leopard (Uncia uncia). Reprod Fertil Dev 1997;9(3):285—95.

[71] Moore HDM, Bush M, Celma M, Garcia A-L, Hartman TD, Hearn JP, et al. Artificial insemination in the Giant panda (Ailuropoda melanoleaca). J Zool 1984;203(2):269—78.

[72] Huang Y, Li D, Zhou Y, Zhou Q, Li R, Wang C, et al. Factors affecting the outcome of artificial insemination using cryopreserved spermatozoa in the giant panda (Ailuropoda melanoleuca). Zoo Biol 2012;31(5):561—73.

[73] Wildt DE, Bush M, Morton C, Morton F, Howard JG. Semen characteristics and testosterone profiles in ferrets kept in long-day photoperiod and the influence of hCG timing and sperm dilution on pregnancy rate after laparoscopic insemination. J Reprod Fertil 1989;86(1):349—58.

[74] Howard JG, Bush M, Morton C, Morton F, Wentzel K, Wildt DE. Comparative semen cryopreservation in ferrets (Mustela putorious furo) and pregnancies after laparoscopic intrauterine insemination with frozen-thawed spermatozoa. J Reprod Fertil 1991;92(1):109—18.

[75] Santymire RM, Livieri TM, Branvold-Faber H, Marinari PE. The black-footed ferret: on the brink of recovery? Adv Exp Med Biol 2014;753:119—34.

[76] Hardin CJ, Liebherr G, Fairchild O. Artificial insemination in chimpanzees. J Reprod Fertil 1975;15(1):132—4.

[77] Gould KG, Martin DE, Warner H. Improved method for artificial insemination in the great apes. Am J Primatol 1985;8(1):61—7.

[78] Cherfas J. Test-tube babies in the zoo. N Sci 1984;16—19.

[79] CBS News. Baby orangutan is best new hope for survival of his species, <https://www.cbsnews.com/news/baby-orangutan-is-best-new-hope-for-survival-of-his-species/>; 2014 [accessed 31.08.19].

[80] Johnston SD, Holt WV. The koala (Phascolarctos cinereus): a case study in the development of reproductive technology in a marsupial. Adv Exp Med Biol 2014;753:171—203.

[81] Howard JG, Wildt DE. Approaches and efficacy of artificial insemination in mustelids and felids. Theriogenology 2009;71(1):130—48.

[82] Howard JG, Lynch C, Santymire RM, Marinari PE, Wildt DE. Recovery of gene diversity using long-term cryopreserved spermatozoa and artificial insemination in the endangered black-footed ferret. Anim Conserv 2016;19(2):102—11.

[83] Brown JL. Reproductive endocrine monitoring of elephants: an essential tool for assisting captive management. Zoo Biol 2000;19(5):347—67.

[84] Allen CD, Burridge M, Mulhall S, Chafer ML, Nicolson VN, Pyne M, et al. Successful artificial insemination in the koala (Phascolarctos cinereus) using extended and extended-chilled semen collected by electroejaculation. Biol Reprod 2008;78(4):661—6.

[85] Australian Geographic. New koala genome bank aims to end chlamydia epidemic, <https://www.australiangeographic.com.au/news/2016/10/new-koala-genome-bank-aims-to-end-chlamydia-epidemic/>; 2016 [accessed 31.08.19].

[86] O'Brien JK, Steinman KJ, Robeck TR. Application of sperm sorting and associated reproductive technology for wildlife management and conservation. Theriogenology 2009;71(1):98—107.

[87] Gee GF. Crane reproductive physiology and conservation. Zoo Biol 1983;2(3):199—213.

[88] Blanco JM, Gee GF, Wildt DE, Donoghue AM. Producing progeny from endangered birds of prey: treatment of urine-contaminated semen and a novel intramagnal insemination approach. J Zoo Wildl Med 2002;33(1):1—7.

[89] Cade TJ, Weaver JD, Platt JB, Burnham WA. The propagation of large falcons in captivity. Raptor Res 1977;11(1-2):28—48.

[90] Brown ME, Singh RP, Pukazhenthi B, Keefer CL, Songsasen N. Cryopreservation effects on sperm function and fertility in two threatened crane species. Cryobiology 2018;82:148—54.

[91] Reuters. South Africa unveils test-tube buffalo, plans IVF rhino, <https://www.reuters.com/article/us-wildlife-buffalo-ivf/south-africa-unveils-test-tube-buffalo-plans-ivf-rhino-idUSKCN11T194>; 2016 [accessed 31.08.19].

[92] Johnston LA, Parrish JJ, Monson R, Leibfried-Rutledge L, Susko-Parrish JL, Northey DL, et al. Oocyte maturation, fertilization and embryo development in vitro and in vivo in the gaur (Bos gaurus). J Reprod Fertil 1994;100(1):131—6.

[93] WCVM Today. World's first IVF bison calves born at WCVM, <https://wcvmtoday.usask.ca/articles/2016/worlds-first-ivf-bison-calves-born-at-wcvm.php>; 2016 [accessed 31.08.19].

[94] Thongphakdee A, Berg DK, Tharasanit T, Thongtip N, Tipkantha W, Punkong C, et al. The impact of ovarian stimulation protocol on oocyte quality, subsequent in vitro embryo development, and pregnancy after transfer to recipients in Eld's deer (Rucervus eldii thamin). Theriogenology 2017;91:134—44.

[95] Smithsonian Conservation Biology Institute. First Eld's deer born from in vitro fertilization with help of Smithsonian Conservation Biology Institute scientists, <https://www.si.edu/newsdesk/releases/first-eld-s-deer-born-vitro-fertilization-help-smithsonian-conservation-biology-institute-s>; 2011 [accessed 31.08.19].

[96] Berg DK, Beaumont SE, Berg MC, Asher GW. Red deer (Cervus elaphus) calves born from in vitro-produced blastocysts fertilized and cultured in deer synthetic oviduct fluid. Reprod Fertil Dev 2004;16(2):222.

[97] Berg DK, Pugh PA, Thompson JG, Asher GW. Development of in vitro embryo production systems for red deer (*Cervus elaphus*): Part 3. In vitro fertilisation using sheep serum as a capacitating agent and the subsequent birth of calves. Anim Reprod Sci 2002;70(1-2):85−98.

[98] Beaumont SE, Berg MC, Strongman K, Saywell DP, Berg DK. Direct-thaw trans-cervical transfer of red deer frozen in vitro blastocysts can result in pregnancies. Reprod Fertil Dev 2005;17(2):244.

[99] Locatelli Y, Cognie Y, Vallet JC, Baril G, Verdier M, Poulin N, et al. Successful use of oviduct epithelial cell coculture for in vitro production of viable red deer (*Cervus elaphus*) embryos. Theriogenology 2005;64(8):1729−39.

[100] The Guardian. Biotechnology lifeline for critically endangered wildlife, < https://www.theguardian.com/environment/2016/jan/18/biotechnology-endangered-wildlife-conservation-species>; 2016 [accessed 31.08.19].

[101] Ptak G, Clinton M, Barboni B, Muzzeddu M, Cappai P, Tischner M, et al. Preservation of the wild European mouflon: the first example of genetic management using a complete program of reproductive biotechnologies. Biol Reprod 2002;66(3):796−801.

[102] Flores-Foxworth G, Coonrod SA, Moreno JF, Byrd SR, Kraemer DC, Westhusin W. Interspecific transfer of IVM IVF-derived red sheep (*Ovis orientalis gmelini*) embryos to domestic sheep (*Ovis aries*). Theriogenology 1995;44(5):681−90.

[103] Donoghue AM, Johnston LA, Seal US, Armstrong DL, Tilson RL, Wolf P, et al. In vitro fertilization and embryo development in vitro and in vivo in the tiger (*Panthera tigris*). Biol Reprod 1990;43(5):733−44.

[104] Pope CE, Gomez MC, Dresser BL. In vitro embryo production and embryo transfer in domestic and non-domestic cats. Theriogenology 2006;66(1):1518−24.

[105] Pope CE, Keller GL, Dresser BL. In vitro fertilization in domestic and non-domestic cats including sequences of early nuclear events, development in vitro, cryopreservation and successful intra- and interspecies embryo transfer. J Reprod Fertil 1993;47(Suppl.):189−201.

[106] Swanson WF. Reproductive biotechnology and conservation of the forgotten felids—the small cats. In: Proceedings of the First International Symposium on Assisted Reproductive Technologies: Conservation & Genetic Management Wildlife. Omaha, NE; 2001. p. 100−120.

[107] Herrick JR, Mehrdadfar F, Campbell M, Levens G, Leiske K, Swanswon WF. Birth of sand cat (*Felis margarita*) kittens following in vitro fertilization and embryo transfer. Biol Reprod 2010;83(Suppl. 1):28.

[108] Pope CE, Gómez MC, Dumas C, MacLean RA, Crichton E, Armstrong D, et al. Birth of black-footed cat kittens after transfer of cryopreserved embryos produced by in vitro fertilization of oocytes with cryopreserved sperm. Reprod Fertil Dev 2011;24(1):173.

[109] Bavister BD, Boatman BE. IVF in nonhuman primates: current status and future directions. In: Wolf DP, Stouffer RL, Brenner RM, editors. In vitro fertilization and embryo transfer in primates. Serono Symposia USA. New York: Springer; 1993.

[110] Kouba AJ, Vance CK. Applied reproductive technologies and genetic resource banking for amphibian conservation. Reprod Fertil Dev 2009;21(6):719−37.

[111] Browne RK, Seratt J, Vance C, Kouba A. Hormonal priming, induction of ovulation and in-vitro fertilization in the endangered Wyoming toad (*Bufo baxteri*). Reprod Biol Endocrinol 2006;4:34. Available from: https://doi.org/10.1186/1477-7827-4-34.

[112] Kouba A, Willis E, Vance C, Hasenstab S, Reichling S, Krebs J, et al. Development of assisted reproduction technologies for the endangered Mississippi gopher frog (*Rana sevosa*) and sperm transfer for in vitro fertilization. Reprod Fertil Dev 2012;24(1):170.

[113] Nakamura Y. Poultry genetic resource conservation using germ cells. J Reprod Dev 2016;62(5):431−7.

[114] Han JY, Lee HC, Park TS. Germline-competent stem cell in avian species and its application. Asian J Androl 2015;17(3):421−6.

[115] van de Lavoir MC, Collarini EJ, Leighton PA, Fesler J, Lu DR, Harriman WD, et al. Interspecific germline transmission of cultured primordial germ cells. PLoS One 2012;7(5):e35664.

[116] Kang SJ, Choi JW, Kim SY, Park KJ, Kim TM, Lee YM, et al. Reproduction of wild birds via interspecies germ cell transplantation. Biol Reprod 2008;79(5):931−7.

[117] Wernery U, Liu C, Baskar V, Guerineche Z, Khazanehdari KA, Saleem S, et al. Primordial germ cell-mediated chimera technology produces viable pure-line Houbara bustard offspring: potential for repopulating an endangered species. PLoS One 2010;5(12):0015824.

[118] Liu C, Khazanehdari KA, Baskar V, Saleem S, Kinne J, Wernery U, et al. Production of chicken progeny (*Gallus gallus domesticus*) from interspecies germline chimeric duck (*Anas domesticus*) by primordial germ cell transfer. Biol Reprod 2012;86(4):101. Available from: https://doi.org/10.1095/biolreprod.111.094409.

[119] Naito M, Tajima A, Tagami T, Yasuda Y, Kuwana T. Preservation of chick primordial germ cells in liquid nitrogen and subsequent production of viable offspring. J Reprod Fertil 1994;102(2):321−5.

[120] Brinster RL, Avarbock MR. Germline transmission of donor haplotype following spermatogonial transplantation. Proc Natl Acad Sci U S A 1994;91(24):11303−7.

[121] Kubota H, Brinster RL. Spermatogonial stem cells. Biol Reprod 2018;99(1):52−74.

[122] Wu X, Goodyear SM, Abramowitz LK, Bartolomei MS, Tobias JW, Avarbock MR, et al. Fertile offspring derived from mouse spermatogonial stem cells cryopreserved for more than 14 years. Hum Reprod 2012;27(5):1249−59.

[123] Silva RC, Costa GM, Lacerda SM, Batlouni SR, Soares JM, Avelar GF, et al. Germ cell transplantation in felids: a potential approach to preserving endangered species. J Androl 2012;33(2):264−76.

[124] Roe M, McDonald N, Durrant B, Jensen T. Xenogeneic transfer of adult quail (*Coturnix coturnix*) spermatogonial stem cells to embryonic chicken (*Gallus gallus*) hosts: a model for avian conservation. Biol Reprod 2013;88(5):129.

[125] Yoshizaki G, Lee S. Production of live fish derived from frozen germ cells via germ cell transplantation. Stem Cell Res 2018;29:103−10.

[126] Lacerda SM, Costa GMJ, de França LR. Biology and identity of fish spermatogonial stem cell. Gen Comp Endocr 2014;207:56−65.

[127] de Siqueira-Silva DH, Dos Santos Silva AP, da Silva Costa R, Senhorini JA, Ninhaus-Silveira A, Veríssimo-Silveira R, et al. Preliminary study on testicular germ cell isolation and transplantation in an endangered endemic species *Brycon orbignyanus* (Characiformes: Characidae). Fish Physiol Biochem 2019;1−10. Available from: https://doi.org/10.1007/s10695-019-00631-8.

[128] Goel S, Reddy N, Mahla RS, Suman SK, Pawar RM. Spermatogonial stem cells in the testis of an endangered bovid: Indian black buck (*Antilope cervicapra L.*). Anim Reprod Sci 2011;126(3):251−7.

[129] Mastromonaco GF, Gonzalez-Grajales LA, Filice M, Comizzoli P. Somatic cells, stem cells, and induced pluripotent stem cells: how do they now contribute to conservation? Adv Exp Med Biol 2014;753:385−427.

[130] Stanton MM, Tzatzalos E, Donne M, Kolundzic N, Helgason I, Ilic D. Prospects for the use of induced pluripotent stem cells in animal conservation and environmental protection. Stem Cell Transl Med 2019;8(1):7−13.

[131] Fink T, Rasmussen JG, Emmersen J, Pilgaard L, Fahlman Å, Brunberg S, et al. Adipose-derived stem cells from the brown bear (*Ursus arctos*) spontaneously undergo chondrogenic and osteogenic differentiation in vitro. Stem Cell Res 2011;7(1):89−95.

[132] Yu J, Thomson JA. Induced pluripotent stem cells. In: Lanza R, Langer R, Vacanti J, editors. Principles of Tissue Engineering. 4th ed. Boston: Academic Press; 2014 [Chapter 30].

[133] Selvaraj V, Wildt DE, Pukazhenthi BS. Induced pluripotent stem cells for conserving endangered species? Nat Methods 2011;8(10):805−7.

[134] Takahashi K, Yamanaka S. Induction of pluripotent stem cells from mouse embryonic and adult fibroblast cultures by defined factors. Cell 2006;126(4):663−76.

[135] Takahashi K, Tanabe K, Ohnuki M, Narita M, Ichisaka T, Tomoda K, et al. Induction of pluripotent stem cells from adult human fibroblasts by defined factors. Cell 2007;131(5):861−72.

[136] Hayashi K, Ogushi S, Kurimoto K, Shimamoto S, Ohta H, Saitou M. Offspring from oocytes derived from in vitro primordial germ cell-like cells in mice. Science 2012;338(6109):971−5.

[137] Friedrich Ben-Nun I, Montague SC, Houck ML, Tran HT, Garitaonandia I, Leonardo TR, et al. Induced pluripotent stem cells from highly endangered species. Nat Methods 2011;8(10):829−31.

[138] Verma PJ, Verma R, Holland MK, Temple-Smith P. Inducing pluripotency in somatic cells from the snow leopard (*Panthera uncia*), an endangered felid. Theriogenology 2012;77(1):220−8 e2.

[139] Verma R, Liu J, Holland MK, Temple-Smith P, Williamson M, Verma PJ. *Nanog* is an essential factor for induction of pluripotency in somatic cells from endangered felids. Biores Open Access 2013;2(1):72−6.

[140] Ramaswamy K, Yik WY, Wang XM, Oliphant EN, Lu W, Shibata D, et al. Derivation of induced pluripotent stem cells from orangutan skin fibroblasts. BMC Res Notes 2015;8:577. Available from: https://doi.org/10.1186/s13104-015-1567-0.

[141] Hildebrandt TB, Hermes R, Colleoni S, Diecke S, Holtze S, Renfree MB, et al. Embryos and embryonic stem cells from the white rhinoceros. Nat Commun 2018;9(1):2589.

[142] Lee Y, Kang E. Stem cells and reproduction. BMB Rep 2019;52(8):482−9.

[143] Gurdon JB, Elsdale TR, Fischberg M. Sexually mature individuals of *Xenopus laevis* from the transplantation of single somatic nuclei. Nature 1958;182(4627):64−5.

[144] Loi P, Ptak G, Barboni B, Fulka Jr J, Cappai P, Clinton M. Genetic rescue of an endangered mammal by cross-species nuclear transfer using post-mortem somatic cells. Nat Biotech 2001;19(10):962−4.

[145] Gomez MC, Pope CE, Giraldo A, Lyons LA, Harris RF, King AL, et al. Birth of African wildcat cloned kittens born from domestic cats. Cloning Stem Cell 2004;6(3):247−58.

[146] Oh HJ, Kim MK, Jang G, Kim HJ, Hong SG, Park JE, et al. Cloning endangered gray wolves (*Canis lupus*) from somatic cells collected post-mortem. Theriogenology 2008;70(4):638−47.

[147] Gougeon A. Human ovarian follicular development: from activation of resting follicles to preovulatory maturation. Annls Endocrinol 2010;71 (3):132−43.

[148] Santos RR, Amorim C, Cecconi S, Fassbender M, Imhof M, Lornage J, et al. Cryopreservation of ovarian tissue: an emerging technology for female germline preservation of endangered species and breeds. Anim Reprod Sci 2010;122(3-4):151−63.

[149] Cleary M, Shaw JM, Jenkin G, Trounson AO. Influence of hormone environment and donor age on cryopreserved common wombat (*Vombatus ursinus*) ovarian tissue xenografted into nude mice. Reprod Fertil Dev 2004;16(7):699−707.

[150] Wolvekamp MC, Cleary ML, Cox SL, Shaw JM, Jenkin G, Trounson AO. Follicular development in cryopreserved common wombat ovarian tissue xenografted to nude rats. Anim Reprod Sci 2001;65(1-2):135−47.

[151] Cleary M, Paris MC, Shaw J, Jenkin G, Trounson A. Effect of ovariectomy and graft position on cryopreserved common wombat (*Vombatus ursinus*) ovarian tissue following xenografting to nude mice. Reprod Fertil Dev 2003;15(6):333−42.

[152] Gunasena KT, Lakey JR, Villines PM, Bush M, Raath C, Critser ES, et al. Antral follicles develop in xenografted cryopreserved African elephant (*Loxodonta africana*) ovarian tissue. Anim Reprod Sci 1998;53(1-4):265−75.

[153] Mattiske D, Shaw G, Shaw JM. Influence of donor age on development of gonadal tissue from pouch young of the tammar wallaby, *Macropus eugenii*, after cryopreservation and xenografting into mice. Reproduction 2002;123(1):143−53.

[154] Wiedemann C, Hribal R, Ringleb J, Bertelsen MF, Rasmusen K, Andersen CY, et al. Preservation of primordial follicles from lions by slow freezing and xenotransplantation of ovarian cortex into an immunodeficient mouse. Reprod Dom Anim 2012;47(Suppl. 6):300−4.

[155] Wiedemann C, Zahmel J, Jewgenow K. Short-term culture of ovarian cortex pieces to assess the cryopreservation outcome in wild felids for genome conservation. BMC Vet Res 2013;9:37. Available from: https://doi.org/10.1186/1746-6148-9-37.

[156] Pukazhenthi B, Comizzoli P, Travis AJ, Wildt DE. Applications of emerging technologies to the study and conservation of threatened and endangered species. Reprod Fertil Dev 2006;18(1-2):77−90.

[157] Shinohara T, Inoue K, Ogonuki N, Kanatsu-Shinohara M, Miki H, Nakata K, et al. Birth of offspring following transplantation of cryopreserved immature testicular pieces and in-vitro microinsemination. Hum Reprod 2002;17(12):3039−45.

[158] Kaneko H, Kikuchi K, Tanihara F, Noguchi J, Nakai M, Ito J, et al. Normal reproductive development of pigs produced using sperm retrieved from immature testicular tissue cryopreserved and grafted into nude mice. Theriogenology 2014;82(2):325–31.

[159] Fayomi AP, Peters K, Sukhwani M, Valli-Pulaski H, Shetty G, Meistrich ML, et al. Autologous grafting of cryopreserved prepubertal rhesus testis produces sperm and offspring. Science 2019;363(6433):1314–19.

[160] Hagedorn MM, Daly JP, Carter VL, Cole KS, Jaafar Z, Lager CVA, et al. Cryopreservation of fish spermatogonial cells: the future of natural history collections. Sci Rep 2018;8(1):6149.

[161] Song Y, Silversides FG. Production of offspring from cryopreserved chicken testicular tissue. Poult Sci 2007;86(7):1390–6.

[162] Liu J, Cheng KM, Silversides FG. Production of live offspring from testicular tissue cryopreserved by vitrification procedures in Japanese quail (*Coturnix japonica*). Biol Reprod 2013;88(5):124.

[163] Higaki S, Kuwata N, Tanaka K, Tooyama I, Fujioka Y, Sakai N, et al. Successful vitrification of whole juvenile testis in the critically endangered cyprinid honmoroko (*Gnathopogon caerulescens*). Zygote 2017;25(5):652–61.

[164] Thuwanut P, Srisuwatanasagul S, Wongbandue G, Tanpradit N, Thongpakdee A, Tongthainan D, et al. Sperm quality and the morphology of cryopreserved testicular tissues recovered post-mortem from diverse wild species. Cryobiology 2013;67(2):244–7.

[165] Thuwanut P, Thongphakdee A, Sommanustweechai A, Siriaroonrat B, Chatdarong K. A case report concerning male gametes rescued from a Siamese Eld's deer (*Rucervus eldii siamensis*): post-thawed testicular and epididymal sperm quality and heterologous zona pellucida binding ability. J Vet Med Sci 2013;75(1):123–5.

[166] Pothana L, Devi L, Goel S. Cryopreservation of adult cervid testes. Cryobiology 2017;74:103–9.

[167] Pothana L, Makala H, Devi L, Varma VP, Goel S. Germ cell differentiation in cryopreserved, immature, Indian spotted mouse deer (*Moschiola indica*) testes xenografted onto mice. Theriogenology 2015;83(4):625–33.

[168] Songsasen N, Comizzoli P. A historical overview of embryo and oocyte preservation in the world of mammalian in vitro fertilization and biotechnology. In: Borini A, Coticchio G, editors. Preservation of human oocytes: from cryobiology science to clinical applications. London: Taylor & Francis Online; 2009. p. 1–11.

[169] Comizzoli P, Songsasen N, Hagedorn M, Wildt DE. Comparative cryobiological traits and requirements for gametes and gonadal tissues collected from wildlife species. Theriogenology 2012;78(8):1666–81.

[170] Rao BS, Mahesh YU, Suman K, Charan KV, Lakshmikantan U, Gibence HR, et al. Meiotic maturation of vitrified immature chousingha (*Tetracerus quadricornis*) oocytes recovered postmortem. Cryobiology 2011;62(1):47–52.

[171] Czarny NA, Rodger JC. Vitrification as a method for genome resource banking oocytes from the endangered Tasmanian devil (*Sarcophilus harrisii*). Cryobiology 2010;60(3):322–5.

[172] Boutelle S, Lenahan K, Krisher R, Bauman KL, Asa CS, Silber S. Vitrification of oocytes from endangered Mexican gray wolves (*Canis lupus baileyi*). Theriogenology 2011;75(4):547–54.

Chapter 8

Reproductive technologies in camelids

Julian A. Skidmore[1], Elizabeth G. Crichton[1], Clara M. Malo[1], Jane L. Vaughan[2], Nisar A. Wani[3] and Muren Herrid[4]

[1]Camel Reproduction Center, Dubai, UAE, [2]Cria Genesis, Victoria, Australia, [3]Reproductive Biotechnology Centre, Dubai, UAE, [4]International Livestock Center, Queensland, Australia

8.1 Introduction

The family Camelidae comprises: The Old World camelids [*Camelus dromedarius* (dromedary: one-humped camel) and *C. bactrianus*, (Bactrian: two-humped camel)], and the New World camelids [otherwise known as South American camelids (SACs)], [*Lama glama* (llama), *Vicugna pacos* (alpaca), *L. guanaco* (guanaco), and *V. vicugna* (vicuna)].

Embryo transfer (ET), artificial insemination (AI), and other related reproductive technologies offer many advantages to commercial animal production to quickly disseminate valuable genetics. These techniques are used routinely in several domestic animals such as horse, cow, pig, and sheep, and more recently have been used successfully for the genetic improvement of camels for milk production and racing. Although methods for the collection and insemination of fresh semen and collection and transfer of fresh, hatched Day 7 blastocysts are now well established with pregnancy rates of around 40%−50% for AI [1,2] and 65%−70% for ET [3−5], there are no reliable methods for commercial application of cryopreserved camelid semen or embryos. Thus, recent studies have focussed on this aspect of camelid reproduction technologies. Interest has also been directed toward establishing in vitro embryo production (IVP) via in vitro fertilization (IVF), recently advancing to incorporating intracytoplasmic sperm injection (ICSI) and somatic cell nuclear transfer (SCNT) to produce embryos in the dromedary camel.

8.2 In vitro embryo production

The technology of embryo production by IVP is still not well-developed in camelids compared with many other domestic species. A contributing factor is the limited availability of oocytes due to the low numbers of females slaughtered. The uniqueness of the reproductive physiology of different species does not allow for direct extrapolation of IVP systems from one species to another. Nevertheless, while much progress awaits this field to establish it routinely in camelids, information has accrued on all steps of this process.

8.2.1 Collection of cumulus oocyte complexes

8.2.1.1 From slaughterhouse ovaries

This method of oocyte collection is restricted to countries and districts where camelids are slaughtered in significant numbers. Ovaries are transported to the laboratory in normal saline at 30−37°C and processed as quickly as possible. Holding ovaries in normal saline at room temperature for up to 12 h is not detrimental to the maturation rate of oocytes, but changes in the temperature of the holding medium during their storage negatively affect their maturation [6]. The cumulus oocyte complexes (COCs) are harvested by aspiration of visible follicles using a needle attached to a syringe or a vacuum pump [7−10], slicing of the whole ovary [11,12], ovarian mincing [13], or by excising the follicles and teasing them apart under a stereomicroscope [14]. Slicing of dromedary camel ovaries yields 6−11 COCs [11], while aspiration of follicles yields 3−7 COCs per ovary [7,8]. In llamas, aspiration of follicles, 2−6 mm in diameter, yields 2−6 oocytes per female [9,10], while mincing the ovaries with a razor blade yields 27 oocytes per female [13]. Aspiration of visible follicles has been the method most commonly employed in camelids because mincing or slicing of

Reproductive Technologies in Animals. DOI: https://doi.org/10.1016/B978-0-12-817107-3.00008-4

ovaries produces a pool of COCs from small follicles that do not mature well and have very low developmental potential after maturation (Wani et al., unpublished data).

8.2.1.2 From live animals

In general, COCs are obtained from superstimulated or unstimulated females through ultrasound-guided transvaginal ovum pick-up (OPU). Ovarian superstimulation, which is induced by eCG or FSH, alone or in combination [3,4,15], increases the number and quality of COCs harvested. For collecting in vivo-matured oocytes, donor camels are injected with GnRH (20 µg of Buserelin) 26 h before the scheduled OPU. For transvaginal ultrasound-guided oocyte aspiration (TUGA) an electronic convex transducer with an attached needle guide is inserted through the vulva and into the cranial-most portion of the vagina. A 17-gauge, 55-cm, single-lumen needle is placed in the needle guide of the ultrasound probe and advanced through the vaginal fornix and into the follicle. Follicular fluid is then aspirated using a regulated aspiration pump and the contents of all follicles >10 mm in diameter are collected into 50 mL conical tubes containing embryo flushing medium supplemented with heparin (10,000 IU/L). Aspirates are transferred to Petri dishes to search for and evaluate the COCs using a stereomicroscope. In dromedary camels, OPU after superstimulation results in about a 90% recovery rate of oocytes from aspirated follicles and yields an average of 12 COCs per donor [16].

Donor SACs may be treated with multiple doses of FSH and/or a single dose of eCG to produce multiple follicles >6 mm [17] for OPU, and oocytes may or may not be matured in vivo prior to harvest. Most commonly, oocytes are harvested using TUGA or by surgical means combining general and/or regional anesthesia and analgesia with flank laparotomy or laparoscopy. The TUGA method has largely been limited to use in llamas due to their larger body size and pelvic capacity, and yields 52%−77% recovery rates of COCs (4−11 oocytes) [10,18−22], but it is also possible in alpacas [23]. Surgical follicular aspiration is a reliable and repeatable method of oocyte harvest in both alpacas and llamas and yields a 55%−80% recovery rate of COCs (up to 18 oocytes) [20,22,24−29].

One study that compared TUGA with the surgical aspiration of oocytes in llamas found that the latter method of collection produced significantly more oocytes of better quality (higher numbers of compact cumulus cells around oocytes) that were more likely to reach metaphase-II stage (M-II) after in vitro maturation (IVM) than the TUGA method of harvest [20].

8.2.2 Maturation of oocytes

8.2.2.1 In vivo oocyte maturation

In the 26−30 h between natural mating and ovulation in camelids, maturation of follicles includes expansion of cumulus cells, a meiotic progression from prophase I to M-II, widening of the perivitelline space, and peripheral migration of cortical granules [21]. As camelids are induced ovulators, a synchronized maturation of oocytes within the follicles can be achieved by inducing an LH surge with exogenous hormones such as a GnRH analog, LH, or hCG followed by timed follicle aspiration. In dromedary camels, the optimal time for oocyte recovery is 24−30 h after GnRH administration [16] whereas for llamas and alpacas it is 20−24 h after GnRH [10,21−23,25−29].

8.2.2.2 In vitro oocyte maturation

The majority of camel oocytes collected from slaughterhouse ovaries or live animals that have not been injected with GnRH before aspiration of follicles, are in the germinal vesicle (GV) stage and need to be cultured in vitro to reach the M-II stage. Suitable COCs with at least two cumulus cell layers and a homogeneous cytoplasm are pooled and washed once in phosphate buffered saline (PBS) with 10% foetal calf serum (FCS), then twice in maturation medium before they are randomly distributed in 4-well culture plates (20−25 COCs/well) containing 400 µL of the maturation medium and cultured at 38.5°C in an atmosphere of 5% CO_2 in air for 28−36 h [30].

Different media have been used for IVM of camelid oocytes including Tissue Culture Medium-199 (TCM-199), Charles Rosenkrans medium (CR-1aa), Hams F-10, and Connaught Medical Research Laboratories medium-1066 [6−14,16,18,19,21,24,26,28,31]. The superiority of TCM-199 has been demonstrated over CR-1aa or modified Connaught Medical Research Laboratories medium-1066 for dromedary oocyte maturation [12,14]. In most studies, the medium is supplemented with 10% FCS, 0.25 mg/mL pyruvate, 50 µg/mL gentamicin, 10 µg/mL FSH, 10 µg/mL LH, and 1 µg/mL estradiol, and in some cases 500 µg/mL cysteamine [7] or L-glutamine and sodium bicarbonate [31,32] have also been added. Supplementation of IVM media with FCS, estrous dromedary serum or bovine serum albumin (BSA) showed that any one of these three protein sources supported oocyte maturation [32]. Supplementation of 20 ng/mL of epidermal growth factor (EGF) to the maturation medium increased the oocyte maturation rate versus media supplemented with

10 ng/mL, 50 ng/mL, or no EGF [32]. The addition of caffeine (10 mM) during the last 6 h of IVM [33], or L-carnitine (0.4 mg/mL) to IVM and in vitro culture (IVC) media [34] helped improve dromedary oocyte maturation and embryo developmental rates.

Initial studies found a maturation time of 32−36 h optimal for dromedary oocytes [11−13,30,32] as extending IVM to 40 h or more resulted in abnormal chromatin configuration and degenerative changes [6]. Ultrastructural studies during IVM showed an increase in the perivitelline space as maturation progressed until 24 h, but no further increase occurred until 36 h of culture. However, further studies on kinetics and ultrastructure of oocytes during nuclear maturation [6,26,35] and their developmental potential after chemical activation suggested the optimal culture time to achieve better maturation and developmental rates to be between 28 and 30 h for camels [36], 30 and 32 h for alpacas [31], and 30 and 36 h for llamas [22].

Khatir et al. [37] compared the developmental competence and viability of dromedary embryos produced by IVM/IVF, to in vivo matured/IVF and in vivo-matured and fertilized oocytes. After ET, all the pregnancies were lost after 90 days except for those derived from the in vivo-matured (both IVF and in vivo -fertilized) oocytes. They hypothesized an incomplete cytoplasmic maturation of the oocytes matured in vitro pointing to the need to focus research on this aspect of oocyte IVM.

Reports on methods of selecting oocytes after IVM as a way to predict their development capacity are few. To circumvent criteria based only on the extrusion of the polar body, Saadeldin et al. [38] proposed a morphological assessment for in vitro-matured dromedary oocytes involving oocyte diameter, zona pellucida thickness, ooplasm diameter, and perivitelline space area and diameter. Their results, while promising, were only preliminary as embryo development ability was not evaluated. Meanwhile, Fathi et al. [39] evaluated Cresyl blue staining and the expression of select transcripts important for folliculogenesis, oocyte development, embryo development, cell cycle regulators, and transcription factors in GV oocytes. Apart from higher maturation, fertilization rates and early embryo development in the oocytes that stained positive to Cresyl blue, the RT-PCR also revealed a higher expression of these markers. Thus, it would seem that Cresyl blue staining is a useful method to predict the developmental capacity of immature dromedary camel oocytes.

During IVM, the degree and pattern of cumulus cell expansion vary between COCs. Based on this observation, Moulavi and Hosseini [40] investigated the relationship between the expansion morphology of cumulus cells (adherent/ nonadherent to the bottom of the culture dish; compact/expanded/dissociated) and oocyte maturation quality in the camel. The adherent/dissociated morphology was correlated with the best oocyte quality in terms of M-II maturation, embryo development, and also other quality parameters [degeneration, reactive oxygen species, mitochondrial potential, zona dissolution time, and peripheral distribution of cortical granules]. In contrast, adherent/compact morphology was correlated with less competent oocytes than those that were nonadherent. Compact or expanded cumulus complexes presented an intermediate quality. It was concluded that cumulus expansion morphology can be used as a non invasive predictive marker of oocyte competence.

8.2.3 In vitro fertilization

For IVF, simple washing, swim-up [8,30,41] and density gradient centrifugation [7,13,21,24,26,42] techniques have been used to isolate highly motile camelid spermatozoa. Percoll sometimes presents the risk of contamination with endotoxins, therefore a sperm selection protocol using Androcoll-E has been adopted in some labs working with llama [22,43] or camel IVF [42].

Capacitation, which occurs naturally in the female reproductive tract, is a prerequisite for the acrosome reaction and required for successful penetration of oocytes in vitro. Most studies of IVF in camelids have followed bovine protocols and included heparin in the fertilization medium to induce sperm capacitation, while a limited number have used Ca-I or caffeine [33,44]. However, one study reported that supplementation of 10 μg/mL heparin to the fertilization medium did not enhance the acrosome reaction and supplementation with 10 μM Ca-I was detrimental in a study that used epididymal spermatozoa of dromedary camels [44]. No study has evaluated the actual concentration, or even need, for any of these agents for successful capacitation of the spermatozoa in any camelid species and Crichton et al. [45] achieved dromedary sperm penetration and decondensation in a heterologous IVF system without them.

Modified Tyrode's medium containing albumin, lactate, and pyruvate (TALP; 7,30,33,36,46,47] has been the general medium of choice although Brackett and Oliphant medium has also been used [48]. Gametes are coincubated at $38°C−39°C$ in a moist atmosphere of 5% CO_2 in air for 20−24 h. A concentration of $0.5−2 \times 10^6$/mL [7,30] has been used successfully for the dromedary. While higher sperm concentrations of $3−4 \times 10^6$/mL can produce better sperm penetration rates and were successful in a llama study [41], they invariably result in polyspermy (Wani, unpublished data).

Fresh [13,21,33,49−51], cooled-stored [30,52] and frozen-thawed [53,54] epididymal spermatozoa as well as fresh [7,20,22,37,46,47] and frozen-thawed [48,55] ejaculated spermatozoa have been successfully used for camelid IVP. A comparative study of the developmental competence of camel oocytes fertilized in vitro by frozen-thawed ejaculated and epididymal spermatozoa found that frozen epididymal spermatozoa fertilized oocytes and produced embryos in vitro better than the ejaculated spermatozoa [48].

Live offspring have been produced by fertilization of in vitro-matured oocytes of abattoir origin in dromedary camels [47] as well as llamas [41].

8.2.4 Intracytoplasmic sperm injection

This procedure involves fertilizing a mature (M-II) oocyte by direct injection of a single spermatozoon into its ooplasm. Embryos have been produced by this technique in both SACs [20,27] and dromedary camels [56]. The matured oocytes are denuded by gentle pipetting in HEPES-TCM 199 prior to ICSI and oocytes with a visible polar body (M-II) are selected for injection of one motile spermatozoon. In camelids, as in cattle and horses, the spermatozoa and the process of injection are not sufficient for triggering oocyte activation [57]; therefore, injected oocytes are activated by exogenous chemical agents. In llamas and dromedary camels, a higher proportion of activated oocytes are obtained after post-injection activation with ionomycine and 6-DMAP [20] or ionomycine and roscovitine [56]. Roscovitine is usually preferred over DMAP because it does not hinder the extrusion of the second polar body and thus ploidy of the embryo is maintained. However, to date, no live offspring have been produced by this technique in any camelid species.

8.2.5 Somatic cell nuclear transfer

The technique of SCNT, used to produce an exact genetic copy of the donor animal, has been successfully applied in Old World camelids with cloned dromedary calves produced from embryos reconstructed with cells from racing champions, winners of beauty competitions, high milk yielders, and elite bulls [58−60]. Cumulus, granulosa, and oviductal cells and skin fibroblasts have all provided donor nuclear material [58−60]. Once a confluent monolayer of cells is obtained it is passaged three to six times, using an enzymatic solution containing 0.25% trypsin and 0.05% EDTA. Then the cells are serum-starved by culturing in Dulbecco's modified Eagles medium (DMEM, Sigma, Aldrich, USA) supplemented with 0.5% FCS, for more than 72 h before being used as donor nuclei. The cell lines can be frozen after the second passage for use at a later date, but have to be thawed and serum-starved before use. No differences have been noticed in embryo development from reconstructs made with actively growing cells, serum-starved cells, or those cultured after confluency, however, live camel calves have only been reported from embryos originating from serum-starved cells [60].

A cloned Bactrian camel calf has also been produced by interspecies SCNT using a dromedary camel as the source for oocytes as well as the surrogate for carrying the pregnancy to term [61]. In SACs, the production of embryos by nuclear transfer has been reported in llamas by fusing llama fibroblast nuclei from a skin biopsy cell line with enucleated oocytes that had been matured in vitro [62]. However, individual transfer of eleven 8−32 cell embryos into the oviduct or uterine horn ipsilateral to the corpus luteum of each synchronized recipient failed to produce any pregnancies. [62].

8.2.6 In vitro culture of embryos

Initially in vitro−produced camelid embryos were cocultured with different cell types, such as oviduct epithelial cells or granulosa cells [7,13], similar to culture methods used for other animal species during their IVP developmental stages [63]. Khatir et al. [7], for example, reported the benefits of coculturing camel zygotes with oviductal rather than granulosa cells in TCM-199. Subsequently, they compared coculture with oviductal cells in TCM-199 and a semidefined medium (mKSOMaa) and found there were no differences in blastocyst formation. However, pregnancies after 60 days were only maintained when mKSOMaa was used [46]. Indeed, the first reported offspring in camelids obtained after ET of in vitro−produced embryos used this system [47] and it has subsequently been employed by others [8]. However, coculture nowadays is generally discouraged because competition between embryos and coculture cells for the nutrients present in the culture medium and cell metabolic waste could negatively impact embryo development [64]. Most studies of SAC have used synthetic oviductal fluid (SOF) with added amino acids [20,22,27,31,41] with a few reports using DMEM-F12 as well [22]. In Old World camelids, TCM 199 [7,30,36] and KSOMaa, [8,47,58,61] with added amino acids, have been the media of choice for embryo culture. The embryos are transferred to medium supplemented with 1% BSA or 10% FCS and cultured at 38.5°C in an atmosphere of 5% CO_2, 5% O_2, and 90% N_2 in air until Day 7. The culture of embryos at 20% O_2 level versus 5% does not seem to affect embryo development in alpacas [31]. Although blastocyst production has been reported in llamas, alpacas, and dromedary and Bactrian camels, only a few studies have reported live offspring from such embryos. In dromedary camels, KSOMaa supplemented with essential and nonessential amino acids was used in the study reporting live offspring from IVF embryos [47] as well as from SCNT embryos [56,58,59,61], while SOF was used in the experiment reporting the birth of the first llama by IVF [41].

8.3 Cryopreservation

8.3.1 Oocyte cryopreservation

Recent research [8] has studied different cryoprotectants (CPAs) and combinations thereof [dimethyl sulfoxide (DMSO), ethylene glycol (EG), and glycerol] and various cryo-techniques (straws or solid surface vitrification) to vitrify dromedary oocytes in the GV stage, confirming their development to blastocysts as equivalent to fresh oocytes. The combination of 25% EG and 25% DMSO was best and both solid surface vitrification and Cryotop use could maintain the intactness of oocytes after vitrification and warming. In a similar study, Moawad et al. [65] also demonstrated the ability of vitrified immature camel oocytes to mature. They compared different CPA combinations but reported success with higher concentrations of 40% EG and 40% DMSO. In contrast, in one study with SACs, only EG was used in the equilibration and vitrification media, and results indicated that increasing the concentration from 25% or 35% to 45% EG in the vitrification media and increasing the exposure time from 30 to 45 s, decreased oocyte viability and cleavage rates [66]. Further research is needed to study the ability of these oocytes to grow to blastocysts in vitro and result in pregnancies following ET.

8.3.2 Embryo cryopreservation

Embryo transfer technologies are used commercially for the genetic improvement of camels for milk production and racing [67] and SACs for fiber and meat production [68]. In camelid industries, there is often a considerable spatial and temporal separation of recipient and donor necessitating the need to freeze embryos so that they can be transported and transferred into recipients at the correct time. Attempts to freeze camelid embryos started in the late 1990s, with the first pregnancies from frozen/thawed llama embryos reported in 2002 [69], and from camels in 2004/2005 [70–72]. Unlike in other species, such as cattle, embryo cryopreservation has not yet been successfully incorporated into an ET program, a reflection of the difficulties associated with developing an effective procedure for camelids. The difficulty of freezing camel embryos might be due to a greater percentage of osmotically inactive intracellular water which would then require larger osmolarity variations, created by a greater concentration of CPAs, to replace the water [73]. In addition, recent research has indicated that camelid embryos may differ from those of some other species in their tolerance of nonpermeating cryoprotectants; for example, Herrid et al. [73] found sucrose to be more toxic than galactose in the warming solutions. Other obstacles to successful freezing of camelid embryos include the lack of zona pellucida (as all embryos retrieved via transcervical flushing in camelids are hatched blastocysts), the large variation in embryo size within any flush [68], the likelihood of lipids in camelid embryos [74] and a lack of a reliable evaluation system for embryo quality post-thaw [73].

8.3.2.1 Controlled-rate, slow-freezing method

Slow-freezing methods use lower concentrations of CPAs and slower freezing rates than vitrification. Embryos are equilibrated in the CPA, loaded into 0.25 mL straws, placed in an embryo freezing machine and seeded at −7°C before being cooled at a controlled rate. Initial studies [75] investigated different CPAs for cryopreserving dromedary camel embryos. Glycerol, EG, propandiol, and DMSO were tested with and without sucrose and, although embryos appeared to survive cryopreservation, in the six treatment groups, immediately post-thawing, only those frozen in 1.5 M EG with (1/8) or without (2/8) 0.2 M sucrose re-expanded after 20 h of IVC. For this reason, ongoing studies focused on using EG, without sucrose, as the CPA. In these studies, embryos were exposed to 1.5 M EG for 1, 5, or 10 min before subjecting them to a cooling rate of −0.5°C/min from −7°C to −33°C and plunging them into liquid nitrogen. Embryos were later thawed at 32°C for 2 min and rehydrated in either holding media (HEPES-buffered Tyrodes medium containing sodium lactate + 3 mg/mL BSA + 10% FCS) or holding media containing 0.2 M sucrose for 5 min or 10 min. Only one pregnancy resulted from embryos exposed to EG for 1 min, suggesting that 1 min was insufficient time for the cryoprotectant to permeate and protect the inner cell mass. Most pregnancies resulted from those embryos exposed to

EG for 10 min, and those embryos expelled into 0.2 M sucrose and incubated for 5 min showed improved pregnancy rates (37%; 7/19) compared with those expelled into 0.2 M sucrose and incubated for 10 min (33%; 4/12), or directly into holding media (25%; 3/12) [70]. This could indicate that a more gradual rehydration procedure using sucrose in the medium to prevent excessive osmotic shock is beneficial to the thawed embryos, but that prolonged exposure is detrimental, perhaps because it interferes with the proper rehydration of the cells or blastocoels.

8.3.2.2 Vitrification

In comparison to slow controlled-rate freezing, vitrification is generally accepted as a simpler and faster technique for cryopreservation of reproductive cells in field conditions and does not require the specialized equipment needed for slow cooling [76−78].

The first pregnancies from successful vitrification of dromedary embryos used two different protocols. Nowshari et al. [72] exposed embryos to EG (7.0 mol/L) with sucrose (0.5 mol/L) in two steps, prior to loading into 0.25 mL straws and plunging them into liquid nitrogen. The embryos were subsequently thawed at 25°C for 10 s, and rehydrated in 0.5 M sucrose in PBS. Three pregnancies resulted from the 49 embryos transferred (6%). Skidmore et al. [71] used a modified version of Aller et al.'s [69] method for vitrifying llama embryos. Embryos were exposed to a more complex vitrification solution (containing 20% glycerol + 20% EG + 0.3 M sucrose + 0.375 M glucose + 3% polyethylene glycol), in three steps, loaded into 0.25 mL straws and plunged into liquid nitrogen. They were then thawed at 25°C for 1 min and rehydrated in decreasing doses of sucrose (0.5 M, 0.25 M, and 0 M sucrose in PBS + 20% FCS), prior to transfer. Pregnancy rates of 38% (8/21) were achieved with smaller Day 6 embryos ($\leq$350 μm) compared with 0% for both the larger Day 7 (0/9) and Day 8 (0/3) embryos because between 50% and 80% of the bigger embryos were fractured or torn after warming. The primary reason for this could be the formation of ice crystals within embryonic cells, which is associated with a slower penetration of cryoprotectants and subsequent incomplete dehydration of larger embryos [76]. Attempts to overcome this problem have adopted a process of artificial shrinkage either by physically removing the blastocoel fluid via needle puncture [79−81] or by incubating embryos in high osmotic solutions [82]. Herrid et al. [73] examined the effects of artificial shrinkage and addition of sucrose to the solutions for vitrifying camel embryos. After 48 h post-thaw survival rates were 51.2% and 80%, respectively; however, no pregnancies developed in either group. This indicated that neither artificial shrinkage nor addition of sucrose to the vitrification media improved the outcome of vitrification for camelid embryos. Similarly, Taylor et al. [83] achieved a 67% (2/3) pregnancy rate in 500−700 μm vitrified llama embryos using a novel coaxial cryoprotectant microinjection system. They initially hyperinflated hatched blastocysts with equilibration solution containing CPAs, then collapsed the embryos prior to vitrification. Blastocysts were also re-inflated with diluent medium after thawing.

A second reason for this lack of success with vitrification of camel embryos could be the toxicity of the cryoprotectants. Toxicities of permeating cryoprotectants have been previously recognized in a variety of animal species; for example, DMSO is widely used for human, cattle, and mouse embryo cryopreservation but has been shown to be toxic to camel blastocysts [70]. There have been fewer studies on the toxicity of nonpermeating CPAs, such as sucrose, trehalose, and galactose, although Arav et al. [84] did show that trehalose was less harmful than sucrose for vitrifying immature cattle oocytes. Herrid et al. [73,85] therefore investigated different combinations of CPAs and sugars for vitrification and rehydration of camel embryos. Embryos were exposed in three steps to a combination of EG and glycerol (final concentration 3.4 M glycerol + 4.6 M EG) before loading them into open pulled straws and plunging into liquid nitrogen. Thawed embryos were rehydrated in decreasing concentrations of either 0.5 M galactose or 0.5 M sucrose solutions (1 min) followed by 0.25 M galactose or sucrose for 5 min. No pregnancies resulted from the embryos rehydrated in sucrose while promising pregnancy rates of 46.1% (6/13) were achieved from vitrified embryos warmed in galactose solutions [73]. Recently, Lutz et al. [86] achieved the first alpaca pregnancy using the embryo vitrification and galactose solution warming method of Herrid et al [73]. This could suggest a possible species-specific toxic effect of sucrose on camelid embryos. Future work needs to continue to develop these protocols to further improve the pregnancy rates from frozen/thawed embryos.

8.3.3 Cryopreservation of semen

Progress in this field has been hampered by difficulties posed by the collection of semen, as males mate in sternal recumbency and have an extended copulation time of 10−20 min. In addition, the viscous semen that is produced makes processing and preservation of the spermatozoa challenging [87−89].

8.3.3.1 Semen collection and extension

Semen can be collected from camels either with a bull artificial vagina (AV), modified by adding a foam imitation cervix, or with a standard bovine electro-ejaculator [EE: 90,91,92]. In general, use of an AV is considered the standard and most repeatable procedure [93,94], obviating the need for pharmacological immobilization of the donor. Ejaculate volume, using an AV, varies greatly among males and ejaculates [92,95,96]. Much depends on the season and frequency of collection [97] as well as the skill of the operator and the willingness of the male to be trained and to respond consistently. Although significant increases were noted for total spermatozoa/ejaculate in EE collections, a study comparing semen collected by AV versus EE indicated no differences between them for most sperm parameters or sperm freezability [92].

The collection of semen from SACs is also difficult. The preferred method uses a modified sheep AV with a silicon liner, artificial cervix, and water-jacketed collection flask [98] wrapped in an electric heating pad with either a receptive female [99] or dummy mount [100−102]. Alternatively, EE under general anesthesia can be used [103]. Differences in males, semen collection methods, and collection frequency will also affect semen parameters [98,99,103−105].

The handling and evaluation of camelid semen are challenging as a result of a number of factors [93]. Camelids produce highly viscous seminal plasma from the combined secretions of their prostate and bulbourethral glands [106]. Initially, spermatozoa are entrapped and oscillatory rather than progressively motile [103], making the assessment of motility difficult. Due to the relatively small testicular size as a percentage of body weight in camelids, ejaculate volumes and sperm concentrations are low: in camels 2−10 + mL, 100−800 million spermatozoa/mL, respectively [96]; in SACs 1−2 mL, 30−300 million sperm/mL, respectively [99,101,103,106]. Camelid semen is alkaline (pH: 7.5−8.5) and hyperosmolar (369−414 mOsm/kg H_2O); some additional characteristics of camelid semen have been described by Wani et al. [107] and Kershaw-Young and Maxwell [106].

It has been hypothesized that the viscosity of camelid semen serves to prevent semen loss from the female reproductive tract following mating and delays release of the spermatozoa from the seminal plasma during the interval ($\sim$24 h) between mating and subsequent ovulation which occurs in response to the seminal plasma protein beta-nerve growth factor [106]. However, liquefaction of camelid semen is required to release motile sperm to enable their evaluation and processing for AI. Liquefaction can occur naturally but can be protracted, therefore efforts have been sought to hasten it by physical and enzymatic means.

Kershaw-Young et al. [108] found the concentration of glycosaminoglycans (GAGs) in alpaca semen to be high, hypothesizing that these substances were responsible for the viscosity. However, while various mucolytic (e.g., α-amylase, collagenase, bromelain, and chymotrypsin) and other enzymes (e.g., hyaluronidase and heparinase) have been the subjects of many studies in camelids to reduce viscosity in seminal plasma [96,100,105,109,110,111,112], there have been implications to acrosome integrity from their use [113−115]. Further studies in alpacas showed GAG-specific enzymes to be only partially effective because the protein mucin 5B is responsible for semen viscosity [105], while the protease papain reduced viscosity without apparent damage to acrosome integrity [114]. Similar favor for papain (over α-amylase and sperm fluid containing bromelain) was found by Monaco et al. [116] and a study by Mal et al. [105,117] identified that guanidine HCL, a protein denaturing agent, also aided liquefaction of dromedary semen without apparently affecting the viability of spermatozoa.

To avoid damage to sperm by enzymatic treatment of semen, alternative successful methods have included liquefaction by physical means: repeated syringing: [102,115], ultrasound treatment [118] and centrifugation. However, separation of the sperm pellet from the seminal plasma after centrifugation resulted in significant loss of sperm in one report [105], and the high centrifugation speed used in another study (18,000 $\times$ g, 15 min) caused significant acrosome damage [119], rendering centrifugation an unsatisfactory method. A combined enzymatic (amylase)-mechanical (syringing) protocol, however, was reported effective for maintaining good sperm for cryopreservation by El-Bahrawy [115]. Wani et al. [107] and Akbar et al. [120] reported liquefaction in an acceptable time interval (60−90 min) by simply extending semen in Tris-based buffers (1:1) with gentle pipetting, while a greater extension of 1:5 in Tris-citrate-fructose buffer also with intermittent gentle pipetting over $\sim$60 min was used for studies by Malo et al. [42,121]. Deen et al. [96] also liquefied semen by dilution (1:1) in Tris buffer which they supplemented with caffeine, finding it to enhance sperm motility compared to the use of 1% α-chymotrypsin.

In terms of semen evaluation, conventional methods to assess camelid sperm quality have been described [94,103,105,122]. Parameters such as motility, concentration, morphology, viability and acrosome status have been assessed, the latter employing both vital and fluorescent dyes, and some limited attention has been directed toward assessing DNA integrity [121,123]. While the development of an IVF test of sperm function has been hampered by a lack of slaughterhouse material for camels, a heterologous in vitro sperm penetration test using zona-free goat oocytes has been used to assess sperm function in fresh and frozen camel semen [42,45,121]. Further research is needed to correlate this in vitro test with conception rates in order to better understand the utility of this test to accurately predict the in vivo fertilizing ability of camelid sperm.

8.3.3.2 Sperm preservation

8.3.3.2.1 Short term

Semen can be extended and sperm adequately maintained by dilution in a variety of Tris, citrate, or lactose-based buffers which have variously been reported as more or less successful depending on the study [1,45,53,54,88,101,107,110,120,121,124−137]. For fresh semen extension, these have included Laciphos, INRA-96, OPTIXcell, Green Buffer (I.M.V. Technologies, L'Aigle, France) Androhep, Biladyl, Triladyl (Minitube, Germany), glucose EDTA, lactose egg-yolk,TALP, Tris-citrate containing fructose and/or glucose. Additionally, research has shown that OPTIXcell, Triladyl, and Green Buffer support chilled dromedary sperm viability and motility during storage of semen at 4°C for 48 h [124,136], and further research by Malo et al. (unpublished) has recorded a pregnancy rate of 66% resulting from dromedary semen cooled for 24 h in Green Buffer. For Bactrian camels, a Tris-based extender, ("SHOTOR"), is almost universally the extender of choice [138].

The inclusion of egg yolk [1,2,105,139] or skimmed milk [128] helps maintain sperm viability. Panahi et al. [140] compared egg yolk from six different avian species and also skimmed camel milk (all 20%) to maintain sperm viability for 24 h at 5°C. While pigeon egg yolk and camel milk provided beneficial effects during chilled storage, in general, chicken egg yolk seems a suitable source of lipoproteins for preventing cold shock during dromedary camel sperm preservation. The choice of the optimal sugar remains uncertain with various sources having been used successfully, namely lactose, mannose, sucrose, fructose, and glucose.

Tris-based buffers and Green Buffer generally seem suitable for extending camel semen as pregnancy rates of 35%−55% have been reported following AI of whole fresh camel semen [1,2]. In llamas, pregnancy rates of 75% (6/8) and 23% (3/13) were achieved when semen was extended with lactose + egg yolk and inseminated at 37°C or after cooling at 5°C for 24 h, respectively [139]. A comprehensive statement on which extender(s) might best serve camelid sperm survival in all aspects of sperm parameters, awaits further research.

8.3.3.2.2 Long term

Success at freezing camelid semen, as measured by pregnancy rates, has been challenging. While researchers studying the Bactrian camel have reported pregnancy rates greater than 90% [141] and Akbar et al. [120] recently reported a 71% successful conception rate in a study for the dromedary, the lack of published data on conception rates resulting from studies of frozen spermatozoa suggest that there has been little success among researchers. Deen et al. [96] and Malo et al. [142] have reported conception rates in dromedaries in the order of 10%. Thus recent research has been directed toward improving this aspect of camelid semen management. Undoubtedly, a large contributor to the problem is inferior sperm cryopreservation resulting from the viscous nature of the seminal plasma which likely prevents cryoprotective agents, such as egg yolk and glycerol, from contacting the sperm plasma membrane and effectively preserving it [143]. As mentioned before, liquefaction of the ejaculate has been addressed in various ways, notably by enzymatic means or alternatively by dilution in appropriate extenders. Studies have either retained [96,120] or removed (by washing) the seminal plasma [121,135] and also looked into returning 10-15% of seminal plasma to sperm to be cryopreserved [144]. However, there is currently no reliable protocol in place for the cryopreservation, insemination, and successful conception resulting from camelid semen.

Much research has focused on identifying optimal cryopreservation buffers. These have included Tris-lactose/glucose/fructose, Triladyl, OPTIXcell, INRA-96, Green/Clear Buffer, SHOTOR, and skim milk-fructose [53,54,102,105,120,129,130,134,135,137,145]. Various CPAs and their concentrations have also been studied: glycerol, EG, alcohols, formamides, and DMSO [102,137,146]. While it is difficult to compare results between different studies, a number of alternatives seem able to successfully protect spermatozoa from cold shock with no conclusion on what might be optimal for more universal application. Glycerol at concentrations of 2%−7% [87,102,105,120,129,137] has been the most widely used.

A number of studies have looked into improving sperm quality prior to or during preservation by applying various agents that have been used successfully for other species. These include the supplementation of sperm with cholesterol (as CLCs), [45,135] and also various antioxidants that might reduce their demise during prolonged handling [142,147−150]. Greater pregnancy success was reported for chilled dromedary sperm cooled in the presence of catalase (500 IU/mL) [149]. Malo et al. [142,150] have also described the beneficial effects of catalase on cryopreserved sperm motility. El-Harairy et al. [151] supplemented epididymal sperm with glutathione or ascorbic acid during chilling (48 h at 5°C) and reported better preservation of the integrity and functionality of the sperm membrane. Pentoxifylline [146] and caffeine have also been used successfully for maintaining motility and acrosome integrity in frozen camel sperm [96,152].

8.3.3.3 Freezing and thawing methods

Semen extended in cryoprotectant is usually frozen in 0.25, 0.5, or 4 mL plastic straws placed in liquid nitrogen vapor at a defined height above the surface of liquid nitrogen for a specified time prior to plunging into liquid nitrogen, but may be frozen in pellets using dry ice [93,105,124,137,142,145]. However, pellets are more difficult to label and store. Most studies of camelid sperm cryopreservation have utilized 0.5 mL straws. Malo et al. [153] have studied rates of freezing (height above liquid nitrogen) and thawing (time and temperature) of camel sperm and concluded that faster rates of both (1 cm above liquid nitrogen vapor, 60°C for 10 s, respectively) are better than slow for the retrieval of maximal numbers of viable sperm. In Bactrian camels, Niasari-Naslaji et al. [154] compared a slower cooling rate of an average of 0.14°C/min with a faster rate of 0.55°C/min and found that progressive forward motility of the spermatozoa cooled at the faster rate was superior to that cooled at the slower rate (47% vs 31%, respectively). However, alpaca sperm viability was better when a fast freezing rate (straws 2 cm above the surface of liquid nitrogen) and a slow thawing rate (60 s in 37°C water bath) were used [143]. Overall, post-thaw sperm motilities of 30%–-40% have been achieved in camelids.

8.3.3.4 Timing of AI, sperm dose, and site of semen deposition

All camelids are induced ovulators, usually ovulating only when mated [155,156]. However, GnRH will reliably induce ovulation when a mature follicle of 1.3−1.7 cm in diameter is present in the ovaries of camels [157], or 0.7−0.9 cm in diameter is present in the ovaries of SACs [158,159]. Ovulation then occurs 28−36 h after GnRH in camels [157,160,161] and 25−29 h after GnRH in SAC [162,163]. However, the timing of AI in relation to ovulation, the number of motile sperm and the site of deposition of semen in the female tract to achieve acceptable conceptions rates are still poorly defined [2].

A 75% pregnancy rate was achieved in llamas when 12 million motile fresh-extended (lactose/egg yolk) spermatozoa were inseminated transcervically 22−24 h after induction of, but prior to, ovulation. Semen was deposited close to the uterotubal junction of the uterine horn ipsilateral to the ovary with the dominant follicle [139]. However, when similarly extended semen was chilled to 5°C for 24 h, insemination of an entire ejaculate (mean 73 million sperm) before ovulation failed to produce any pregnancies. Insemination of an entire, similarly extended, ejaculate (76−210 million sperm) 26−30 h after induction of ovulation, and 2 h after ovulation was confirmed by ultrasound, resulted in a 23% pregnancy rate [139].

Skidmore et al. [2] investigated pregnancy rates following transcervical AI in camels using fresh-extended semen (Green Buffer/egg yolk). When 150, 80, and 40 million motile sperm were deposited at the uterotubal junction of the uterine horn ipsilateral to the ovary containing the dominant follicle 24 h after induction of ovulation, 43%, 40%, and 7% of females conceived, respectively. In contrast, when the same doses were deposited in the uterine body just cranial to the cervix, 53%, 7%, and 0% females conceived, respectively.

8.4 Camel tissue culture, isolation of germline stem cells

In contrast to its use in human medicine, the application of tissue culture techniques as a technology in livestock species, is still in the early stages of research.

8.4.1 Reproductive stem cells/precursor cells

8.4.1.1 In vitro spermatogenesis from spermatogonial stem cells

In camels, various growth factors, matrix substrates, and serum-free supplements have been investigated to develop a defined system for culturing putative camel spermatogonial stem cells (SSCs). Using two types of culture media (DMEM and STEMPRO), supplemented with knockout serum replacement (KSR) and supplements B27 and GDNF, putative dromedary SSCs can be cultured on gelatin-coated flasks for five passages. Smears prepared from each passage have identified SSCs with DBA-FITC (DBA: *Dolichos biflorus* agglutinin, a glycoprotein that specifically binds to gonocytes or Type A spermatogonia). Under optimized culture conditions consisting of DMEM + 15% KSR supplement, 30%–40% DBA-positive cells were present in the first three passages, however, these decreased to less than 10% at the 4th and 5th passages, indicating a need to further optimize a reliable culture system for camel SSCs (2017, Herrid and Skidmore, unpublished data).

8.4.1.2 Spermatogonial stem cell transplantation

In contrast to in vitro spermatogenesis, SSC transplantation is emerging as a novel reproductive technology with application in animal breeding systems [164,165]. Testicular stem cells from a genetically elite individual transplanted into another can develop and produce a surrogate male—an animal that produces the functional sperm of the original elite individual. One of the major advantages of this technology in livestock species is that the recipient testes do not require an immune suppression procedure to allow donor cell survival in heterologous transplantation [166–168].

In comparison to other livestock species, such as cattle and sheep, the isolation efficiency of a two-step enzymatic digestion protocol is low and difficult in camels, due to the large amount of white connective tissue and the rich testicular vascular supply [169]. On average, 20.8 ± 1.8 million spermatogonial cells can be isolated per gram of testis tissue, with a range from 10.6-25.8 million. However, density gradient centrifugation with either Percoll (Sigma, U.S.A.) or Bovicoll (provided by Prof Jane Morrell, Swedish University of Agricultural Sciences) can significantly increase the number of spermatogonia in the single-cell suspension [169].

The first recipient camel to produce donor-origin sperm was reported by Herrid et al. [169]. These authors attempted to deplete the germ cells in the recipient male camel by injecting DBA directly into the rete testis. Then, 4–6 weeks after treatment, 51–170 million cells from an unenriched single-cell suspension from the donor were injected into the rete testis of each testicle. This resulted in spermatozoa of donor origin being produced by the recipient male and indicated that, as in other mammals, testicular germ cell transplantation in camels is feasible. However, the functionality of those sperm produced in the recipient's testis has not yet been confirmed via natural mating and live birth.

Conclusions

Ongoing research over recent years has led to many advances in the application of assisted reproductive techniques in camelids. For example, live offspring have been produced from fertilization of in vitro-matured oocytes from camels and llamas and the technique of SCNT has been successfully applied to Old World camelids with the birth of both dromedary and Bactrian cloned camel calves.

There have been significant advances in the success of vitrification of camelid embryos although the stage-dependent (small embryos survive vitrification better than large embryos) sensitivity of camel embryos is a major obstacle to a practical application of the vitrification method. Further refinements of these protocols are therefore required before these techniques can become commercially viable. While considerable advances have also been made in attempts to cryopreserve and inseminate frozen sperm, this reproductive technology remains a challenge in camelids. Conclusive results are hampered by the huge variability among males, their ejaculates, and the large number of different studies that mostly have not had access to sufficient numbers of animals from which to draw firm conclusions. Although post-thaw motilities of 35-40% have been achieved, conception rates from frozen semen are low and a universal protocol for the reliable long-term preservation of camelid semen and its timely insemination relative to (induced) ovulation are subjects for ongoing research. It is very encouraging that the first recipient to produce donor-origin sperm after spermatogonial stem cell transplantation has been reported, but further research is required to refine the protocols until the functionality of the donor sperm can be confirmed.

References

[1] Anouassi A, Adnani M, Raed EL. Artificial insemination in the camel requires induction of ovulation to achieve pregnancy. In: Allen WR, Higgins AJ, editors. Proceedings of first international camel conference. UK: R&W Publications (Newmarket) Ltd; 1992. p. 175–8.

[2] Skidmore JA, Billah M. Comparison of pregnancy rates in dromedary camels (*Camelus dromedarius*) after deep intra-uterine versus cervical insemination. Theriogenology 2006;66:292–6.

[3] McKinnon AO, Tinson AH, Nation G. Embryo transfer in dromedary camels. Theriogenology 1994;41:145–50.

[4] Skidmore JA, Billah M, Allen WR. Investigation of factors affecting pregnancy rate after embryo transfer in the dromedary camel. Reprod Fertil Dev 2002;14:109–16.

[5] Anouassi A, Tibary A. Development of a large commercial camel embryo transfer program: 20 years of scientific research. Anim Reprod Sci 2013;136:211–21.

[6] Wani NA, Nowshari MA. Kinetics of nuclear maturation and effect of holding ovaries at room temperature on in vitro maturation of camel (*Camelus dromedarius*) oocytes. Theriogenology 2005;64:75—85.

[7] Khatir H, Anouassi A, Tibary M. Production of dromedary (*Camelus dromedarius*) embryos by IVM and IVF and co-culture with oviductal or granulosa cells. Theriogenology 2004;62:1175—85.

[8] Fathi M, Moawad AR, Badr MR. Production of blastocysts following in vitro maturation and fertilization of dromedary camel oocytes vitrified at the germinal vesicle stage. PLoS One 2018;13(3):e0194602.

[9] M.R. Del Campo, M.X. Donoso, C.H. Del Campo, R. Rojo, C. Barros, R. Mapletoft, In vitro maturation of llama (*Lama glama*) oocytes. In: *Proceedings of the 12th International Congress on Animal Reproduction*; 1992. p. 324—326.

[10] Ratto M, Berland M, Huanca W, Singh J, Adams GP. In vitro and in vivo maturation of llama oocytes. Theriogenology 2005;63:2445—57.

[11] Abdoon ASS. Factors affecting follicular population, oocyte yield and quality in camels (*Camelus dromedarius*) ovary with special reference to maturation time in vitro. Anim Reprod Sci 2001;66:71—9.

[12] Torner H, Heleil B, Alm H, Ghoneim IM, Srsen V, Kanitz W, et al. Changes in cumulus oocyte complexes of pregnant and non-pregnant camels (*Camelus dromedarius*) during maturation in vitro. Theriogenology 2003;60:977—87.

[13] Del Campo MR, Donosos MX, Del Campo CH, Berland M. In vitro fertilization and development of llama (*Lama glama*) oocytes using epididymal spermatozoa and oviductal cell co-culture. Theriogenology 1994;41:1219—29.

[14] Nowshari MA. The effect of harvesting technique on efficiency of oocyte collection and different maturation media on the nuclear maturation of oocytes in camels (*Camelus dromedarius*). Theriogenology 2005;63:2471—81.

[15] Anouassi A, Ali A. Embryo transfer in camels (*Camelus dromedarius*). In: Saint-Martin G, editor. Proceedings of the workshop "Is it possible to improve the reproductive performance of the camel?". Paris: CIRAD-EMVT; 1990. p. 327—32.

[16] Wani NA, Skidmore JA. Ultrasonographic-guided retrieval of in vivo matured oocytes after super-stimulation in dromedary camel (*Camelus dromedarius*). Theriogenology 2010;74:436—42.

[17] Ratto MH, Silva ME, Huanca W, Huanca T, Adams GP. Induction of superovulation in South American camelids. Anim Reprod Sci 2013;136:164—9.

[18] Brogliatti GM, Palasz AT, Rodriguez-Martinez H, Mapletoft RJ, Adams GP. Transvaginal collection and ultrastructure of llama (*Lama glama*) oocytes. Theriogenology 2000;54:1269—79.

[19] Ratto MH, Berland M, Adams GP. Ovarian superstimulation and ultrasound-guided oocyte collection in llamas. Theriogenology 2002;57:590.

[20] Sansinena MJ, Taylor SA, Taylor PJ, Schmidt EE, Denniston RS, Godke RA. In vitro production of llama (*Lama glama*) embryos by intracytoplasmic sperm injection: effect of chemical activation treatments and culture conditions. Anim Reprod Sci 2007;99:342—53.

[21] Berland MA, von Baer A, Parraguez V, Morales P, Adams GP, Silva ME. In vitro fertilization and embryo development of cumulus oocyte complexes collected by ultrasound guided follicular aspiration in llamas treated with gonadotropin. Reprod Fertil Dev 2011;22:288—9.

[22] Trasorras VL, Giuliano S, Chaves MG, Neild D, Agüero A, Carretero M. In vitro embryo production in llamas (*Lama glama*) from in vivo matured oocytes with raw semen processed with Androcoll-E using defined embryo culture media. Reprod Domest Anim 2012;47:562—7.

[23] J.L. Vaughan, Control of ovarian follicular growth in the alpaca (*Lama pacos*). PhD thesis, Central Queensland University, Australia; 2001.

[24] Gomez G, Gattto MH, Berland M, Wolter M, Adams GP. Superstimulation response and oocytes collection in alpacas. Theriogenology 2002;57:584.

[25] Miragaya MH, Chaves MG, Capdevielle EF, Ferrer MS, Pinto M, Rutter B, et al. In vitro maturation of llama (*Lama glama*) oocytes obtained surgically using follicle aspiration. Theriogenology 2002;57:731.

[26] Ratto MH, Gomez C, Berland M, Adams GP. Effect of ovarian superstimulation on COC collection and maturation in alpacas. Anim Reprod Sci 2007;97:246—56.

[27] Conde PA, Herrera C, Trasorras VL, Giuliano SM, Director A, Miragaya MH. In vitro production of llama (*Lama glama*) embryos by IVF and ICSI with fresh semen. Anim Reprod Sci 2008;109:298—308.

[28] Trasorras V, Chaves M, Miragaya M, Pinto M, Rutter B, Flores M. Effect of eCG Superstimulation and buserelin on cumulus-oocyte complexes recovery and maturation in llamas (*Lama glama*). Reprod Domest Anim 2009;44:359—64.

[29] Trasorras V, Baca Castex C, Alonso A, Giuliano S, Santa Cruz R, Arraztoa C, et al. First llama (*Lama glama*) pregnancy obtained after in vitro fertilization and in vitro culture of gametes from live animals. Anim Reprod Sci 2014;148:83—9.

[30] Wani NA. In vitro embryo production in camel (*Camelus dromedarius*) from in vitro matured oocytes fertilized with epididymal spermatozoa stored at 4°C. Anim Reprod Sci 2009;111:69—79.

[31] Ruiz J, Santayana RP, Mendoza MJ, Landeo JL, Huamán E, Ticllancuri F. Effect of oocyte maturation time, sperm selection method and oxygen tension on in vitro embryo development in alpacas. Theriogenology 2017;95:127—32.

[32] Wani NA, Wernery U. Effect of different protein supplementations and epidermal growth factor on in vitro maturation of dromedary camel (*Camelus dromedarius*) oocytes. Reprod Domest Anim 2010;45:189—93.

[33] Fathi M, Seida AA, Sobhy RR, Darwish GM, Badr MR, Moawad AR. Caffeine supplementation during IVM improves frequencies of nuclear maturation and subsequent preimplantation development of dromedary camel oocytes following IVF. Theriogenology 2014;81(9):1286—92.

[34] Fathi M, El-Shahat KH. L-carnitine enhances oocyte maturation and improves in vitro development of embryos in dromedary camels (*Camelus dromedaries*). Theriogenology 2017;104:18—22.

[35] Kafi M, Mesbah F, Nili H, Khalili A. Chronological and ultra structural changes in camel (*Camelus dromedarius*) oocytes during in vitro maturation. Theriogenology 2005;63:2458—70.

[36] Wani NA. Chemical activation of in vitro matured dromedary camel (*Camelus dromedarius*) oocytes: optimization of protocols. Theriogenology 2008;69:591—602.

[37] Khatir H, Anouassi A, Tibary A. Quality and developmental ability of dromedary *(Camelus dromedarius)* embryos obtained by IVM/IVF, in vivo matured/IVF or in vivo matured/fertilized oocytes. Reprod Domest Anim 2007;42(3):263—70.

[38] Saadeldin IM, Swelum AA, Yaqoob SH, Alowaimer AN. Morphometric assessment of in vitro matured dromedary camel oocytes determines the developmental competence after parthenogenetic activation. Theriogenology 2017;95:141—8.

[39] Fathi M, Ashry M, Salama A, Badr MR. Developmental competence of dromedary camel *(Camelus dromedarius)* oocytes selected using brilliant cresyl blue staining. Zygote 2017;25(4):529—36.

[40] Moulavi F, Hosseini SM. Diverse patterns of cumulus cell expansion during in vitro maturation reveal heterogeneous cellular and molecular features of oocyte competence in dromedary camel. Theriogenology 2018;119:259—67.

[41] Landeo L, Mendoza J, Manrique L, Taipe E, Molina R, Conteras J, et al. First Llama born by in vitro fertilization. Reprod Fertil Dev 2017;29:188.

[42] Malo C, Crichton EG, Morrell JM, Pukazhenthi BS, Skidmore JA. Single layer centrifugation of fresh dromedary camel semen improves sperm quality and in vitro fertilization capacity compared with simple sperm washing. Reprod Domest Anim 2017;52:1097—103.

[43] Santa Cruz R, Carretero M, Arraztoa C, Neild D, Trasorras V, Miragaya M. Use of Androcoll-E™ for selecting llama sperm. Preliminary results. Vet 2010;12:292.

[44] Wani NA, Nowshari MA. Effect of heparin and calcium ionophore on acrosome reaction in epididymal spermatozoa of dromedary camel *(Camelus dromedarius)*. Reprod Fertil Dev 2005;17:250—1.

[45] Crichton EG, Malo C, Pukazhenthi BS, Nagy P, Skidmore JA. Evaluation of cholesterol-treated dromedary camel sperm function by heterologous IVF and AI. Anim Reprod Sci 2016;174:20—8.

[46] Khatir H, Anouassi A, Tibary A. In vitro and in vivo developmental competence of dromedary *(Camelus dromedarius)* embryos produced in vitro using two culture systems (mKSOMaa and oviductal cells). Reprod Domest Anim 2005;40(3):245—9.

[47] Khatir H, Anouassi A. The first dromedary *(Camelus dromedarius)* offspring obtained from in vitro matured, in vitro fertilized and in vitro cultured abattoir-derived oocytes. Theriogenology 2006;65:1727—36.

[48] Scholkamy TH, El Badry DA, Mahmoud KGM. Developmental competence of dromedary camel oocytes fertilized in vitro by frozen thawed ejaculated and epididymal spermatozoa. Iran J Vet Res 2016;17:253—8.

[49] N.A. Wani, M.A. Nowshari, U. Wernery, Storage of camel *(Camelus dromedarius)* epididymal spermatozoa in different extenders and at different temperatures. In: *38th annual meeting of the society for the study of reproduction*, Quebec; 2005, W 693.

[50] Moawad AR, Darwish GM, Badr MR, El-Wishy AB. In vitro fertilization of dromedary camel *(Camelus dromedarius)* oocytes with epididymal spermatozoa. Reprod Fertil Dev 2012;24:192—3.

[51] Trasorras VL, Giuliano S, Chaves MG, Baca Castex C, Carretero M, Negro V. In vitro production of llama embryos. Vet 2010;12:296.

[52] Badr MR, Abdel-Malak MG. In vitro fertilization and embryo production in dromedary camel using epididymal spermatozoa. Glob Vet 2010;4:271—6.

[53] A.S. Abdoon, O.M. Kandil, F. Pizzi, F. Turri, A. El Atrash, H.A. Sabra, Effect of semen extender on cryopreservation and fertilization rates of dromedary camel epididymal spermatozoa. In: *Scientific conference in camel research and production*. Khartoum, Sudan; April 17th—18th, 2013. p. 75—82.

[54] El-Badry DA, Mohamed RH, El-Metwally HA, Abo Al-Naga TR. The effect of some cryoprotectants on dromedary camel frozen—thawed semen. Reprod Domest Anim 2017;52:522—5.

[55] El-Sayed A, Sayed HA, El-Hassanein EE, Murad H, Barkawi AH. Effect of epidermal growth factor on in vitro production of camel *(Camelus dromedaries)* embryos by using frozen semen. Egypt J Anim Prod 2012;49:39—45.

[56] Wani NA, Hong SB. Intracytoplasmic sperm injection (ICSI) of in vitro matured oocytes with stored epididymal spermatozoa in camel *(Camelus dromedarius)*: effect of exogenous activation on in vitro embryo development. Theriogenology 2018;113:44—9.

[57] Perreault SD, Barbee RR, Elstein KH, Zucker RM, Keefer CL. Interspecies differences in the stability of mammalian sperm nuclei assessed in vivo by sperm microinjection and in vitro by flow cytometry. Biol Reprod 1988;39:157—67.

[58] Wani NA, Wernery U, Hassan FA, Wernery R, Skidmore JA. Production of the first cloned camel by somatic cell nuclear transfer. Biol Reprod 2010;82:373—9.

[59] Wani NA, Hong S, Vettical BS. Cytoplast source influences development of somatic cell nuclear transfer (SCNT) embryos in vitro but not their development to term after transfer to synchronized recipients in dromedary camels *(Camelus dromedarius)*. Theriogenology 2018;118:137—43.

[60] Wani NA, Hong SB. Source, treatment and type of nuclear donor cells influences in vitro and in vivo development of embryos cloned by somatic cell nuclear transfer in camel *(Camelus dromedarius)*. Theriogenology 2018;106:186—91.

[61] Wani NA, Vettical BS, Hong SB. First cloned Bactrian camel *(Camelus bactrians)* calf produced by interspecies somatic cell nuclear transfer: a step towards preserving the critically endangered wild Bactrian camels. PLoS One 2017;12(5):0177800. <https://doi.org/10.1371/journal.pone.0177800>.

[62] Sansinena MJ, Taylor SA, Taylor PJ, Denniston RS, Godke RA. Production of nuclear transfer llama *(Lama glama)* embryos from in vitro matured llama oocytes. Cloning Stem Cell 2003;5:191—8.

[63] Eyestone WH, First NL. Co-culture of early cattle embryos to the blastocyst stage with oviductal tissue or in conditioned medium. J Reprod Fertil 1989;85:715—20.

[64] Bavister BD. Culture of preimplantation embryos: facts and artifacts. Hum Reprod Update 1995;1:91—148.

[65] Moawad M, Hussein HA, Abd El-Ghani M, Darwish G, Badr M. Effects of cryoprotectants and cryoprotectant combinations on viability and maturation rates of *Camelus dromedarius* oocytes vitrified at germinal vesicle stage. Reprod Domest Anim 2019;54(1):108—17.

[66] Ruiz J, Landeo L, Mendoza J, Artica M, Correa JE, Silva M, et al. Vitrification of in vitro mature alpaca oocyte: effect of ethylene glycol concentration and time of exposure in the equilibration and vitrification solutions. Anim Reprod Sci 2013;143:72−8.

[67] Skidmore JA. The use of some assisted reproductive technologies in old world camelids. Anim Reprod Sci 2019. Available from: https://doi.org/10.1016/j.anireprosci.2019.06.001.

[68] Vaughan J, Mihm M, Wittek T. Factors influencing embryo transfer success in alpacas: a retrospective study. Anim Reprod Sci 2013;136:194−204.

[69] Aller JF, Rebuffi GE, Cancino AK, Alberio RH. Successful transfer of vitrified llama (*Lama glama*) embryos. Anim Reprod Sci 2002;73(1-2):121−7.

[70] Skidmore JA, Billah M, Loskutoff NM. Developmental competence in vitro and in vivo of cryopreserved, hatched blastocysts from the dromedary camel (*Camelus dromedarius*). Reprod Fertil Dev 2004;16:605−9.

[71] Skidmore JA, Billah M, Loskutoff NM. Comparison of two different methods for the vitrification of hatched blastocysts from the dromedary camel (*Camelus dromedarius*). Reprod Fert Dev 2005;17:523−7.

[72] Nowshari MA, Ali SA, Saleem S. Offspring resulting from transfer of cryopreserved embryos in camel (*Camelus dromedarius*). Theriogenology 2005;63(9):2513−22.

[73] Herrid M, Billah M, Skidmore JA. Successful pregnancies from vitrified embryos in the dromedary camel: Avoidance of a possible toxic effect of sucrose on embryos. Anim Reprod Sci 2017;187:116−23.

[74] von Baer A, Del Campo M. Vitrification and cold storage of llama (*Lama glama*) hatched blastocysts. Theriogenology 2002;57:489.

[75] Skidmore JA, Loskutoff NM. Developmental competence in vitro and in vivo of cryopreserved expanding blastocysts from the dromedary camel (*Camelus dromedarius*). Theriogenology 1999;51:293 Abs.

[76] Edgar DH, Gook DA. A critical appraisal of cryopreservation (slow cooling versus vitrification) of human oocytes and embryos. Hum Reprod Update 2012;18(5):536−54.

[77] Fasano G, Fontenelle N, Vannin AS, Biramane J, Devreker F, Englert Y, et al. A randomized controlled trial comparing two vitrification methods versus slow-freezing for cryopreservation of human cleavage stage embryos. J Assist Reprod Genet 2014;31(2):241−7.

[78] Vajta G. Vitrification of the oocytes and embryos of domestic animals. Anim Reprod Sci 2000;60-61:357−64 Review.

[79] Mukaida T, Oka C, Goto T, Takahashi K. Artificial shrinkage of blastocoeles using either a micro-needle or a laser pulse prior to the cooling steps of vitrification improves survival rate and pregnancy outcome of vitrified human blastocysts. Hum Reprod 2006;21(12):3246−52.

[80] Choi YH, Velez IC, Riera FL, Roldán JE, Hartman DL, Bliss SB, et al. Successful cryopreservation of expanded equine blastocysts. Theriogenology 2011;76:143−52.

[81] Diaz F, Bondiolli K, Paccamonti D, Gentry GT. Cryopreservation of Day 8 equine embryos after blastocyst micromanipulation and vitrification. Theriogenology 2016;85(5):894−903.

[82] Barfield JP, McCue PM, Squires EL, Seidel Jr. GE. Effect of dehydration prior to cryopreservation of large equine embryos. Cryobiology 2009;59(1):36−41.

[83] Taylor PJ, Taylor S, Sansinema MJ, Godke RA. Llama (*Lama glama*) pregnancies from vitrified/warmed blastocysts using a novel coaxial cryoprotectant microinjection system. Reprod Fertil Dev 2005;18:164 (Abs 112).

[84] Arav A, Shehu D, Mattioli M. Osmotic and cytotoxic study of vitrification of immature bovine oocytes. J Reprod Fertil 1993;99:353−8.

[85] Herrid M, Billah M, Malo C, Skidmore JA. Optimization of a vitrification protocol for hatched blastocysts from the dromedary camel (*Camelus dromedarius*). Theriogenology 2016;85(4):585−90.

[86] Lutz J, Johnson S, Duprey K, Taylor P, Vivanco H, Ponce-Salaar M, et al. 7 Pregnancy from a vitrified-warmed alpaca pre-implantation embryo. Reprod Fert Dev 2020;32(2):128.

[87] Bravo PW, Skidmore JA, Zhao XX. Reproductive aspects and storage of semen in Camelidae. Anim Reprod Sci 2000;62:173−93.

[88] Bravo PW, Moscoso V, Alarcon V, Ordonez C. Ejaculatory process and related semen characteristics. Arch Androl 2002;48(1):65−72.

[89] Adams GP, Ratto MH, Collins CW, Bergfelt DR. Artificial insemination in South American camelids and wild equids. Theriogenology 2009;71(1):166−75.

[90] M. Abdel-Raouf, M.A. El-Naggar, Studies on reproduction in camels (*Camelus dromedarius*). VI. Properties and constituents of ejaculated semen. In: *VIIIth international congress on animal reproduction and artificial insemination*. Cracow; 1978. p. 862−65.

[91] Tingari MD, El-Manna MMM, Rahim ATA, Ahmed AK, Hamad MH. Studies on camel semen. I: electroejaculation and some aspects of semen characteristics. Anim Reprod Sci 1986;12:213−22.

[92] Mostafa TH, Abdel-Salaam AM, El-Badry DE. Abear M. Freezability and DNA integrity of dromedary camel spermatozoa in semen collected by artificial vagina and electro-ejaculator. Egypt J Anim Prod 2014;51(2):145−55.

[93] Tibary A, Anouassi A. Artificial breeding and manipulation of reproduction in camelidae. In: Tibary A, editor. Theriogenology in camelidae: anatomy, physiology, pathology and artificial breeding. Actes Edition. Institut Agronomique et Veterinaire Hassan II. Abu Dhabi Printing and Publishing Company; 1997. p. 413−52.

[94] Skidmore JA, Morton KM, Billah M. Artificial insemination in dromedary camels. Anim Reprod Sci 2013;136(3):178−86.

[95] El-Hassanein EE. An invention for easy semen collection from dromedary camels, El-Hassanein camel dummy. Available from: www.ivis.org In: Skidmore JA, Adams GP, editors. Recent advances in camelid reproduction. Ithaca, NY: International Veterinary Information Service; 2003.

[96] Deen A, Vyas S, Sahani MS. Semen collection, cryopreservation and artificial insemination in the dromedary camel. Anim Reprod Sci 2003;77:223−33.

[97] Al Bulushi S, Manjunatha BM, Bathgate R, Rickard JP, de Graaf SP. Effect of semen collection frequency on the semen characteristics of dromedary camels. Anim Reprod Sci 2018;197:145−53.

[98] Morton KM, Thomson PC, Bailey KJ, Evans G, Maxwell WMC. Quality parameters for alpaca (*Vicugna pacos*) are affected by semen collection procedure. Reprod Domest Anim 2010;45:637−43.

[99] Urquieta B, Flores P, Munoz C, Bustos-Obregon E, Garcia-Huidobro J. Alpaca semen characteristics under free and directed mounts during a mating period. Anim Reprod Sci 2005;90:329−39.

[100] Bravo PW, Callo M, Garnica J. The effect of enzymes on semen viscosity in llamas and alpacas. Small Rumin Res 2000;38:91−5.

[101] Vaughan JL, Galloway DB, Hopkins DA. Artificial insemination in alpacas (*Lama pacos*). Canberra: *Rural Industries Research and Development Corporation*; 2003.

[102] Santiani A, Huanca W, Sapana R, Huanca T, Sepulveda N, Sanchez R. Effects on the quality of frozen-thawed alpaca (*Lama pacos*) semen using two different cryoprotectants and extenders. Asian J Androl 2005;7:303−9.

[103] Giuliano S, Director A, Gambarotta M, Trasorras V, Miragaya M. Collection method, season and individual variation on seminal characteristics in the llama (*Lama glama*). Anim Reprod Sci 2008;104:359−69.

[104] Bravo PW, Flores D, Ordonez C. Effect of repeated collection on semen characteristics of alpacas. Biol Reprod 1997;57(3):520−4.

[105] Morton KM, Vaughan JL, Maxwell CWM. Continued development of artificial insemination technology in alpacas. Canberra: *Rural Industries Research and Development Corporation*; 2008.

[106] Kershaw-Young CM, Maxwell WM. Seminal plasma components in camelids and comparisons with other species. Reprod Domest Anim 2012;47(Suppl 4):369−75.

[107] Wani NA, Billah M, Skidmore JA. Studies on liquefaction and storage of ejaculated dromedary camel (*Camelus dromedarius*) semen. Anim Reprod Sci 2008;109:309−18.

[108] Kershaw-Young CM, Evans G, Maxwell WMC. Glycosaminoglycans in the accessory sex glands, testes, and seminal plasma of alpaca and ram. Reprod Fertil Dev 2012;24:362−9.

[109] El-Bahrawy KA, El-Hassanein EE. Effect of different mucolytic agents on viscosity and physical characteristics of dromedary camel semen. Alex J Agric Res 2009;54:1−6.

[110] Ghoneim IM, Al-Eknah MM, Waheed MM, Alhaider AK, Homeida AM. Effect of some extenders and enzymes on semen viscosity and sperm viability in dromedary camels (*Camelus dromedarius*). J Camel Pract Res 2010;17:85−9.

[111] Giuliano S, Carretero M, Gambarotta M, Neild D, Trasorras V, Pinto M, et al. Improvement of llama (*Lama glama*) seminal characteristics using collagenase. Anim Reprod Sci 2010;118:98−102.

[112] Shekhar C, Vyas S, Purohit GN, Patil NV. Use of collagenase type-1 to improve the seminal characteristics of dromedary camel semen. Eur J Vet Med 2012;1(1):17−27.

[113] El-Bahrawy KA. Cryopreservation of dromedary camel semen supplemented with α-amylase enzyme. J Camel Pract Res 2010;17:211−16.

[114] Kershaw-Young CM, Stuart C, Evans G, Maxwell WMC. The effect of glycosaminoglycan enzymes and proteases on the viscosity of alpaca seminal plasma and sperm function. Anim Reprod Sci 2013;138:261−7.

[115] El-Bahrawy KA. Influence of enzymatic and mechanical liquefaction of seminal plasma on freezability of dromedary camel semen. World Vet J 2017;7:108−16.

[116] Monaco D, Fatnassi M, Padalino B, Hammadi M, Khorchani T, Lacalandra GM. Effect of alpha-amylase, papain, and spermfluid® treatments on viscosity and semen parameters of dromedary camel ejaculates. Res Vet Sci 2016;105:5−9.

[117] G. Mal, S. Vyas, A. Srinivasan, N.V.R. Patil, K.M.L. Pathak, Studies on liquefaction time and proteins involved in the improvement of seminal characteristics in dromedary camels (*Camelus dromedarius*). Scientifica 2016;2016:4659358. <https://doi.org/10.1155/2016/4659358>.

[118] Rateb SA. Ultrasound-assisted liquefaction of dromedary camel semen. Small Rumin Res 2016;141:48−55.

[119] El-Bahrawy KA. Effect of seminal plasma centrifugation for viscosity elimination on cryopreservation of dromedary camel semen. Nat Sci 2010;8(9):196−202.

[120] Akbar SJ, Hassan SM, Ahmad M. Studies on semen processing, cryopreservation and artificial insemination in dromedary camel. Int J Anim Sci 2018;2:1018−23.

[121] Malo C, Crichton EG, Morrell JM, Pukazhenthi BS, Johannisson A, Splan R, et al. Colloid centrifugation of fresh semen improves post-thaw quality of cryopreserved dromedary camel spermatozoa. Anim Reprod Sci 2018;192:28−34.

[122] Skidmore JA, Malo CM, Crichton EG, Morrell JM, Pukazhenthi BS. An update on semen collection, preservation and artificial insemination in the dromedary camel (*Camelus dromedarius*). Anim Reprod Sci 2018;194:11−18.

[123] Carretero MI, Giuliano SM, Casaretto CI, Gambarotta MC, Neild DM. Evaluation of the effect of cooling and of the addition of collagenase on llama sperm DNA using toluidine blue. Andrologia 2012;44(Suppl 1):239−47.

[124] Sieme H, Merkt H, Musa E, Badreldin H, Willmen T. Liquid and deep freeze preservation of camel semen using different extenders and methods. In: Saint-Martin G, editor. Proceedings Unite de Coordination pour l'Elevage Camelin. Workshop: "Is it possible to improve the reproductive performance of the camel?". Paris: CIRAD-EMVT; 1990. p. 273−84.

[125] Vyas S, Goswami P, Rai AK, Khanna ND. Use of Tris and lactose extenders in preservation of camel semen at refrigerated temperature. Indian Vet J 1998;75(9):810−12.

[126] Skidmore JA, Billah M, Allen WR. Using modern reproductive technologies such as embryo transfer and artificial insemination to improve the reproductive potential of dromedary camels. Rev Elev Med Vet Pays Trop 2000;53(2):97−100.

[127] Deen A, Vyas S, Jain M, Sahani MS. Refrigeratory preservation of camel semen. J Camel Pract Res 2004;11:137−9.

[128] Bergeron A, Manjunath P. New insights towards understanding the mechanisms of sperm protection by egg yolk and milk. Mol Reprod Dev 2006;73:1338−44.

[129] K.A. El-Bahrawy, E.E. El-Hassanien, A.Z. Fateh El-Bab, M.M. Zeitoun, Semen characteristics of the male camel and its freezability after dilution in different extenders. In: *Proceedings of the international scientific camel conference*, El-Qaseem; 2006. p. 2037−2053.

[130] El-Hassanein EE. Freezability of camel semen collected by "El-Hassanein camel dummy" and diluted in two steps in sucrose and/or Tris-based extenders. Vet Med J Giza 2006;54:29−46.

[131] Waheed M, Ghoneim IM, Al-Eknah MM, Al-Haider AK. Effect of extenders on motility and viability of chilled-stored camel spermatozoa. (*Camelus dromedarius*). J Camel Pract Res 2010;17:217−20.

[132] Morton KM, Gibb Z, Bertoldo M, Maxwell CWM. Effect of diluent, dilution rate and storage temperature on longevity and functional integrity of liquid stored alpaca (*Vicugna pacos*) semen. J Camelid Sci 2009;2:15−25.

[133] K.M. Morton, M. Billah, J.A. Skidmore, Artificial insemination of dromedary camels with fresh and chilled semen: effect of diluent and sperm dose, preliminary results. In: *Proceedings of the eighth international symposium on reproduction in domestic ruminants*; 2010. p. 493 (Abs).

[134] Kutty CI, Koroth A. Collection, evaluation, processing and preservation of semen from dromedary camels (*Camelus dromedarius*). In: Juhasz J, Nagy P, Skidmore JA, Malo CM, editors. Proceedings of the ICAR satellite meeting on camelid reproduction. Vancouver; 2012. p. 57−60.

[135] Crichton EG, Pukazhenthi BS, Billah M, Skidmore JA. Cholesterol addition aids the cryopreservation of dromedary camel (*Camelus dromedarius*) spermatozoa. Theriogenology 2015;83:168−74.

[136] Al-Bulushi S, Manjunatha BM, Bathgate R, de Graaf SP. Effect of different extenders on sperm motion characteristics, viability and acrosome integrity during liquid storage of dromedary camel semen. Anim Reprod Sci 2016;169:127 (Abs).

[137] Malo C, Crichton EG, Skidmore JA. Optimization of the cryopreservation of dromedary camel semen: cryoprotectants and their concentration and equilibration times. Cryobiology 2017;74:141−7.

[138] Niasari-Naslaji A, Mosaferi S, Bahmani N, Gharahdaghi AA, Abarghani A, Ghanbari A. Effectiveness of a tris-based extender (SHOTOR diluent) for the preservation of Bactrian camel (*Camelus bactrianus*) semen. Cryobiology 2006;53:12−21.

[139] Giuliano SM, Chaves MG, Trasorras VL, Gambarotta M, Neild D, Director A, et al. Development of an artificial insemination protocol in llamas using cooled semen. Anim Reprod Sci 2012;131:204−10.

[140] Panahi F, Niasari-Naslaji A, Seyedasfari F, Ararooti T, Razavi K, Moosavi-Movaheddi AA. Supplementation of tris-based extender with plasma egg yolk of six avian species and camel skim milk for chilled preservation of dromedary camel semen. Anim Reprod Sci 2017;184:11−19.

[141] Zhao XX, Huang YM, Chen BX. Artificial insemination with deep-frozen semen in the bactrian camel (*Camelus bactrianus*). In: Department of Animal Husbandry and Health, Ministry of Agriculture and Chinese Association of Agricultural Science Societies, editor. Research progress in animal industry and animal products processing, the first China international annual meeting on agricultural and technology. Beijing: China Agricultural Scientech Press; 1996. p. 237−42.

[142] Malo C, Grundin J, Morrell JM, Skidmore JA. Male dependent improvement in post-thaw dromedary camel sperm quality after addition of catalase. Anim Reprod Sci 2019; in press.

[143] C. Stuart, The development of assisted reproductive technologies in camelid, especially the alpaca (*Vicugna pacos*). PhD thesis. University of Sydney, 2016.

[144] Kershaw-Young CM, Maxwell WM. The effect of seminal plasma on alpaca sperm function. Theriogenology 2011;76:1197−206.

[145] Morton KM, Bathgate R, Evans G, Maxwell WMC. Cryopreservation of epididymal alpaca (*Vicugna pacos*) sperm: a comparison of citrate-based, tris-based and lactose-based diluents and pellets and straws. Reprod Fertil Dev 2007;19:792−6. Available from: https://doi.org/10.1071/rd07049 (2007).

[146] Abdel-Salaam AM. Freezability of camel spermatozoa as affected by cryoprotective agents and equilibration periods added with pentoxifylline. Egypt J Basic App Physiol 2013;12(1):17−48.

[147] Azam S, Had N, Khan NU, Hadi SM. Antioxidant and pro-oxidant properties of caffeine, theobromide and xanthine. Med Sci Monit 2003;9:325−30.

[148] Santiani A, Evangelista S, Valdivia M, Risopatron J, Sanchez R. Effect of the addition of two superoxide dismutase analogues (Tempo and Tempol) to alpaca semen extender for cryopreservation. Theriogenology 2013;79:842−6.

[149] Medan MS, Absy G, Zeidan AE, Khalil MH, Khalifa HH, Abdel-Salaam AM, et al. Survival and fertility rate of cooled dromedary camel spermatozoa supplemented with catalase enzyme. J Reprod Dev 2008;54(1):84−9.

[150] C. Malo, L. Soederstroem, B. Elwing, J.A. Skidmore, J.M. Morrell, Effect of antioxidants and thawing rates on the quality of cryopreserved camel sperm. In: 33rd Annual Meeting A.E.T.E. Bath, U.K. 2017.

[151] El-Harairy MA, El-Razek IMA, Abdel-Khalek EA, Shamiah SM, Zaghloul HK, Khalil WA. Effect of antioxidants on the stored dromedary camel epididymal sperm characteristics. Asian J Anim Sci 2016;10(2):147−53.

[152] Zeidan AE. Semen quality, enzymatic activities and penetrating ability of spermatozoa into she-camel cervical mucus as affected by caffeine addition. J Camel Pract Res 2002;9:153−61.

[153] Malo C, Elwing B, Soederstroem L, Lundeheim N, Morrell JM, Skidmore JA. Effect of different freezing rates and thawing temperatures on cryosurvival of dromedary camel spermatozoa. Theriogenology 2019;125:43−8.

[154] Niasari-Naslaji A, Mosaferi S, Bahman N, Gerami A, Gharahdaghi AA, Abarghani A, et al. Semen cryopreservation in Bactrian camel (*Camelus bactrianus*) using SHOTOR diluent: effects of cooling rates and glycerol concentrations. Theriogenology 2007;68:618−25.

[155] Fernandez-Baca S, Madden DHL, Novoa C. Effect of different mating stimuli on induction of ovulation in the alpaca. J Reprod Fertil 1970;22:261−7.

[156] Musa BE, Abusineina ME. The oestrous cycle of the camel (*Camelus dromedarius.*). Vet Rec 1978;102:556−7.

[157] Skidmore JA, Billah M, Allen WR. The ovarian follicular wave pattern and induction of ovulation in the mated and non-mated one-humped camel (*Camelus dromedarius*). J Reprod Fertil 1996;106:185−92.

[158] Adams GP, Sumar J, Ginther OJ. Effects of lactational and reproductive status on follicular waves in llamas. J Reprod Fert 1990;90:535−45.

[159] Bravo PW, Fowler ME, Stabenfeldt GH, Lasley BL. Ovarian follicular dynamics in the llama. Biol Reprod 1990;43:579−85.

[160] Marie M, Anouassi A. Mating induced luteinizing hormone surge and ovulation in the female camel (*Camelus dromedarius*). Biol Reprod 1986;35:792−8.

[161] Marie M, Anouassi A. Induction of luteal activity and progesterone secretion in the non-pregnant one-humped camel (*Camelus dromedarius*). J Reprod Fertil 1987;80:183−92.

[162] Bourke DA, Kyle CE, McEvoy TG, Young P, Adam CL. Recipient synchronisation and embryo transfer in South American camelids. Theriogenology 1995;43:171.

[163] Stuart CC, Vaughan JL, Kershaw-Young CM, Wilkinson J, Bathgate R, de Graaf SP. Effects of varying doses of beta-nerve growth factor on the timing of ovulation, plasma progesterone concentration and corpus luteum size in female alpacas (*Vicugna pacos*). Reprod Fertil Dev 2015;27:1181−6 2015.

[164] Hill JR, Dobrinski I. Male germ cell transplantation in livestock. Reprod Fertil Dev 2006;18:13−18.

[165] Herrid M, Davey RJ, Hutton K, Colditz IG, Hill JR. A comparison of methods for preparing enriched populations of bovine spermatogonia. Reprod Fertil Dev 2009;21:393−9.

[166] Honaramooz A, Megee SO, Dobrinski I. Germ cell transplantation in pigs. Biol Reprod 2002;66(1):21−8.

[167] Honaramooz A, Behboodi E, Blash S, Megee SO, Dobrinski I. Germ cell transplantation in goats. Mol Reprod Dev 2003;64(4):422−8.

[168] Herrid M, Vignarajan S, Davey R, Dobrinski I, Herrid M, McFarlane JR. Application of testis germ cell transplantation in breeding systems of food producing species: a review. Anim Biotechnol 2013;24:293−306.

[169] Herrid M, Nagy P, Juhasz J, Morrell JM, Billah M, Khazanehdari K. Donor sperm production in heterologous recipients by testis germ cell transplantation in the dromedary camel. Reprod Fertil Dev 2018. Available from: https://doi.org/10.1071/RD18191 Oct 12.

Chapter 9

Reproductive technologies in companion animals

G.C. Luvoni, M.G. Morselli and M. Colombo

Department of Health, Animal Science and Food Safety "Carlo Cantoni" - Università degli Studi di Milano, Milan, Italy

9.1 Artificial insemination

Artificial Insemination (AI) in dogs began in the 18th century with the experiments of Lazzaro Spallanzani [1], a pioneer in assisted reproduction technologies (ARTs), when there were still beliefs of "spontaneous generation" and embryo development as a result of growth of preformed parts (theory of preformationism). From that time, with the progress of science and the in-depth study of gametes, AI became a common practice first in large animals and then in dogs, although the first artificial vagina was developed in 1914 for dogs by Amantea, an Italian professor of the Medical School, and it was later modified and adopted for other animals. While in other species, as ruminants, it is the method of choice for semen collection, in dogs ejaculated spermatozoa are easily obtained by digital manipulation rather than using the artificial vagina.

Currently, breeders and owners are very well prepared on the topic and require this procedure for different reasons: to avoid stress for animal transportation, to avoid sanitary risks due to physical contact among animals, to obtain conception in dogs that are unable to naturally mate for anatomical or pathological reasons, or to obtain progeny from old or even dead individuals.

It follows that, along with the diffusion of AI, procedures for semen preservation were also investigated. Seager in 1969 [2] reported the first successful pregnancy from frozen-thawed dog semen, and today short-term preservation by chilling at $4°C-5°C$ or long-term preservation in liquid nitrogen is a common way to extend spermatozoa survival for a postponed use.

Deposition of spermatozoa in the female genital system can be done more or less close to the oviductal site of fertilization, and the choice is based on the survival time of gametes. Since fresh spermatozoa are characterized by a long survival, they can be deposited into the cranial vagina with an insemination pipette and, due to their motility and to uterine contractions, they can reach the oviducts, where fertilization occurs. On the other hand, chilled or, even more, frozen-thawed spermatozoa are stressed by cold temperatures and their survival decreases according to the temperature of preservation (few days in chilled and few hours in frozen-thawed semen). Therefore an intrauterine insemination is highly advisable for chilled semen, but mandatory for frozen semen. The relative inaccessibility to the uterine cavity from outside due to the anatomical conformation of the vagina (narrow and with a prominent dorsal medial fold) and the cervical *os* orientation (toward the floor of the vagina) requires specific equipment and techniques. Indeed, intrauterine insemination is commonly performed by endoscopic-assisted transcervical catheterization or a surgical approach (laparotomy/laparoscopy) [3,4], which however raises ethical concerns and it is not allowed in all the countries because of its invasiveness [5].

To ensure high rates of successful pregnancy, the optimal time for insemination must be identified. Differently from those animals, as ruminants, characterized by a short estrus, in which performing insemination before ovulation is not a challenge, the bitch has a peculiar reproductive physiology that includes a long estrus period and ovulation of primary oocytes, at the beginning of the first meiotic division. Subsequent stages of meiotic maturation are resumed in the oviduct and take 2-3 days to achieve the metaphase II (MII) stage [6]. This means that to maximize fertility rates,

Reproductive Technologies in Animals. DOI: https://doi.org/10.1016/B978-0-12-817107-3.00009-6

particularly in the case of short survival time of the preserved male gametes, insemination should be performed after ovulation, when oocytes are prone to be fertilized. This issue makes the monitoring of optimal time for insemination harder than in the cow, and the determination of progesterone concentration [which initial rise is concurrent with the luteinizing hormone (LH) surge] is the preferred way to identify the fertile period [3].

Different factors affect the efficiency of AI, as site of deposition, time, and number of inseminations, as well as concentration and quality of fresh or preserved semen. Therefore it is hard to talk about average efficiency of the technique, but it is generally known that it may reach 80% with fresh semen by intravaginal deposition and 70%−75% with frozen semen by intrauterine insemination [7,8]. Intrauterine insemination can also be adopted for fresh semen when the quality of the ejaculate is below the standard.

Recent progress in AI includes the use of sex-sorted fresh or frozen spermatozoa placed respectively into the uterus by transcervical endoscopic catheterization or surgically into the oviducts [9,10]. Birth of healthy puppies of predicted sex was obtained and new perspectives in dog breeding were opened especially for working dogs, such as guide dogs, military, and police dogs.

Insemination with ejaculated spermatozoa is not the only possibility to obtain progeny. Epididymal spermatozoa also represent a source of male germplasm that can be obtained from isolated testicles of animals orchiectomized for medical purposes or that unexpectedly die. Even in the case of erectile dysfunction or ejaculation failure due to pathological conditions or presence of an azoospermic ejaculate, epididymal spermatozoa can be retrieved and used for AI as demonstrated in dogs with fresh, chilled, and frozen epididymal gametes that resulted in the birth of puppies [11].

In cats, due to the anatomical features of the genital organ and, in most cases, to the scarcely collaborative temperament, collection of semen with an artificial vagina is not a common procedure, and it can only be accomplished in trained males. In not trained males electroejaculation in general anesthesia or urethral catheterization after administration of medetomidine [12] has been proposed.

Although the identification of optimal time for insemination is not as difficult as in dogs, since queens are induced ovulators, a pharmacological treatment [one or two injections of human chorionic gonadotropin (hCG)] must be planned at the right time during estrus to cause ovulation [13].

Intravaginal deposition of fresh and frozen semen and the first documented successful AI were reported in the 1970s [14,15]. However, for cryopreserved semen, intrauterine insemination should be performed and the technique is even more complicated than in dogs. Besides the anatomical conformation, the small dimension of the vagina (diameter lower than 2 mm) requires the use of very small catheters [16,17,18]. Recently, a human semirigid sialendoscope has been proposed for endoscopic transcervical insemination of queens [18], but it is clear that this procedure is not commonly adopted. The alternative is the surgical insemination, with the related ethical problems, that resulted in 70%−75% of pregnancy rates in small groups of queens [19,20]. Epididymal semen has also been used in the cat and after freezing and surgical intrauterine or intratubal insemination, kittens were born [21,22].

9.2 Embryo production in vitro

The in vitro embryo production (IVEP) in companion animals started more than 40 years ago [23,24] and from then on, several studies focused on in vitro maturation (IVM), in vitro fertilization (IVF), and embryo development were published. The ultimate goal is to produce embryos that can be transferred to obtain progeny or that can be cryopreserved for the safeguard of biodiversity. Cats and dogs are valuable models for endangered species and cats can also be used as recipients for wild animals' embryos [25,26].

In domestic cats, good results have been obtained in terms of IVM (60%), cleavage (60%), and cleaved embryos reaching the morula/blastocyst stage (50%), and successful pregnancies with the birth of live kittens after embryo transfer were also documented [27]. Nevertheless, the ideal environment to improve IVEP to the same extent than in other species is still under investigation, and the identification of physical and chemicals factors determining the acquisition of oocyte developmental competence during in vitro culture (IVC) seems to be the key factor for success [27]. Basic or enriched, home-made or commercial media were tested to understand the real requirements of gametes and embryos in in vitro conditions [27]. Along the years "one-step" cultures have been replaced by sequential media, in order to better mimic the dynamic conditions that reproductive cells meet during their migration in the reproductive system [28,29]. The role of protein supplementations, sugars, ions, amino acids, hormones, and growth factors was investigated to determine the best composition of culture media. For example, nowadays there is agreement on the beneficial effects of follicle-stimulating hormone and LH during IVM and on the use of bovine serum albumin for IVM and at the beginning of embryo culture. Serum addition should be limited only to later stages of development, as well as the use of essential amino acids to stimulate blastocyst formation [27].

In addition, immunostaining approaches and molecular biology studies were applied to understand how the acquisition of developmental competence and the gene expression profile of oocytes and embryos during culture occur [30−33]. To mimic more faithfully the biological follicular niche, cocultures with companion cells (somatic cells or cumulus oophorus-oocyte complexes) in two-dimensional or three-dimensional systems were also evaluated in our studies [27,34]. Ongoing studies are looking at the creation of an artificial ovary in which multiple living cells are connected with each other on the same support (chip) to reproduce in vivo conditions [35].

In dogs, most of the ARTs advancements have been hindered by the peculiar reproductive physiology of bitches, which complicates the creation of ideal in vitro conditions [36,37]. Due to many factors (i.e., the estrus phase of the female from which oocytes are collected, the IVC timing, the cultural supplementation [36]), in vitro meiosis resumption is only accomplished by 30% of recovered oocytes [26]. To improve maturation outcomes, extensive studies were performed to acquire basic knowledge about the in vivo endocrine control and the physiological environment of oocytes [38]. Trying to mimic the in vivo hormonal effects, the application of sequential media and prolonged IVM (culture in presence of hCG for the first 48 hours and without in the last 48 hours [39]) gave the highest rate of full nuclear maturation (approximately 40% MII). Further studies focused on the distribution of cortical granules during IVM in biphasic systems confirmed the close connection between culture time, hormonal supplementation, and bitches' reproductive status [40,41]. Oocytes cultured in isolated ligated oviduct, as in our study [42], or in coculture with homologous or heterologous somatic cells (oviductal or granulosa cells, embryonic fibroblasts, or liver cells), gave variable results [36−38,43−46]. Despite all these efforts, very few in vitro−derived embryos have been produced. The latest very encouraging result, which gives some hopes for canine ARTs, was reported in 2015: in vivo−matured oocytes were in vitro fertilized and, after transfer of vitrified four-cell stage embryos into recipients' oviducts, live pups were successfully obtained [47].

As well as in the feline species, molecular biology investigations on canine oocytes were introduced to improve ARTs success. Different approaches were applied, such as the proteomic analysis of tubal fluid or the genetic expression of specific molecules and growth factors during in vivo and IVM. For example, our study [48], together with others [49,50], allowed the detection and the quantification of target genes, as the growth differentiation factor-9 (GDF-9) and the bone morphogenetic protein-15 (BMP-15), giving more information on the regulation of the molecular mechanisms that trigger the acquisition of oocyte developmental competence and meiosis resumption. Very recently, the addition of these factors to the culture medium was proved to be beneficial for the acquisition of full nuclear maturation [51].

9.3 Cryopreservation of gametes, embryos, and gonadal tissue

Cryopreservation of gametes, embryos, and gonadal tissue might significantly contribute to the preservation of genetic material of valuable individuals, or, as already mentioned for semen, might facilitate international exchange of germplasm. Different cells or set of cells, as embryos or tissues, have different characteristics of resilience to nonphysiological temperatures and osmotic stress and permeability to cryoprotectant agents (CPAs). Therefore several procedures (slow freezing, vitrification, etc.) and protocols (types, concentrations and combinations of CPAs, time of exposure, etc.) have been proposed for cryostorage in different species, including cats and dogs.

9.3.1 Spermatozoa

Spermatozoa are highly specialized cells, different from somatic cells. Their cytoplasm is very limited, many common organelles are absent, and specialized structures, such as the acrosome and the proximal centriole, with a crucial role in the fertilization process, are instead present.

During cryopreservation, the cell is exposed to subzero temperatures and different strategies to limit cryodamages (e.g., membranes destabilization, changes in the lipid composition, modifications of acrosome status), including the dilution in extenders with egg yolk, the addition of cryoprotectants, and a controlled temperature decrease, have been tested [52].

Egg yolk is commonly added to chilling extender (20% v/v) because its content of phospholipids, cholesterol, and low-density lipoproteins protects the cellular membranes. Cryoprotectants modulate the rate of cellular dehydration, but since they are toxic, especially at room temperature, the addition to the diluted semen after an equilibration period at chilling temperature is needed. This is the reason for the two-step dilution applied in some freezing procedures, as the well-known Uppsala system for dog and cat semen [8,13]. The two-step system consists of a first dilution at room temperature with the extender supplemented with a low concentration of glycerol as cryoprotectant, and a second dilution, after chilling to +4°C/5°C, with the extender supplemented with a higher dose of glycerol to reach its final concentration.

Dog semen is generally frozen in straws with a gradual decrease of temperature by exposure to liquid nitrogen vapors or in pellets by using dry ice. Results of AI have been previously reported in this review and successful pregnancy rates justify the widespread use of frozen semen in canine breeding.

Cryopreservation of epididymal spermatozoa is also an option to preserve genetic material, but it has been documented in dogs [11] and cats [52] that their resilience to cold damages is lower than ejaculated spermatozoa. Even testicular spermatozoa, which are not able to spontaneously fertilize oocytes for their immobility and immaturity, can be cryopreserved [53] for potential use by intracytoplasmic sperm injection (ICSI).

These findings indicate that an optimal freezing protocol has not been defined yet and, although cryopreservation of epididymal spermatozoa in dogs and cats brought encouraging results, further investigations aimed at protecting gametes from freezing injuries are still needed.

Cat breeders are only recently approaching cryoprocedures, mainly due to the difficulties of semen collection from tomcats, although exhaustive reviews and protocols for the cryopreservation of ejaculated and epididymal gametes are already available [54]. One of the most commonly used protocols has been modified from a protocol that was originally developed for preservation of canine spermatozoa [13].

As previously mentioned, live kittens were obtained after AI with frozen ejaculated and epididymal gametes [17,19,20,22,55], even though this ART is not as diffuse in cats as in dogs.

Innovative methods of long-term preservation, such as vitrification and freeze-drying, are currently under investigation in several animal species, including dogs and cats.

Vitrification was developed as an alternative to conventional freezing in order to avoid formation of ice crystals. It requires the direct immersion of the cells into liquid nitrogen, suspended in a viscous solution with high concentration of cryoprotectants. The rapid decrease of the temperature causes the glass transition of extra- and intracellular liquids, thus preventing the formation of ice crystals.

The procedure is easy and quick, but not free of dangers for spermatozoa. Highly concentrated cryoprotectants are toxic and the "glassy" state is quite fragile. However, with the use of only sucrose, a nonpermeable cryoprotectant, encouraging results in terms of sperm motility and viability after warming were obtained in dog semen [56].

Very recently, cleaved embryos (35%) were successfully produced after IVF of cat oocytes with vitrified ejaculated sperm [57], further suggesting a possible application of this new technology to small carnivores' semen.

Lyophilization, or freeze-drying, is a preservation method in which frozen material is dried by sublimation of ice. Its main purpose is the storage at room temperature or +4°C of the samples and then their reconstitution with the addition of water [58]. With this method, costs for semen storage and shipping would be significantly reduced, because liquid nitrogen would no longer be necessary. However, for the loss of motility, fertilization could only be obtained in vitro by ICSI.

Application of freeze-drying to dog and cat spermatozoa is still at an experimental level. Studies are currently focused on the definition of protocols preserving DNA integrity, which is severely damaged by this method [59,60]. In the cat, cleaved embryos were obtained after ICSI with freeze-dried spermatozoa, but severe anomalies of spermatozoa were also observed [61,62,63].

9.3.2 Oocytes and embryos

Oocytes are more sensitive than embryos to cold injuries, which for female gametes include cytoskeleton disorganization, chromosome and DNA abnormalities, spindle disintegration, plasma membrane disruption, and premature cortical granule exocytosis with consequent hardening of the zona pellucida. This is due to their morphological characteristics such as the low surface area/volume ratio, the presence of a proteinaceous coat (zona pellucida) and the surrounding cumulus cells. Moreover, the domestic cat oocyte has a high content of lipid droplets in the ooplasm [52]. Therefore the permeability of the cell and the time of osmotic balance with the cryoprotectant solution are affected by the structure of cumulus-oocyte complexes and their response to physical and chemical events occurring during freezing may affect their quality. Despite the extreme sensitivity of oocytes to chilling, significant advances have been achieved, particularly in feline species.

Firstly described by Luvoni and colleagues on immature (germinal vesicle, GV) and mature (MII) cat oocytes [64,65], slow freezing and vitrification procedures were applied for fertility preservation aims [52,66]. Significant advancements are represented by successful pregnancies and live births after the transfer of in vitro−derived embryos from vitrified GV or in vitro−matured (MII) oocytes [66], and kittens obtained by transferring cryopreserved in vitro−produced embryos [67]. However, an optimal cryopreservation procedure for oocytes or embryos has not been defined yet and several studies were aimed at the reduction of the cryoinjuries and improvements of maturation and embryonic developmental rates, for instance with the use of enriched culture conditions after warming, as our recent study investigated [68]. The attention is now focused on the investigation of the intracellular mechanisms affected by the cold-induced damages that, consequently, negatively influence the oocyte developmental competence. For instance, the inhibition of specific molecules that regulate apoptosis [69], or the block of transmembrane channels components [70], brought some improvements in IVM or cleavage of cryopreserved oocytes. Recently, one study on proteomic analysis revealed a different expression profile of 258 proteins after IVM between control (fresh) and vitrified-warmed oocytes, highlighting which molecules are involved and mainly stressed during cryoprocedures [71]. Again, the evaluation of the lipid droplet phase transition on cryopreserved cat oocytes and embryos, evaluated by Raman spectroscopy, provided new interesting insights [72] on small carnivores' oocyte "freezability."

As in spermatozoa, the possibility of cat oocyte preservation at supra-zero temperature was investigated with the aim of facilitating storage and transportation without the need of liquid nitrogen. Desiccation, storage at 4°C, rehydration, and transplantation into fresh conspecific cytoplasts allowed the resumption of meiosis of air-dessicated GV [73], whereas preliminary results of microwave-assisted drying were reported in 2015 [74].

In dogs, due to the poor in vitro developmental competence of the oocytes, experimental research devoted to female germplasm cryopreservation is still limited and from the first study published in 2008 [75], only few attempts were made to preserve the viability and the membrane integrity of immature canine cumulus-oocyte complexes. On the other hand, cryopreservation of in vivo−produced embryos, although also poorly investigated, allowed the birth of live puppies from vitrified embryos [76,77].

9.3.3 Gonadal tissues

In gonadal tissues, the presence of different cell types with variable cryotolerance makes their cryopreservation more challenging than gametes alone. In cats, the maintenance of follicle viability and enclosed oocyte ability to resume meiosis after slow freezing or vitrification of ovarian tissue have been documented [66]. Cryopreservation of cat ovarian tissue followed by xenotransplantation to severe combined immunodeficient (SCID) mice has also been reported. Although only few follicles survived after freezing and transplantation, some retained the ability to resume growth from early to more advanced stages [78]. Perfecting protocols (different CPAs, time of exposure, temperature, dimensions of gonadal tissue fragments, etc.) for the improvement of follicular functional integrity after warming is the main aim of recent studies on this topic.

In dogs, follicular viability and functionality after slow freezing and vitrification of ovarian fragments were evaluated, and the preservation of tissue normal histology was obtained. However, after xenotransplantation of preserved tissue into mice, follicular survival and growth were variable [79−82].

Testicular tissue, as a reservoir of germ cells, can be stored to widen the availability of male genetic resources. Freezing of feline tissue needs further investigations in order to identify the best protocols and conditions to retrieve, after thawing, useful gametes for IVEP (by ICSI) or viable fragments able to resume the spermatogenesis after transplantation.

The goal of fertility preservation was achieved when the birth of kittens after the transfer of frozen-thawed embryos produced by ICSI with testicular sperm retrieved from frozen tissue was successfully reported in 2012 [83].

9.4 Innovative approaches

In the near future, the most striking progresses in small animal reproductive biotechnologies could occur with the widespread application of somatic cloning or with the development of in vitro gametogenesis, which would also include the use of stem cells technologies to produce gametes from embryonic stem cells, spermatogonial or oogonial progenitors, or induced pluripotent stem cells (iPSCs) [84]. In this context, although expensive and still experimental, these techniques would be applied on the domestic species as models to design protocols for other purposes. For instance, cloning allows the propagation of animals, even if endangered or extinct, and together with in vitro gametogenesis it might help to maximize the genetic resources available to produce offspring of valuable individuals or endangered feline or canine species. Instead, somatic cell nuclear transfer (SCNT) could also be used for the production of transgenic animals (e.g., as disease models for biomedical research) and could help us to widen our knowledge of the developmental biology of domestic carnivores.

9.4.1 Cloning and somatic cell nuclear transfer

In domestic animals, two main techniques are used to produce individuals that are genetically indistinguishable: embryo splitting and nuclear transfer. Whereas there are no reports of embryo splitting to produce genetically identical individuals in the domestic cat and dog, SCNT was employed in both these species, even if the efficiency is low due to intrinsic features of this technology.

In the domestic cat, the first cloned kitten was obtained at the end of 2001 [85], transferring a cumulus cell into the perivitelline space of an enucleated oocyte activated by electrofusion. Since then, some studies demonstrated the fertility of cloned cats, the creation of transgenic animals by SCNT, the transmission of the transgene to the offspring, and the ability of clones to be re-cloned. Recently the interest faded, and the last study attempting to improve the efficiency of a crucial step of the technique (i.e., the electrofusion of the cells) was published some years ago [86].

However, domestic cat oocytes have been successfully used for interspecies cloning with wild felids [87] and that seems to be the main interest, nowadays.

With some years of delay, in 2005, even the first cloned dog was born fusing an adult skin fibroblast with an in vivo−matured oocyte [88]. The same authors proved the fertility of clones and the possibility to re-clone them, but because of the peculiarities of bitch reproductive physiology, dog cloning remains a challenge, especially for what concerns the source of mature oocytes that have to be matured in vivo and collected with some difficulties [89,90]. To demonstrate that cloning could be used to preserve endangered canids, interspecies SCNT was employed, using dog oocytes, to produce individuals of threatened species such as the wolf [91], but the interest for dog cloning is wider. Besides the generation of pets and elite working dogs with excellent abilities, such as the assistance of disabled people, drug detection, and rescue activity, the creation of human models of disease is worth the effort since dogs share the environment and some diseases with humans. As a proof of concept, transgenic dogs expressing fluorescent proteins, also in an inducible manner, were generated (the first in 2009 [92]), but it seems that we are close to have a Duchenne muscular dystrophy model created by SCNT of engineered donor cells [90], and the same method could be applied for other diseases.

9.4.2 In vitro gametogenesis

In vitro gametogenesis is aimed at inducing germ cell development in laboratory conditions, providing a better understanding of reproductive cell biology, giving the chance to create transgenic offspring and, potentially, supplying an additional (and abundant) source of gametes to be used for fertility preservation purposes and embryo production. Most of the studies have been conducted on the mouse model, and in the domestic cat and dog the knowledge on stem cells is still scarce.

For instance, embryonic stem cells (ESC) can be maintained in culture, genetically modified, and then injected into a blastocyst to generate chimeric offspring, but in the domestic cat there is a lack of ESC lines, which limits their applications [84]. In this species, only ESC-like cells have been obtained [93], and the generation of iPSC has not been achieved (although it has been in wild felids [94]). Instead, in dogs, ESC with typical features have been established [95], as well as different lines of iPSC (since 2010 [96]), also capable of differentiation. However, there are no reports of the use of these cell types for in vitro gametogenesis in any of these species, but a successful attempt was made in the dog with other stem cells (i.e., adipose mesenchymal) that were able to differentiate into primordial germ cell-like cells, even though further differentiation into gametes was not investigated [97].

Germ cells themselves, which are not terminally differentiated, could also be useful for artificial gametogenesis. Spermatogonial stem cells (SSCs), whose "stemness" is reduced compared to ESC or iPSC, have already been characterized and used for transplantation in vivo with some success (cat [98], dog [99]); they could allow to produce offspring with a chosen genetic makeup, from another individual (e.g., to propagate the genome of wild animals) or genetically engineered (dog [100]). Differently, for the female counterpart, there are only few studies on oogonial stem cells (mostly in mice), and we are still far from their application for gametogenesis in domestic carnivores.

In both species, the study on in vitro oogenesis is limited at the IVC of follicles, isolated or within the ovarian cortex [101]. Different strategies have been experimented, especially in recent years, with the addition of several compounds (mainly hormones or growth factors) or the modification of culture conditions (e.g., scaffold encapsulation, osmotic pressure, gas atmosphere), with different results also depending on the follicle developmental stage. Follicle viability and growth were reported, and steroidogenesis and gene expression profile were investigated, but we are still far from obtaining kittens and puppies from these gametes. Similarly, with regard to IVC of testicular tissue there are even fewer reports, and in the domestic cat testicular tissue fragments have been cultured without obtaining germ cells differentiation [102].

The development of germ cells in vitro is still a challenge, but these studies may open the way to the creation of artificial gametes from other cell types and could finally revolutionize ARTs, not only in felids and in canids.

References

[1] Foote RH. The history of artificial insemination: selected notes and notables. J Anim Sci 2002;80:1−10. Available from: https://doi.org/10.2527/animalsci2002.80e-suppl_21a.

[2] Seager SWJ. Successful pregnancies utilizing frozen dog semen. AI Dig 1969;17:26.

[3] Mason SJ. Current review of artificial insemination in dogs. Vet Clin North Am Small Anim Pract 2018;48:567−80. Available from: https://doi.org/10.1016/j.cvsm.2018.02.005.

[4] Romagnoli S, Lopate C. Transcervical artificial insemination in dogs and cats: review of the technique and practical aspects. Reprod Domest Anim 2014;49:56−63. Available from: https://doi.org/10.1111/rda.12395.

[5] England GCW, Millar KM. The ethics and role of AI with fresh and frozen semen in dogs. Reprod Domest Anim 2008;43:165−71. Available from: https://doi.org/10.1111/j.1439-0531.2008.01157.x.

[6] Concannon PW. Reproductive cycles of the domestic bitch. Anim Reprod Sci 2011;124:200−10. Available from: https://doi.org/10.1016/j.anireprosci.2010.08.028.

[7] Hollinshead FK, Hanlon DW. Factors affecting the reproductive performance of bitches: a prospective cohort study involving 1203 inseminations with fresh and frozen semen. Theriogenology 2017;101:62−72. Available from: https://doi.org/10.1016/j.theriogenology.2017.06.021.

[8] Linde-Forsberg C, Forsberg M. Fertility in dogs in relation to semen quality and the time and site of insemination with fresh and frozen semen. J Reprod Fertil Suppl 1989;39:299−310.

[9] Meyers MA, Burns G, Arn D, Schenk JL. Birth of canine offspring following insemination of a bitch with flow-sorted spermatozoa. Reprod Fertil Dev 2008;20:213.

[10] Wei Y-F, Chen F-L, Tang S-S, Mao A-G, Li L-G, Cheng L-G, et al. Birth of puppies of predetermined sex after artificial insemination with a low number of sex-sorted, frozen−thawed spermatozoa in field conditions. Anim Sci J 2017;88:1232−8. Available from: https://doi.org/10.1111/asj.12763.

[11] Luvoni GC, Morselli MG. Canine epididymal spermatozoa: a hidden treasure with great potential. Reprod Domest Anim 2017;52. Available from: https://doi.org/10.1111/rda.12820.

[12] Zambelli D, Prati F, Cunto M, Iacono E, Merlo B. Quality and in vitro fertilizing ability of cryopreserved cat spermatozoa obtained by urethral catheterization after medetomidine administration. Theriogenology 2008;69:485−90. Available from: https://doi.org/10.1016/j.theriogenology.2007.10.019.

[13] Axnér E. Updates on reproductive physiology, genital diseases and artificial insemination in the domestic cat. Reprod Domest Anim 2008;43:144−9. Available from: https://doi.org/10.1111/j.1439-0531.2008.01154.x.

[14] Sojka NJ, Jennings LL, Hamner CE. Artificial insemination in the cat (*Felis catus* L.). Lab Anim Care 1970;20:198−204.

[15] Platz CC, Follis T, Demorest N, Seager SWJ. Semen collection, freezing and insemination in the domestic cat. Proceedings of eighth international congress on animal reproduction and artificial insemination 1976;1053−6.

[16] Zambelli D, Cunto M. Transcervical artificial insemination in the cat. Theriogenology 2005;64:698−705. Available from: https://doi.org/10.1016/j.theriogenology.2005.05.020.

[17] Chatdarong K, Axnér E, Manee-In S, Thuwanut P, Linde-Forsberg C. Pregnancy in the domestic cat after vaginal or transcervical insemination with fresh and frozen semen. Theriogenology 2007;68:1326−33. Available from: https://doi.org/10.1016/j.theriogenology.2007.07.022.

[18] Zambelli D, Bini C, Cunto M. Endoscopic transcervical catheterization in the domestic cat. Reprod Domest Anim 2015;50:13−16. Available from: https://doi.org/10.1111/rda.12442.

[19] Tsutsui T, Mizutani T, Matsubara Y, Toyonaga M, Oba H, Hori T. Surgical intrauterine insemination with cat semen cryopreserved with Orvus ES Paste or sodium lauryl sulfate. J Vet Med Sci 2011;73:259−62. Available from: https://doi.org/10.1292/jvms.09-0568.

[20] Villaverde AISB, Melo CM, Martin I, et al. Comparison of efficiency between two artificial insemination methods using frozen-thawed semen in domestic cat (*Felis catus*). Artif Insemdomest Cats Anim Reprod Sci 2009;114:434−42. Available from: https://doi.org/10.1016/j.anireprosci.2008.10.008.

[21] Tsutsui T, Wada M, Anzai M, Hori T. Artificial insemination with frozen epididymal sperm in cats. J Vet Med Sci 2003;65:397−9. Available from: https://doi.org/10.1292/jvms.65.397.

[22] Toyonaga M, Sato Y, Sasaki A, Kaihara A, Tsutsui T. Artificial insemination with cryopreserved sperm from feline epididymides stored at 4 °C. Theriogenology 2011;76:532−7. Available from: https://doi.org/10.1016/j.theriogenology.2011.03.005.

[23] Hamner CE, Jennings LL, Sojka NJ. Cat (*Felis catus* L.) spermatozoa require capacitation. J Reprod Fertil 1970;23:477−80 <https://doi.org/10.1530/jrf.0.0230477>.

[24] Mahi CAYR. Maturation and sperm penetration of canine ovarian oocytes in vitro. J Exp Zool 1976;196:189−96.

[25] Comizzoli P, Holt WV. Recent advances and prospects in germplasm preservation of rare and endangered species. Adv Exp Med Biol 2014;753:331−56. Available from: https://doi.org/10.1007/978-1-4939-0820-2_14.

[26] Van Soom A, Rijsselaere T, Filliers M. Cats and dogs: two neglected species in this era of embryo production in vitro? Reprod Domest Anim 2014;49:87−91. Available from: https://doi.org/10.1111/rda.12303.

[27] Luvoni GC, Colombo M, Morselli MG. The never-ending search of an ideal culture system for domestic cat oocytes and embryos. Reprod Domest Anim 2018;53:110−16. Available from: https://doi.org/10.1111/rda.13331.

[28] Herrick JR, Bond JB, Magarey GM, Bateman HL, Krisher RL, Dunford SA, et al. Toward a feline-optimized culture medium: impact of ions, carbohydrates, essential amino acids, vitamins, and serum on development and metabolism of in vitro fertilization-derived feline embryos relative to embryos grown in vivo. Biol Reprod 2007;76:858−70. Available from: https://doi.org/10.1095/biolreprod.106.058065.

[29] Pope CE, Gómez MC, Kagawa N, Kuwayama M, Leibo SP, Dresser BL. In vivo survival of domestic cat oocytes after vitrification, intracytoplasmic sperm injection and embryo transfer. Theriogenology 2012;77:531−8. Available from: https://doi.org/10.1016/j.theriogenology.2011.08.028.

[30] Jewgenow K, Fickel J. Sequential expression of zona pellucida protein genes during the oogenesis of domestic cats. Biol Reprod 1999;60:522−6. Available from: https://doi.org/10.1095/biolreprod60.2.522.

[31] Filliers M, Goossens K, Van Soom A, Merlo B, Pope CE, De Rooster H, et al. Gene expression profiling of pluripotency and differentiation-related markers in cat oocytes and preimplantation embryos. Reprod Fertil Dev 2012;24:691−703. Available from: https://doi.org/10.1071/RD11068.

[32] Lee PC, Wildt DE, Comizzoli P. Proteomic analysis of germinal vesicles in the domestic cat model reveals candidate nuclear proteins involved in oocyte competence acquisition. MHR Basic Sci Reprod Med 2018;24:14−26. Available from: https://doi.org/10.1093/molehr/gax059.

[33] Phillips TC, Wildt DE, Comizzoli P. Increase in histone methylation in the cat germinal vesicle related to acquisition of meiotic and developmental competence. Reprod Domest Anim 2012;47:210−14. Available from: https://doi.org/10.1111/rda.12052.

[34] Morselli MG, Luvoni GC, Comizzoli P. The nuclear and developmental competence of cumulus−oocyte complexes is enhanced by three-dimensional coculture with conspecific denuded oocytes during in vitro maturation in the domestic cat model. Reprod Domest Anim 2017;52. Available from: https://doi.org/10.1111/rda.12850.

[35] Nagashima JB, El Assal R, Songsasen N, Demirci U. Evaluation of an ovary-on-a-chip in large mammalian models: species specificity and influence of follicle isolation status. J Tissue Eng Regen Med 2018;12:e1926−35. Available from: https://doi.org/10.1002/term.2623.

[36] Luvoni GC, Chigioni S, Allievi E, Macis D. Factors involved in vivo and in vitro maturation of canine oocytes. Theriogenology 2005;63:41−59. Available from: https://doi.org/10.1016/j.theriogenology.2004.03.004.

[37] Luvoni GC, Chigioni S, Beccaglia M. Embryo production in dogs: from in vitro fertilization to cloning. Reprod Domest Anim 2006;41:286−90. Available from: https://doi.org/10.1111/j.1439-0531.2006.00704.x.

[38] Chastant-Maillard S, Viaris De Lesegno C, Chebrout M, Thoumire S, Meylheuc T, Fontbonne A, et al. The canine oocyte: uncommon features of in vivo and in vitro maturation. Reprod Fertil Dev 2011;23:391−402. Available from: https://doi.org/10.1071/RD10064.

[39] De Los Reyes M, De Lange J, Miranda P, Palominos J, Barros C. Effect of human chorionic gonadotrophin supplementation during different culture periods on in vitro maturation of canine oocytes. Theriogenology 2005;64:1−11. Available from: https://doi.org/10.1016/j.theriogenology.2004.11.008.

[40] Apparicio M, Alves A, Pires-Butler E, Ribeiro A, Covizzi G, Vicente W. Effects of hormonal supplementation on nuclear maturation and cortical granules distribution of canine oocytes during various reproductive stages. Reprod Domest Anim 2011;46:896−903. Available from: https://doi.org/10.1111/j.1439-0531.2011.01761.x.

[41] Apparicio M, Mostachio GQ, Motheo TF, Alves AE, Padilha L, Pires-Butler EA, et al. Distribution of cortical granules and meiotic maturation of canine oocytes in bi-phasic systems. Reprod Fertil Dev 2015;27:1082−7. Available from: https://doi.org/10.1071/RD14022.

[42] Luvoni GC, Chigioni S, Allievi E, Macis D. Meiosis resumption of canine oocytes cultured in the isolated oviduct. Reprod Domest Anim 2003;38:410−14. Available from: https://doi.org/10.1046/j.1439-0531.2003.00457.x.

[43] No J, Zhao M, Lee S, Ock SA, Nam Y, Hur TY. Enhanced in vitro maturation of canine oocytes by oviduct epithelial cell co-culture. Theriogenology 2018;105:66−74. Available from: https://doi.org/10.1016/j.theriogenology.2017.09.002.

[44] Abdel-Ghani MA, Shimizu T, Asano T, Suzuki H. In vitro maturation of canine oocytes co-cultured with bovine and canine granulosa cell monolayers. Theriogenology 2012;77:347−55. Available from: https://doi.org/10.1016/j.theriogenology.2011.08.007.

[45] Enginler SO, Sandal AI, ÖZdaş ÖB, Arici R, Ertürk E, Çinar EM, et al. The effect of oviductal cells on in vitro maturation of canine oocytes in different culture media. Turkish J Vet Anim Sci 2013;38:14−19. Available from: https://doi.org/10.3906/vet-1306-57.

[46] Saikhun J, Sriussadaporn S, Thongtip N, Pinyopummin A, Kitiyanant Y. Nuclear maturation and development of IVM/IVF canine embryos in synthetic oviductal fluid or in co-culture with buffalo rat liver cells. Theriogenology 2008;69:1104−10. Available from: https://doi.org/10.1016/j.theriogenology.2008.01.024.

[47] Nagashima JB, Sylvester SR, Nelson JL, Cheong SH, Mukai C, Lambo C, et al. Live births from domestic dog (*Canis familiaris*) embryos produced by in vitro fertilization. PLoS One 2015;10:1−13. Available from: https://doi.org/10.1371/journal.pone.0143930.

[48] Morselli MG, Loiacono M, Colombo M, Michele M, Luvoni GC. Nuclear competence and genetic expression of growth differentiation factor-9 (GDF-9) of canine oocytes in 3D culture. Reprod Domest Anim 2018;53:117−24. Available from: https://doi.org/10.1111/rda.13336.

[49] De Los Reyes M, Rojas C, Hugo V, Palomino J. Expression of growth differentiation factor 9 (GDF-9) during in vitro maturation in canine oocytes. Theriogenology 2013;80:587−96. Available from: https://doi.org/10.1016/j.theriogenology.2013.06.001.

[50] Hashimoto O, Takagi R, Yanuma F, Doi S, Shindo J, Endo H, et al. Identification and characterization of canine growth differentiation factor-9 and its splicing variant. Gene 2012;499:266−72. Available from: https://doi.org/10.1016/j.gene.2012.03.003.

[51] Garcia P, Aspee K, Ramirez G, Dettleff P, Palomino J, Peralta OA, et al. Influence of growth differentiation factor 9 and bone morphogenetic protein 15 on in vitro maturation of canine oocytes. Reprod Domest Anim 2019;54:373−80. Available from: https://doi.org/10.1111/rda.13371.

[52] Luvoni GC. Gamete cryopreservation in the domestic cat. Theriogenology 2006;66:101−11. Available from: https://doi.org/10.1016/j.theriogenology.2006.03.012.

[53] Chatdarong K, Thuwanut P, Morrell JM. The development of cat testicular sperm cryopreservation protocols: effects of tissue fragments or sperm cell suspension. Theriogenology 2016;85:200−6. Available from: https://doi.org/10.1016/j.theriogenology.2015.09.030.

[54] Buranaamnuay K. Protocols for sperm cryopreservation in the domestic cat: a review. Anim Reprod Sci 2017;183:56−65. Available from: https://doi.org/10.1016/j.anireprosci.2017.06.002.

[55] Tsutsui T, Tanaka A, Takagi Y, Nakagawa K, Fujimoto Y, Murai M, et al. Unilateral intrauterine horn insemination of frozen semen in cats. J Vet Med Sci 2000;62:1247−51. Available from: https://doi.org/10.1292/jvms.62.1247.

[56] Sánchez R, Risopatrón J, Schulz M, Villegas J, Isachenko V, Kreinberg R, et al. Canine sperm vitrification with sucrose: effect on sperm function. Andrologia 2011;43:233−41. Available from: https://doi.org/10.1111/j.1439-0272.2010.01054.x.

[57] Swanson WF, Bateman HL, Vansandt LM. Urethral catheterization and sperm vitrification for simplified semen banking in felids. Reprod Domest Anim 2017;52:255−60. Available from: https://doi.org/10.1111/rda.12863.

[58] Gil L, Olaciregui M, Luño V, Malo C, González N, Martínez F. Current status of freeze-drying technology to preserve domestic animals sperm. Reprod Domest Anim 2014;49:72−81. Available from: https://doi.org/10.1111/rda.12396.

[59] Watanabe H, Asano T, Abe Y, Fukui Y, Suzuki H. Pronuclear formation of freeze-dried canine spermatozoa microinjected into mouse oocytes. J Assist Reprod Genet 2009;26:531−6. Available from: https://doi.org/10.1007/s10815-009-9358-y.

[60] Olaciregui M, Luño V, Gonzalez N, De Blas I, Gil L. Freeze-dried dog sperm: dynamics of DNA integrity. Cryobiology 2015;71:286−90. Available from: https://doi.org/10.1016/j.cryobiol.2015.08.001.

[61] Ringleb J, Waurich R, Wibbelt G, Streich WJ, Jewgenow K. Prolonged storage of epididymal spermatozoa does not affect their capacity to fertilise in vitro-matured domestic cat (*Felis catus*) oocytes when using ICSI. Reprod Fertil Dev 2011;23:818−25. Available from: https://doi.org/10.1071/RD10192.

[62] Patrick JL, Elliott GD, Comizzoli P. Structural integrity and developmental potential of spermatozoa following microwave-assisted drying in the domestic cat model. Theriogenology 2017;103:36−43. Available from: https://doi.org/10.1016/j.theriogenology.2017.07.037.

[63] Tsujimoto Y, Kaneko T, Yoshida T, Kimura K, Inaba T, Sugiura K, et al. Development of feline embryos produced using freeze-dried sperm. Theriogenology 2020. Available from: https://doi.org/10.1016/j.theriogenology.2020.02.021.

[64] Luvoni GC, Pellizzari PBM. Effects of slow and ultrarapid freezing on morphology and resumption of meiosis in immature cat oocytes. J Reprod Fertil Suppl 1997;51:93−8.

[65] Luvoni GC, Pellizzari P. Embryo development in vitro of cat oocytes cryopreserved at different maturation stages. Theriogenology 2000;53:1529−40. Available from: https://doi.org/10.1016/S0093-691X(00)00295-8.

[66] Luvoni G. Cryosurvival of ex situ and in situ feline oocytes. Reprod Domest Anim 2012;47:266−8. Available from: https://doi.org/10.1111/rda.12040.

[67] Gómez MC, Pope CE, Harris R, Mikota S, Dresser BL. Development of in vitro matured, in vitro fertilized domestic cat embryos following cryopreservation, culture and transfer. Theriogenology 2003;60:239−51. Available from: https://doi.org/10.1016/S0093-691X(03)00004-9.

[68] Colombo M, Morselli MG, Tavares MR, Apparicio M, Luvoni GC. Developmental competence of domestic cat vitrified oocytes in 3D enriched culture conditions. Animals 2019;. Available from: https://doi.org/10.3390/ani9060329.

[69] Arayatham S, Tiptanavattana N, Tharasanit T. Effects of vitrification and a Rho-associated coiled-coil containing protein kinase 1 inhibitor on the meiotic and developmental competence of feline oocytes. J Reprod Dev 2017;63:511−17. Available from: https://doi.org/10.1262/jrd.2017-004.

[70] Snoeck F, Szymanska KJ, Sarrazin S, Ortiz-Escribano N, Leybaert L, Van Soom A. Blocking connexin channels during vitrification of immature cat oocytes improves maturation capacity after warming. Theriogenology 2018;122:144−9. Available from: https://doi.org/10.1016/j.theriogenology.2018.09.011.

[71] Turathum B, Roytrakul S, Changsangfa C, Sroyraya M, Tanasawet S, Kitiyanant Y, et al. Missing and overexpressing proteins in domestic cat oocytes following vitrification and in vitro maturation as revealed by proteomic analysis. Biol Res 2018;51:1−12. Available from: https://doi.org/10.1186/s40659-018-0176-5.

[72] Okotrub KA, Mokrousova VI, Amstislavsky SY, Surovtsev NV. Lipid droplet phase transition in freezing cat embryos and oocytes probed by Raman Spectroscopy. Biophys J 2018;115:577−87. Available from: https://doi.org/10.1016/j.bpj.2018.06.019.

[73] Graves-Herring JE, Wildt DE, Comizzoli P. Retention of structure and function of the cat germinal vesicle after air-drying and storage at supra-zero temperature. Biol Reprod 2013;88:139. Available from: https://doi.org/10.1095/biolreprod.113.108472.

[74] Elliott GD, Lee P-C, Paramore E, Van Vorst M, Comizzoli P. Resilience of oocyte germinal vesicles to microwave-assisted drying in the domestic cat model. Biopreserv Biobank 2015;13:164−71. Available from: https://doi.org/10.1089/bio.2014.0078.

[75] Abe Y, Lee DS, Kim SKSH. Vitrification of canine oocytes. J Mamm Ova Res 2008;25:32−6.

[76] Suzuki H. Cryopreservation of canine embryos and resulting pregnancies. Reprod Domest Anim 2012;47:141−3. Available from: https://doi.org/10.1111/rda.12068.

[77] Hori T, Ushijima H, Kimura T, Kobayashi M, Kawakami E, Tsutsui T. Intrauterine embryo transfer with canine embryos cryopreserved by the slow freezing and the Cryotop method. J Vet Med Sci 2016;78:1137−43. Available from: https://doi.org/10.1292/jvms.16-0037.

[78] Bosch P, Hernandez-Fonseca HJ, Miller DM, Wininger JD, Massey JB, Lamb SV, et al. Development of antral follicles in cryopreserved cat ovarian tissue transplanted to immunodeficient mice. Theriogenology 2004;61:581−94.

[79] Wakasa I, Hayashi M, Abe Y, Suzuki H. Distribution of follicles in canine ovarian tissues and xenotransplantation of cryopreserved ovarian tissues with even distribution of follicles. Reprod Domest Anim 2017;52:219−23. Available from: https://doi.org/10.1111/rda.12857.

[80] Suzuki H, Ishijima T, Maruyama S, Yanagimoto Ueta Y, Abe Y, Saitoh H. Beneficial effect of desialylated erythropoietin administration on the frozen-thawed canine ovarian xenotransplantation. J Assist Reprod Genet 2008;25:571−5. Available from: https://doi.org/10.1007/s10815-008-9271-9.

[81] Commin L, Buff S, Rosset E, Galet C, Allard A, Bruyere P, et al. Follicle development in cryopreserved bitch ovarian tissue grafted to immunodeficient mouse. Reprod Fertil Dev 2012;24:461−71. Available from: https://doi.org/10.1071/RD11166.

[82] Ishijima T, Kobayashi Y, Lee D-S, Ueta YY, Matsui M, Lee J-Y, et al. Cryopreservation of canine ovaries by vitrification. J Reprod Dev 2006;52:293−9. Available from: https://doi.org/10.1262/jrd.17080.

[83] Tharasanit T, Buarpung S, Manee-In S, Thongkittidilok C, Tiptanavattana N, Comizzoli P, et al. Birth of kittens after the transfer of frozen-thawed embryos produced by intracytoplasmic sperm injection with spermatozoa collected from cryopreserved testicular tissue. Reprod Domest Anim 2012;47:305−8. Available from: https://doi.org/10.1111/rda.12072.

[84] Travis A, Kim Y, Meyers-Wallen V. Development of new stem cell-based technologies for carnivore reproduction research. Reprod Domest Anim 2009;44:22−8. Available from: https://doi.org/10.1111/j.1439-0531.2009.01396.x.

[85] Shin T, Kraemer D, Pryor J, Liu L, Rugila J, Howe L, et al. A cat cloned by nuclear transplantation. Nature 2002;415:859. Available from: https://doi.org/10.1038/nature723.

[86] Do LTK, Terazono MWT, Taniguchi YSM, Kazuki TTY, Otoi KKMOT. Effects of duration of electric pulse on in vitro development of cloned cat embryos with human artificial chromosome vector. Reprod Domest Anim 2016;1039−43. Available from: https://doi.org/10.1111/rda.12766.

[87] Gómez MC, Pope CE. Cloning endangered Felids by interspecies somatic cell nuclear transfer. Methods Mol Biol 2015;1330:133−52. Available from: https://doi.org/10.1007/978-1-4939-2848-4_13.

[88] Lee BC, Kim MK, Jang G, Oh HJ, Yuda F, Kim HJ, et al. Dogs cloned from adult somatic cells. Nature 2005;436:641. Available from: https://doi.org/10.1038/436641a.

[89] Lee S, Oh H, Kim M, Kim G, Setyawan E, Ra K, et al. Dog cloning — no longer science fiction. Reprod Domest Anim 2018;53:133−8. Available from: https://doi.org/10.1111/rda.13358.

[90] Oh H, Ra K, Kim M, Kim G, Setyawan E, Lee S, et al. The promise of dog cloning. Reprod Fertil Dev 2017;30:1−7. Available from: https://doi.org/10.1071/RD17375.

[91] Kim MK, Jang G, Oh HJ, Yuda F, Kim HJ, Hwang WS, et al. Endangered wolves cloned from adult somatic cells. Cloning Stem Cell 2007;9:130−7. Available from: https://doi.org/10.1089/clo.2006.0034.

[92] Hong SG, Kim MK, Jang G, Oh HJ, Park JE, Kang JT, et al. Generation of red fluorescent protein transgenic dogs. Genesis 2009;47:314−22. Available from: https://doi.org/10.1002/dvg.20504.

[93] Yu X, Jin G, Yin X, Cho S, Jeon J, Lee S, et al. Isolation and characterization of embryonic stem-like cells derived from in vivo-produced cat blastocysts. Mol Reprod Dev 2008;75:1426−32. Available from: https://doi.org/10.1002/mrd.20867.

[94] Verma R, Holland MK, Temple-Smith P, Verma PJ. Inducing pluripotency in somatic cells from the snow leopard (*Panthera uncia*), an endangered felid. Theriogenology 2012;77:220−8. Available from: https://doi.org/10.1016/j.theriogenology.2011.09.022.

[95] Vaags AK, Rosic-Kablar S, Gartley CJ, Zheng YZ, Chesney A, Villagómez DAF, et al. Derivation and characterization of canine embryonic stem cell lines with in vitro and in vivo differentiation potential. Stem Cell 2009;27:329−40. Available from: https://doi.org/10.1634/stemcells.2008-0433.

[96] Shimada H, Nakada A, Hashimoto Y, Shigeno K, Shionoya Y, Nakamura T. Generation of canine induced pluripotent stem cells by retroviral transduction and chemical inhibitors. Mol Reprod Dev 2010;77:2. Available from: https://doi.org/10.1002/mrd.21117.

[97] Wei Y, Fang J, Cai S, Lv C, Zhang S, Hua J. Primordial germ cell−like cells derived from canine adipose mesenchymal stem cells. Cell Prolif 2016;49:503−11. Available from: https://doi.org/10.1111/cpr.12271.

[98] Silva RC, Costa GMJ, Lacerda SMSN, Batlouni SR, Soares JM, Avelar GF, et al. Germ cell transplantation in felids: a potential approach to preserving endangered species. J Androl 2012;33:264−76. Available from: https://doi.org/10.2164/jandrol.110.012898.

[99] Kim Y, Turner D, Nelson J, Dobrinski I, McEntee M, Travis AJ. Production of donor-derived sperm after spermatogonial stem cell transplantation in the dog. Reproduction 2008;136:823−31. Available from: https://doi.org/10.1530/rep-08-0226.

[100] Harkey M, Asano A, Zoulas M, Torok-Storb B, Nagashima J, Travis A. Isolation, genetic manipulation, and transplantation of Canine spermatogonial stem cells: progress toward transgenesis through the male germ line. Reproduction 2013;146:1−27. Available from: https://doi.org/10.1530/REP-13-0086.

[101] Songsasen N, Comizzoli P, Nagashima J, Fujihara M, Wildt D. The domestic dog and cat as models for understanding the regulation of ovarian follicle development in vitro. Reprod Domest Anim 2012;47:13−18. Available from: https://doi.org/10.1111/rda.12067.

[102] Silva AF, Escada-Rebelo S, Amaral S, Tavares RS, Schlatt S, Ramalho-Santos J, et al. Can we induce spermatogenesis in the domestic cat using an in vitro tissue culture approach? PLoS One 2018;13:1−14. Available from: https://doi.org/10.1371/journal.pone.0191912.

Chapter 10

Reproductive technologies in laboratory animals

Takehito Kaneko[1] and Wiebke Garrels[2]

[1]Division of Fundamental and Applied Sciences, Graduate School of Science and Engineering, Iwate University, Morioka, Japan, [2]Institute for Laboratory Animal Science and Central Animal Facility, Hannover Medical School, Hannover, Germany

10.1 Cryopreservation in mice and rats

The most frequently used mammalian laboratory animals are mice and rats, and due to their short generation interval and their high reproductive performance they are ideal animal models for basic and applied research (Table 10.1). The cryopreservation of rodent gametes and embryos is an important basic method for artificial reproductive technologies (ART), as well as maintenance of valuable lines.

Cryopreservation employs low temperatures (typically $-196°$C/liquid nitrogen) to preserve structurally and functionally intact gametes and embryos for extended periods of time. Due to ice crystal formation, unprotected freezing is normally lethal for cells, and this aspect led to the development of cryoprotectants to replace/reduce the cellular water, and alternatively to vitrification procedures. For mice and rats the cryopreservation of early cleavage stage embryos, spermatozoa, and ovarian tissue are the most commonly applied methods. A cryopreserved stock allows to protect valuable rodent lines (wild type and transgenic ones) against genetic drift, inbreeding depression, outbreak of a disease, environmental disaster, or failures of the facility equipment [1]. Animal welfare is an important aspect, and transportation of frozen embryos, spermatozoa, or ovaries instead of living animals is preferred [2]. In addition, the costs of a cryopreserved stock are much lower, compared to those for maintaining living animals, which may not be permanently required. The costs for the maintenance of a mice colony are a mixed calculation of costs for various items, equipment, energy, food, cages, etc. The Jackson Laboratory has calculated the costs for one strain per year to be around 10,000 USD, while in Germany is between 6900 and 14,500 Euros depending on the number of cages. Commercial companies, following the initial costs for purchasing cryopreservation equipment, offer their service for around 3000 USD for cryopreservation and 1000 USD for sperm freezing, whereas the costs for rats are higher, and the storage per year is around 150 USD. At the same time, several universities have in-house facilities and offer the services at reduced rates.

Before starting the cryopreservation, it is important to design the most suitable approach for the preferred strain, since the reproductive characteristics may vary broadly between strains [3]. Hormonal superovulation, following synchronization of donors, will lead to a higher number of oocytes or embryos, while reducing the animal number as low as possible. For superovulation of mice, different strategies mainly based on the use of pregnant mare serum gonadotropin (PMSG) and human chorionic gonadotropin (hCG) have been developed. The dosage and timing of application followed by overnight breeding are the most important aspects (Table 10.2). In our laboratories, we use 7 International Units (IU) of PMSG and hCG for all strains with a black C57BL/6 background. In strains with a BALB/c background we reduce the dosage to 5 IU, and in outbreed strains such as NMRI we increase the dosage up to 10 IU of both hormones. The timing of the hormonal application depends on the light cycle in the animal room. Commonly, light cycles are characterized by 12 hours day and 12 hours night intervals. In this case it is ideal to inject (i.p.) the prepuberal females (35−42 days of age) with PMSG at noon, followed by hCG injection 46−48 hours later. In some strains, such as 129S1/SvImJ, the outcome of high-quality embryos can be improved with the administration of gonadotropin-releasing hormone 24 hours before PMSG application [4]. In the event aged mice (>6 months of age) are going to be used, the ovulation rate can be improved by a cetrorelix acetate injection (gonadotropin releasing hormone antagonist) prior to ovarian stimulation [5]. In some strains, it might be helpful to synchronize the mice (without superovulation)

Reproductive Technologies in Animals. DOI: https://doi.org/10.1016/B978-0-12-817107-3.00010-2

TABLE 10.1 Reproductive physiology of laboratory animals.

	Mouse		Rat	
	Female	Male	Female	Male
Sexual maturity (days)	28–29	28–35	50–72	
Breeding maturity (days)	56–70		90–100	
Estrous cycle (days)	4–5	–	4–6	–
Gestation time (days)	18–21	–	20–23	–
Litter size	3–12		6–12	
Weaning age (days)	18–21		18–21	

TABLE 10.2 Strains used and hormone doses for superovulation.

Strain	Hormone treatment (PMSG and hCG) (IU)
C57BL/6J	7
BALB/cJ	5
FVB/N HanZtm	7
NMRI	10

 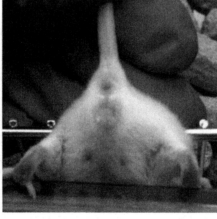

FIGURE 10.1 Plug check: (Left) Female without visible plug; (Right) female with prominent plug.

by the application of cloprostenol (synthetic analog of Prostaglandin $F_{2\alpha}$) and progesterone, which is a common strategy in large animals as well [6]. A relative new superovulation strategy for mice is based on an antiinhibin antibody in combination with hCG [7,8], and a fivefold increase in the number of embryos can be reported. A main disadvantage is that the quality of the embryos may be compromised, but it still is a suitable technique that is used for in vitro fertilization (IVF) or to get offspring from genetically engineered mice [9].

Successful copulation of synchronized females is controlled after overnight mating by the so-called plug check (Fig. 10.1). For best copulation rates superovulated females should be grouped with one male [10].

In rats, superovulation is more complicated, unstable and quite often very expensive. Some protocols are based on timed PMSG and hCG application [11]. Armstrong and Opavsky used an osmotic pump in which they applied a purified follicle-stimulating hormone (FSH), although this method is not usable in a lot of strains [12]. The protocol was modified with an injection of luteinizing hormone-releasing analog to synchronize the animals, followed by the

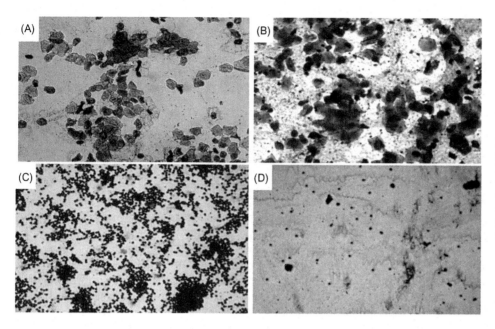

FIGURE 10.2 Vaginal smears for rat cycle diagnosis. (A) Proestrus: duration around 12 hours (mostly nucleated cells, some keratinized epithelial cells, and sporadic leucocytes). (B) Estrus: duration around 12 hours (mostly keratinized epithelial cells, which form clusters at the end of the estrus). (C) Metestrus: duration around 24 hours (dominating cell type is leucocytes, and some are epithelial cells, with and without a nucleus). (D) Diestrus: duration around 48 hours (mucus as a typical sign, some leucocytes and some nucleated epithelial cells).

implantation of the mini-pump with FSH [13]. In our laboratories, we use a modification of this protocol: three days after the implantation of the mini-pump, the rats are treated with a single injection of hCG followed by overnight breeding. Vaginal smears after breeding are performed (Fig. 10.2), because in rats the vaginal plug quite often is not visible. Therefore thanks to the vaginal smears the correct cycle stages can be diagnosed and consequently the number of killed animals can be reduced [14].

10.1.1   Embryo freezing

The first successful freezing method for mice embryos was established in 1972 [15]. During the last decades, refined protocols and different cryoprotectants were developed. There are two basic principles for cryopreservation: slow freezing with cryoprotective substances and vitrification. With slow freezing protocols, the cellular water is replaced by a cryoprotectant, and ice crystal formation is induced by specific cooling programs. Instead, vitrification leads to a glass-like solidification, without ice crystal formation. The exchange of cellular water by a cryoprotectant in the slow freezing process and the washing out after thawing make these protocols relatively time-consuming, whereas the vitrification typically is rather fast. For the slow freezing, a commonly applied protocol uses a continuous cooling of embryos down to $-192°C$. To achieve constant freezing rates of around 0.5°C/minute, programmable freezing machines are commercially available. A modification of the continuous freezing is the so-called two-step method, where the embryos are frozen down to $-32°C$ and then transferred directly into liquid nitrogen. For vitrification, embryos are mixed with the cryoprotectants, and then they are plugged in a carrier directly into liquid nitrogen.

10.1.1.1   Calculation of required donors

How many donors are necessary to establish a sufficient frozen embryo bank? The calculation of the needed animals is influenced by several factors such as age and genetic background [16]. The most commonly used background strains are C57BL/6, 129, and BALB/c of which BALB/c has the lowest revitalization rate [17]. Rall et al. [18] developed an equation to be used for the calculation of the required donors to obtain a sufficient number of embryos:

$$S = \frac{KE}{(SE \times GE)}$$

where S is the required donors, KE is the planned number of cryopreserved embryos, SE is the number of plugged females following the superovulation process, and GE is the number of embryos per each donor.

The last two parameters are readily available following the first superovulation trials. Revitalization rates of rat embryos are much lower compared to those of mice, but the described equation gives an approximation as well.

10.1.1.2 Freezing of oocytes and postfertilization embryos

All the protocols mentioned before are suitable for freezing oocytes, zygotes, two-cell, 8-cell or blastocyst stages, but the success is variable. Oocytes are quite difficult to freeze, because the metaphase II structure is quite complex [19]. There are different two-step protocols, such as embryos frozen with dimethyl sulfoxide (DMSO) as a cryoprotectant [15,20], and the so-called Jackson protocol, which uses 1,2-propanediol (PrOH) and sucrose as cryoprotectants [21].

In our laboratories, we prefer the freezing of two-cell embryos, because they are not so sensitive to the freezing process. For the two-step freezing protocol, a controlled-rate [20] freezer is necessary. We use two different two-step protocols, of which the first one is the modified two-step method [22]. In this protocol, the embryos are frozen with 1.5 M PrOH in phosphate buffered 1 (PB1), and mini-straws are used as carriers. The freezing process starts at 0°C, cools down to −6°C followed by seeding. Seeding is a mechanical stimulus with a precooled (in liquid nitrogen at −196°C) metallic stamp. This is important for starting the ice crystal formation away from the embryos. The cooling rate down to −32°C takes place in steps of −0.5°C/minute followed by transfer into liquid nitrogen. The thawing process is time-consuming because a stepwise dilution of the cryoprotectant is mandatory.

The second protocol is used by the Jackson Laboratory and several other companies and institutes [23]. Start temperature is set at −7°C, the seeding takes place after 5 minutes followed by cooling rate down to −32°C with 0.5°C/minute. The method is simple, effective, and can be applied to cryopreserve embryos from the oocyte to the blastocyst stage and the survival rates are quite often as high as 90%. Thawing is simple because all of the needed elements are within the straws, making this very useful for distribution of cryopreserved embryos.

An additional possibility is the vitrification of embryos. Vitrification induces the formation of a glass state during freezing and avoids the formation of ice crystals by working with high osmolar concentrations of cryoprotectant molecules, such as ethylene glycol (EG), dimethyl sulfoxide (DMSO) and sucrose followed by transfer into liquid nitrogen with a carrier, such as cryotop [24,25]. An advantage of this technique is that no slow cooling or a programmable freezer is necessary and so this technique is very fast. To avoid cryodamage to the embryos, they must be stored below −130°C. There is a protocol that overcomes this problem by using a high osmolality vitrification solution containing 42.5% (v/v) ethylene glycol, 17.3% (w/v) Ficoll, and 1.0 M sucrose. Most (80%) embryos cryopreserved in this solution survive at −80°C for at least 30 days. Normal mice were recovered even after intercontinental transportation in a conventional dry ice package for 2−3 days, indicating that special containers such as dry shippers with liquid nitrogen vapor are unnecessary [26]. An exact and quick thawing process is necessary; otherwise, a detrimental effect caused by the intracellular ice formation is observed [27]. Vitrification is suitable for all stages of mouse embryonic development, although the best rates were observed at four- and eight-cell stages of development [24].

In rats, vitrification is a better method for embryo cryopreservation [28]. On average, more than 90% of the embryos survive the freezing procedure and can be transferred to foster mothers [11].

10.2 Freezing of spermatozoa

10.2.1 Freezing of mice spermatozoa

In farm animals, especially cattle, cryopreservation of sperm cells is an efficient technology, which has found widespread commercial applications. Mice and rat spermatozoa, however, are very sensitive against osmotic changes and most cryoprotectants are toxic at room temperature to the sperm cells [29]. The freezing of sperm cells in mice started in the early 1990s, where the technique of using skim milk and raffinose [30] improved the success rates. It has been shown that there is a huge difference among strains, with regard to the fertilization competence of sperm cells after thawing. If you want to have a safe backup of your mouse strain, a test IVF-revitalization is necessary. The first protocol from the 1990s for cryopreserving mouse sperm cells is based on the use of 18% raffinose and 3% skim milk as cryoprotectant agents (CPAs) and has been widely used since 1990 [31]. There are mainly three protocols for freezing mouse sperm: (1) the pellet method [30], (2) the Jackson method [17] (Fig. 10.3), and (3) the CARD method [32]. Unfortunately, the reliability and efficiency of using frozen-thawed sperm for IVF are highly variable and depend on the genetic background [33]. The addition of monothioglycerol (MTG, an antioxidant) or glutamine (Glu) to the skim milk improved the survival rate of the frozen-thawed sperm cells [17,34]. Mouse sperm cryopreserved in skim milk with MTG has significantly better postthaw progressive motility. No significant difference in fertilization rate between sperm cryopreserved in cryovials as compared to cryostraws was reported [34].

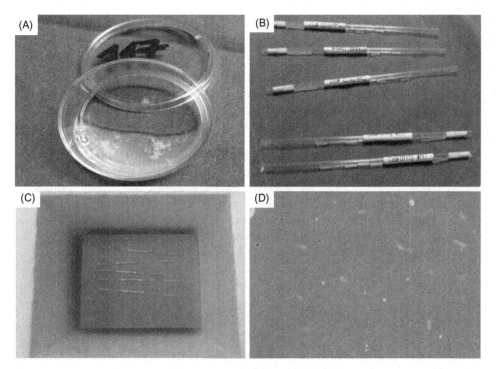

(A) (B)

(C) (D)

FIGURE 10.3 Sperm freezing by Jackson method. (A) Prepared epididymis in CPA, (B) filled straws with five pellets each, (C) cooling down in gas phase of liquid nitrogen for 10 minutes, and (D) Sperm in CPA.

10.2.1.1   Isolation of spermatozoa from donor males

The male mice should be at least 8 weeks old, and they should have not been used for breeding for at least a week. Electroejaculation or chemical ejaculation is possible in mice, but on account of animal welfare these methods are not commonly used. Normally the males are euthanized after preparation of the two caudae epididymis. These are then transferred to the CPA, followed by a short incubation to allow the sperm cells to swim out. When the males must be preserved, then the surgical sampling of the epididymal sperm becomes an option [35].

10.2.1.2   Freezing of spermatozoa

In the pellet method by Takeshima et al. [30], raffinose and skim milk are used as cryoprotectants. After mixing sperm cells and cryoprotectant solution, the sperm solution is pipetted into holes in a dry ice block. The dry ice block is prepared with a special punch to make small holes of 5 mm diameter, which are deep enough for 50 μL of the sperm solution. The temperature of the dry ice is around $-70°C$. After the solution has frozen, the pellets are placed into a precooled cryovial and are then transferred into liquid nitrogen. This method is very quick and cheap, because no special equipment is needed. But the sperm quality after thawing is not so good compared to the Jackson method, and therefore the fertilization rate is quite low, often not more than 30%.

In the Jackson method [17] the CPA is improved with alpha-monothioglycerol, which is an antioxidant and increases the fertilization rate [23]. After preparation, the sperm cells are transferred to French straws, sealed, and then cooled in gas phase of the liquid nitrogen ($-150°C$) followed by transfer and storage into liquid nitrogen ($-196°C$). In the so-called CARD method L-glutamine is added to the cryoprotectant, with the effect of enhancing the protective action of the CPA [32].

10.2.1.3   Freezing of rat spermatozoa

Scarce data on freezing rat sperm cells are available. Egg yolk—based freezing media are promising approaches [36]. It is important that the age of the male is around 20 weeks. The protocol is based on egg yolk and lactose monohydrate as CPAs. The recovery of the spermatozoa is similar to mice. After filling the straws at room temperature sperm cells are cooled down in a programmable freezer with a rate of $-0.5°C/minute$ to $-5°C$. After 5 minutes of incubation, they are then transferred into the vapor of liquid nitrogen and after 15 minutes they are directly transferred into liquid nitrogen [37]. If no programmable freezer is available, it is possible to cool the sperm cells down in a normal fridge to $4°C$. A successful production of living offspring was reported after intrauterine inseminations of frozen-thawed spermatozoa [37] and intracytoplasmic sperm injection (ICSI) [38,39].

10.3 Freeze-dry preservation of gametes

The drying of gametes is the ultimate preservation technique (Fig. 10.4). The freeze-drying of sperm has been well studied and has been applied as a long-term preservation technique replacing ordinary cryopreservation methods. The advantage of freeze-drying is that sperm cells can be stored for a long term in a refrigerator (4°C) without using liquid nitrogen and dry ice. Furthermore, easy and safe transportation and short-term preservation at ambient temperature are possible. Various animal species have already been obtained from oocytes fertilized with freeze-dried sperm. Due to the complete loss of motility following freeze-drying, ICSI is required to achieve fertilization and development into offspring.

The freeze-drying of sperm has been studied for long time. The first attempt to freeze-dry sperm cells was performed in 1949 [40]. In this report, 50% of freeze-dried fowl sperm cells that were stored for 2 hours at room temperature regained their motility after rehydration. However, in recent scientific literature there is no evidence that sperm cells regain motility after freeze-drying in any animal species. The freeze-drying of mammalian sperm cells was well studied by Yanagimachi et al. [41]. The pronucleus was formed in the oocytes injected with freeze-dried hamster sperm cells that were immotile after rehydration. Subsequently, mouse offspring was obtained from an oocyte injected with immotile sperm that had been freeze-dried with culture medium and stored for 3 months at 4°C [42].

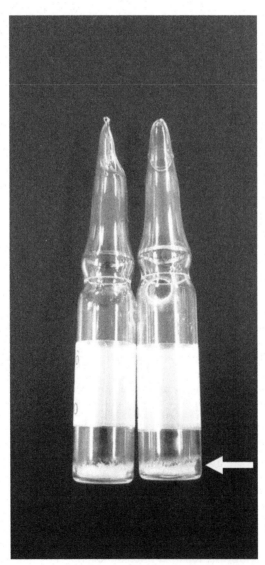

FIGURE 10.4 Freeze-dried spermatozoa in glass ampoules. The arrow shows the dried sperm cells at the bottom of the ampoules.

Fertility was increased by improvement of the preservation buffer [43—46], storage temperature [47—49], and sperm condition [50,51]. Freeze-drying of sperm cells has been successful in various mammals, and stable results have been obtained in mouse and rat as well. In fact, offspring could be obtained from oocytes fertilized with freeze-dried mouse and rat sperm stored at 4°C for at least 5 years [51,52]. Furthermore, the same offspring showed normal health, life expectancy, and reproductive ability [53]. The efficiency of freeze-drying of sperm cells has made possible and easier the transportation of genetic resources in biobanking worldwide [54,55].

10.4 Cryopreservation of ovaries

Ovary cryopreservation of mice and rats, in combination with an ovary transplantation to isogenic, or immunodeficient recipients can complement embryo and sperm cryopreservation programs [56—58]. One clear advantage is that only few females are needed, and the method is fast and cheap. In principle ovaries of all postnatal developmental stages are freezable, as long as some primordial follicles are present. A disadvantage is that only maternal genome is represented. In addition, due to the large size of tissue involved, an increased risk of freezing injuries and a high risk of transmission of pathogens have to be taken into consideration.

Mouse studies have shown that fertility can be reestablished in more than 50% of the female recipients, leading to the conclusion that ovary cryopreservation can be a useful option for banking mouse germplasm, when embryos or sperm cells cannot be used or are not cost-effective [56].

The ovary is a complex structure with different cell types, thus it is important that the CPA is permeating quickly to all layers because this represents the limiting factor in the cryopreservation of tissues and organs [59]. The risk of DNA damage is high due to the different stages in oogenesis [60].

10.4.1 Freezing of mouse ovaries

Harp and Sztein were the first to successfully freeze ovaries, followed by the revitalization process and the generation of living offspring [56,61]. The best results are obtained when the females are around 3—7 weeks old. The females are killed with carbon dioxide and the ovaries quickly prepared, while removing the bursa and the fat tissue. If the ovaries are too big, they can be halved. The CPA contains 1.5 M PrOH and is supplemented with 10% fetal calf serum. Each ovary is transferred into a cryovial with 200 μL CPA. Equilibration takes place for 10 minutes at room temperature followed by a 45-minute incubation on wet ice. Afterward the cryovials are transferred into a programmable freezer at −6°C. After 5 minutes of incubation, seeding takes place and the ovaries are cooled down with a rate of −0.5°C/minute to −80°C. After incubation for ten minutes at −80°C, they are then transferred into liquid nitrogen. Vitrification protocols for mouse ovarian tissue show the best morphological preservation when a combination of ethylene glycol and DMSO is used [62].

10.4.2 Freezing of rat ovaries

Dorsch et al. were able to adapt the protocol to rat ovaries [63], with a freezing procedure similar to mice. The ovaries are cut in four pieces, so that a transplantation to four recipients is possible, and the pieces are transferred into 500 μL CPA.

In rats the protocols are quite similar to the mouse. In our laboratories we prefer the slow freezing with a concentration of 1.5 M DMSO, but there are studies that show that the quality depends on the concentration of the cryoprotectant used, rather than the freezing protocol (e.g., slow freezing vs vitrification) [57].

10.4.3 Thawing and transfer of ovaries in mice and rat

Cryovials are taken out of liquid nitrogen tank and thawed at room temperature. When all ice crystals have disappeared, the CPA must be removed from the ovaries and 500 μL fresh PB1 (recipe online available) is added on top of the ovaries. The rehydration takes around 10 minutes, and before transplantation, the PB1 is changed. Ideally the recipients of the ovaries are coisogenic to the donors to avoid an immune reaction. However, quite often the genetic background is not clear and therefore the ovaries must be transferred into immunodeficient recipients. Following narcosis, the recipients are ovariectomized followed by a transfer of the new ovarian tissue to the bursa, which can be closed with two stitches [64]. Immunodeficient rat strains often have a low milk production, and commonly foster mothers have to be synchronized to raise the pups.

10.4.4   Calculation of needed time to obtain a new breeding nucleus

The cryopreservation of the ovaries is a quick procedure and takes around 1 or 2 days, depending on the number of animals, whereas another day is needed for thawing and transplantation. It takes around 1 month until the recipients start cycling and can be used for breeding. Breeding and pregnancy take around 21 days. The born pups need around 80 days to reach breeding maturity, and then at least one backcrossing is necessary to get to produce homozygote animals. From the above, it can be concluded that the process is comparable to the IVF procedure and it takes around 200 days.

10.5   In vitro fertilization with murine and rat gametes

IVF is an important procedure for the production of embryos, which can be used for rederivation of a strain. Quite often it is realized too late that a breeding depression occurs in a strain, and the males are old and show no breeding behavior anymore. In such cases, IVF is the best choice to rescue the strain. As a source of sperm cells, different possibilities have been developed: (1) fresh sperm cells from the epididymis after killing the male, (2) puncture of the epididymis in narcosis [35], whenever only one male is available and need to be maintained, and (3) frozen sperm cells. A big advantage is that for a backup with frozen sperm cells you only need sperm cells from around six males, whereas from one male typically 10 straws with five pellets are frozen. A disadvantage of IVF is that it does not work well for strains carrying multiple transgenic modifications, because for oocyte donation the unmodified background strain is employed, resulting in segregation of the different genetic modifications. Thus a long breeding procedure may be required to recreate a particular strain. With regard to the number of animals needed, IVF is not really the best option, because to perform an IVF session several donors from the background strain are needed and the efficiency following superovulation varies. In particular, the C57BL/6 strain is characterized by low ovulation rates, and in this case the use of inhibin within the protocol is really helpful [7−9]. Typically, around 80 females of the C57BL/6 strain are necessary for the donation of oocytes.

In the IVF procedure, the timing of the steps involved is of paramount importance. In our facility we have a 7 a.m. to 7 p.m. day cycle and 7 p.m. to 7 a.m. night cycle, respectively. Under these conditions, a program has to be decided and made with regard to the superovulation protocol, the harvesting of the oocytes, and carrying out IVF, followed by washing and embryo transfer usually after overnight culture. Nowadays, the shipment of cryopreserved sperm cells has become a standard procedure. To rederivate animals resulting from IVF session to a high specific pathogen free barrier the risk of transferring unwanted bacteria or viruses is much higher when compared to an ordinary embryo transfer. In our laboratories, we keep foster mothers in a special gnotocage [65], and the sperm cells are tested first, before the foster mothers are allowed to be transferred in the barrier. It is important to know the genetic background of the sperm cells donor, and if unknown, it is advisable to perform a background analysis. In fact, males with an unclear background make it difficult to perform experiments considering the long time required for the needed backcrossing. It has been shown that the method of choice for euthanizing males, may affect the efficiency of the IVF procedure [66]. In most laboratories, mice and rats are euthanized by carboxygen inhalation, and a recent study has shown that this method has an influence on the embryonic developmental rate when compared to cervical dislocation [67].

Different protocols incorporating methyl-α-cyclodextrin (MBCD) and reduced glutathione (GSH) concentration have been reported to improve recovery of cryopreserved mouse sperm cells for IVF, belonging to a C57BL/6 (J and N) genetic background [33]. Li et al. performed a study where they correlated sperm motility with fertilization rate and compared the efficiency of different IVF protocols. High linear correlation between sperm fertilization rate and progressive motility was found [33]. High amounts of CPA were reported to impair both sperm capacitation and fertilization. It was concluded that the efficiency of IVF using cryo-recovered mouse sperm cells in media containing MBCD and GSH can be predicted from sperm progressive motility [33]. At present, the IVF protocol using MBCD and GSH is routinely used [68−70]. In this protocol, the sperm cells are added after thawing to a long flat MBCD drop. MBCD enhances the sperm capacitation and enriches the progressive motile sperm cells in the periphery. Following COCs recovery, they are transferred to an MBCD droplet containing GSH. After incubation, the sperm cells from the periphery of the MBCD droplet are added to the oocytes, and after 5 hours of coincubation/fertilization COCs are washed in MBCD droplets and cultured overnight [69]. In the second method, the so-called rescued IVF in MBCD [33], the thawed sperm cells are washed and centrifuged in Cook medium with MTG, the supernatant is discharged and the pellet is resuspended in MBCD medium. The sperm suspension is used to make long flat droplets, and the remaining of the procedure is followed as previously detailed. Commercial media for IVF are quite expensive. The MBCD and GSH method for IVF is preferable because all media can be prepared in the lab. After overnight culture two-cell embryos are transferred into the oviduct of pseudo-pregnant females on day 0.5 postcoitum. It is also possible to transfer directly zygotes, although checking all embryos for evidence of the two polar bodies is time-consuming.

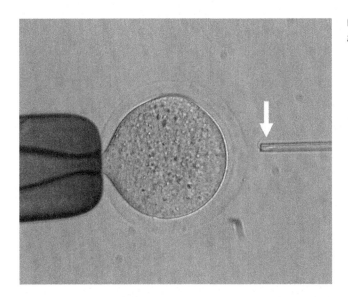

FIGURE 10.5   ICSI. Arrow indicating the sperm head on the tip of the glass pipette.

10.6   Intracytoplasmic sperm injection

ICSI represents a powerful fertilization tool in various animal species [71] including humans [72]. ICSI is also applied as routine technique for efficient embryo production in laboratory animals. The oocytes are fertilized by direct injection of a spermatozoon drawn into a thin glass pipette attached to a micromanipulator (Fig. 10.5). Using ICSI, embryos can be produced from oocytes injected with immotile [73] and immature [74] sperm cells. The establishment of ICSI has dramatically increased the fertility potential of sperm cells.

The first successful reports of mammalian ICSI was reported by Uehara and Yanagimachi [75,76]. These reports showed that normal pronuclear development was achieved in oocytes injected with hamster spermatozoa. Offspring from various mammals have been reported since then [77].

The first successful mouse ICSI was reported by Kimura and Yanagimachi in 1995 [78], following the first successful human ICSI in 1992 [72]. Mouse oocytes are quite sensitive and vulnerable to physical stress; in fact the oolemma of the mouse oocyte is easily broken by injection of sperm cell using conventional sharp glass pipette. Physical damage by injection using glass pipette was dramatically decreased by development of the piezo pulse-driven micromanipulator [78]. With this approach a flat-cut thin glass pipette can be used, and a small hole with minimal damage to the oolemma with a piezo pulse is made. The survival rate of mouse oocytes after ICSI with piezo pulse-driven micromanipulator was in fact significantly increased. C57BL/6 mouse is one of the popular strains for production of genetically engineered animals. The ICSI using this strain has also been continuously improved [79,80] and currently used as a routine technique.

It is difficult to inject rat sperm cells into oocytes because the sperm head has a unique structure [81], and oocytes in addition are more sensitive to physical damage when compared to mouse oocytes [82]. No oocytes survived following injection using a glass pipette with a diameter wide enough to aspirate the whole sperm head. Survival was obtained though, by injection of a single sperm head hanging on the tip of a glass pipette with a narrow tip [38,55].

It has also been possible to produce offspring through ICSI using oocytes fertilized by freeze-dried sperm cells [55], testicular sperm [50,51], and round immotile spermatid [74].

10.7   Production of transgenic animals

Transgenic animals are an important tool for scientific work. The first successful generation of transgenic mice was achieved by microinjection of a DNA solution into the pronucleus of a zygote in 1980 [83]. The first transgenic mice with a transgene-mediated phenotype change (increased growth due to ectopic growth hormone expression) were produced in 1982 [84]. Transgenic mouse and rat models are nowadays a standard tool for molecular research. Mouse and rat models have the big advantage of a short generation interval compared to large animal models (Fig. 10.6). Furthermore, there are a lot of tools available to produce transgenic models (Fig. 10.7). The most important constraint in using laboratory animals as model animals is given by the limitations caused by the anatomic and physiological differences with humans.

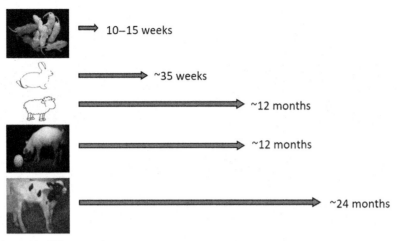

FIGURE 10.6 Generation interval in different species.

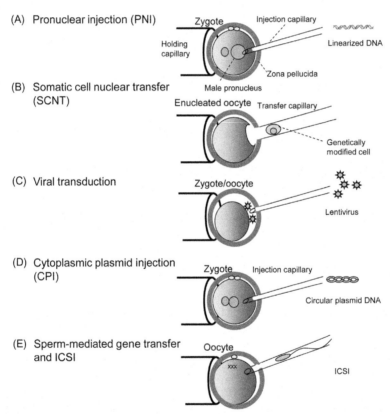

FIGURE 10.7 Validated methods for genetic engineering in mammalian oocytes and zygotes. (A) Pronuclear injection (PNI): Linearized DNA molecules are injected into the male pronucleus by a finely pulled glass capillary. Typically, 1%−5% of the treated zygotes result in transgenic offspring; however, high rates of (transgene) mosaic animals, concatemeric integrations, and transgene silencing should be expected. Instead of linearized DNA, circular transposon plasmids or designed nuclease components [DNA, RNA, ribonucleoprotein (RNP) complexes] can be injected. (B) Somatic cell nuclear transfer (SCNT): First, the genome of metaphase II oocytes is removed (enucleation), then a somatic cell or a purified nucleus is transplanted to reconstruct a cloned embryo. Typically the somatic cells are genetically modified and characterized before SCNT. Due to the low developmental capacity of the reconstructed embryos, only 1%−5% develop to offspring and deformed offspring are commonly found. (C) Viral transfection: Disarmed, recombinant lenti- or retroviruses are injected into the perivitelline space. Subsequently, they infect the embryo and integrate their reverse-transcribed genome. Typically 30%−80% of the offspring are transgenic; however, reduced offspring numbers, a high mosaicism rate, silencing of the proviruses, and animals carrying multiple integrations are commonly found. (D) Cytoplasmic plasmid injection (CPI): simplified version of the PNI methods. Since the vulnerable pronucleus membrane is not touched, the developmental rates are increased. Circular expression plasmids are injected into the cytoplasm, where they transiently persist as episomal entities. By employing transposon systems (such as Sleeping Beauty), active enzyme-catalyzed transgene integration of single copy transgenes can be achieved. By employing gene-editing compounds (expression plasmids, RNAs, or CRISPR/Cas9 ribonucleoprotein complexes) targeted modifications of specific genes are possible. Up to 10% of the treated zygotes may result in genetically modified offspring. (E) Sperm-mediated gene transfer and intracytoplasmic sperm injection (ICSI): The membranes of sperm cells are damaged by freezing, drying, or NaOH treatment, and incubated with DNA. The spermatozoa are then no longer mobile and must be injected into oocytes by ICSI. Up to 10% of the treated zygotes may result in transgenic offspring; however, the same drawbacks as for the PNI methods exist.

10.7.1   Microinjection

The standard method for microinjection is the injection of linearized DNA into the pronuclei. This technique has one big disadvantage, as the integration of DNA occurs randomly in the genome. The ratio of protein-coding genes in the genome is only 3% and thus the integration may take place in regions that are not suitable for transgene expression. With this technique the success rate to produce transgenic animals is quite low, 1%−4% of microinjected embryos [85]. With a simplified microinjection method of circular DNA in the cytoplasm of the zygotes, a new more efficient method for transient DNA expression from episomally persistent plasmids was established and the success rate was between 40% and 60% [86]. The transgene expression was transient because after around 10 cell divisions the plasmid was so diluted, not to be detectable anymore, although this method of choice is ideal for investigation of the early embryonic development. Nowadays it is possible to use this approach for active DNA integration employing transposon systems, or for genome editing approaches using CRISPR/Cas or other nucleases [87].

10.7.2   Microinjection procedure

For microinjection to take place (Fig. 10.8), zygotes are needed. Following superovulation and mating, plugged females are sacrificed and oviducts prepared. Sometimes, when zygotes are still enclosed in cumulus cells, they may stick together and the oviduct has to be open with two small forceps to recover them. To remove the cumulus cells, a treatment with hyaluronidase is necessary. The embryos are analyzed under the microscope and only the ones with two polar bodies are used for microinjection. For handling the embryos under air condition PB1 or M2 can be used. The prepared plasmids are filled in an injection pipette and the pipette is connected to a microinjector. The microinjection can be performed in a small Petri dish, although in our lab we prefer to use a siliconized glass plate. Prewarmed medium (37°C) is placed on the glass plate (around 500 μL) and the zygotes are transferred into this droplet. Following microinjection, zygotes are transferred in droplets of M2, which are covered with oil to avoid changes in the osmolality due to evaporation.

The injected zygotes can be transferred directly to foster mothers (day 0.5) or they can be cultured to assess the developmental competence.

10.7.3   Electroporation

The production of genome-edited animals by modifying targeted genes is required to foster studies for the benefit of animals and humans as well. Although microinjection is nowadays the gold standard method for the production of genome-edited animals, it requires a micromanipulator and sophisticated technical skills to prevent cell damage. Furthermore, microinjection is not convenient when many cells must be assessed simultaneously, because the endonucleases must be injected into the embryos one at a time.

Endonucleases have been introduced in cultured cells by electroporation, although this approach has not been found suitable for their introduction into mammalian embryos. In fact, electroporation causes damage to the embryos due to

FIGURE 10.8   Typical working place for microinjection.

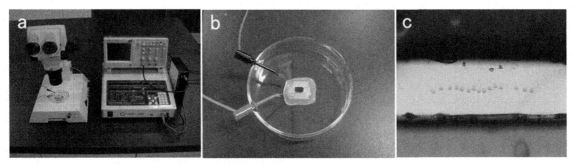

FIGURE 10.9   (A) Super electroporator NEPA21. (B) Petri dish with platinum plate electrodes. (C) Pronuclear-stage embryos are placed in a line between metal plates in a glass chamber filled with buffer carrying the endonucleases.

the conventional needed electric pulses and by the use of acid Tyrode's solution just before electroporation, with the intent to increase the possibility of endonucleases introduction [88,89]. In addition, this procedure affects subsequent embryonic development and implantation [90,91].

The technique for animal knockout system by electroporation (TAKE) method, was developed in order to efficiently introduce endonucleases into mammalian embryos. A new electroporation device, NEPA21 (Nepa Gene, Chiba, Japan), is characterized by a three-step electrical pulse system (Fig. 10.9) [92,93] that greatly reduces the damage to the embryos. The first pulse, the poring pulse, makes micro-holes in the zona pellucida and the oolemma of the developing embryos. The second pulse, the transfer pulse, transfers the endonucleases into the cytoplasm of the embryos. The third pulse, the polarity-changed transfer pulse, increases the opportunity of introducing the endonucleases into the embryos [94]. This technique can also be used with intact embryos and no weakening of the zona pellucida. Many valuable mouse and rat new strains have been already produced using the TAKE method. This technique has revolutionized the method for the introduction of endonucleases, and it is now being applied for the introduction of endonucleases into embryos within the oviduct.

Conclusion

In this chapter, the current status of old and the most recently developed reproductive technologies in laboratory animals has been discussed, with a clear understanding of a continuing refinement and a higher efficiency to be witnessed in the near future. This will contribute to an improvement of animal welfare through the avoidance of unneeded studies and the reduction in the number of laboratory animals.

References

[1]   Lyon MF, Whittingham DG, Glenister P. Long-term storage of frozen mouse embryos under increased background irradiation. Ciba Found Symp 1977;273−90.

[2]   Kenyon J, Guan M, Bogani D, Marschall S, Raspa M, Pickard A, et al. Transporting mouse embryos and germplasm as frozen or unfrozen materials. Curr Protoc Mouse Biol 2014;4:47−65.

[3]   Luo C, Zuniga J, Edison E, Palla S, Dong W, Parker-Thornburg J. Superovulation strategies for 6 commonly used mouse strains. J Am Assoc Lab Anim Sci 2011;50:471−8.

[4]   Vasudevan K, Sztein JM. In vitro fertility rate of 129 strain is improved by buserelin (gonadotropin-releasing hormone) administration prior to superovulation. Lab Anim 2012;46:299−303.

[5]   Sotomaru Y, Kanda A, Nobukiyo A. Effect of cetrorelix acetate administration on ovarian stimulation in aged mice. In: 14th FELASA congress; 2019. Abstract PC28.

[6]   Pallares P, Gonzalez-Bulnes A. A new method for induction and synchronization of oestrus and fertile ovulations in mice by using exogenous hormones. Lab Anim 2009;43:295−9.

[7]   Hasegawa A, Mochida K, Matoba S, Yonezawa K, Ohta A, Watanabe G, et al. Efficient production of offspring from Japanese wild-derived strains of mice (*Mus musculus* molossinus) by improved assisted reproductive technologies. Biol Reprod 2012;86:1−7.

[8]   Mochida K, Hasegawa A, Otaka N, Hama D, Furuya T, Yamaguchi M, et al. Devising assisted reproductive technologies for wild-derived strains of mice: 37 strains from five subspecies of *Mus musculus*. PLoS One 2014;9:e114305.

[9]   Takeo T, Nakagata N. Superovulation using the combined administration of inhibin antiserum and equine chorionic gonadotropin increases the number of ovulated oocytes in C57BL/6 female mice. PLoS One 2015;10:e0128330.

[10]   Peters AG, Festing MF. NIH/Ola: a highly productive inbred strain of laboratory mouse. Lab Anim 1985;19:320−7.

[11] Taketsuru H, Kaneko T. Efficient collection and cryopreservation of embryos in F344 strain inbred rats. Cryobiology 2013;67:230−4.

[12] Armstrong DT, Opavsky MA. Superovulation of immature rats by continuous infusion of follicle-stimulating hormone. Biol Reprod 1988;39:511−18.

[13] Rouleau AM, Kovacs PR, Kunz HW, Armstrong DT. Decontamination of rat embryos and transfer to specific pathogen-free recipients for the production of a breeding colony. Lab Anim Sci 1993;43:611−15.

[14] Cora MC, Kooistra L, Travlos G. Vaginal cytology of the laboratory rat and mouse: review and criteria for the staging of the estrous cycle using stained vaginal smears. Toxicol Pathol 2015;43:776−93.

[15] Whittingham DG, Leibo SP, Mazur P. Survival of mouse embryos frozen to -196 degrees and -269 degrees c. Science 1972;178:411−14.

[16] Schmidt PM, Hansen CT, Wildt DE. Viability of frozen-thawed mouse embryos is affected by genotype. Biol Reprod 1985;32:507−14.

[17] Ostermeier GC, Wiles MV, Farley JS, Taft RA. Conserving, distributing and managing genetically modified mouse lines by sperm cryopreservation. PLoS One 2008;3:e2792.

[18] Rall WF, Schmidt PM, Lin X, Brown SS, Ward AC, Hansen CT. Factors affecting the efficiency of embryo cryopreservation and rederivation of rat and mouse models. ILAR J 2000;41:221−7.

[19] Landel CP. Cryopreservation of mouse gametes and embryos. Methods Enzymol 2010;476:85−105.

[20] Wood MJ, Farrant J. Preservation of mouse embryos by two-step freezing. Cryobiology 1980;17:178−80.

[21] Renard JP, Babinet C. High survival of mouse embryos after rapid freezing and thawing inside plastic straws with 1-2 propanediol as cryoprotectant. J Exp Zool 1984;230:443−8.

[22] Dorsch M. Cryopreservation of preimplantation embryos and gametes and associated methods. Lab Mouse 2012;675−93.

[23] Taft R. Mouse embryo cryopreservation by slow freezing. Cold Spring Harb Protoc 2018;5.

[24] Ghandy N, Karimpur Malekshah AA. Which stage of mouse embryos is more appropriate for vitrification? Int J Fertil Steril 2017;10:357−62.

[25] Inna N, Sanmee U, Saeng-Anan U, Piromlertamorn W, Vutyavanich T. Rapid freezing versus cryotop vitrification of mouse two-cell embryos. Clin Exp Reprod Med 2018;45:110−15.

[26] Mochida K, Hasegawa A, Li MW, Fray MD, Kito S, Vallelunga JM, et al. High osmolality vitrification: a new method for the simple and temperature-permissive cryopreservation of mouse embryos. PLoS One 2013;8:e49316.

[27] Jin B, Mochida K, Ogura A, Hotta E, Kobayashi Y, Ito K, et al. Equilibrium vitrification of mouse embryos. Biol Reprod 2010;82:444−50.

[28] Taketsuru H, Kaneko T. Tolerance to vitrification of rat embryos at various developmental stages. Cryobiology 2018;84:1−3.

[29] Willoughby CE, Mazur P, Peter AT, Critser JK. Osmotic tolerance limits and properties of murine spermatozoa. Biol Reprod 1996;55:715−27.

[30] Takeshima T, Nakagata N, Ogawa S. Cryopreservation of mouse spermatozoa. Jikken Dobutsu 1991;40:493−7.

[31] Tada N, Sato M, Yamanoi J, Mizorogi T, Kasai K, Ogawa S. Cryopreservation of mouse spermatozoa in the presence of raffinose and glycerol. J Reprod Fertil 1990;89:511−16.

[32] Takeo T, Sztein J, Nakagata N. The card method for mouse sperm cryopreservation and in vitro fertilization using frozen-thawed sperm. Methods Mol Biol 2019;1874:243−56.

[33] Li MW, Glass OC, Zarrabi J, Baker LN, Lloyd KC. Cryorecovery of mouse sperm by different IVF methods using MBCD and GSH. J Fertil In Vitro 2016;4.

[34] Li MW, Vallelunga JM, Kinchen KL, Rink KL, Zarrabi J, Shamamian AO, et al. IVF recovery of mutant mouse lines using sperm cryopreserved with MTG in cryovials. Cryo Lett 2014;35:145−53.

[35] Del Val GM, Robledano PM. In vivo serial sampling of epididymal sperm in mice. Lab Anim 2013;47:168−74.

[36] Varisli O, Scott H, Agca C, Agca Y. The effects of cooling rates and type of freezing extenders on cryosurvival of rat sperm. Cryobiology 2013;67:109−16.

[37] Nakatsukasa E, Inomata T, Ikeda T, Shino M, Kashiwazaki N. Generation of live rat offspring by intrauterine insemination with epididymal spermatozoa cryopreserved at -196 degrees c. Reproduction 2001;122:463−7.

[38] Hirabayash M, Kato M, Aoto T, Sekimoto A, Ueda M, Miyoshi I, et al. Offspring derived from intracytoplasmic injection of transgenic rat sperm. Transgenic Res 2002;11:221−8.

[39] Kaneko T, Kimura S, Nakagata N. Offspring derived from oocytes injected with rat sperm, frozen or freeze-dried without cryoprotection. Theriogenology 2007;68:1017−21.

[40] Polge C, Smith AU, Parkes AS. Revival of spermatozoa after vitrification and dehydration at low temperatures. Nature 1949;164:666.

[41] Katayose H, Matsuda J, Yanagimachi R. The ability of dehydrated hamster and human sperm nuclei to develop into pronuclei. Biol Reprod 1992;47:277−84.

[42] Wakayama T, Yanagimachi R. Development of normal mice from oocytes injected with freeze-dried spermatozoa. Nat Biotechnol 1998;16:639−41.

[43] Kusakabe H, Szczygiel MA, Whittingham DG, Yanagimachi R. Maintenance of genetic integrity in frozen and freeze-dried mouse spermatozoa. Proc Natl Acad Sci USA 2001;98:13501−6.

[44] Kaneko T, Whittingham DG, Yanagimachi R. Effect of pH value of freeze-drying solution on the chromosome integrity and developmental ability of mouse spermatozoa. Biol Reprod 2003;68:136−9.

[45] Kaneko T, Nakagata N. Improvement in the long-term stability of freeze-dried mouse spermatozoa by adding of a chelating agent. Cryobiology 2006;53:279−82.

[46] Nakai M, Kashiwazaki N, Takizawa A, Maedomari N, Ozawa M, Noguchi J, et al. Effects of chelating agents during freeze-drying of boar spermatozoa on DNA fragmentation and on developmental ability in vitro and in vivo after intracytoplasmic sperm head injection. Zygote 2007;15:15−24.

[47] Kaneko T, Nakagata N. Relation between storage temperature and fertilizing ability of freeze-dried mouse spermatozoa. Comp Med 2005;55:140−4.

[48] Kawase Y, Araya H, Kamada N, Jishage K, Suzuki H. Possibility of long-term preservation of freeze-dried mouse spermatozoa. Biol Reprod 2005;72:568−73.

[49] Kusakabe H, Tateno H. Characterization of chromosomal damage accumulated in freeze-dried mouse spermatozoa preserved under ambient and heat stress conditions. Mutagenesis 2011;26:447−53.

[50] Kaneko T, Whittingham DG, Overstreet JW, Yanagimachi R. Tolerance of the mouse sperm nuclei to freeze-drying depends on their disulfide status. Biol Reprod 2003;69:1859−62.

[51] Kaneko T, Serikawa T. Successful long-term preservation of rat sperm by freeze-drying. PLoS One 2012;7:e35043.

[52] Kaneko T, Serikawa T. Long-term preservation of freezedried mouse spermatozoa. Cryobiology 2012;64:211−14.

[53] Li MW, Willis BJ, Griffey SM, Spearow JL, Lloyd KC. Assessment of three generations of mice derived by ICSI using freeze-dried sperm. Zygote 2009;17:239−51.

[54] Kaneko T. The latest improvements in the mouse sperm preservation. In: Lewandoski M, editor. Methods in molecular biology, 1092. New York, NY: Springer; 2014. p. 357−65.

[55] Kaneko T. Simple sperm preservation by freeze-drying for conserving animal strains. In: Shondra M, editor. Chromosomal mutagenesis, 2nd ed., Methods in molecular biology, 1239. New York, NY: Springer; 2015. p. 317−29.

[56] Sztein J, Sweet H, Farley J, Mobraaten L. Cryopreservation and orthotopic transplantation of mouse ovaries: new approach in gamete banking. Biol Reprod 1998;58:1071−4.

[57] Milenkovic M, Diaz-Garcia C, Wallin A, Brannstrom M. Viability and function of the cryopreserved whole rat ovary: comparison between slow-freezing and vitrification. Fertil Steril 2012;97:1176−82.

[58] Sztein J. Mouse ovary cryopreservation. Cold Spring Harb Protoc 2017;3.

[59] Bedaiwy MA, Falcone T. Ovarian tissue banking for cancer patients: reduction of post-transplantation ischaemic injury: intact ovary freezing and transplantation. Hum Reprod 2004;19:1242−4.

[60] Bouquet M, Selva J, Auroux M. Cryopreservation of mouse oocytes: mutagenic effects in the embryo? Biol Reprod 1993;49:764−9.

[61] Harp R, Leibach J, Black J, Keldahl C, Karow A. Cryopreservation of murine ovarian tissue. Cryobiology 1994;31:336−43.

[62] Ghavami M, Mohammadnejad D, Beheshti R, Solmani-Rad J, Abedelahi A. Ultrastructural and morphalogical changes of mouse ovarian tissues following direct cover vitrification with different cryoprotectants. J Reprod Infertil 2015;16:138−47.

[63] Dorsch MM, Wedekind D, Kamino K, Hedrich HJ. Cryopreservation and orthotopic transplantation of rat ovaries as a means of gamete banking. Lab Anim 2007;41:247−54.

[64] Dorsch M, Wedekind D, Kamino K, Hedrich HJ. Orthotopic transplantation of rat ovaries as a tool for strain rescue. Lab Anim 2004;38:307−12.

[65] Basic M, Bleich A. Gnotobiotics: past, present and future. Lab Anim 2019;53:232−43.

[66] Wuri L, Agca C, Agca Y. Euthanasia via CO_2 inhalation causes premature cortical granule exocytosis in mouse oocytes and influences in vitro fertilization and embryo development. Mol Reprod Dev 2019;86:825−34.

[67] Conlee KM, Stephens ML, Rowan AN, King LA. Carbon dioxide for euthanasia: concerns regarding pain and distress, with special reference to mice and rats. Lab Anim 2005;39:137−61.

[68] Choi YH, Toyoda Y. Cyclodextrin removes cholesterol from mouse sperm and induces capacitation in a protein-free medium. Biol Reprod 1998;59:1328−33.

[69] Takeo T, Hoshii T, Kondo Y, Toyodome H, Arima H, Yamamura K, et al. Methyl-beta-cyclodextrin improves fertilizing ability of C57BL/6 mouse sperm after freezing and thawing by facilitating cholesterol efflux from the cells. Biol Reprod 2008;78:546−51.

[70] Bath ML. Inhibition of in vitro fertilizing capacity of cryopreserved mouse sperm by factors released by damaged sperm, and stimulation by glutathione. PLoS One 2010;5:e9387.

[71] Kaneko T. Sperm freeze-drying and micro-insemination for biobanking and maintenance of genetic diversity in mammals. Reprod Fertil Dev 2016;28:1079−87.

[72] Palermo G, Joris H, Devroey P, Van Steirteghem AC. Pregnancies after intracytoplasmic injection of single spermatozoon into an oocyte. Lancet 1992;340:17−18.

[73] Goto K, Kinoshita A, Takuma Y, Ogawa K. Fertilisation of bovine oocytes by the injection of immobilised, killed spermatozoa. Vet Rec 1990;127:517−20.

[74] Kimura Y, Yanagimachi R. Mouse oocytes injected with testicular spermatozoa or round spermatids can develop into normal offspring. Development 1995;121:2397−405.

[75] Uehara T, Yanagimachi R. Microsurgical injection of spermatozoa into hamster eggs with subsequent transformation of sperm nuclei into male pronuclei. Biol Reprod 1976;15:467−70.

[76] Uehara T, Yanagimachi R. Behavior of nuclei of testicular, caput and cauda epididymal spermatozoa injected into hamster eggs. Biol Reprod 1977;16:315−21.

[77] Kaneko T. Simple gamete preservation and artificial reproduction of mammals using micro-insemination techniques. Reprod Med Biol 2014;14:99−105.

[78] Kimura Y, Yanagimachi R. Intracytoplasmic sperm injection in the mouse. Biol Reprod 1995;52:709−20.

[79] Kaneko T, Ohno R. Improvement in the development of oocytes from C57BL/6 mice after sperm injection. J Am Assoc Lab Anim Sci 2011;50:33—6.

[80] Sakamoto W, Kaneko T, Nakagata N. Use of frozen-thawed oocytes for efficient production of normal offspring from cryopreserved mouse spermatozoa showing low fertility. Comp Med 2005;55:136—9.

[81] Benson JD, Woods EJ, Walters EM, Critser JK. The cryobiology of spermatozoa. Theriogenology 2012;78:1682—99.

[82] Dozortsev D, Wakaiama T, Ermilov A, Yanagimachi R. Intracytoplasmic sperm injection in the rat. Zygote 1998;6:143—7.

[83] Gordon JW, Scangos GA, Plotkin DJ, Barbosa JA, Ruddle FH. Genetic transformation of mouse embryos by microinjection of purified DNA. Proc Natl Acad Sci USA 1980;77:7380—4.

[84] Palmiter RD, Brinster RL, Hammer RE, Trumbauer ME, Rosenfeld MG, Birnberg NC, et al. Dramatic growth of mice that develop from eggs microinjected with metallothionein-growth hormone fusion genes. Nature 1982;300:611—15.

[85] Wall RJ, Rexroad CE, Powell A, Shamay A, McKnight R, Hennighausen L. Synthesis and secretion of the mouse whey acidic protein in transgenic sheep. Transgenic Res 1996;5:67—72.

[86] Iqbal K, Barg-Kues B, Broll S, Bode J, Niemann H, Kues W. Cytoplasmic injection of circular plasmids allows targeted expression in mammalian embryos. Biotechniques 2009;47:959—68.

[87] Garrels W, Ivics Z, Kues WA. Precision genetic engineering in large mammals. Trends Biotechnol 2012;30:386—93.

[88] Grabarek JB, Plusa B, Glover DM, Zernicka-Goetz M. Efficient delivery of dsRNA into zona-enclosed mouse oocytes and preimplantation embryos by electroporation. Genesis 2002;32:269—76.

[89] Peng H, Wu Y, Zhang Y. Efficient delivery of DNA and morpholinos into mouse preimplantation embryos by electroporation. PLoS One 2012;7:e43748.

[90] Bronson RA, McLaren A. Transfer to the mouse oviduct of eggs with and without the zona pellucida. J Reprod Fertil 1970;22:129—37.

[91] Modliński JA. The role of the zona pellucida in the development of mouse eggs in vivo. J Embryol Exp Morphol 1970;23:539—47.

[92] Kaneko T, Sakuma T, Yamamoto T, Mashimo T. Simple knockout by electroporation of engineered endonucleases into intact rat embryos. Sci Rep 2014;4:6382.

[93] Kaneko T, Mashimo T. Simple genome editing of rodent intact embryos by electroporation. PLoS One 2015;10:e0142755.

[94] Kaneko T. Genome editing in mouse and rat by electroporation. In: Hatada I, editor. Genome editing in animals: Methods and protocols, Methods in molecular biology, 1630. New York, NY: Springer; 2017. p. 81—9.

Chapter 11

Standard and innovative reproductive biotechnologies for the development of finfish farming

E. Figueroa[1], L. Sandoval[4], O. Merino[3], J. Farías[2], J. Risopatrón[3] and I. Valdebenito[4]

[1]Nucleus of Research in Food Production, Department of Biological and Chemical Sciences, Faculty of Natural Resources, Catholic University of Temuco, Temuco, Chile, [2]Chemical Engineering Department, Faculty of Engineering and Science, University of La Frontera, Temuco, Chile, [3]Center of Excellence of Biotechnology in Reproduction (BIOREN-CEBIOR), Faculty of Medicine, University of La Frontera, Temuco, Chile, [4]Nucleus of Research in Food Production, Department of Agricultural and Aquaculture Sciences, Faculty of Natural Resources, Catholic University of Temuco, Temuco, Chile

11.1 Introduction

Within fish research and production, the term biotechnology includes broodstock management, induction of oocytes and spermatozoa maturation, in vitro gamete manipulation, DNA manipulation, controlling the sex of the populations bred, and embryo development. Broadly speaking, many of these technologies focus on the early stages of the life cycle of the species of commercial interest, through gamete and chromosome manipulation, cryopreservation, transgenesis, production of chimeras, cloning and controlling the phenotypic expression of sex [1]. The present chapter offers a conceptual analysis of the principal biotechnologies used in the production of fish of commercial interest, particularly Atlantic salmon (*Salmo salar*), silver or coho salmon (*Oncorhynchus kisutch*), and rainbow trout (*Oncorhynchus mykiss*). It also describes biotechnologies applied to the production of other freshwater species such as tilapia, catfish, and carp, and marine fish such as cobia, bass, and sea bream. These biotechnologies focus on a number of areas: artificial manipulation of the photoperiod to obtain out-of-season spawning (advanced or delayed), use of hormone treatments with extracts of carp or salmon hypophysis [human chorionic gonadotropin (hCG), luteinizing hormone-releasing hormone analogue (LH-RHa), and gonadotropin-releasing hormone analogue (GnRHa)] to synchronize final maturation of the oocyte, acceleration of sexual maturity and/or increasing the volume of semen produced by males, and cold storage of semen (up to several weeks) using different media enriched with antioxidants, antibiotics, and salts. The cryopreservation of semen in liquid nitrogen, where sperm cells have an almost indefinite lifetime, is used on a massive scale in some species like salmonids and marine species. Oocytes can be stored for up to 48 hours, and experimentally for up to 4 days. In slow-growing cold-water species such as salmonids, the number of days of embryo development is regularly reduced or prolonged, by increasing or reducing the water temperature during incubation. Monosex populations are frequently produced in species where one sex is preferred for its productive yield (e.g., females in salmonids or males in tilapias). The survival of polyploid specimens allows commercial production of triploid (3n) or tetraploid (4n) specimens by temperature shock close to 28°C or pressure shock around 10,000 psi. In 3n groups, females do not reach sexual maturity or express secondary sexual characteristics, resulting in better yield per carcass and greater growth. On the other hand, 4n groups are used for the production of interploid (3n) populations by crossing a 2n female with a 4n male. A wide variety of polyploid groups can be produced experimentally. The present review will discuss only those which are most developed in fish farming worldwide or which offer the best prospects for future application.

Reproductive Technologies in Animals. DOI: https://doi.org/10.1016/B978-0-12-817107-3.00011-4

11.2 The control of sexual maturation

11.2.1 Modification of the spawning season

In fish, as in all vertebrates, reproduction is regulated by the brain through the gonadotropin-releasing hormone (GnRH) produced in the hypothalamus [1−5]. This decapeptide stimulates production of gonadotropins (GtH) by the hypophysis or pituitary gland. In some fish, dopamine is a negative inhibitor of GtH production and release by the hypophysis [3,5,6]. This gland controls reproduction through two gonadotropic hormones: the follicle-stimulating hormone (FSH or GtH I), which regulates vitellogenesis in females and spermatogenesis in males, and the luteinizing hormone (LH or GtH II), which is responsible for controlling final maturation of the oocyte in females [5].

In salmonids, sexual maturation is controlled synergically by the combination of photoperiod and temperature, the environmental factors that trigger the hormone cycles that control reproduction. Vitellogenesis is facilitated by photoperiods with long days combined with "high" temperatures (around 15°C). Final maturation of the oocytes and spawning require photoperiods with short days and temperatures below 10°C. These basic requirements can be maintained in production, in fact where low temperatures exist, it is easy to manipulate the photoperiod at little cost by prolonging or shortening the number of hours of light per day (Fig. 11.1A and B), achieving out-of-season spawning and even two spawnings in a year. This is done particularly with rainbow trout, but more and more experiments are being carried out with Atlantic salmon, catfish, cyprinids, and some marine species such as sole, cobia, bass, sea bream, tuna, and eel [7−13].

11.2.2 Induction of sexual maturation

Many species kept in artificial culture systems suffer severe reproductive dysfunctions. The most common is for the females to start gonadal development normally, but never reach final maturation of the oocyte, ovulation and/or spawning. The males, although much more resistant to the environmental conditions of fish farming and the consequent stressing effects caused, produce a smaller volume of semen, or semen of poor quality. Manipulating certain environmental factors, such as photoperiod, temperature, salinity, tank volume, and vegetation in the substrate, can often improve gamete quality. In some species, however, hormone treatments are the only option for achieving successful reproductive processes, for example, in sole and some species of carp [14]. In catadromous species, such as the European eel (*Anguilla anguilla*), this type of biotechnology is starting to produce its first fruits with the induction of sexual maturation in captivity, since in the wild full sexual maturity in this species occurs at depths of thousands of meters, and replicating these conditions in a fish farm is not yet possible [15−17].

In recent years, a range of hormones and associated technologies have been used successfully in numerous species [18−21]. These methods were first developed in 1930 with the injection of raw extract of hypophysis from mature fish (with high levels of GtH), followed by the injection of salmon or carp GtH to induce spawning. Today, various powerful synthetic compounds of the gonadotropin-releasing hormone (GnRHa or LH-RHa) are used, applied by simple injection in physiological serum or in pellets (slow-release systems) in concentrations around 10 μg/kg of fish (Fig. 11.2A−D), in a single or double dose. They are also used in association with a dopamine inhibitor such as domperidone or pimozide, a method that has solved numerous problems in the synchronization of final sexual maturity in various species [22]. Extracts of salmon or carp hypophysis in isolation, or in combination with injections of hCG, continue to be used in the production of carp and some Amazon species such as pacú (*Piaractus mesopotamicus*), tambaquí (*Colossoma macropomum*), surubí (*Pseudoplatystoma fasciatum* and *Pseudoplatystoma corruscans*) and South American catfish (*Rhamdia quelen*). Experimental work has been done on hormones such as kisspeptin, prostaglandins,

FIGURE 11.1 (A) Maintenance system of rainbow trout broodstock (*Oncorhynchus mykiss*) under natural photoperiod. (B) Maintenance system of rainbow trout broodstock under artificial photoperiod.

(A)

(B)

FIGURE 11.2 (A) Commercial product (Ovaprim) to sexual maturation induction based on GnRH plus a dopamine inhibitor. (B) Application of a hormonal dose in Atlantic salmon (*Salmo salar*) through intraperitoneal injection (IP). (C) Hormonal implants (Ovaplant) used in salmon industry for sexual maturity induction. (D) Application of hormonal implants in salmon coho (*Oncorhynchus kisutch*) through IP.

hormone baths (for small, scaleless fish), and direct injections into the hypophysis (carp) [23]. So far they have been successful on a small scale, but there is no doubt that some of these products could become available in the market for mass use in the medium term.

11.3   In vitro gamete management

In vitro gamete preservation (Fig. 11.3A−F) is assuming an important role in the production of ova and the genetic management of broodstock in salmon farming. Long-term storage techniques have not yet produced a viable solution for storing ova; for spermatozoa, however, various storage techniques have been developed from cryopreservation (Fig. 11.4A−F) to low-temperature storage [24].

The need to keep or preserve the semen of various fish species outside the testicle arises from:

- asynchrony in full maturity between males and females [24,25];
- the increasing demand to optimize the management and efficiency of reproductive facilities [26,27];
- the need to carry out genetic studies and preserve the genetic heritage of species in danger of extinction [27,28];
- the requirement to store semen while awaiting the ichthyo-pathological diagnosis to which broodstock are subjected to avoid vertical transmission of diseases;
- the need to transport gametes when the broodstock are located at a distance from the culture centers where fertilization and incubation are carried out.

11.3.1   Semen diluents

These products are isotonic saline solutions that mimic the medium in which the spermatozoa exist inside the testicle; they are also called "extenders," "immobilizing media," or "dilutors." They are regularly used in the artificial management of gametes in mammals [29].

Diluents for fish sperm cells are based principally on the chemical composition of the seminal plasma of each species, since spermatozoa only need to be kept immobile for a short time. Other compounds can also be added, such as some type of buffer and glucose [30]. The chemical composition of these compounds is very variable, even those recommended for use in the same species. Table 11.1 shows some examples of the diluents used in salmonid species.

Some researchers have studied the use of a diluent that mimics the content of the seminal plasma of fish of commercial importance, principally trout and salmon. Truscott and Idler [35], report the motility of Atlantic salmon spermatozoa

FIGURE 11.3 (A) Atlantic salmon male broodstock (*Salmo salar*). (B) Milt extraction through cannula in *S. salar*. (C) Motility level evaluation. (D) Oxygen injection to milt in cold storage. (E) Stripping in female rainbow trout (*Oncorhynchus mykiss*). (F) Artificial fertilization of rainbow trout gametes.

after 6 days' storage with a diluent consisting principally of sodium, potassium, and calcium chlorides, fructose, and water at pH 7.3. Erdahl and Graham [30] stored trout semen for 24 hours using a diluent in which the principal components were sodium, potassium, and magnesium chlorides and glucose, with good fertilization results. A review carried out by Linhart et al. [37] reports storage of *Polyodon spathula* semen for short periods using a saline solution of sodium chloride at 0.9%. Conte et al. [38], stored white sturgeon semen (*Acipenser transmuntanus*) for 14 days at 4°C. Bromage and Roberts [24] reported that the provision of oxygen and/or the addition of antibiotics, combined with low storage temperatures, prolonged the viability period of spermatozoa. Similar results were found by Valdebenito [39] for storing *Galaxias maculatus* (Galaxiidae) semen for 3 days without a significant loss in fertilizing capacity.

Works with rainbow trout semen have shown the effectiveness of a combination of penicillin and streptomycin for semen storage over prolonged periods [26]. Antibiotics are commonly used to protect mammal semen during storage but have been little used in salmonids.

McNiven et al. [25] diluted rainbow trout semen in nonaqueous media with fluorocarbons, achieving storage for up to 37 days with acceptable levels of motility. So long as the semen is kept in a sperm diluent, the temperature and luminosity must be controlled. Storage is ideally carried out in the dark at temperatures of 0°C–4°C [28,34,40,41]. Gaseous oxygen is frequently injected to further prolong the survival and viability of the spermatozoa. Gordon et al. [40] recommended a ratio of 1:120 between the volumes of the stored semen and the gas covering it; they also indicated that the height of the semen column should not exceed 6 mm. This technique allows semen to be stored for several days with good viability. Semen is frequently diluted with sperm diluent at a ratio of 1:2. The pH of the storage medium is an important factor that determines the movement capacity of the spermatozoon after activation. Woolsey et al. [32] found

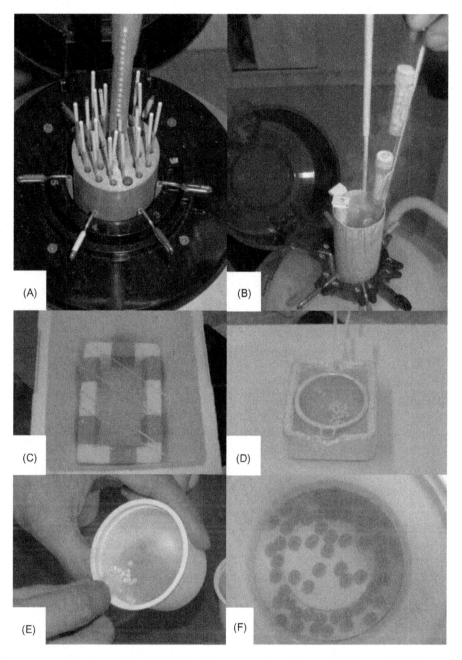

FIGURE 11.4 System of (A) straws and (B) cryotubes used in salmonid milt cryopreservation. (C) Cryopreservation process. (D) vitrification. (E) "In vitro" fertilization with vitrified sperm. (F) Evaluation of fertilizing capacity in Atlantic salmon gametes.

that rainbow trout semen stored for 2 hours at pH 8.5 presented a higher percentage of motile spermatozoa than semen stored at pH 7.1.

The use of sperm diluents solves the problem of keeping semen outside the testicle, allowing producers to extend the period during which the spermatozoa maintain their fertilizing capacity. This is a great help to ova producers, particularly in salmonid farming where the gametes need to be preserved for several days while waiting for the results of male screening, or transported for several hours when the broodstock are kept at a distance from the hatcheries.

11.3.2 Sperm activators

There are several works reporting fish semen subjected to different extra-testicular environments, in order to determine the factors that control their activation and inhibition, viability and behavior. Some of the solutions used in these

TABLE 11.1 Sperm diluents used for semen storage of salmonids.

Component (mM)	1	2	3	4	5 (g/L)	6	7 (g/2 L)
NaCl	80	80	110	103	5.16	110	11.7
KCl	40	40	28.3	40	1.64	28.3	5.1
CaCl$_2$	0.1	0.1	–	1.0	0.143	1.8	0.2
NaHCO$_3$	–	–	–	–	1.00	–	–
NaH$_2$PO$_4$	–	–	–	–	0.41	–	–
Na$_2$HPO$_4$	–	–	–	–	–	–	0.5
Tris	30	30	0.02	20	–	0.02	–
MgSO$_4$	–	–	1.1	–	0.223	1.1	–
MgSO$_2$	–	–	–	0.8			–
MgCl$_2$	–	–	–	–	–	–	0.4
CaCl	–	–	1.8	–	–	–	–
Fructose	–	–	–	–	1.00	–	–
Glucose	–	–	–	–	–	–	20
Citric acid	–	–	–	–	–	–	0.2
Bicine (5.3 g/100 mL)	–	–	–	–	–	–	20 mL
KOH (1.27 g/100 mL)	–	–	–	–	–	–	20 mL
pH	9.2	6.5–8.5	9.0	7.8	7.3	9	7.8
Osmolarity (mOsm)	–	–	–	–	226	–	310

1 = Cosson et al. [31]; 2 = Woolsey et al. [32]; 3 = Billard [33]); 4 = Lahnsteiner et al. [34]; 5 = Truscott and Idler [35]; 6 = Billard and Jalabert [36]; 7 = Erdahl and Graham [30].

investigations could be classified as "activating solutions" since they are intended to prolong the time and intensity of sperm motility and thus their fertilizing capacity.

In salmonids, the ovarian or coelomic fluid that accompanies the ova during spawning is a very effective natural sperm activator [32]; consequently, many sperm activators used in these species are based on mimicking its composition.

Erdahl and Graham [30] studied the activation of teleost fish semen in order to increase the number of ova that can be fertilized, and to this aim they used freshwater, although with poor results. Valdebenito [42] evaluated the effect of an activation solution that included caffeine on the motility of rainbow trout spermatozoa, finding that motility levels increased significantly as compared to the control (river water) with a concentration of 3.5 mM of caffeine. Saline solutions have also been widely studied as a possible activation medium, producing better results than freshwater, but with more limited motility periods [30]. Goodall et al. [43] examined some of the factors which regulate the activation and duration of motility in *Sillago ciliata* semen, in a study oriented toward the fertilizing capacity of spermatozoa using solutions containing glucose and chlorides of sodium and potassium at different osmolarities. They found that when activation media containing chlorides of sodium and potassium at osmolarities of 600 mOsm/kg were used, the motility of the spermatozoa increased.

Ohta et al. [44] used saline solutions based on the ionic components of the seminal plasma of *Anguilla japonica*, modifying the concentrations of some components, such as potassium, bicarbonate, and pH, to determine which factors promote motility in this species. They found that when the concentrations of potassium and bicarbonate were increased, and the pH of the solutions raised to the range from 7.8 to 8.7, the percentage of motile spermatozoa increased.

In production, the salmonid fish farms installed in Chile regularly use saline solutions or sperm activators in a proportion of around 20%−30% of the volume of ova for the purpose of fertilization (personal observation). This prolongs the duration of sperm motility, the intensity of flagellar movement, and finally the fertilizing capacity of the semen. Table 11.2 shows the composition of some sperm activators used in fish farming.

TABLE 11.2 Sperm activators used for activation of sperm motility in salmonids.

Component (mM)	1	2	3
NaCl	125	125	155
KCl	0	0	3.1
CaCl$_2$	0.1	0.1	–
Tris	30	30	0.02
MgSO$_4$	–	–	1.3
CaCl	–	–	3.4
pH	9.2	8.5	9.0
Others	–	1.0	–

1 = Cosson et al. (1999); 2 = Woolsey et al. [32]; 3 = Billard (1983).

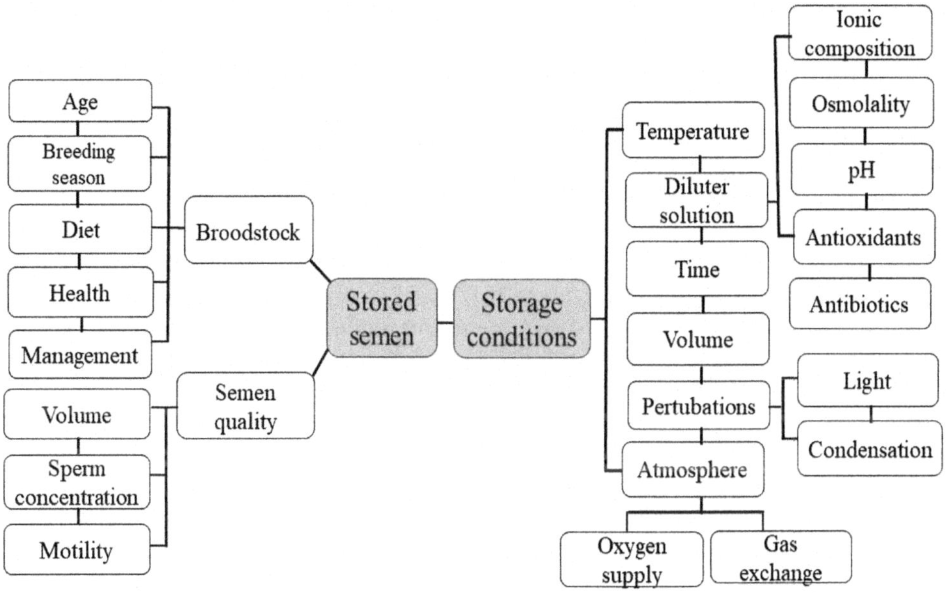

FIGURE 11.5 Factors that can influence on cold storage of fish semen [47].

11.3.3 Short-term storage of sperm in fish

The fish farming industry is dependent on in vitro fertilization for breeding purposes. In this context, reproductive management is essential, implying collection and manipulation of gametes in such a way as to ensure their quality and maximize the results of fertilization [45]. Among the reproductive technologies used for gamete conservation, short-term sperm storage is a common practice in aquaculture, allowing the use of gametes to be maximized. It is a viable, inexpensive, and simple procedure to perform, and provides a solution to problems such as broodstock maintenance and asynchronic sexual maturity [46,47]. However, it has been shown that multiple factors can influence the success of sperm storage [45,47] (Fig. 11.5). In fish, it is known that cold storage of semen decreases sperm quality in parameters such as motility, viability, plasma membrane integrity, mitochondrial function, and DNA structural integrity; it also increases the production of reactive oxygen species (ROS) [48–54]. All these factors are directly related to sperm fertilizing capacity and offspring quality [55,56]. Analyses of sperm function provide important information for the optimization of storage protocols and successful in vitro fertilization. Different effects caused by in vitro gamete storage may be species-specific and/or related to differences in protocols [50].

Sperm motility decreases with storage time [48,52,57–59]; one possible cause is the significant decrease recorded in the adenosine triphosphate (ATP) content [57,60]. On the other hand, a higher concentration of ATP has been observed in spermatozoa when the semen is stored in dilution [52,61]. This suggests that dilution of stored semen allows the sperm to consume less ATP in the maintenance of vital functions during storage and to present more favorable conditions for ATP recovery, with beneficial effects on their motility parameters. In general, a longer semen storage period is possible in marine species than in freshwater fish [53,61,62]. Sperm motility is one of the most widely used parameters to determine semen quality. It can be assessed in an optical microscope (subjective evaluation) [63] or by computer-assisted sperm analysis (CASA/objective evaluation) [63,64]. Changes in fish sperm motility occur both naturally over the breeding season (in vivo) and also during in vitro storage of spermatozoa [50,59,65].

The main cause of the loss of sperm function has been associated with the effects of oxidative stress (OS) caused by the generation of ROS. These accelerate cell aging during short-term storage, inducing changes that include membrane fluidity alterations and mitochondrial and DNA damage [48,53,66,67]. Fish spermatozoa are vulnerable to OS attack [68]. The sperm cell is particularly susceptible to oxidative damage due to the lipid composition of the sperm membrane, with a high content of unsaturated fatty acids, and due to its lack of a cytoplasm, which is also a source of antioxidants [69,70]. When sperm cells are being prepared for cryopreservation, the seminal plasma must be removed. This fluid contains enzymatic and nonenzymatic antioxidants that scavenge and neutralize excessive ROS, avoiding OS and deterioration of sperm function [71,72]. Supplementation of the storage medium with antioxidants such as uric acid, methionine, sodium alginate, butylated hydroxytoluene (BHT), ascorbic acid, trolox, and polyphenol has been used to increase semen quality during short-term storage. This procedure improves sperm motility and membrane integrity and decreases lipid peroxidation and superoxide anion production in the sperm [51,58,67,73,74].

The mitochondrion plays a fundamental role in fish sperm motility and its fertilizing potential. Nevertheless, it has been shown that short-term storage of fish semen induces a significant decrease in sperm mitochondrial function [50−52] and structure [54], leading to a decrease in ATP and low motility [52,61]. OS damages the mitochondrial function, increasing ROS production and creating a vicious circle in which the mitochondria can be both the generator and the victim of oxidative damage [75].

Different species have developed different strategies to protect sperm DNA: one is to increase the condensation level of chromatin during spermatogenesis, reaching a different degree depending on the species and the degree of cell maturation [76]; a second strategy is the presence of antioxidants in the seminal plasma [77]. Different factors (developmental/environmental) to which the cell is exposed could promote damage to sperm cell DNA [78]. Several studies have described that fish sperm cells suffer DNA damage during cold storage [51,53,79]. Sperm with damaged genetic material may preserve their fertilization capacity; however, alterations may occur later during embryo development, affecting the offspring [80,81]. During cold storage, nitrogenous bases, particularly the guanine in DNA, are the main targets of ROS attack due to their low redox potential. ROS action generates 8-oxoguanine (8-oxo G), one of the commonest lesions observed [79,82]. This is a sensitive and specific indicator of oxidative damage to DNA [83], destabilizing DNA structure and resulting in localized strand breaks [67].

Storage conditions must be optimized for each species individually, considering the factors that influence sperm cell quality (motility, viability, mitochondrial membrane potential, ROS production, ATP content, and DNA integrity) during storage in order to preserve the fertility potential of sperm stored in vitro.

11.3.4 Sperm cryopreservation in fish of interest

Developing gamete conservation programs will be very useful in future reproduction studies of fish of commercial interest. Gamete cryopreservation is an important tool for fish reproduction and of great interest for fish farming. It is currently used routinely in reproduction laboratories [84−87].

Gametes can be cryopreserved by both slow and fast (ultrafast) freezing and by vitrification at the temperature of liquid nitrogen (−196°C) [88−91]. These techniques offer various advantages: conservation of genetic material from wild animals originating from remote locations or where access is difficult, elimination of problems caused by asynchronic gonad maturity between males and females; use of gametes from animals selected for improvement programs or subjected to genetic manipulation (tetraploids, clones, transgenic material), reduction in the costs and risks of transporting live animals, and establishment of hybridization programs using fish with different reproductive periods [92−96].

The first work on freezing fish semen was carried out in 1953 by Blaxter [97], who crossed different populations of Atlantic herring (*Clupea harengus*) that spawned at different times of year [92,98]. On this basis, semen cryopreservation protocols have since been developed experimentally in more than 200 species of fish of interest [96,99].

Cryopreservation of marine fish semen normally presents better results for motility rate and fertilizing capacity after thawing in species such as sea bass (*Dicentrarchus labrax*) and turbot (*Scophthalmus maximus*), with mean embryo survival of 80%, than in freshwater species such as Atlantic salmon, rainbow trout, and carp that present embryo survival of around 65% with cryopreserved semen [92,96,100]. These results are probably related to the fact that spermatozoa from marine species are adapted to high osmotic pressures [90].

In recent years, due to the rapid development of aquaculture and the conservation problems facing some fish species, cryopreservation has played an important role in freezing gametes to protect fish of high biological and economic value. Among salmonids these include rainbow trout *(O. mykiss)*, brown trout *(Salmo trutta)*, brook trout *(Salvelinus fontinalis)* and charr *(Salvelinus alpinus)* [55,101]. Other species of commercial interest which need protection are the sturgeons: beluga *(Huso huso)*, sterlet *(Acipenser ruthenus)*, pallid *(Scaphirhynchus albus)*, Siberian, European, and Russian *(Aythya baeri, Acipenser sturio, Acipenser gueldenstaedtii)*, and shortnose *(Acipenser brevirostrum)* [102]; catfish: African and European catfish *(Clarias gariepinus, Silurus glanis)*; and common *(Cyprinus carpio)* and silver carp *(Hypophthalmichthys molitrix)* [103–106].

The first step in developing a cryopreservation protocol is to choose the best composition for the diluent, generally a saline or glucose solution with suitable osmolality containing a cryoprotection agent [107]. A characteristic of paramount importance in a diluent is that it should not initiate sperm motility [96] and that it should be sterile and remain stable throughout storage. The purpose of this solution is to dilute the semen, which in some fish presents high viscosity and low volume. In recent years, most of the diluents used in marine fish have contained salts, small quantities of monosaccharide or disaccharide sugars (glucose, saccharose, fructose, lactose, trehalose, etc.) and protein (bovine serum albumin, BSA) with pH of 7.0–8.2 and osmolality of 205–400 mOsm/kg (for further details of composition, see Magnoti et al. [108]). The object of adding an easily metabolized sugar to the diluent is to enable the spermatozoa to obtain energy, which is found naturally in the seminal plasma of mammals, insects, nematodes, and fish [109–111]. Antioxidant substances such as ascorbic acid, vitamin E, reduced glutathione, reduced methionine, carnitine, and uric acid can be added to eliminate free radicals of the superoxide, peroxide, and hydroxide types that lead to peroxidation of phospholipids in the plasma membrane, causing death of the sperm cell [112,113]. Their effectiveness in freshwater teleost fish was shown by Lahnsteiner [68], Lahnsteiner et al. [77], Lahnsteiner and Mansour [73], and Ubilla and Valdebenito [74].

Lahnsteiner et al. [77] and Lahnsteiner and Mansour [73] mention that uric acid is the principal antioxidant found in the semen of burbot *(Lota lota)*, perch *(Perca fluviatilis)*, bleak *(Alburnus alburnos)*, and brown trout *(S. trutta)* and that it can increase sperm motility and membrane integrity. Liu et al. [114] suggest that antioxidant protection for mammal spermatozoa is also related to proteins from seminal plasma, and Sasaki et al. [115] attribute it to low molecular weight protein fractions. In fish, the antioxidant capacity of seminal plasma is insufficient to prevent peroxidation of the lipids during freezing [91].

The function of cryoprotectants is to protect the cell against the formation of intracellular ice crystals and excessive dehydration, which produce irreversible damage in the plasma and nuclear membranes (damage to DNA) and alterations in cell organelles [78]. Cryoprotectants can be divided into two groups depending on their intracellular or extracellular action. Those in the first group are capable of permeating the cell membrane; they are generally of low molecular weight (less than 400 kDa), not electrolytic and highly soluble in water [116,117]. The most common compounds for semen cryopreservation are: dimethyl sulfoxide (DMSO), dimethyl acetamide (DMA), glycerol (Gly), 1,2-propanediol, ethylene glycol (EG), methyl glycol (2-methoxyethanol), propylene glycol (PG), methanol (MeOH), and 2,3-butanediol (BD). However, most of these substances are toxic and their concentration and equilibrium time (interaction with the cell) may have negative effects on sperm physiology, such as osmotic shock or biochemical problems (energy metabolism) [85]. According to Suquet et al. [92] and Liu et al. [98], DMSO is considered to produce good results when used as a cryoprotectant in most fish species. Within studies published over the last 15 years, DMSO has also been tested and approved in most marine species studied, as has the cryoprotectant propylene glycol (PG). The second group consists of nonpermeable substances of high molecular weight, which may be monosaccharide, disaccharide, or polysaccharide sugars, macromolecules such as polyvinyl pyrrolidone (PVP) and hydroxyethyl starch (HES) [116], lipoproteins or proteins derived from milk, egg-yolk, and vegetable oils [118,119]. These macromolecules have two functions: to increase the osmolality of the extracellular space, causing dehydration of the cells during freezing, and to prevent excessive osmotic dilatation during thawing [117].

Cryopreservation protocols are classified as slow or fast depending on the cooling rate as a function of the concentration of the cryoprotectants used. Various slow-freezing protocols are used in fish spermatozoa [120]. When *O. mykiss* spermatozoa were frozen in a solution of sucrose 0.6 M + 10% DMSO, good fertilization percentages were obtained (81% and 83%); however, the percentages in other species were lower. Thus to improve the fertilizing capacity of frozen spermatozoa, some variants were introduced. These include different protocols and cryoprotectants such as propylene glycol, ethylene glycol, DMSO, sucrose, methanol, and glycerol [121,122]. Different freezing rates can be programmed, for example, 40 seconds between 0°C and 2°C and 5 minutes at −79°C, and then immersion in liquid nitrogen at −196°C [118]. The addition of albumin (BSA fraction V) to the cryoprotectant solution gives the highest rates of fertilization (95.3%) in some salmonid species. Salte et al. [123] reported similar results in the rainbow trout, with a fertilization rate higher than 80%, with semen diluted at 1:3 or 1:5 in a cold extender containing methanol as the cryoprotectant. The aliquots were transferred to 0.5 mL straws, balanced at 4°C for 10 minutes, and then frozen in racks in a programmable freezer. The freezing curve followed a freezing rate of 30°C/minute in the range −14°C to −110°C. For thawing, Cabrita et al. [124] tested different temperatures and thawing rates by immersing straws of different volumes directly in a thermo-regulated bath, with variable results for motility and postthawing fertilization rate.

The vitrification method consists of direct immersion of small volumes of semen in liquid nitrogen; it has been applied successfully in the spermatozoa of fish such as rainbow trout and Atlantic salmon [89−91,125,126]. However, to date only Cuevas-Uribe et al. [90] have applied the technique to the semen of marine fish. Vitrification does not require specialized equipment and can easily be applied in procedures in production plants and in fieldwork in areas of difficult access. The basic difference between vitrification and slow cryopreservation is that during direct immersion in liquid nitrogen the viscosity of the medium is increased and the water molecules in the sample cannot become positioned so as to form a crystalline structure, but constitute an amorphous form of solid (vitrified) water [117]. As a rule, the most recommended system for vitrification is a mixture of cryoprotectants with different characteristics. For example, human serum albumin + 0.25 M of sucrose is used successfully as a nonpermeable cryoprotectant in the vitrification of human sperm by ultra-freezing [127,128]. Cuevas-Uribe et al. [90] included eight cryoprotectants in different concentrations and found that mixing permeable and nonpermeable substances produced better results than using just one cryoprotectant in a high concentration. In these experiments the mixture of 15% DMSO + 15% EG + 10% Gly + 1% X-1000 + 1% Z-1000 resulted in the highest rate of vitrification, motility, and membrane integrity in the semen of red snapper (*Lutjanus campechanus*), spotted seatrout (*Cynoscion nebulosus*), and red drum (*Sciaenops ocellatus*).

Studies into cryopreservation of the semen of freshwater and marine fish have intensified in recent years. However, most of this work deals only with technical aspects, modifying diluents and cryoprotectants and reporting the effects on sperm motility and fertility. Some studies present more in-depth work to determine the effects on cell physiology and morphology caused by these chemical substances by freezing and vitrification; for further details on alterations to sperm functions see Figueroa et al. [87]. In future, cellular and molecular techniques will have to be used to characterize cryodamage. New additives will also come into use for diluents, such as antioxidants and organic acids, and new cryoprotectants such as macromolecular seminal plasma used as a nonpermeable cryoprotectant in cryopreservation and vitrification [87].

Another important aspect is the cryopreservation of fish oocytes and embryos, which are highly sensitive to low temperatures; the conditions involved are specific to every species and cell type [129−131]. Due to the large size of their cells, fish embryos are very big, in general bigger than those of most mammals. The diameter of a human zygote is approximately 100 μm, whereas a fish zygote (>1 mm diameter) is around 1000 times bigger. The result of this larger size is that fish embryos have a low surface:volume ratio and lower membrane permeability to water and cryoprotectant solutions. Fish oocytes enjoy some advantages over other species, especially in the first development stages [132−134]; the optimization of protocols and use of antifreeze proteins could improve the short- and long-term storage of fish oocytes and embryos [135,136,137].

11.4   Control of sex

Some fish species are dioic, and others are simultaneous or consecutive hermaphrodites; among the latter, species have been recorded, which are protogynous and gynandrous, among other forms of sex [138]. The sex of fish is determined genetically; however, its phenotypic expression is regulated by a series of environmental factors such as temperature and social factors [20,138−140]. In various species of fish, such as trout, tilapias, bass, and sea bream, and in some crustaceans, the animals of one sex have better productive characteristics than those of the other. These characteristics may include faster growth, late maturation, or both. For example, a large percentage of male salmonids mature before they enter saltwater, and on average they mature 1 year earlier than females. The secondary changes caused by maturation (development of the hooked lower jaw, formation of a kind of hump in the dorsal region, loss of orange flesh color, cessation of growth, reduction of their osmoregulatory capacity in seawater, and increased mortality), reduce the animals' market value and oblige the producer to harvest them before they reach their full growth potential. In females on the other hand these problems do not occur or are much less intense.

Techniques for modifying the phenotypic expression of sex in fish are extensively used in fish farms by indirect methods to produce "all-female" populations. These populations are produced in large numbers through the production of sex-reversed females (Fig. 11.6A), whose phenotypic sex expression is altered by the administration of male

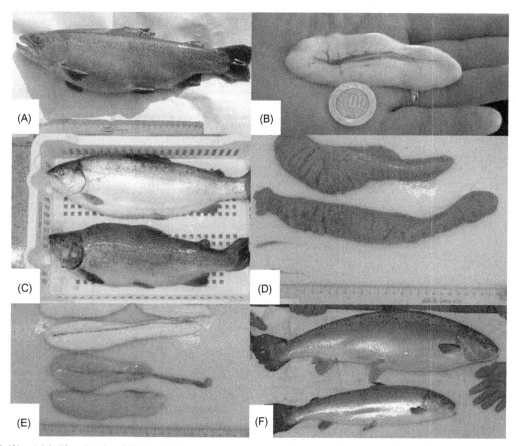

FIGURE 11.6 (A) Adult "functional male" of rainbow trout (*Oncorhynchus mykiss*). (B) Testicle (without duct) extracted from a rainbow trout "functional-male." (C) Triploid specimens of 2-year-old rainbow trout, female (top) and male (bottom). (D) Ovaries from a normal female (top) and triploid (bottom). (E) Testicles from a normal male (top) and triploid (bottom). (F) Atlantic salmon specimens (*Salmo salar*) transgenic (top) and normal (bottom). *Picture taken from http://www.greenpeace.*

hormones (regularly 17α-methyltestosterone) to specimens in prehatching stages (Pacific salmon) or during first feeding (trout and Atlantic salmon). These specimens are genetically female (XX) but phenotypically male. Once full sexual maturity is reached (after at least 2 years' rearing), all the spermatozoa produced carry the X chromosome; for this reason the semen obtained from these "neo-males" is called "female semen." If it is used to fertilize a normal female, the offspring will be "all-female" (Fig. 11.6B). Experimentally, gynogenetic or androgenetic groups have been produced by radiating the semen or oocytes with ultraviolet light (Fig. 11.7) [141–143].

In tilapias, the most profitable individuals are the males. They mature later than the females and expend less energy in reproduction; also the females incubate the embryos in their mouths and protect the larvae in the buccal cavity after hatching (which stops them from feeding); thus the males grow more quickly and produce a larger, better quality fillet than the females. MacIntosh and Little [144] report that "all-male" populations are produced in tilapias by sex reversal by administration of male hormones (regularly 17α-methyltestosterone) to young fish; this procedure produces interspecific hybrids in which the females and males are homogametic (XX and ZZ, respectively), or else "super males" (YY) obtained by sexual reversal to female of an XY male. When this phenotypic female (which is genetically male) breeds with a normal male, it will produce offspring formed of 25% females (XX), 50% males (XY), and 25% super males (YY). The latter will produce 100% male offspring if crossed with a normal female [143], and therefore they are carefully guarded in fish farms (Fig. 11.7).

11.5 Production of sterile fishes

In many aquatic species (salmonids, soles, bass, sea bream, etc.), polyploid specimens are viable and those with uneven sets of chromosomes are regularly sterile. The fact that the specimens cultivated in fish farming are sterile is sometimes

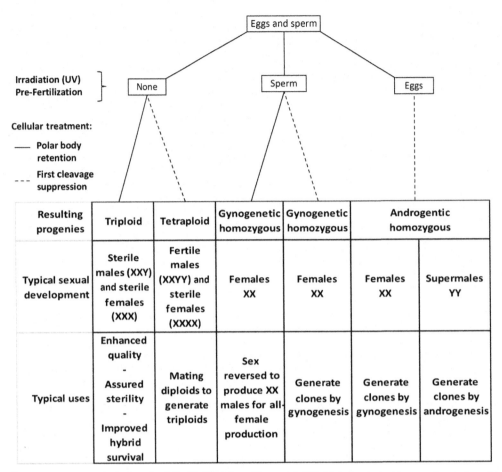

FIGURE 11.7   Chromosome set manipulations for fish. *Modified from Thorgaard GH. Biotechnological approaches to broodstock management (Chapter 4). In: Bromage N, Roberts R, editors. Broodstock management and egg and larval quality. Cambridge: Blackwell Science Ltd.; 1995. p. 76–93.*

an advantage for production as greater growth occurs, or else the specimens are used in repopulation programs in protected areas in order to prevent the introduction of exotic species (Fig. 11.6).

In fish, triploidy (3n) is induced by preventing the release of the second polar corpuscle by chemical, physical, or electric treatments [143]. In salmonids, temperature shocks (at approximately 28°C) (Fig. 11.8A) and/or pressure shocks (at 10,000 psi) are very widely used, applied to the eggs approximately 30 minutes after fertilization (Fig. 11.8B). These methods achieve percentages of triploidy, which are regularly over 90%. In adult salmonids, triploid males are aneuploid and present the same signs of sexual maturity as normal males (the skin grows darker, jaw becomes deformed, the flesh loses carotenoids, and the immune system is depressed), (Fig. 11.6C and E). However, the females do not develop gonads as their hypothalamic-pituitary-gonadal axis is inactive. This means that they retain their prepubic appearance throughout their lives, with silver skin, absence of gonadal development, and high levels of muscle pigmentation, and expend no energy on developing gonads (Fig. 11.6C and D), and therefore they grow around 20% larger than a normal female. The disadvantages of triploid groups include the lower resistance to management conditions and disease and the higher percentage of deformed specimens. For this reason, tetraploid (4n) specimens are produced in some species (e.g., rainbow trout), obtained by the application of thermal shock (warm in cold-water species and cold in warm-water species), chemical shock with mitotic poisons such as colchicine or cytochalasin B, pressure shock, or electric shock during the first cleavage of the embryo to prevent the separation of the first nuclei (2n), forming a tetraploid genome (4n) by their fusion [143]. Males and females are fertile and produce 2n gametes; when crossed with normal diploid specimens (of n gametes) they produce 3n offspring. These interploid triploids present higher survival and similar behavior to a normal (2n) specimen as they are not subjected to physical or chemical shocks (Fig. 11.9) [143].

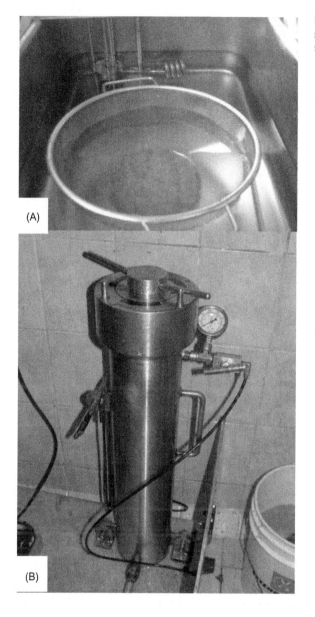

FIGURE 11.8 (A) Thermic shocks application to recently fertilized eggs from rainbow trout (*Oncorhynchus mykiss*). (B) Pressure system for triploidy induction.

Due to the productive advantage only present in triploid females, this technique is regularly applied in monosex "all-female" populations obtained with one of the methods described above. The resulting populations are 100% sterile, with no signs of sexual maturity (producing a fillet of better quality that fetches a higher price), and can be harvested at any time of year.

11.6   Regulation of ontogenetic development time

Being poikilothermic species, fish regulate their metabolism and development as a function of the ambient temperature. All species have one "optimum" thermal preference for growth and another for reproduction. In salmonids, the fastest growth rates occur with temperatures around 16°C; reproduction (spawning, fertilization, and embryo development) on the other hand requires temperatures below 10°C. This means that their ontogenetic development is not measured in time, but in the heat accumulated daily by specimens, measured in accumulated thermal units (ATU) or degree days (°day) which reflect the sum of the daily mean temperatures in which the embryos are incubated. An Atlantic salmon embryo requires about 800 ATU (or °day) to start the feeding process; so if it is reared in temperatures of 8°C, it will require approximately 100 days before it starts feeding, but if it is kept at 4°C it will need practically twice as long because its development will be slower. This allows the fish farmer to accelerate or delay embryo development by

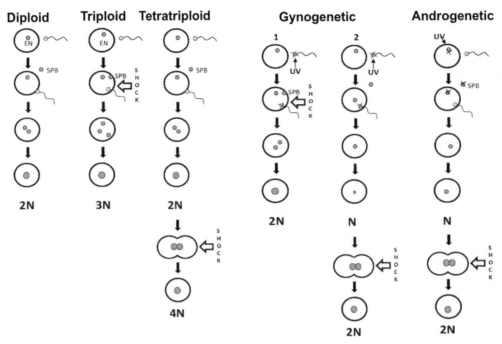

FIGURE 11.9 General description of the treatments with gametes and embryos for the manipulation of chromosome sets. Techniques for the production of triploid, tetraploid, gynogenetic, and androgenetic progeny compared to diploid progeny (details described in the text). *Modified from Tave D. Genetics for fish hatchery managers. 2nd ed. New York: Van Nostrand Reinhold; 1993. p. 415.*

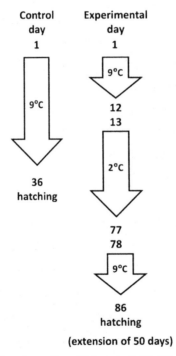

FIGURE 11.10 Effects of cold-water (2°C) incubation of eggs (from 12 days postfertilization) on time to hatching in rainbow trout embryos. *Taken from Bromage N, Cumaranatunga R. Egg production in the rainbow trout. In: Muir JF, Roberts RJ, editors. Recent advances in aquaculture, vol. 3. London and Sydney/Portland, Oregon; 1988. p. 63–138.*

raising or lowering the temperature of the water in which fishes are kept. Salmonids are regularly incubated at around 8°C, which is the optimum temperature. However, the water temperature of embryos with approximately 100 UTA is frequently reduced (slowly) to 2°C to delay their development by up to 2 months, while the fish farmer waits for a buyer. The more months of development the embryo has already passed through, the lower its tolerance to temperature changes, and the changes must be gradual to reduce bone alterations and malformations in the alevins (Fig. 11.10) [145].

11.7   Hybridization

Intraspecific or interspecific hybrids are frequently produced in pursuit of hybrid vigor. Although this is not practiced on a large scale, Tave [143] and Thorgaard [141] describe the advantages and disadvantages of interspecific hybrid production in salmonids (for further details see Tave [143]).

11.8   Experimental technologies

Substitutes in teleost fish are produced by inducing germline chimeras. To do this, the primordial germ cells (PGCs) are isolated from the donor embryo and transplanted to the host embryos. The hosts become germline chimeras if the transplanted PGCs migrate successively to the genital ridge and differentiate into functional gametes. Thus the donor's genotypes can be restored in the next generation. Fish seed production is expected to be made more efficient by substitute production using germline chimeras between two different species with differing biological properties. For example, the life cycle of the fish could be severely shortened if a species with a shorter life cycle is used as a substitute father and produces donor gametes. Other uses of this biotechnology would be the protection of species in danger of extinction and the production of gametes of big fish (e.g., tuna) in tanks in the form of chimeras easily kept in captivity. Furthermore, if the PGCs of the larvae distributed in nature are collected and reproduced by means of germline chimeras, an enormous genetic biodiversity could be kept as stock seedlings for rearing aquatic animals in fish farms [146–148].

Production of transgenic fish in fish farms has already made it possible to commercialize transgenic fish carrying the enzyme luciferase, which produces chemoluminescence in aquarium fishes. Some years ago the Food and Drug Administration (FDA) authorized the production of transgenic salmon (AquAdvantage) that had an exogenous growth hormone incorporated into their genome, producing very large specimens in a short time. This technology has been developed for various species such as coho salmon, Atlantic salmon, tilapia, and rainbow trout [148]. Generating these specimens has not been easy in fish as the hardness of the chorion and the cell size make it difficult for the transgenic fish to incorporate the exogenous gene into its genome and gonads, making it a great challenge for genetic engineering researchers [149].

11.9   Conclusion

Some of the technologies described in this chapter are already in frequent, large-scale use in fish farms around the world. Monosex and/or triploid populations are biotechnological tools used permanently in the production of salmonids and some marine species such as bass and sea bream. There is little doubt that in the medium term we will see mass use of the cryopreservation of oocytes and embryos and the production of chimeras and/or transgenic organisms, once the consumer population adapts to these new biotechnologies for fish production.

Acknowledgements

This work was supported by the Fund for Promotion of Scientific and Technological Development CONICYT, FONDECYT / POSTDOCTORAL (N° 3180765); FONDECYT REGULAR (N°1180387) and Internal Research Project of the Catholic University of Temuco (N₀ 4136-2018).

References

[1]   Donaldson EM. The role of biotechnology in sustainable aquaculture. In: Bardach JE, editor. Sustainable aquaculture. Wiley & Sons, Inc; 1997. p. 101–26.

[2]   Harvey BJ, Hoar WS. Theory and practice of induced breeding in fish. IDRC-TS2le. Ottawa: International Development Research Centre; 1979. 48 p.

[3]   Zohar Y. Gonadotropin releasing hormone in spawning induction in teleosts: basic and applied considerations. In: Zohar Y, Breton B, editors. Reproduction in fish. Basic and applied aspects in endocrinology and genetics. Paris: INRA Press; 1988. p. 47–62.

[4]   Montero M, Dufour S. Gonadotropin-releasing hormones (GnRH) in fishes: evolutionary data on their structure, localization, regulation, and function. Zool Stud 1996;35(3):149–60.

[5]   Peter RE, Yu KL. Neuroendocrine regulation of ovulation in fishes: basic and applied aspects. Rev Fish Biol Fish 1997;7(2):173–97.

[6]   Redding JM, Patiño R. Reproductive physiology. In: Evans DH, editor. The physiology of fishes. Boca Raton, FL; 1993, p. 503–534.

[7]   Carrillo M, Zanuy S, Prat F, Cerda J, Ramos J, Mañanos E, et al. Sea bass (*Dicentrarchus labrax*). In: Bromage N, Roberts R, editors. Broodstock management and egg and larval quality. Cambridge: Blackwell Science Ltd; 1995. p. 38–168.

[8] Asturiano JF, Pérez L, Garzón DL, Peñaranda DS, Marco-Jiménez F, Martínez-Llorens S, et al. Effect of different methods for the induction of spermiation on semen quality in European eel. Aquac Res 2005;36(15):1480–7.

[9] Asturiano JF, Marco-Jiménez F, Pérez L, Balasch S, Garzón DL, Peñaranda DS, et al. Effects of hCG as spermiation inducer on European eel semen quality. Theriogenology 2006;66:1012–20.

[10] Mylonas CC, Fostier A, Zanuy S. Broodstock management and hormonal manipulations of fish reproduction. Gen Comp Endocrinol 2010;165:516–34.

[11] Lokman PM, Wylie MJ, Downes M, Di Biase A, Damsteegt EL. Artificial induction of maturation in female silver eels, *Anguilla australis:* The benefits of androgen pre-treatment. Aquaculture 2015;437:111–19.

[12] Lokman PM, Damsteegt EL, Wallace J, Downes M, Goodwin SL, Facoory LJ, et al. Dose-responses of male silver eels, *Anguilla australis*, to human chorionic gonadotropin and 11-ketotestosterone in vivo. Aquaculture 2016;463:97–105.

[13] Mylonas CC, Duncan NJ, Asturiano JF. Hormonal manipulations for the enhancement of sperm production in cultured fish and evaluation of sperm quality. Aquaculture 2017;472:21–44.

[14] Rothbard S, Yaron Z. Carps (Cyprinidae). In: Bromage N, Roberts R, editors. Broodstock management and egg and larval quality. Cambridge: Blackwell Science Ltd; 1995. p. 321–52.

[15] Di Biase A, Lokman PM, Govoni N, Casalini A, Emmanuele P, Parmeggiani A, et al. Co-treatment with androgens during artificial induction of maturation in female eel, *Anguilla anguilla:* effects on egg production and early development. Aquaculture 2017;479(June):508–15.

[16] Mordenti O, Emmanuele P, Casalini A, Lokman PM, Zaccaroni A, Di Biase A, et al. Effect of aromatable androgen (17-methyltestosterone) on induced maturation of silver European eels (*Anguilla*): oocyte performance and synchronization. Aquac Res 2018;49(1):442–8.

[17] Herranz-Jusdado JG, Rozenfeld C, Morini M, Pérez L, Asturiano JF, Gallego V. Recombinant vs purified mammal gonadotropins as maturation hormonal treatments of European eel males. Aquaculture 2019;501:527–36.

[18] Donaldson EM, Yamazaki F, Dye H, Philleo WW. Preparation of gonadotropin from salmon (*Oncorhynchus tshawytscha*) pituitary glands. Gen Comp Endocrinol 1972;1972(18):469–81.

[19] Donaldson EM, Hunter GA. Induced final maduration, ovulation and spermiation in cultured fishes Part B In: Hoar WS, Randall DJ, Donaldson EM, editors. Fish physiology. Reproduction, vol. 9. Orlando: Academic Press; 1983. p. 351–403.

[20] Donaldson EM, Devlin RH, Piferrer F. Solar II. Hormones and sex control in fish with particular emphasis on salmon. Asian Fish Sci 1996;9:1–8.

[21] Mylonas CC, Zohar Y. Endocrine regulation and artificial induction of oocyte maturation and spermiation in basses of the genus Morone. Aquaculture 2001;202(3–4):205–20.

[22] Zohar Y, Harel M, Hassin S, Tandler A. Gilt-head sea bream (*Sparus aurata*) Chapter 5. In: Bromage N, Roberts R, editors. Broodstock management and egg and larval quality. Cambridge: Blackwell Science Ltd; 1995. p. 94–117.

[23] Mikolajczyk T, Roelants I, Sokolowska M, Gorka R, Epler P. On "pituitary" injection: a novel method for the in vivo investigation of secretory function of the pituitary gland in the common carp. Cyprinus carpio L Aquacult Fish Manage 1994;25(4):409–18.

[24] Bromage N, Roberts R. Broodstock management and egg and larval quality. Cambridge: Blackwell Science Ltd; 1995. p. 424.

[25] McNiven MA, Gallant RK, Richardson GF. Fresh storage of rainbow trout (*Oncorhynchus mykiss*) semen using a non-aqueosus medium. Aquaculture 1993;109(1):71–82.

[26] Stoss J, Holtz W. Susscesful storage of chilled rainboi trout (*Salmo gairdneri*) spermatozoa for up to 34 days. Aquaculture 1983;31(2–4):269–74.

[27] Christensen JM, Tiersch TR. Refrigerated storage of channel catfish sperm. J World Aquac Soc 1996;27(3):340–3.

[28] Stoss J, Refstie T. Short-term storage and cryopreservation of milt from Atlantic Salmon and Sea Trout. Aquaculture 1983;30(1–4):229–36.

[29] Sánchez A, Rubilar J. Obtención de cachorros mediante inseminación artificial con semen canino refrigerado. Primera descripción en Chile. Arch Med Vet 2001;33(1):105–10.

[30] Erdahl AW, Graham EF. Fertility of teleost semen as affected by dilution and storage in a seminal plasma-mimicking medium. Aquaculture 1987;60(3–4):311–21.

[31] Cosson J, Billard R, Cibert C, Dreanno C, Suquet M. Ionic factors regulating the motility of fish sperm. In: Gagnon C, editor. The Male Gamete: From basic knowledge to clinical applications. Paris, France: Cache River Press; 1999. p. 161–86.

[32] Woolsey J, Holcomb M, Cloud JG, Ingermann RL. Sperm motility in the steelhead *Oncorhynchus mykiss* (Walbaum): influence of the composition of the incubation and activation media. Aquac Res 2006;37(3):215–23.

[33] Billard R. Effects of coelomic and seminal fluids and various saline diluents on the fertilizing ability of spermatozoa in the rainbow trout, Salmo gairdneri. Reproduction 1983;68(1):77–84.

[34] Lahnsteiner F, Mansour N, Berger B. Seminal plasma proteins prolong the viability of rainbow trout (*Oncorynchus mykiss*) spermatozoa. Theriogenology 2004;62(5):801–8.

[35] Truscott B, Idler DR. An improved extender for freezing Atlantic salmon spermatozoa. Journal of the Fisheries Board of Canada 1969;26(12):3254–8.

[36] Billard R, Jalabert B. L'insdmination artificielle de la truite (*Salmo gairdneri* Richardson). Ann. Biol. anim. Bioch. Biophys 1974;14:601–10.

[37] Linhart O, Mims SD, Shelton WL. Motility of spermatozoa from shovelnose sturgeon and paddlefish. J Fish Biol 1995;47(5):902–9.

[38] Conte F, Doroshov S, Lutes P. Hatchery Manual for the White Sturgeon, Acipenser transmontanus, with Application to other North American Acipenseridae. Publication #3322, Coop. Extension, Div. Agric. Nat. Resources, Univ. California, Davis 1988; 104 pp.

[39] Valdebenito I. Estudio de los parámetros reproductivos del puye (*Galaxias maculatus*, Jenyns, 1842) (OSMERIFORMES:GALAXIIDAE) bajo condiciones de cultivo experimental en el Sur de Chile. Tesis doctoral de la Universidad de Las Palmas de Gran Canaria, España; 2004. p. 211.

[40] Gordon MR, Klotins KC, Campbell VM, Cooper MM. Farmer salmon broodstock management. Vancouver: B.C. Research; 1987. p. 145.

[41] Vladic T, Järvi T. Sperm motility and fertilization time span in Atlantic salmon and brown trout - the effect of water temperature. J Fish Biol 1997;50(5):1088–93.

[42] Valdebenito I. Efecto de la cafeína en la motilidad y fertilidad espermática de trucha arcoiris (*Oncorhynchus mykiss*). Inf Tecn 2007;18(2):61–5.

[43] Goodall J, Blackshaw A, Capra M. Factors affecting the activation and duration of motility of the spermatozoa of the summer whiting (*Sillago ciliata*). Aquaculture 1989;77(2–3):243–50.

[44] Ohta H, Tanaka H, Kagawa H, Okuzawa K, Iinuma N. Artificial fertilization using testicular spermatozoa in the Japanese eel *Anguilla japonica*. Fish Sci 1997;63(3):393–6.

[45] Beirão J, Boulais M, Gallego V, O'Brien JK, Peixoto S, Robeck TR, et al. Sperm handling in aquatic animals for artificial reproduction. Theriogenology 2019;133:161–78.

[46] Bobe J, Labbé C. Chilled storage of sperm and eggs. In: Cabrita E, Robles V, Herráez MP, editors. Methods in reproductive aquaculture: marine and freshwater species. CRC Press, Taylor and Francis Group; 2009. p. 219–31.

[47] Contreras P, Dumorné K, Ulloa-Rodríguez P, Merino O, Figueroa E, Farías JG, et al. Effects of short-term storage on sperm function in fish semen: A review. Rev Aquac 2019;1–17.

[48] Shaliutina A, Hulak M, Gazo I, Linhartova P, Linhart O. Effect of short-term storage on quality parameters, DNA integrity, and oxidative stress in Russian (*Acipenser gueldenstaedtii*) and Siberian (*Acipenser baerii*) sturgeon sperm. Anim Reprod Sci 2013;139(1–4):127–35.

[49] Aguilar-Juárez M, Ruiz-Campos G, Paniagua-Chávez CG. Cold storage of the sperm of the endemic trout *Oncorhynchus mykiss* nelsoni: a strategy for short-term germplasm conservation of endemic species. Rev Mex Biodivers 2014;85(1):294–300.

[50] Trigo P, Merino O, Figueroa E, Valdebenito I, Sánchez R, Risopatrón J. Effect of short-term semen storage in salmon (*Oncorhynchus mykiss*) on sperm functional parameters evaluated by flow cytometry. Andrologia 2015;47(4):407–11.

[51] Merino O, Figueroa E, Cheuquemán C, Valdebenito I, Isachenko V, Isachenko E, et al. Short-term storage of salmonids semen in a sodium alginate-based extender. Andrologia 2017;49(5):1–5.

[52] Contreras P, Ulloa P, Merino O, Valdebenito I, Figueroa E, Farías J, et al. Effect of short-term storage on sperm function in Patagonian blenny (*Eleginops maclovinus*) sperm. Aquaculture 2017;481(June):58–63.

[53] Risopatrón J, Merino O, Cheuquemán C, Figueroa E, Sánchez R, Farías JG, et al. Effect of the age of broodstock males on sperm function during cold storage in the trout (*Oncorhynchus mykiss*). Andrologia 2018;50(2):1–7.

[54] Díaz R, Lee-Estevez M, Quiñones J, Dumorné K, Short S, Ulloa-Rodríguez P, et al. Changes in Atlantic salmon (*Salmo salar*) sperm morphology and membrane lipid composition related to cold storage and cryopreservation. Anim Reprod Sci 2019;204(March):50–9.

[55] Lahnsteiner F, Berger B, Weismann T, Patzner R. The influence of various cryoprotectants on semen quality of the rainbow trout (*Oncorhynchus mykiss*) before and after cryopreservation. J Appl Ichthyol 1996;12(2):99–106.

[56] Pérez-Cerezales S, Martínez-Páramo S, Beirão J, Herráez MP. Fertilization capacity with rainbow trout DNA-damaged sperm and embryo developmental success. Reproduction 2010;139(6):989–97.

[57] Aramli MS, Kalbassi MR, Nazari RM, Aramli S. Effects of short-term storage on the motility, oxidative stress, and ATP content of Persian sturgeon (*Acipenser persicus*) sperm. Anim Reprod Sci 2013;143(1–4):112–17.

[58] Sarosiek B, Dryl K, Kucharczyk D, Zarski D, Kowalski RK. Motility parameters of perch spermatozoa (*Perca fluviatilis* L.) during short-term storage with antioxidants addition. Aquac int 2014;22(1):159–65.

[59] Bernáth G, Csenki Z, Bokor Z, Várkonyi L, Molnár J, Szabó T, et al. The effects of different preservation methods on ide (*Leuciscus idus*) sperm and the longevity of sperm movement. Cryobiology 2018;81(September 2017):125–31.

[60] Dziewulska K, Rzemieniecki A, Domagała J. Motility and energetic status of Atlantic salmon (*Salmo salar* L.) sperm after refrigerated storage. J Appl Ichthyol 2010;26(5):668–73.

[61] Ulloa-Rodríguez P, Contreras P, Dumorné K, Lee-Estevez M, Díaz R, Figueroa E, et al. Patagonian blenny (*Eleginops maclovinus*) spermatozoa quality after storage at 4 °C in Cortland medium. Anim Reprod Sci 2018;197(August):117–25.

[62] Gallego V, Pérez L, Asturiano JF, Yoshida M. Relationship between spermatozoa motility parameters, sperm/egg ratio, and fertilization and hatching rates in pufferfish (*Takifugu niphobles*). Aquaculture 2013;416–417:238–43.

[63] Gallego V, Asturiano JF. Sperm motility in fish: technical applications and perspectives through CASA-Mot systems. Reprod Fertil Dev 2018;30(6):820–32.

[64] Fauvel C, Suquet M, Cosson J. Evaluation of fish sperm quality. J Appl Ichthyol 2010;26(5):636–43.

[65] Babiak I, Ottesen O, Rudolfsen G, Johnsen S. Chilled storage of semen from Atlantic halibut, *Hippoglossus hippoglossus* L. I: Optimizing the protocol. Theriogenology 2006;66(9):2025–35.

[66] Agarwal A, Virk G, Ong C, du Plessis SS. Effect of oxidative stress on male reproduction. World J Mens Health 2014;32(1):1–17.

[67] Cabrita E, Martínez-Páramo S, Gavaia PJ, Riesco MF, Valcarce DG, Sarasquete C, et al. Factors enhancing fish sperm quality and emerging tools for sperm analysis. Aquaculture 2014;432:389–401.

[68] Lahnsteiner F. The role of free amino acids in semen of rainbow trout *Oncorhynchus mykiss* and carp *Cyprinus carpio*. J Fish Biol 2009;75(4):816–33.

[69] Koppers AJ, Garg ML, Aitken RJ. Stimulation of mitochondrial reactive oxygen species production by unesterified, unsaturated fatty acids in defective human spermatozoa. Free Radic Biol Med 2010;48(1):112−19.

[70] Agarwal A, Makker K, Sharma R. Clinical relevance of oxidative stress in male factor infertility: An update. Am J Reprod Immunol 2008;59 (1):2−11.

[71] Martínez-Páramo S, Diogo P, Dinis MT, Herráez MP, Sarasquete C, Cabrita E. Incorporation of ascorbic acid and α-tocopherol to the extender media to enhance antioxidant system of cryopreserved sea bass sperm. Theriogenology 2012;77(6):1129−36.

[72] Agarwal A, Majzoub A. Laboratory tests for oxidative stress. Indian J Urol 2017;33(3):199−206.

[73] Lahnsteiner F, Mansour N. A comparative study on antioxidant systems in semen of species of the Percidae, Salmonidae, Cyprinidae, and Lotidae for improving semen storage techniques. Aquaculture 2010;307(1−2):130−40.

[74] Ubilla A, Valdebenito I. Use of antioxidants on rainbow trout *Oncorhynchus mykiss* (Walbaum, 1792) sperm diluent: effects on motility and fertilizing capability. Lat Am J Aquat Res 2011;39(2):338−43.

[75] Ferramosca A, Pinto Provenzano S, Montagna DD, Coppola L, Zara V. Oxidative stress negatively affects human sperm mitochondrial respiration. Urology 2013;82(1):78−83.

[76] Lewis JD, Saperas N, Song Y, Zamora MJ, Chiva M, Ausió J. Histone H1 and the origin of protamines. Proc Natl Acad Sci U S A 2004;101 (12):4148−52.

[77] Lahnsteiner F, Mansour N, Plaetzer K. Antioxidant systems of brown trout (*Salmo trutta* f. fario) semen. Anim Reprod Sci 2010;119 (3−4):314−21.

[78] Cabrita E, Robles V, Rebordinos L, Sarasquete C, Herráez MP. Evaluation of DNA damage in rainbow trout (*Oncorhynchus mykiss*) and gilthead sea bream (*Sparus aurata*) cryopreserved sperm. Cryobiology 2005;50(2):144−53.

[79] Pérez-Cerezales S, Martínez-Páramo S, Cabrita E, Martínez-Pastor F, de Paz P, Herráez MP. Evaluation of oxidative DNA damage promoted by storage in sperm from sex-reversed rainbow trout. Theriogenology 2009;71(4):605−13.

[80] Emery BR, Carrell DT. The effect of epigenetic sperm abnormalities on early embryogenesis. Asian J Androl 2006;8(2):131−42.

[81] De Mello F, Garcia JS, Godoy LC, Depincé A, Labbé C, Streit DP. The effect of cryoprotectant agents on DNA methylation patterns and progeny development in the spermatozoa of *Colossoma macropomum*. Gen Comp Endocrinol 2017;245:94−101.

[82] Aitken RJ, Jones KT, Robertson SA. Reactive oxygen species and sperm function-in sickness and in health. J Androl 2012;33(6):1096−106.

[83] Cooke C, Podmore I, Herbert K, Lunec J, Thomas S. UV mediated DNA damage and its assessment. Toxicol Ecol N 1996;3:101−9.

[84] Carolsfeld J, Harvey B, Godinho HP, Zaniboni-Filho E. Cryopreservation of sperm in Brazilian migratory fish conservation. J Fish Biol 2003;63(2):472−89.

[85] Tiersh TR. Process pathways for cryopreservation research, application and commercialization. In: Tiersch TR, Green CC, editors. Cryopreservation in aquatic species. 2nd Edition Louisiana: World Aquaculture Society; 2011. p. 646−71.

[86] Maria AN, Carneiro PCF. Fish semen cryopreservation in Brazil: state of the art and future perspectives. Cienc Anim 2012;22(1):124−31.

[87] Figueroa E, Lee-Estevez M, Valdebenito I, Watanabe I, Oliveira RPS, Romero J, et al. Effects of cryopreservation on mitochondrial function and sperm quality in fish. Aquaculture 2019;511:634190.

[88] Oliveira AV, Viveiros ATM, Maria AN, Freitas RTF, Izaú ZA. Sucess of cooling and freezing of pirapitinga (*Brycon nattereri*) semen. Arq Bras Med Vet Zootec 2007;59(6):1509−15.

[89] Merino O, Sánchez R, Risopatrón J, Isachenko E, Katkov II, Figueroa E, et al. Cryoprotectant-free vitrification of fish (*Oncorhynchus mykiss*) spermatozoa: First report. Andrologia 2012;44((SUPPL.1):390−5.

[90] Cuevas-Uribe R, Chesney EJ, Daly J, Tiersch TR. Vitrification of sperm from marine fish: effect on motility and membrane integrity. Aquac Res 2015;46(7):1770−84.

[91] Figueroa E, Merino O, Risopatrón J, Isachenko V, Sánchez R, Effer B, et al. Effect of seminal plasma on Atlantic salmon (*Salmo salar*) sperm vitrification. Theriogenology 2015;83(2):238−45.

[92] Suquet M, Dreanno C, Fauvel C, Cosson J, Billard R. Cryopreservation of sperm in marine fish. Aquac Res 2000;31(3):231−43.

[93] Linhart O, Rodina M, Flajshans M, Gela D, Kocour M. Cryopreservation of European catfish *Silurus glanis* sperm: sperm motility, viability, and hatching success of embryos. Cryobiology 2005;51(3):250−61.

[94] Tiersch TR, Yang H, Jenkins JA, Dong Q. Sperm cryopreservation in fish and shellfish. Soc Reprod Fertil Suppl 2007;65(April):493−508.

[95] Maria NA, Azevedo HC, Carneiro PCF. Crioconservação de sêmen de peixes no contexto do agronegócio da piscicultura. In: Tavares-Dias M, editor. Manejo e sanidade de peixes em cultivo. Amapá, Brasil: Embrapa Amapá; 2009. p. 47−63.

[96] Gwo J-C. Cryopreservation of sperm of some marine fishes. In: Tiersch TR, Green CC, editors. Cryopreservation in aquatic species. 2nd Edition Baton Rouge. Lousiana: World Aquaculture Society; 2011. p. 459−81.

[97] Blaxter JHS. Sperm storage and cross-fertilization of spring and autumn spawning herring. Nature 1953;172:1189−90.

[98] Liu Q, Li J, Zhang S, Ding F, Xu X, Xiao Z, et al. An efficient methodology for cryopreservation of spermatozoa of red seabream, *Pagrus major*, with 2-mL cryovials. J World Aquac Soc 2006;37(3):289−97.

[99] Billard R, Zhang T. Technique of genetic resources banking fish. In: Watson PF, Holt WV, editors. Cryobanking the genetic resource: wildlife conservation for the future. London: Taylor and Francis; 2001. p. 143−70.

[100] Scott AP, Baynes SM. A review of the biology, handling and storage of salmonid spermatozoa. J Fish Biol 1980;17(6):707−39.

[101] Cabrita E, Alvarez R, Anel L, Rana KJ, Herraez MP. Sublethal damage during cryopreservation of rainbow trout sperm. Cryobiology 1998;37 (3):245−53.

[102] Horváth Á, Wayman WR, Dean JC, Urbányi B, Tiersch TR, Mims SD, et al. Viability and fertilizing capacity of cryopreserved sperm from three North American acipenseriform species: a retrospective study. J Appl Ichthyol 2008;24(4):443−9.

[103] Horva Â, Urba B. The effect of cryoprotectants on the motility and fertilizing capacity of cryopreserved African catfish *Clarias gariepinus* (Burchell 1822) sperm. Aquac Res 2000;31:317−24.

[104] Alvarez B, Arenal A, Fuentes R, Pimentel R, Abad Z, Pimentel E. Use of post-thaw silver carp (*Hypophtalmichthys molitrix*) spermatozoa to increase hatchery productions. In: Cabrita E, Robles V, Herráez MP, editors. Methods in reproductive aquaculture: marine and freshwater species. Biology series. CRC Press (Taylor and Francis group); 2008. p. 345−50.

[105] Maisse G, Ogier de Balny B, Labbé C. Cryopreservation of testicular sperm from European catfish (*Silurus glanis*). In: Cabrita E, Robles V, Herráez MP, editors. Methods in reproductive aquaculture: marine and freshwater species. Biology series. CRC Press (Taylor and Francis group); 2008. p. 397−401.

[106] Viveiros ATM, Komen J. Semen cryopreservation of the African catfish, *Clarias gariepinus*. In: Cabrita E, Robles V, Herráez MP, editors. Methods in reproductive aquaculture: marine and freshwater species. Biology series. CRC Press (Taylor and Francis group); 2008. p. 403−7.

[107] Viveiros ATM, Orfão LH, Nascimento AF, Corrêa FM, Caneppele D. Effects of extenders, cryoprotectants and freezing methods on sperm quality of the threatened Brazilian freshwater fish pirapitinga-do-sul *Brycon opalinus* (Characiformes). Theriogenology 2012;78(2):361−8.

[108] Magnotti C, Cerqueira V, Lee-Estevez M, Farias JG, Valdebenito I, Figueroa E. Cryopreservation and vitrification of fish semen: a review with special emphasis on marine species. Rev Aquac 2018;10(1):15−25.

[109] Gregory RW. Occurrence of frutose in trout seminal plasma. Trans Am Fish Soc 1968;97(2):203−4.

[110] Pourkhazaei F, Ebrahimi E, Ghaedi A. Arginine effects on biochemical composition of sperm in rainbow trout, *Oncorhynchus mykiss*. Aquac Res 2017;48(7):3464−71.

[111] Bozkurt Y, Öğretmen F, Erçin U, Yildiz Ü. Seminal plasma composition and its relationship with physical spermatological parameters of Grass carp (*Ctenopharyngodon idella*) semen: with emphasis on sperm motility. Aquac Res 2008;39(15):1666−72.

[112] Ciereszko A, Dabrowski K. Sperm quality and ascorbic acid concentration in rainbow trout semen are affected by dietary vitamin C: an across-season study. Biol Reprod 1995;52(5):982−8.

[113] Anghel A, Zamfirescu S, Dragomir C, Nadolu D, Elena S, Florica B. The effects of antioxidants on the cytological parameters of cryopreserved buck semen. Rom Biotechnol Lett 2010;15((SUPPL.3):26−32.

[114] Liu X, Robinson GW, Gouilleux F, Groner B, Hennighausen L. Cloning and expression of Stat5 and an additional homologue (Stat5b) involved in prolactin signal transduction in mouse mammary tissue. Proc Natl Acad Sci U S A 1995;92(19):8831−5.

[115] Sasaki T, Wiedeman H, Matzner M, Chu M-L, Timpl R. Expression of fibulin-2 by fibroblastos and deposition with fibronectin into a fibrilar matrix. J Cell Sci 1996;190:2895−904.

[116] Bakhach J. The cryopreservation of composite tissues: principles ans recente advancement on cryopreservation of diferente type of tissues. Organogenesis 2009;5(3):119−26.

[117] Cuevas-Uribe R, Tiersch TR. Non-equilibrium vitrification: an introduction and review of studies done in fish. In: Tiersch TR, Green CC, editors. Cryopreservation in aquatic species. 2nd Edition Baton Rouge, Louisiana: World Aquaculture Society; 2011. p. 309−24.

[118] Babiak I, Glogowski J, Goryczko K, Dobosz S, Kuzminski H, Strzezek J, et al. Effect of extender composition and equilibration time on fertilization ability and enzymatic activity of rainbow trout cryopreserved spermatozoa. Theriogenology 2001;56(1):177−92.

[119] Cabrita E, Anel L, Herraéz MP. Effect of external cryoprotectants as membrane stabilizers on cryopreserved rainbow trout sperm. Theriogenology 2001;56(4):623−35.

[120] Steinberg H, Hedder A, Baulain R, Holtz W. Cryopreservation of rainbow trout (*Oncorhynchus mykiss*) semen in straws. In: Proceedings of the fifth international symposium on the reproductive physiology of fish. University of Texas at Austin; 1995. p 147.

[121] Babiak I, Glogowski J, Dobosz S, Kuzminski H, Goryczko K. Semen from rainbow trout produced using cryopreserved spermatozoa is more suitable for cryopreservation. J Fish Biol 2002;60(3):561−70.

[122] Cabrita E, Robles V, Herraéz P. Sperm quality assessment. In: Cabrita E, Robles V, Herráez P, editors. Methods in reproductive aquaculture: marine and freshwater species. Boca Raton, FL: CRC Press Taylor and Francis Group; 2009. p. 93−147.

[123] Salte R, Galli A, Falaschi U, Fjalestad KT, Aleandri R. A protocol for the on-site use of frozen milt from rainbow trout (*Oncorhynchus mykiss* Walbaum) applied to the production of progeny groups: comparing males from different populations. Aquaculture 2004;231(1−4):337−45.

[124] Cabrita E, Robles V, Alvarez R, Herráez MP. Cryopreservation of rainbow trout sperm in large volume straws: application to large scale fertilization. Aquaculture 2001;201(3−4):301−14.

[125] Merino O, Risopatrón J, Sánchez R, Isachenko E, Figueroa E, Valdebenito I, et al. Fish (*Oncorhynchus mykiss*) spermatozoa cryoprotectant-free vitrification: stability of mitochondrion as criterion of effectiveness. Anim Reprod Sci 2011;124(1−2):125−31.

[126] Figueroa E, Risopatrón J, Sánchez R, Isachenko E, Merino O, Isachenko V, et al. Spermatozoa vitrification of sex-reversed rainbow trout (*Oncorhynchus mykiss*): effect of seminal plasma on physiological parameters. Aquaculture 2013;372−375:119−26.

[127] Isachenko E, Isachenko V, Weiss JM, Kreienberg R, Katkov II, Schulz M, et al. Acrosomal status and mitochondrial activity of human spermatozoa vitrified with sucrose. Reproduction 2008;136(2):167−73.

[128] Isachenko V, Sanchez R, Rahimi G, Mallmann P, Isachenko E, Merzenich M. Cryoprotectant-free vitrification of spermatozoa: Fish as a model of human. Andrologia 2019;51(1):1−7.

[129] Zhang T, Liu XH, Rawson DM. Effects of methanol and developmental arrest on chilling injury in zebrafish (*Danio rerio*) embryos. Theriogenology 2003;59(7):1545−56.

[130] Valdez DM, Miyamoto A, Hara T, Edashige K, Kasai M. Sensitivity to chilling of medaka (*Oryzias latipes*) embryos at various developmental stages. Theriogenology 2005;64(1):112−22.

[131] Streit Jr DP, Godoy LD, Ribeiro RP, Fornari DC, Digmayer M, et al. Cryopreservation of embryos and oocytes of South American fish species. Recent advances in cryopreservation. London: Intech; 2014. p. 45−58. Chap. 3.

[132] Zhang T, Rawson DM. Cryo-conservation in fish: the potentials and challenges. Cryobiology 2007;55(3):354.

[133] Tsai S, Lin C. Advantages and applications of cryopreservation in fisheries science. Braz Arch Biol Technol 2012;55(3):425−33.

[134] Godoy LC, Streit Jr DP, Zampolla T, Bos-Mikich A, Zhang T. A study on the vitrification of stage III stage zebrafish (*Danio rerio*) ovarian follicles. Cryobiology 2013;67(3):347−54.

[135] Robles V, Cabrita E, Fletcher GL, Shears MA, King MJ, Herráez MP. Vitrification assays with embryos from a cold tolerant sub-arctic fish species. Theriogenology 2005;64(7):1633−46.

[136] Figueroa E, Lee-Estévez M, Valdebenito I, Farías JG, Romero J. Potential biomarkers of DNA quality in cryopreserved fish sperm: impact on gene expression and embryonic development. Rev Aquac 2018;1−10.

[137] Robles V, Valcarce DG, Riesco MF. The use of antifreeze proteins in the cryopreservation of gametes and embryos. Biomolecules 2019;9 (5):1−12.

[138] Piferrer F, Beaumont A, Falguière JC, Flajšhans M, Haffray P, Colombo L. Polyploid fish and shellfish: production, biology and applications to aquaculture for performance improvement and genetic containment. Aquaculture 2009;293(3−4):125−56.

[139] Donaldson EM, Hunter A. Sex control in fish with particular reference to salmonids. Can J Fish Aquat Sci 1982;39(1):99−110.

[140] Devlin RH, Nagahama Y. Sex determination and sex differentiation in fish: An overview of genetic, physiological, and environmental influences. Aquaculture 2002;208(3−4):191−364.

[141] Thorgaard GH. Biotechnological approaches to broodstock management. Chapter 4. In: Bromage N, Roberts R, editors. Broodstock management and egg and larval quality. Cambridge: Blackwell Science Ltd; 1995. p. 76−93.

[142] Richter CJJ, Eding EH, Verreth JAJ, Fleuren WLG. In: Bromage N, Roberts R, editors. Broodstock management and egg and larval quality. Cambridge: Blackwell Science Ltd; 1995. p. 241−76.

[143] Tave D. Genetics for fish hatchery managers. 2nd Edition New York: Van Nostrand Reinhold; 1993. p. 415.

[144] MacIntoch DJ, Little DC. Nile tilapia (*Oreochoromis niloticus*). In: Bromage N, Roberts R, editors. Broodstock management and egg and larval quality. Cambridge: Blackwell Science Ltd; 1995. p. 277−320.

[145] Bromage N, Cumaranatunga R. Egg production in the rainbow trout. In: Muir JF, Roberts RJ, editors. Recent advances in aquaculture, vol. 3. London and Sydney/Portland, Oregon; 1988, p. 63−138.

[146] Yamaha E, Saito T, Goto-Kazeto R, Arai K. Developmental biotechnology for aquaculture, with special reference to surrogate production in teleost fishes. J Sea Res 2007;58(1):8−22.

[147] Golpour A, Siddique MAM, Siqueira-Silva DH, Pšenička M. Induced sterility in fish and its potential and challenges for aquaculture and germ cell transplantation technology: A review. Biologia (Pol) 2016;71(8):853−64.

[148] De Siqueira-Silva DH, Saito T, dos Santos-Silva AP, da Silva Costa R, Psenicka M, Yasui GS. Biotechnology applied to fish reproduction: tools for conservation. Physiol Biochem 2018;44(6):1469−85.

[149] Rathipriya A, Lakshmi DK. Developmental biotechnology for aquaculture, with special reference to production of fish by surrogate technology. J Aquacult Trop 2016;31(3−4):47.

Chapter 12

Assisted reproductive technologies in nonhuman primates

Shihua Yang

College of Veterinary Medicine, South China Agricultural University, Guangzhou, P.R. China

12.1 Introduction

The majority of available studies with regard to the development and implementation of assisted reproductive technologies (ARTs) in nonhuman primates (NHPs) focus on the macaque genus. The duration of menstrual cycle in macaques is about 25–32 days, with an average of 28 days, and the bleeding period ranges from 1 to 4 days. The cycle is generally divided into a follicular and a luteal phase. The first period of the cycle lasts from 8 to 14 days, in which follicles grow, mature, and secrete estrogen, coincident with the presence of a large dominant follicle, whereas the second period of the cycle, dominated by the presence of a corpus luteum, lasts on average 15 days. About 70%–80% of adult females menstruate in one cycle during a fertile season, and about 25% of them will be characterized by having either short or long cycles. As a result in fact, some macaque monkeys do not ovulate, due to luteinization of the dominant follicle, occurring in 10%–15% of cycles in the course of their reproductive season. The uterus in macaques is pear-shaped, similar to humans, and the cervix is narrow and curved. The gestation period is about 160 days and a single child is usually delivered. Therefore, thanks to such reproductive physiological similarities to humans, ARTs have had a quick development.

At present, the most common ARTs employed in NHPs include semen collection and freezing, superovulation, in vitro fertilization, and embryo culture and transfer to recipients. In recent years, following the study and implementation of ARTs in NHPs, a number of breakthroughs have been reported in biomedicine, avoiding thus the use of human reproductive cells and the relative restrictions imposed by laws and the common ethical sense among the public at large [1]. With regard to the application of more complex reproductive technologies, somatic cell nuclear transfer (SNCT) (cloning) has been achieved in macaque monkey [2], and the use of induced pluripotent stem cells (iPSCs) successfully established, and chimerism obtained in rhesus monkey [3]. In addition, in other studies, stem cells have also been differentiated into haploid pluripotent stem cells [4], testicular tissue cultured in vitro, and derived sperm cells have been reported to be able to fertilize mature ova and to obtain offspring [5]. Furthermore, more than 10 gene-edited macaques and relative disease models have been developed by using ARTs and gene editing technology, such as SHANK3-knockout monkeys [6], MECP2-knockout monkeys [7], BMAL1-knockout monkeys [2], among others. It is expected in the future, that ARTs applied to macaques could result in a significant contribution, not only for the benefit of a more efficient reproductive success in macaques themselves, but also for the establishment of a more closely related animal model for human diseases as well as biomedicine research.

12.2 Semen collection and cryopreservation

The availability of high-quality semen in sufficient quantities is one of the key factors in research and development of NHP reproduction and reproductive technology, and an excellent semen collection method is the premise for obtaining a large volume of high-quality sperm.

Reproductive Technologies in Animals. DOI: https://doi.org/10.1016/B978-0-12-817107-3.00012-6

12.2.1 Sperm collection methods

Currently, the approach to collect sperm from NHPs includes (1) releasing sperm from epididymis and vas deferens in vitro, (2) electro-ejaculation, and (3) vibratory stimulation. Releasing sperm from epididymis in vitro is a simple but damaging and nonreusable method and is used to collect semen from nature or dead macaques. By contrast, noninvasive sperm collection methods such as electro-ejaculation and vibratory stimulation can be adopted to obtain plenty of semen safely and repeatedly for production and scientific research.

12.2.1.1 Electro-ejaculation

The electro-ejaculation method, which is widely used to collect sperm from NHPs, includes the direct application of electrodes to the penis (probe electro-ejaculation method (PEM)) or else inside the rectum (rectal probe electro-ejaculation method (REM)).

Mastroianni and Manson in 1963 [8], successfully applied for the first-time PEM for semen collection in macaque monkeys. To our days, PEM has developed into a mature and commonly used semen collection method. It works by using two conductors of a certain width, as electrodes. The electrodes are wrapped up in moist absorbent cotton twine around the tip of the glans and the base of penis separately and linked with the positive and negative poles of the electromagnetic pulse generator, respectively, when PEM is conducted. The penis is therefore the target of the intensity of electrical stimulation, resulting in the excitation of local nerves and leading to ejaculation.

The first use of REM in monkeys was reported in 1965 by Weisbroth and Young who described a successful ejaculation of Old World macaques through the use of rectal probes [9]. With the development of REM over time, available devices for electric stimulation have been used to collect semen from more than 30 NHPs strains in a safe, simple, and effective manner [10]. Similarly to other animal species receiving the same stimulation, when REM is performed, the rectal probe carrying the electrode is inserted into the rectum, resulting in the excitation by electric current of the low-level nerve center and the nerve endings near the ampulla of the vas deferens, leading to the contraction of the vas deferens themselves and the final expulsion of semen in the epididymis together with the seminal plasma [11].

12.2.1.2 Vibratory stimulation

The vibratory stimulation method, also known as penile vibratory stimulation method (PVSM), is a new technology that has been applied in recent years, and had been used in the past to collect semen from marmoset by Schneiders et al. in 2004 [12]. In macaques, with the use of a monkey collar and chair to control the animals in a safe manner, the risks caused by the needed anesthesia and the consequent adverse effect on sperm collection have been avoided. In addition, this method of vibratory stimulation works effectively and mimics closely the natural mating conditions leading to the induced ejaculation.

12.2.1.3 Comparison of semen collection methods

Differences between PEM and REM methods for semen collection in macaques are evident [13–16]. Although the time needed for semen collection time by REM is shorter and the rate of success is high, semen quality parameters obtained by PEM are better, including semen weight and density, sperm cell motility, acrosome integrity, and abnormality rate. It has been proved that the lower motility of sperm cells recovered by REM is due to the high content of urea in the seminal plasma, mixed by the contraction of ureters caused by electrical stimulation.

As rhesus and cynomolgus monkeys are very sensitive to the surrounding environment and are highly alert and aggressive when treated without anesthesia, semen collection methods are often performed in anesthesia. However, the conditions created in the course of anesthesia and the process of electrical stimulation can cause a certain degree of safety risk. This is true especially whenever anesthesia is administered often in the course of repeated semen collection, and electrical stimulation can, under such conditions, easily cause harm and pain to the body of animals. For example, a too high applied current will burn the contact rectal mucosa or penis skin [14,17]. The emergence of PVSM technology, which is operatively simple, noninvasive, and painless in the process of semen collection, has solved these problematic issues. However, following PVSM stimulation in a number of adult male cynomolgus monkeys, the efficiency in collecting semen is very low.

12.2.2 Semen liquefaction

Generally semen from primates will condense following collection. Human semen is always completely liquefied in 30 minutes at room temperature; however, in many NHPs, especially macaque, it is only partly liquefied. In macaques, semen is usually collected in a 50-mL centrifuge tube and incubated for 30 minutes at 37°C in order to be partially liquefied. In the current practice, liquefied semen is collected and added to nine parts of similar volume of Tyrode's albumin lactate pyruvate (TALP)-HEPES washing semen and washed twice by $200 \times g$ centrifugation. Finally the supernatant is removed and the precipitated sperm is collected and gently mixed and left until use.

12.2.3 Semen cryopreservation

Cryopreservation of NHP sperm cells has been studied for more than 30 years in baboon, macaque, squirrel monkey, cynomolgus monkey, Japanese monkey, lemur, Tibetan macaque, chimpanzee, gorilla, and marmoset. However, in very few studies researchers have tried to perform artificial insemination using frozen-thawed sperm cells. Macaque sperm cell cryopreservation is relatively successful, using mainly yolk for anticold shock and glycerin as cryoprotectant, and has been widely used in production centers and for scientific research [18].

12.3   Superovulation

In order to obtain a higher number of mature oocytes from both ovaries, overcoming thus the barrier imposed by the physiological constraints of the animal species, exogenous reproductive hormones must be administered to NHPs in order to stimulate accessory ovarian follicular development. Following maturation of hormone-stimulated follicles, semiinvasive (endoscopic procedure) or invasive surgical procedure has to be performed in order to collect the matured oocytes. The procedure of ovarian stimulation has gone through a slower process of research, development, and application to NHPs, when compared to production or laboratory animals. Again, the barrier imposed by the observance of strict regulations on the extension of research and methodology that can be implemented in NHPs [1] results in working protocols and superovulation schemes whose efficiency is still far from optimal.

12.3.1 Superovulation protocols

Several reports on the superovulation of cynomolgus monkeys, rhesus monkeys, marmosets, squirrel monkeys, African green monkeys, and gorillas are available (Table 12.1). The protocols used in these species are similar. The use of follicle-stimulating hormone (FSH) alone can induce a high number of follicle development. Following administration of gonadotropin-releasing hormone (GnRH), luteinizing hormone (LH), or human chorionic gonadotropin (hCG), the number of oocytes collected by superovulation is not significantly increased, although, whether a difference in oocyte development and pregnancy outcome exists among protocols, has not been reported so far.

Different superovulation protocols in rhesus monkeys have been established and tested over the years, but only three of them remain to these days the main choice: (1) in 1995 [34] rhesus monkeys were treated with GnRH for 7 days in the middle and terminal period of the last menstrual cycle. From the first day of menstruation, 30 IU of FSH were given twice daily for 6 days and 30 IU FSH and 30 IU LH were administered twice daily for additional 3 days, followed by injection of 1000 IU hCG the day after. Oocytes were collected via surgical procedure 25−32 hours following hCG administration; (2) in 1989 Don Wolf and his colleagues [28] developed a different protocol in which female monkeys from 1 to 3 days of menstruation were intramuscularly administered 30 IU FSH twice daily for 8 days, followed by injection of 1000 IU hCG the day after. Then, 28−32 hours following hCG administration, oocytes were collected from follicles with a diameter of 0.5 mm or more from both the ovaries. (3) Weizhi Ji and Shihua Yang and colleagues in 2007 [31], developed an alternative superovulation scheme based on the second of the protocols described above, with the first dose of FSH unchanged, but with the subsequent doses of FSH halved. As a result, the number and quality of recovered oocytes were improved. This last one proved to be an efficient protocol also in cynomolgus monkeys [35].

To reduce labor and use of reproductive hormones, sustained release of hormones has also been evaluated. Polyvinyl pyrrolidone (PVP), as a safe food additive, is used as a sustained release agent in pharmaceuticals. FSH when dissolved with 30% PVP solution can effectively complete superovulation [36], cutting down nearly 70% of hormone costs and reducing nearly half of the workload. Moreover, a similar number of oocytes can be obtained, when compared to more traditional approaches, and both fertilization and blastocyst rates are not significantly affected. In addition, exogenous hormones dissolved in glycerol solution can decrease the frequency of administration and prolong the interval in the course of the superovulation protocol [37]. However, the fertility of oocytes obtained by these slow-release treatment methods needs to be further explored if development to term of the same oocytes has to be considered.

TABLE 12.1 Superovulation protocols on NHPs.

	Protocol	Oocytes collected	Oocytes maturation (%) at collection	Blastocyst (%)	References
Marmosets	$PGF_{2\alpha}$/hCG	17	65	61	[19]
	$PGF_{2\alpha}$/rhFSH/hCG	14	-	34	[20]
	rhFSH/hCG	23	90	67	[21]
Squirrel monkey	FSH/hCG	-	37	42	[22]
	PMSG	-	-	-	[23]
African green monkey	GnRHa/hFSH/hCG	19	36.8	58.6	[24]
	GnRHa/eCG/hCG	18	66.7	77.8	[24]
Cynomolgus monkey	hMG/hCG	19	83	64	[25]
	GnRHa/rhFSH/hCG	31	38.7	-	[24]
	rhFSH/hCG	27–48	50–65	40–55	[16]
	GnRHa/eCG/hCG	18	38.9	-	[24]
Rhesus monkey	PMSG/hCG	21	66	66	[26]
	hFSH/hMG/hCG	25	64	51	[27]
	hFSH/hLH/hCG	18	63	25–75	[28]
	hMG/FSH + LH	11	29	24	[29]
	pFSH/hCG	20	71	48	[30]
	rhFSH/hCG	32	73	64	[31]
Baboon	CC	5.5	64	39	[32]
	CC/PMSG	11.3	92	39	[32]
	hMG	6.8	72	44	[32]
Gorillas	hFSH/hCG	8	-	-	[33]

CC, Clomiphene citrate.

In the final stage of superovulation, hCG is used to induce follicular oocyte maturation, and GnRH is also used to induce final oocyte maturation. In general, oocytes are usually collected 32–35 hours following hCG injection. A time interval longer than 36 hours from hCG treatment to oocyte recovery may cause some large follicles to spontaneously ovulate.

12.3.2   Influence of seasonal breeding on superovulation

In rhesus monkeys, the follicular developmental capacity and the level of reproductive hormones are significantly different between breeding and nonbreeding seasons. In the nonbreeding season, that is, from April to September each year, the rhesus monkey spontaneous ovulation and mating ability are low [38–40]. Likewise, when the reproductive function is physiologically reduced, the response to exogenous gonadotropin stimulation is also poor [41]. According to data derived from a collection of superovulations in adult rhesus monkeys, in the early stage of the reproductive season (September–October), the average number of oocytes collected was significantly lower than the middle of the reproductive season (November–December) and the last period (January–March). No difference in the number of oocytes collected and the maturation competence was found between the middle and the last period of the reproductive season. Therefore according to this available recorded data, the reproductive function is only partly restored at the beginning of the reproductive season, and this aspect has to be taken into consideration when applying superovulation protocols. Differently, cynomolgus monkeys have a similar reproductive function across all seasons considered, as confirmed also by a similar efficiency of superovulation among the seasons. The proportion of monkeys with regular menstrual cycles, though, is variable among seasons, and the implementation of superovulation protocols is affected by the actual presence of menstrual blood among animals.

12.3.3   The effect of age on superovulation

In general, the right age for macaques to undergo superovulation treatment is between 4 and 16 years, and the effect of age on the efficiency of hormonal treatment needs to be considered. It has been reported that macaques prior to puberty and in premenopausal stage have a reduced response to superovulatory treatment, resulting in few recovered oocytes of lower quality, when compared to puberal and adult macaques. The superovulation effect on monkeys at the time of menarche is close to the efficiency in adult monkeys, and however, it gradually decreases from the age of 12 onward [42]. From the above, it can be stated that the best time for obtaining the most successful superovulatory treatment is between 4 and 10 years of age.

12.3.4   Repeated superovulation

Due to limited resources devoted to NHPs research, repeated superovulation is a common practice. Studies have shown that macaques can undergo repeated superovulation up to three times in a year and that the number of oocytes collected and their developmental potential are not different among sessions [43]. In addition, a repeatability in the response following superovulation has been reported within single animals. Poor responses to the first treatment tend to be consistently seen in follow-up trials. Therefore macaques with better superovulation response should be selected for repeated superovulation [16].

12.3.5   Other influential factors

From our previous research it is clear that beginning of the superovulatory treatment between 1 and 4 days after menstruation brings similar consistent results. Another factor influencing the efficiency of superovulatory treatment is the live weight of donor monkeys. Both obese (heavier than 7 kg) and lightweight (less than 4 kg) rhesus monkeys have often suffered from reduced efficiency following hormonal treatment for superovulation, characterized also by lowered development potential of recovered oocytes. In a recent study, the recovery rate of metaphase II oocytes from 4.5 to 5.0 kg cynomolgus monkeys was higher than cynomolgus monkeys of different range of weights [44]. Therefore when selecting oocyte donors, the range in body weight should be between 4 and 7 kg in rhesus monkeys and 3 and 5 kg in cynomolgus monkeys. At the beginning of superovulation in NHPs, menstrual blood should be present and should mark the start of the superovulatory treatment. Sometimes though, disorder in the menstrual cycle of macaques may occur, causing poor results. For example, long-distance transportation of animals and diarrheal conditions are some of the factors affecting the menstrual cycle and leading to poor response following ovarian stimulation. In addition, the presence of a functional corpus luteum and high level of progesterone should also be considered as possible causes of unsatisfactory results following superovulation.

12.4   Fertilization and embryo culture

Similarly as described for production animals, both in vivo and in vitro fertilization approaches are implemented in NHPs. The former includes natural mating and artificial insemination, whereas the latter can be performed by either fertilizing matured oocytes with capacitated sperm cells or by intracytoplasmic sperm injection (ICSI).

12.4.1   In vivo fertilization

Pregnancy rate of macaques following natural mating is about 40% [45]. As for artificial insemination, sperm suspension (containing approximately 2×10^6 fresh or cryopreserved-thawed sperm in 100 μL medium) is prepared and injected 1 day before and 1 day after estrogen surge during a single menstrual cycle for each female. Pregnancy rate ranges from 4% to 40% [2,46], and the reason for such variable and low success rate is probably to be found in the presence of five to six intraluminal folds that do not allow the introduction of the insemination pipette for the deposition of semen into the uterus [47].

12.4.2 In vitro fertilization

Following capacitation in TALP medium supplemented with 0.1 mM sodium pyruvate and 0.3% bovine serum albumin (BSA), fertilization is performed in 100 mL droplets of fertilized culture medium (TALP + 2% fetal calf serum (FCS) + 3 mg/mL fraction V BSA), where 1×10^6 spermatozoa/mL and mature oocytes are coincubated at 37°C in a humidified atmosphere containing 5% CO_2 in air. The presence of two pronuclei is considered a mark for successful fertilization and the recorded rates are different for cynomolgus monkeys (75%−90%), rhesus monkey (68%) [48], and *Callithrix jacchus* (57%) [49]. In addition, in this last species, a small amount of polyspermic fertilization has also been reported [5,49].

ICSI can be considered an efficient method for fertilization in NHPs and has been successfully used in assisted reproduction to overcome the inherent problems associated with the lack of zona pellucida, azoospermia, polyspermism, and asthenospermia. However, when sperm cells cannot be re recovered due to idiopathic azoospermia, round spermatids can be collected from testes of sexually mature males, representing this approach the only available chance to achieve fertilization. In cynomolgus monkey round spermatids are 10−12 μm in diameter and therefore easily distinguished from primary spermatocytes. In a previous study, the round spermatids plasma membrane was broken, and the nucleus together with small amount of cytoplasm injected into the ooplasm, reaching a fertilization rate of approximately 69% [50].

12.5 Embryo culture

Not only the quality of the embryo, but the health of the future newborn may significantly be affected by improper culture conditions in vitro. Some options are available in in vitro culture: (1) surrogate mothers may be used and embryos cultured for a temporary period of time in the oviduct of the foster animal. In some studies, rabbit fallopian tubes have been used for culturing squirrel monkey early embryos, although this approach has been found to be quite complex and not suitable for large-scale culture and embryo research in developmental biology [51], and (2) coculture system, where embryos are cultured on a feeder layer of cells of different origin. In previous studies, buffalo rat liver cells in CMRL-1066 medium supplemented with 10% FCS were used, and the blastocyst rate achieved was 22.7%. If supplementation of fetal bovine serum was increased to 20%, then blastocyst rate was also improved up to 37% [2,52]. A blastocyst rate of 30% was reached when fertilized oocytes after ICSI were cultured in TALP medium supplemented with 0.3% BSA, and after development to two-cell stage, the cells were transferred to CMRL medium containing 10% fetal bovine serum. Nowadays, the most common approach is to culture embryos in HECM-9 with 10% fetal bovine serum, reaching a higher blastocyst rate [53]. Other alternatives include the adoption of human embryonic medium (COOK IVF medium), used for cynomolgus embryo culture, with a blastocyst rate of about 20% [54], while with G1/G2 medium-high blastocyst rates could not be obtained [55].

12.6 Embryo transfer

A successful synchronization must be in place in order to foster the development to term of an embryo at the blastocyst stage that has been transferred into a suitable recipient. Many other components play a vital role in the establishment of a pregnancy following embryo transfer (ET). In the case of NHPs, four main important aspects have to be considered: (1) embryo quality screening and selection, (2) recipient screening and selection, (3) methodology of ET, and (4) early pregnancy diagnosis and live birth.

12.6.1 Embryo quality screening and selection

Despite the available tests devised to assess the quality of embryos, their viability cannot currently be confirmed with certainty due to the absence of a reliable assessment system in place. One of the most efficient methods on embryo diagnosis entails the technique of blastomere biopsy, mainly used though to detect embryos bearing genetic defects, and currently adopted in human prenatal genetic diagnosis. Therefore, nowadays and with regard to NHPs, the only methodology employed to assess embryo quality and predict the potential of embryo development and consequently the future probability of implantation is based on the visual assessment of embryo morphology. From the above, and likewise for other animal species, the only available parameters to be taken into consideration for selection are the time of development to a particular stage, the uniformity in size and shape of the embryo, and the relatively transparent color of the blastomeres. In clinical practice, in order to obtain higher pregnancy rates, the number of transplanted embryos is usually increased, although this approach can bring unwanted multiple pregnancies, increased abortion rate, and a reduced live birth rate.

12.6.2   Selection of recipients

Other than some general basic consideration in the selection of NHPs recipient, such as the use of young animals characterized by a healthy reproductive status, from reported results in macaques the following factors should also be considered [56]:

1. Age and body weight. Generally, the most suitable recipients are females with an appropriate body weight of 4−8 kg for rhesus monkey and 3−6 kg for cynomolgus monkey and an age between 5 and 12 years. Lighter weights may give indication of animals in poor nutrition, whereas heavier weights in females may affect implantation of the transferred embryos due to possible endocrine and metabolic disorders.
2. Parity in recipients. It is always better to use recipients that have already delivered babies. However, due to the limitations of resources in monkey research, it is often difficult to rely on a sufficient number of recipient animals. Therefore whenever recipient healthy females with regular menstrual cycle, characterized by a recorded ovulation or in the early stage of corpus luteum development, are available, they are usually selected for transfer.
3. Superovulation. In our experience, superovulated macaques are characterized by normal menstrual bleeding and achieve pregnancy when mated with males. In fact, in our trials, transfer of embryos has been performed on macaques just following the end of the hormonal stimulation and pregnancies have been obtained.
4. Endometrial requirements. We have retrospectively analyzed the size of the uterus and the ovaries, serum E2/P4 levels, the length of the menstrual cycle, the timing of follicular development, and ovulation of the dominant follicle before transfer. When considering pregnancy rates, however, no differences were reported when considering all the above variables following ET. Nevertheless, an important indicator for the screening of ET recipients is the ratio between endometrium and myometrium in cynomolgus and rhesus monkeys having normal ovulation and menstrual cycles [16].
5. Synchronization. The stage of embryonic development must be synchronized with the recipient's uterus. Some studies have suggested that the period of ET window may vary from preovulation of follicles (1 day before ovulation) to early luteal stage (1−4 days after ovulation) [57]. With regard to the stage of embryo development, it is generally believed that embryos from one-cell stage to blastocyst should be transplanted into timed and synchronized uterus or slightly late uterus, with ovulation as the reference. For example, two-cell stage embryos can be transplanted into recipient on day 1 after ovulation, and even blastocyst can be transferred into the uterus on days 1−4 after ovulation.

12.6.3   Embryo transfer

Since the first successful ET in rhesus monkey in 1984, a great progress in the implementation of this technology has been made. At the very start, embryos were used to be transferred into the uterus or the oviducts of recipients by abdominal surgery, and the resulting pregnancy rates were not satisfactory. Although transfer of embryos through vagina and cervix with ultrasound aid is successful in marmoset, this is not feasible in macaques due to the anatomical layout and constraints of the cervix. Nowadays, through laparoscopic surgical procedure, high-resolution intra-abdominal endoscopy is widely and successfully used, and selected embryos are transferred into the oviducts (2−3 cm in depth) of recipient female monkeys following conventional anesthetic procedure [31].

12.6.4   Confirmation of clinical pregnancy

For confirmation of pregnancy following ET, recipient females are checked for menstrual bleeding, assayed for E2 and P4 hormonal values, and monitored by B-ultrasound. Both, a continuous rise of progesterone values and the detection by ultrasound of the early embryonic sac from 15 days after ET are an indication of early pregnancy. Ultrasound will confirm a successful pregnancy from 25 to 35 days after ET by detecting a fetal heartbeat.

12.7   Multipregnancy treatments

Macaques are monotocous species, although with the implementation of ARTs, the number of transplanted embryos is usually more than one. Therefore multiple embryo implantation and pregnancy can occur, resulting in increased abortion rate and reduced live birth rate. To counteract this negative aspect, a technology for multifetal pregnancy reduction (MPR) has been developed [58]. Briefly, a 0.5−1 mL of 10% potassium chloride solution is injected into the fetal heart of smaller fetuses or to the ones positioned closer to the abdominal wall, resulting in the death of the targeted fetus. MPR is safer and more efficient when implemented at 1 month of pregnancy. Alternatively, cesarean section can be performed, preceded by a continuous daily pregnancy monitoring. Whenever the monkey carrying multiple pregnancies displays signs of childbirth, either early in gestation or near the physiological end of pregnancy, then a cesarean section should be performed immediately. This will effectively reduce maternal dystocia and improve mother-child survival.

12.8   Cloning

As with all other animal species used for cloning, each step involved in NHP cloning such as oocyte enucleation, nuclear donor cell transfer, oocyte activation, and epigenetic regulation has been researched.

12.8.1   Oocyte enucleation

The enucleation procedure using Hoechst 33342 staining and ultraviolet exposure [59] may affect oocyte and embryo viability, although the adoption of different approaches [60,61] has improved blastocyst rate up to 24%. The Oosight spindle imaging system allows real-time microscopy and manipulation during enucleation in primate oocytes, with the result of reducing damage to the oocytes and improving blastocyst rates [62].

12.8.2   Donor somatic cell transfer and oocyte activation

In 1997, live monkeys were generated using nuclear transfer of early blastomeres [63]. In the 21st century, the exploration of SNCT began in NHP. Cumulus cells [64], amniotic epithelial cells [64], and skin fibroblasts [59–61,63–67] were tested as nuclear donor cells. Subsequently, donor cells (fetal and adult fibroblasts commonly used) were introduced into the perivitelline space of enucleated oocytes, and then cell membrane fusion by electrofusion [59,60], SeV (an extract from Sendai virus) [68,69] or hemagglutinin virus of Japan envelope [70] could be achieved. Following cell fusion, reconstructed embryos were commonly activated with ionomycin and 6-dimethylaminopurine [61,62]. Two embryonic stem cell lines were successfully isolated from SCNT embryos in 2007 [71]. As oocytes are difficult to obtain, interspecific SCNT is widely used for NHP cloning experiments, although blastocyst development rate is low [72,73].

12.8.3   Developmental potential of somatic cell nuclear transfer

Early studies had pointed out that SCNT in NHPs might prove difficult and unachievable, mainly due to abnormal expression of nuclear mitotic apparatus in reconstructed embryos [74]. Some studies have shown that SCNT reconstructed embryos underwent inappropriate reprogramming [75] and abnormal epigenetic modifications [36]. The treatment of trichostatin A (TSA), a histone deacetylase inhibitor that can reduce the activity of histone deacetylase, thus resulting in an increased level of histone acetylation, could improve the efficiency of SCNT in several species. Furthermore, reprogramming-resistant region, which was enriched for histone 3 lysine 9 trimethylation (H3K9me3) modification, was identified in SCNT embryos and Kdm4d can significantly improve it together with the efficiency of SCNT in mouse [76]. Therefore, in 2018, SCNT using fetal fibroblasts with injection of *Kdm4d* mRNA into oocytes and treatment with TSA, yielded two live SCNT monkeys for the first time in the world, named Zhong Zhong and Hua Hua [70], and later clones of the *BMAL1* mutant macaque monkey were produced by using the same procedure [77].

12.9   Induced pluripotent stem cells

As a milestone technology in biomedicine, the technology delivering iPSCs has a prospect for a very broad development. This technology avoids not only the problem of immune rejection and the shortage of experimental materials, but also overcomes ethical and legal issues. However, problems related to low induction efficiency, high carcinogenic risk, and the efficiency in the process of inducing iPSCs are still there. Although the results in establishing NHPs iPSC are not as advanced as in rodents, the technology is gradually improving. In 2008, for the first time, monkey iPSC cells were generated by retrovirus-mediated transduction of four transcription factors OCT4, SOX2, KLF4, and c-MYC into fibroblasts of rhesus monkeys. The iPSCs produced were detected to differentiate into the three embryonic cell types in vitro. The iPSCs of rhesus monkey were then injected into non-obese diabetic/severe combined immunodeficient mice, and teratomas were formed in 6–8 weeks, confirming the totipotency of rhesus monkey iPSC [78]. In 2015, by using the PiggyBac transposable system of the six transcription factors (OCT3/4, SOX2, c-MYC, KLF4, NANOG, LIN28), induction of iPSCs was accomplished without exogenous gene insertion and completely unmodified, with the possibility to pass generated iPSCs for more than 80 times and forming teratomas [79]. In addition, many studies have demonstrated that NHP iPSCs could be induced to differentiate into pigment epithelial cells [80], insulin-producing cells [81], endothelial cell systems [82], etc.

12.10 Summary

ARTs in NHPs are tools of paramount importance not only for the improvement of reproductive efficiency. In fact, they can also offer a strategic approach and provide support for the study and development of basic and applied research for the benefit of human population health, biomedicine, and the establishment of animal models for major human diseases. In this last regard, the etiology of many human diseases can be understood from the genetic evaluations of embryos. That is the reason why studies on primate genomics and gene mutation are instrumental for a wide comprehension of genetic diseases and acquire importance of scientific and medical value. As human embryos, due to ethical and law constraints, cannot be studied to the full extent, NHPs become indispensable animals to explore the genetic basis of neurological and metabolic disorders, among others. However, a number of problems related to the efficient application of ARTs to NHPs are still there, such as the synchronization of the reproductive cycle and its length and the low efficiency in pregnancy rates following ET. Despite a number of shortfalls, the efforts in overcoming the issues raised by the implementation of ARTs in NHPs have become a commitment of a large number of researchers worldwide.

References

[1] Bavister BD. ARTs in action in nonhuman primates: symposium summary-advances and remaining issues. Reprod Biol Endocrinol 2004;2:43.
[2] Liu Z, Cai Y, Wang Y, Nie Y, Zhang C, Xu Y, et al. Cloning of macaque monkeys by somatic cell nuclear transfer. Cell 2018;174:245.
[3] Fang R, Liu K, Zhao Y, Li H, Zhu D, Du Y, et al. Generation of naive induced pluripotent stem cells from rhesus monkey fibroblasts. Cell Stem Cell 2014;15:488−97.
[4] Sagi I, Chia G, Golan-Lev T, Peretz M, Weissbein U, Sui L, et al. Derivation and differentiation of haploid human embryonic stem cells. Nature 2016;532:107−11.
[5] Sato T, Katagiri K, Gohbara A, Inoue K, Ogonuki N, Ogura A, et al. In vitro production of functional sperm in cultured neonatal mouse testes. Nature 2011;471:504−7.
[6] Zhou Y, Sharma J, Ke Q, Landman R, Yuan J, Chen H, et al. Atypical behaviour and connectivity in SHANK3-mutant macaques. Nature 2019;570:326−31.
[7] Chen Y, Yu J, Niu Y, Qin D, Liu H, Li G, et al. Modeling rett syndrome using TALEN-edited MECP2 mutant cynomolgus monkeys. Cell 2017;169:945−955.e10.
[8] Mastroianni LJ, Manson WJ. Collection of monkey semen by electroejaculation. Proc Soc Exp Biol Med 1963;112:1025−7.
[9] Weisbroth S, Young FA. The collection of primate semen by electro-ejaculation. Fertil Steril 1965;16:229−35.
[10] Gould KG, Warner H, Martin DE. Rectal probe electroejaculation of primates. J Med Primatol 1978;7:213−22.
[11] Jinhuai H, Weiguo L, Quanfu L, et al. Electric stimulus collecting semen test in Taihang monkey. J Henan Norm Univ (Nat Sci) 1998;26 (1):75−7.
[12] Schneiders A, Sonksen J, Hodges JK. Penile vibratory stimulation in the marmoset monkey: a practical alternative to electro-ejaculation, yielding ejaculates of enhanced quality. J Med Primatol 2004;33:98−104.
[13] Gould KG, Mann DR. Comparison of electrostimulation methods for semen recovery in the rhesus monkey (*Macaca mulatta*). J Med Primatol 1988;17:95−103.
[14] Matsubayashi K. Comparison of the two methods of electroejaculation in the Japanese monkey (*Macaca fuscata*). Jikken Dobutsu 1982;31:1−6.
[15] VandeVoort CA. High quality sperm for nonhuman primate ART: production and assessment. Reprod Biol Endocrinol 2004;2:33.
[16] Ma Y, Li J, Wang G, Ke Q, Qiu S, Gao L, et al. Efficient production of cynomolgus monkeys with a toolbox of enhanced assisted reproductive technologies. Sci Rep 2016;6:25888.
[17] Shangchuan Y, Weizhi J, Jianchun C, et al. The use of improved penile electroejaculation in rhesus, tibetan and assamese macaques and study on the parameters of their semen. Zool Res 1994.
[18] Si W, Zheng P, Tang X, He X, Wang H, Bavister BD, et al. Cryopreservation of rhesus macaque (*Macaca mulatta*) spermatozoa and their functional assessment by in vitro fertilization. Cryobiology 2000;41:232−40.
[19] Lopata A, Summers PM, Hearn JP. Births following the transfer of cultured embryos obtained by in vitro and in vivo fertilization in the marmoset monkey (*Callithrix jacchus*). Fertil Steril 1988;50:503−9.
[20] Marshall VS, Browne MA, Knowles L, Golos TG, Thomson JA. Ovarian stimulation of marmoset monkeys (*Callithrix jacchus*) using recombinant human follicle stimulating hormone. J Med Primatol 2003;32:57−66.

[21] Grupen CG, Gilchrist RB, Nayudu PL, Barry MF, Schulz SJ, Ritter LJ, et al. Effects of ovarian stimulation, with and without human chorionic gonadotrophin, on oocyte meiotic and developmental competence in the marmoset monkey (*Callithrix jacchus*). Theriogenology 2007;68:861−72.

[22] Kuehl TJ, Dukelow WR. Maturation and in vitro fertilization of follicular oocytes of the squirrel monkey (*Saimiri sciureus*). Biol Reprod 1979;21:545−56.

[23] Schuler AM, Westberry JM, Scammell JG, Abee CR, Kuehl TJ, Gordon JW. Ovarian stimulation of squirrel monkeys (Saimiri boliviensis boliviensis) using pregnant mare serum gonadotropin. Comp Med 2006;56:12−16.

[24] Shimozawa N, Okada H, Hatori M, Yoshida T, Sankai T. Comparison of methods to stimulate ovarian follicular growth in cynomolgus and African green monkeys for collection of mature oocytes. Theriogenology 2007;67:1143−9.

[25] Balmaceda JP, Pool TB, Arana JB, Heitman TS, Asch RH. Successful in vitro fertilization and embryo transfer in cynomolgus monkeys. Fertil Steril 1984;42:791−5.

[26] Bavister BD, Boatman DE, Leibfried L, Loose M, Vernon MW. Fertilization and cleavage of rhesus monkey oocytes in vitro. Biol Reprod 1983;28:983−99.

[27] Lanzendorf SE, Gliessman PM, Archibong AE, Alexander M, Wolf DP. Collection and quality of rhesus monkey semen. Mol Reprod Dev 1990;25:61−6.

[28] Wolf DP, Vandevoort CA, Meyer-Haas GR, Zelinski-Wooten MB, Hess DL, Baughman WL, et al. In vitro fertilization and embryo transfer in the rhesus monkey. Biol Reprod 1989;41:335−46.

[29] Morgan PM, Boatman DE, Bavister BD. Relationships between follicular fluid steroid hormone concentrations, oocyte maturity, in vitro fertilization and embryonic development in the rhesus monkey. Mol Reprod Dev 1990;27:145−51.

[30] Schramm RD, Bavister BD. Use of purified porcine follicle-stimulating hormone for ovarian stimulation of macaque monkeys. Theriogenology 1996;45:727−32.

[31] Yang S, He X, Hildebrandt TB, Jewgenow K, Goeritz F, Tang X, et al. Effects of rhFSH dose on ovarian follicular response, oocyte recovery and embryo development in rhesus monkeys. Theriogenology 2007;67:1194−201.

[32] Fourie FR, Snyman E, van der Merwe JV, Grace A. Primate in vitro fertilization research: preliminary results on the folliculogenic effects of three different ovulatory induction agents on the chacma baboon, *Papio ursinus*. Comp Biochem Physiol A Comp Physiol 1987;87:889−93.

[33] Lanzendorf SE, Holmgren WJ, Schaffer N, Hatasaka H, Wentz AC, Jeyendran RS. In vitro fertilization and gamete micromanipulation in the lowland gorilla. J Assist Reprod Genet 1992;9:358−64.

[34] Zelinski-Wooten MB, Hutchison JS, Hess DL, Wolf DP, Stouffer RL. Follicle stimulating hormone alone supports follicle growth and oocyte development in gonadotrophin-releasing hormone antagonist-treated monkeys. Hum Reprod 1995;10:1658−66.

[35] Huang Y, Ding C, Liang P, Li D, Tang Y, Meng W, et al. HBB-deficient Macaca fascicularis monkey presents with human beta-thalassemia. Protein Cell 2019;10:538−42.

[36] Yang S, He X, Hildebrandt T, Zhou Q, Ji W. Superovulatory response to a low dose single-daily treatment of rhFSH dissolved in polyvinylpyrrolidone in rhesus monkeys. Am J Primatol 2007;69:1278−84.

[37] Livesey JH, Roud HK, Metcalf MG, Donald RA. Glycerol prevents loss of immunoreactive follicle-stimulating hormone and luteinizing hormone from frozen urine. J Endocrinol 1983;98:381−4.

[38] Dailey RA, Neill JD. Seasonal variation in reproductive hormones of rhesus monkeys: anovulatory and short luteal phase menstrual cycles. Biol Reprod 1981;25:560−7.

[39] Walker ML, Wilson ME, Gordon TP. Endocrine control of the seasonal occurrence of ovulation in rhesus monkeys housed outdoors. Endocrinology 1984;114:1074−81.

[40] Hutz RJ, Dierschke DJ, Wolf RC. Seasonal effects on ovarian folliculogenesis in rhesus monkeys. Biol Reprod 1985;33:653−9.

[41] Nusser KD, Mitalipov S, Widmann A, Gerami-Naini B, Yeoman RR, Wolf DP. Developmental competence of oocytes after ICSI in the rhesus monkey. Hum Reprod 2001;16:130−7.

[42] Yang S, He X, Niu Y, Hildebrandt TB, Jewgenow K, Goeritz F, et al. Ovarian response to gonadotropin stimulation in juvenile rhesus monkeys. Theriogenology 2009;72:243−50.

[43] Yang S, Shen Y, Niu Y, Hildebrandt TB, Jewgenow K, Goeritz F, et al. Effects of rhFSH regimen and time interval on ovarian responses to repeated stimulation cycles in rhesus monkeys during a physiologic breeding season. Theriogenology 2008;70:108−14.

[44] Kim JS, Yoon SB, Jeong KJ, Sim BW, Choi SA, Lee SI, et al. Superovulatory responses in cynomolgus monkeys (*Macaca fascicularis*) depend on the interaction between donor status and superovulation method used. J Reprod Dev 2017;63:149−55.

[45] Klooster KL, Burruel VR, Meyers SA. Loss of fertilization potential of desiccated rhesus macaque spermatozoa following prolonged storage. Cryobiology 2011;62:161−6.

[46] Sánchez-Partida LG, Maginnis G, Dominko T, Martinovich C, McVay B, Fanton J, et al. Live rhesus offspring by artificial insemination using fresh sperm and cryopreserved sperm. Biol Reprod 2000;63:1092−7.

[47] Czaja JA, Eisele SG, Goy RW. Cyclical changes in the sexual skin of female rhesus: relationships to mating behavior and sucessful artificial insemination. Fed Proc 1975;34:1680−4.

[48] Wolf DP, Thomson JA, Zelinski-Wooten MB, Stouffer RL. In vitro fertilization-embryo transfer in nonhuman primates: the technique and its applications. Mol Reprod Dev 1990;27:261−80.

[49] Wilton LJ, Marshall VS, Piercy EC, Moore HD. In vitro fertilization and embryo development in the marmoset monkey (*Callithrix jacchus*). J Reprod Fertil 1993;97:481−6.

[50] Ogonuki N. Pregnancy by the tubal transfer of embryos developed after injection of round spermatids into oocyte cytoplasm of the cynomolgus monkey (*Macaca fascicularis*). Hum Reprod 2003;18:1273–80.

[51] Yunhan M, Shihua Y. Advances in genetically modified Chinese macaques. Prog Biochem Biophys 2014;41:1089–98.

[52] Yamasaki J, Iwatani C, Tsuchiya H, Okahara J, Sankai T, Torii R. Vitrification and transfer of cynomolgus monkey (*Macaca fascicularis*) embryos fertilized by intracytoplasmic sperm injection. Theriogenology 2011;76:33–8.

[53] Wolf DP, Thormahlen S, Ramsey C, Yeoman RR, Fanton J, Mitalipov S. Use of assisted reproductive technologies in the propagation of rhesus macaque offspring. Biol Reprod 2004;71:486–93.

[54] Curnow EC, Pawitri D, Hayes ES. Sequential culture medium promotes the in vitro development of *Macaca fascicularis* embryos to blastocysts. Am J Primatol 2002;57:203–12.

[55] Wolf DP. Assisted reproductive technologies in rhesus macaques. Reprod Biol Endocrinol 2004;2:37.

[56] Ji W. Reproduction and breeding of rhesus monkey. Peking Sci Press 2013;133–4.

[57] Chen Y, Niu Y, Yang S, He X, Ji S, Si W, et al. The available time window for embryo transfer in the rhesus monkey (*Macaca mulatta*). Am J Primatol 2012;74:165–73.

[58] Zhuo Y, Feng S, Huang S, Chen X, Kang Y, Si C, et al. Transabdominal ultrasound-guided multifetal pregnancy reduction in 10 cases of monkeys. Biol Reprod 2017;97:758–61.

[59] Simerly C, Navara C, Hwan Hyun S, Chun Lee B, Keun Kang S, Capuano S, et al. Embryogenesis and blastocyst development after somatic cell nuclear transfer in nonhuman primates: overcoming defects caused by meiotic spindle extraction. Dev Biol 2004;276:237–52.

[60] Zhou Q, Yang SH, Ding CH, He XC, Xie YH, Hildebrandt TB, et al. A comparative approach to somatic cell nuclear transfer in the rhesus monkey. Hum Reprod 2006;21:2564–71.

[61] Mitalipov SM, Zhou Q, Byrne JA, Ji WZ, Norgren RB, Wolf DP. Reprogramming following somatic cell nuclear transfer in primates is dependent upon nuclear remodeling. Hum Reprod 2007;22:2232–42.

[62] Ng SC, Chen N, Yip WY, Liow SL, Tong GQ, Martelli B, et al. The first cell cycle after transfer of somatic cell nuclei in a non-human primate. Development 2004;131:2475–84.

[63] Meng L, Ely JJ, Stouffer RL, Wolf DP. Rhesus monkeys produced by nuclear transfer. Biol Reprod 1997;57:454–9.

[64] Okahara-Narita J, Tsuchiya H, Takada T, Torii R. Cloned blastocysts produced by nuclear transfer from somatic cells in cynomolgus monkeys (*Macaca fascicularis*). Primates 2007;48:232–40.

[65] Sparman M, Dighe V, Sritanaudomchai H, Ma H, Ramsey C, Pedersen D, et al. Epigenetic reprogramming by somatic cell nuclear transfer in primates. Stem Cell 2009;27:1255–64.

[66] Niu Y, Yang S, Yu Y, Ding C, Yang J, Wang S, et al. Impairments in embryonic genome activation in rhesus monkey somatic cell nuclear transfer embryos. Cloning Stem Cell 2008;10:25–36.

[67] Mitalipov SM, Yeoman RR, Nusser KD, Wolf DP. Rhesus monkey embryos produced by nuclear transfer from embryonic blastomeres or somatic cells. Biol Reprod 2002;66:1367–73.

[68] Sparman ML, Tachibana M, Mitalipov SM. Cloning of non-human primates: the road "less traveled by". Int J Dev Biol 2010;54:1671–8.

[69] Tachibana M, Sparman M, Sritanaudomchai H, Ma H, Clepper L, Woodward J, et al. Mitochondrial gene replacement in primate offspring and embryonic stem cells. Nature 2009;461:367–72.

[70] Liu Z, Cai Y, Wang Y, Nie Y, Zhang C, Xu Y, et al. Cloning of macaque monkeys by somatic cell nuclear transfer. Cell 2018;172:881–887.e7.

[71] Byrne JA, Pedersen DA, Clepper LL, Nelson M, Sanger WG, Gokhale S, et al. Producing primate embryonic stem cells by somatic cell nuclear transfer. Nature 2007;450:497–502.

[72] Kwon D, Koo OJ, Kim MJ, Jang G, Lee BC. Nuclear-mitochondrial incompatibility in interorder rhesus monkey-cow embryos derived from somatic cell nuclear transfer. Primates 2016;57:471–8.

[73] Kwon DK, Kang JT, Park SJ, Gomez MN, Kim SJ, Atikuzzaman M, et al. Blastocysts derived from adult fibroblasts of a rhesus monkey (*Macaca mulatta*) using interspecies somatic cell nuclear transfer. Zygote 2011;19:199–204.

[74] Simerly C, Dominko T, Navara C, Payne C, Capuano S, Gosman G, et al. Molecular correlates of primate nuclear transfer failures. Science 2003;300:297.

[75] Simerly CR, Navara CS. Nuclear transfer in the rhesus monkey: opportunities and challenges. Cloning Stem Cell 2003;5:319–31.

[76] Matoba S, Liu Y, Lu F, Iwabuchi KA, Shen L, Inoue A, et al. Embryonic development following somatic cell nuclear transfer impeded by persisting histone methylation. Cell 2014;159:884–95.

[77] Liu Z, Cai Y, Liao Z, Xu Y, Wang Y, Wang Z, et al. Cloning of a gene-edited macaque monkey by somatic cell nuclear transfer. Natl Sci Rev 2019;6:101–8.

[78] Liu H, Zhu F, Yong J, et al. Generation of induced pluripotent stem cells from adult rhesus monkey fibroblasts. Cell Stem Cell 2008;3 (6):587–90.

[79] Debowski K, Warthemann R, Lentes J, et al. Non-viral generation of marmoset monkey iPS cells by a six-factor-in-one-vector approach. PLoS One 2015;10(3):e118424.

[80] Okamoto S, Takahashi M. Induction of retinal pigment epithelial cells from monkey iPS cells. Investig Ophthalmol Vis Sci 2011;52(12):8785.

[81] Zhu FF, Zhang PB, Zhang DH, et al. Generation of pancreatic insulin-producing cells from rhesus monkey induced pluripotent stem cells. Diabetologia 2011;54(9):2325–36.

[82] Thoma EC, Heckel T, Keller D, et al. Establishment of a translational endothelial cell model using directed differentiation of induced pluripotent stem cells from cynomolgus monkey. Sci Rep 2016;6:35830.

Chapter 13

Reproductive technologies in avian species

Judit Barna, Barbara Végi, Krisztina Liptói and Eszter Patakiné Várkonyi

National Centre for Biodiversity and Gene Conservation (NCBGC), Institute for Farm Animal Gene Conservation, Gödöllő, Hungary

13.1 Artificial insemination

The importance of artificial insemination (AI) in birds applies to both ordinary breeding programs and avian conservation programs. By the implementation of AI, the reproductive efficiency of economically important species (poultry, ratites) as well as the preservation of endangered birds can be achieved [1]. AI was first successfully performed in birds about a century ago when Ivanov [2] produced fertile chicken eggs using semen recovered from the ductus deferens [3]. This assisted breeding technique is now an integral part to commercial turkey, chicken, duck, goose, and guinea fowl production and has also been adapted to produce progenies in more than 40 types of nondomestic birds such as various waterfowls, raptors, cranes, passerines, psittacines, and ratites [4]. AI is a labor-intensive process that requires bird capture, restraint, and special pen design. The risk of injury both to bird and handler has also to be considered, as well as possible transfer of diseases in the process of capturing and handling of domestic and nondomestic birds. In establishing an AI program, it is very important to integrate the knowledge of the sex-specific differences in reproductive anatomy and copulatory strategy since each step of the procedure varies species by species [5]. Due to the various seasonal effects on reproduction in birds, the timing of AI implementation is also very important and has to be taken into account.

AI is the manual transfer of semen mostly into the female's vagina or in special cases into the magnum by laparoscopy. It consists basically of a three-step procedure: (1) collecting semen from the male; (2) semen handling and dilution, evaluation, and storage; and (3) introducing the prepared semen into the female.

13.1.1 Semen collection

The copulatory organ of birds is the phallus that, due to anatomical variation of the phallic region, may vary significantly in size and structure [6]. Therefore the techniques of semen collection in birds cannot be uniform. In contrast to ratites and waterfowls, characterized by a phallus protrudens, Galliformes (chicken, turkey, guinea fowl, and quail) copulatory organ (phallus nonprotrudens) consists of folds and bulges that make contact with the females' cloaca at mating. Usually three different methods can be used: (1) a "cooperative" approach in imprinted birds-of-prey and some cases in ratites where the birds voluntarily copulate on special devices in response to a behavioral stimulus without any handling, stress or trauma [7,8]; (2) electro-ejaculation, practiced in duck and goose [9], pigeons [10], and psittacines [11], with the disadvantage of requiring anesthesia, frequent semen contamination with urine, and sometimes serious injuries on tissues; and (3) finally the most practical and frequently used procedure being the dorso-abdominal massage [12] with various modification adopted on the different species (Fig. 13.1). The massage technique is performed as follows: the male is restrained by an assistant or by a special restraining apparatus, followed by gentle stroking of the back region from behind the wing toward the tail and at the same time the ventral abdominal region. Among avian species, the turkey prefers more massage at the vent region. Most males respond with phallic tumescence, at which time the handler gently squeezes the cloaca. To get the first drop of semen, the personnel performing this operation applies gentle pressure to the dorsal lip of the cloaca. Then the cloacal stimulation is repeated until the remaining semen is collected. Semen collection takes usually 5–10 seconds. The collection device can be a cup, a glass tube, Eppendorf tube, funnel, capillary tube, or an artificial vagina (for ratites see Ref. [8]). In most species,

Reproductive Technologies in Animals. DOI: https://doi.org/10.1016/B978-0-12-817107-3.00013-8

193

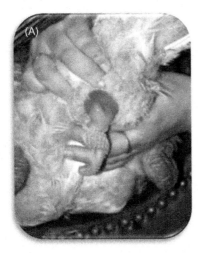  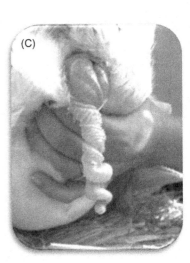

FIGURE 13.1 Sperm collection—domesticated species: everted phalluses following dorso-abdominal massage and cloacal stimulation: (A) gander (*Anser anser*), (B) guinea fowl (*Numida meleagris*), and (C) drake (*Anas platyrhynchos domestica*).

to perform the procedure two people are needed, while in the industrial breeding of turkey or fowl special devices are available for restraining the males, so that the personnel can work without direct help of other people. Additional adaptation of this method has been devised for several species such as waterfowl, ratites, guans and tinamous characterized by penis-like copulatory organs [13]. For birds with small amount of semen, the operator has to evert the phallus early at collection and a suction device has to be used to avoid losing semen on the phallic surface [14], and sometimes it is also recommended to add a drop of diluent to the semen or cloaca region in order to avoid the drying of the semen. Definitely, for different wild birds some peculiar restraining apparatus needs to be developed [4,15]. In all semen collecting methods the best and most consistent semen quality can be achieved only when males have been rightly conditioned and trained for semen collection.

A detailed description of semen collection of various nondomestic birds has been reviewed by Gee et al. [5]. Usually, the frequency of semen collections is between two and three times a week, although it has to be considered that breeds within a given species can respond differently to the frequency [16].

13.1.2 Techniques of semen handling and quality of semen

Following ejaculation, a visual examination of freshly collected semen is important. The collected semen should be clean, pearly white, and viscous in most cases. Watery semen and semen contaminated with blood, feces, or urates debris should not be used for insemination.

The volume of semen can vary from a few microliters to several milliliters, and although larger semen volumes are collected from larger birds, not always large birds produce large semen samples. This is dependent on a number of variables, such as the method of semen collection, the amount of accessory fluids added to the semen at ejaculation, the frequency and technique of collection, the stage of the breeding season, the timing of collection during the day, and lastly but equally important, the personnel trained to collect the semen. A variety of pipettes and syringes are available for measuring the small amount of semen samples. Gee et al. [17] used the white- and red-cell diluting pipettes for measuring a volume of 20−100 µL of semen samples.

The pH of semen is species-specific and varies from 6.0 to 8.0. Semen pH is more alkaline for psittacine species compared to raptors, and the pH of falcon semen is more acidic compared to eagles. Moreover, there is a species-specific wider pH range out of which 90% sperm remains immotile [18]. This information is crucial when developing and optimizing diluents for each species, as well as for understanding how pH changes can influence motility [19]. There are reliable pH meters available in the market, which can be used with very small sperm samples, avoiding the loss of valuable ejaculates when working with critically endangered species. Urine contamination usually pushes the pH into the more alkaline range.

The expected osmolality of semen should be close to blood plasma (generally around 270−350 mOsm/kg). Several kinds of osmometers are available for accurate determinations, requiring only a very small amount of semen. It is known that avian sperm is more resistant to changes in osmolality compared to mammalian species, and in order to develop specific diluents for each avian species it is of paramount importance the knowledge of values of normal semen osmolality [18].

13.1.2.1 Morphology of avian spermatozoa

The shape of avian spermatozoa varies from the simple to a complex helical type with exterior ribbon-like membrane and relatively long tail. Usually, the more primitive species have the simple form and the more recently evolved passerines the more complex form [20]. The exception to the rule is the round to slightly flattened sperm of American kestrel [21]. Sperm size and the ratios of the various structures of the spermatozoa are species-specific.

During handling of avian semen, it is important to be aware of sperm size and morphology to avoid cell injury and to detect abnormalities such as giant or small cells [22], swollen head and tail, multiple or broken and coiling tails, and droplets. Certainly, the normal morphology is in positive correlation to fertility [23]. These parameters can be determined by using vital staining methods (see Section 13.1.2.3). A higher level of abnormal and immature spermatozoa can usually be found in ejaculates from the early stages of the breeding season.

13.1.2.2 Sperm concentration

The concentration of spermatozoa in ejaculates is also typically species-specific. In domestic species the average values in million/μL are 2.5−5.5 in domestic fowl, Peking drakes, and guinea fowls; 6−10 in turkey toms; and as low as 0.2−1.5 in ganders, while in nondomestic species semen is much less concentrated [5]. In order to achieve a high fertility rate, sperm concentration is an important parameter for both domestic and nondomestic bird species. The oldest but the most reliable method is the simple sperm count with a hemocytometer. The cytometer counts are the basis of all other measurements using a standard dilution curve with ascending sperm concentration.

In a well-equipped laboratory the most commonly used indirect techniques to determine sperm concentration are the optical density (OD, photometry) and the packed cell volume (also referred to as a spermatocrit). For the measurement of OD, several types of photometers are available. Sperm concentration is obtained using a conversion factor or previously derived standard curve, by comparing and graphically plotting sperm cells to blood hematocrit values. Semen is aspirated into micro-hematocrit tubes and centrifuged in a hematocrit centrifuge until the sperm cells are tightly packed (10 minutes).

The percentage of packed sperm cells relative to the original semen volume in the micro-tube is then determined. Sperm concentration is also derived using a conversion factor or previously derived standard curve by comparing and graphically plotting varying ascending sperm concentrations from cytometer counts to corresponding spermatocrit readings (for detailed protocols to determine sperm concentration and the derivation of standard curves, see Ref. [24]).

13.1.2.3 Sperm viability

Traditionally, in order to assess sperm viability, stain exclusion assays are used, such as aniline blue-eosin, eosin-nigrosine or eosin-fast green. Integrity of the plasma membrane is shown by the ability of a viable cell to exclude the dye, whereas the dye will diffuse passively into sperm cells when encountering damaged plasma membranes. The stained smears can be viewed under oil immersion objective of light microscope and the percentage of live-dead sperm ratio as well as the spermatozoa with abnormal morphology can be determined. Usually 100−200 cells are counted and the percentages of live cells calculated. Under a bright-field microscope the viable sperm cells remain pearly white, while eosin will stain nonviable cells with a pink color. The aniline blue or nigrosine serves as a background to differentiate nonviable from viable sperm cells. After staining the smears it is advisable to dry them quickly using a cold-air hair drier, since in a humid environment the slides may dry slowly and the cells die before the eosin has dried. The stain works well for most avian species and the aviculturists can prepare the slides under most field conditions; however, in some species dead cells also exclude the eosin stain or fail to stain reliably. This phenomenon was found in the Budgerigar [25] and the American kestrel [21]. Therefore this staining process should be previously tested for each species. A variety of fluorescent dyes are also available such as ethidium bromide, which penetrate dead cells but not live cells [24].

In contrast to this technique, a more expensive but more sensitive method is the use of flow cytometry that sorts the live and dead cells after staining with fluorescent stains SYBR-14 and propidium iodide. SYBR-14 labels live sperm with green fluorescence, and propidium iodide labels dead sperm with red fluorescence [26,27]. The hypoosmotic swelling (HOS) test is a simple, inexpensive technique, which has been adapted to several species [28]. The HOS test predicts membrane integrity by determining the ability of the sperm membrane to maintain equilibrium between the sperm cell and its environment. Influx of the fluid due to hypoosmotic stress causes the sperm tail to coil and balloon or "swell." A higher percentage of swollen sperm cells indicates the presence of sperm having a functional and intact plasma membrane [29].

13.1.2.4 Sperm motility and mobility

Motility is one of the most important features of a fertile spermatozoon. It reflects several structural and functional competences as well as essential aspects of sperm metabolism. Sperm motility can be progressive (forward direction) and nonprogressive (random movement or oscillations). Motility as "mass motility" can be determined subjectively (hanging drop technique) [24,30] or even in a capillary tube using light microscope with low magnification, which is cheap, quick, and reliable if the technician is a well-trained person, although it is less correlated with fertility. Moreover, the rate of dilution, the type of extender (Table 13.1), the temperature, the pH and osmolality, as well as the observer can influence motility scores [37]. The percentage of sperm cells moving in a forward direction can be scored as follows: 0 = no motile sperm; 1 = less than 25% motile; 2 = 25%−49% motile; 3 = 50%−74% motile; 4 = more than 75% motile. In most cases of nondomestic birds only the latter method can be used under field condition.

The sperm mobility is a slightly different concept. Mobility defines the ability of sperm cells to move progressively against a viscous medium (Accudenz) at 41°C. With the help of the sperm mobility assay [38,39] you can measure an individual male ability to produce highly mobile sperm that are more likely to fertilize ovum than males producing less mobile sperm. This parameter is very useful for selection of the most fertile males to be used in AI of domestic species [37].

TABLE 13.1 Composition of the most frequently used diluents for poultry semen.

Chemical components	1	2	3	4	5	6
	Lake	BPSE	BHSV	ASG	LKS	LK
	Fowl	Fowl	Fowl	Turkey	Fowl	Goose
Na glutamate	1.92	0.867	2.85	1.52	1.92	1.4
Fructose	0.8	0.5		−	0.8	0.2
Glucose	−		0.5	0.6		0.7
Inositol			0.25			0.7
Mg acetate	0.08		0.07	0.08		
Mg chloride		0.034				
K acetate	0.5		0.5	−	0.5	
Na acetate		0,43				
K citrate	−	0.064		0.128		0.14
K monophosphate		0.065				
K hydrogen phosphate		1.27				
Na dihydrogen phosphate						0.21
Disodium hydrogen phosphate						0.98
Protamine sulfate					0.032	0.02
BES	−			3.05		
TES		0.195				
Polyvinyl pyrrolidone	0.3			−	0.3	0.1
NaOH (1 N)	−			5.8 mL		
pH	7.1	7.5		7.1	6.85	7.8
Osmolality (mOsm/kg)	340	333		325		380−400
References	[31]	[32]	[33]	[34]	[35]	[36]

The figures are in g constituent/100 mL unless otherwise noted.

13.1.2.5   Other functional tests

Usage of other tests can be important in the case of frozen-thawed avian sperm assessment. For the location of species-specific mitochondria and the evaluation of mitochondrial activity two fluorescent probes can be used: Rhodamine 123 (R123) and MitoTrack that have been successfully used in many species. R123 is a potentiometric membrane dye, which selectively stains functional mitochondria; therefore unstained cells do not contain functional mitochondria [40]. Similarly, by adopting MitoTrack, functional mitochondria can be selectively labeled [41].

The acrosome of the spermatozoa is an acidic secretory organelle filled with hydrolytic enzymes. The role of this structure is the maintenance of spermatozoa ability to penetrate and fuse with the ovum plasma membrane. Acrosome integrity can be effectively evaluated using peanut agglutinin, the more reliable fluorescence probe for this purpose in avian sperm. Using this lectin, acrosome-intact sperm cells will not bind the lectin and thus can be distinguished from acrosome-reacted sperm cells that will bind the lectin [42].

13.1.2.6   Multifunctional tests

The technique "sperm−egg interaction assay" [43] has also been used in domestic and nondomestic bird research. The assay uses the chicken egg inner perivitelline layer (IPVL) to detect sperm activity (motility, penetration, and acrosome reaction at the same time). The number of holes made by the sperm cell per unit area of IPVL is viewed under dark-field optics at $100 \times$ magnification. The IPVL samples for the assay are separated from the outer perivitelline layer (OPVL) by acid hydrolysis. The assay can be used for ranking individual males on the basis of their semen quality and is especially recommended for monitoring the quality of stored semen [43,44]. Although the use of the chicken IPVL assay has been explored for some endangered avian species as well, major modifications are required, since some avian species unrelated to the Galliformes are unable to penetrate chicken IPVL [18,45].

The more advanced and objective method of semen assessment is the computer-assisted sperm analysis (CASA), for which a number of products are nowadays available. This system allows the assessment of many motility parameters of individual spermatozoa [46], which are better correlated to the efficiency of assisted reproductive technologies [47]. Blesbois et al. [48] proved that parameters assessed with CASA are correlated with fertility of frozen-thawed chicken spermatozoa. Another important advantage is the immediate measurement of sperm cell concentration, total number of sperm cells in the ejaculate, morphological traits, live and dead cell ratio, and the automated calculation of number of insemination units from one ejaculate. Nevertheless, CASA systems need standardization and validation before assessment [47]. Unfortunately, the type and depth of the used chamber, the number of fields analyzed, the temperature during analysis, and the protocol of semen preparation can all influence the results. Only after the complete optimization of technical settings, the comparison between intra- and interlaboratory results can be meaningful, regardless of the instruments employed [49].

13.1.3   Short-term or "liquid" semen storage

Without dilution, avian semen can be stored without significant decrease in sperm quality only for less than 1 hour. The short-term storage (at 4°C−5°C for 6−48 hours) of avian semen can be used for transport or assessment of semen quality in vitro. Diluents are buffered saline solutions based mostly on the composition of seminal plasma and used to extend the concentrated semen, maintaining the viability of spermatozoa and maximize the number of females that can be inseminated. However, it is important to note that an amount higher than 30%−40% of spermatozoa considered viable after 24-hour storage is considered "unfit" to survive in the oviduct [50]. It means that conventional viability tests are unable to detect some kinds of damaged sperm cells. Therefore, in order to reach high fertility rate for gene conservation purposes, the use of fresh semen is a better choice than liquid stored semen.

There are many diluents available for various poultry semen, both laboratory recipes and commercially available products [51,52]. There is no standard diluent either for domesticated or nondomesticated species, but the most prominent anionic constituent of avian seminal plasma, glutamic acid, is a standard component of avian diluents. The most important factors are the ones that will be instrumental in maintaining the correct range of pH (6.5−7.3), osmolality (250−460 mOsm/kg) and provide energy source (glucose, fructose). From the 1970s until now, several varieties of poultry semen extenders have been developed, and among them the compositions of the most commonly used ones are shown in Table 13.1.

Important to the development of diluents and storage for poultry semen is the physiological differences and metabolic requirements of spermatozoa of different species. For example, fowl spermatozoa can tolerate both aerobic and anaerobic environments in vitro, while turkey spermatozoa require high level of oxygen to survive [53,54].

Generally, the usage of poultry semen extenders for nondomestic species is a good start, but the aviculturist needs to adjust these conditions to the requirements of the species of interest. Therefore several extenders used with domestic poultry may support most avian semen samples (Table 13.2). Adjusting the pH and osmolality of the domestic species as a buffer to the ejaculated pH and osmolality of the nondomestic species may improve in vitro sperm quality. Specific diluent for ratites was developed on the basis of the macro-mineral (OS1, [106]) and micro-mineral (OS2, [107]) composition of ostrich seminal plasma. In the case of emu, the short-term storage at 10°C proved to be more effective for

TABLE 13.2 Most commonly used extenders, dilution rates, and precooling time in different poultry species.

Species	Extenders	Dilution rate	Precooling	References
Domestic fowl	Lake	1:2–3	5°C, 15–45 min	[55–63]
	Tselutin	1:1–2	5°C, 20 min	[35,57,64,65]
	Modified Lake	1:2–3	5°C, 20–30 min	[66,67]
	BPSE	1:2–5	5°C, 15–120 min	[68–71]
	BHSV	1:2.5–4	2°C–5°C, 25–30 min	[69,72,73]
	HS-1	1:1–2	5°C, 30–60 min	[74,75]
	HEPES	1:2		[76]
	Minnesota Avian Extender		5°C–6°C, 45–60 min	[77,78]
	Nabi	1:3	4°C, 120 min	[79]
Turkey	BPSE	1:2–5	4°C–5°C, 5–120 min	[80–82]
	Lake	1:1–5	5°C, 5–30 min	[34,83]
	Macpherson's Extender	1:5	5°C, 5 min	[83]
	Blumberger	1:2–4	5°C, 30 min	[84–86]
	IGGKPh	1:4	2°C–5°C, 40 min	[87]
	Tselutin	1:1–4	4°C, 25–60 min	[88–90]
	ASG	1:1	5°C, 15 min	[34]
Guinea fowl	IGGK	1:3	4°C, 15 min	[91]
	Lake	1:1	3°C, 25 min	[92]
	Tselutin	1:2	2°C, 5 min	[92]
	BHSV	1:2	4°C, 15 min	[93]
Goose	Tselutin	1:3	2°C–4°C, 40 min	[35]
	B-26	1:1	10°C–20°C, 10–20 min	[94]
	EK	1:0.5–2	4°C–5°C, 5–15 min	[36,95–98]
	IGGK	1:2	5°C, 10 min	[99]
	Tai	1:2–3	10 min, ambient	[99,100]
Pekin duck	BPSE	1:1	Not specified	[101]
	Tselutin	1:4	–	[35]
	IGGKP	1:1	120 min	[102]
	Lake	1:4	5°C, 10 min	[103]
Cairina moschata	Blumberger	1:1–2	–	[104]
	HIA-1	1:3	–	[105]
	AU	1:3	–	[105]

sperm survival than at 5°C [45]. Spermatozoa of guinea fowl semen, extended with a phosphate buffer, have retained motility for up to 10 days [108], but for most birds the spermatozoa survive for no more than 24−48 hours. According to the available literature and experience, although sperm cells may survive for many hours, fertility with fresh semen usually exceeds that of stored semen, and antibiotic treatments in the extender may help maintain sperm cells in contaminated samples [109,110]. It is advised to avoid vaginal insemination with contaminated samples, while instead placing the semen in the cloaca near the opening to the oviduct [5].

13.1.4 Long-term semen storage—cryopreservation of avian spermatozoa

The investigations of bird semen cryopreservation began a few decades ago. The first experiment to freeze poultry semen was made in 1939, although it was unsuccessful [111]. Polge et al. [112] worked on semen freezing using fructose as a cell protector, but the real key event, however, came from the accidental discovery of Polge himself [113], revealing the highly active cell protective effect of glycerol against freezing. More intensive studies on semen freezing of birds began only in the 1980s [55,114−117]. It has been proven that avian spermatozoa are much more sensitive to deep-freezing than sperm cells of mammalian species, due to differences in sperm membrane structure, low levels of cytoplasm, and other physiological and reproductive features [118]. A different tolerance of avian spermatozoa from different species is known and therefore, in the development of freezing protocols it should be taken into account that different species require different semen diluents, cryoprotectants, and cooling rates [119].

In spite of a wide literature focused on deep-freezing of avian sperm, nowadays there is yet no standardized method for cryopreservation. Efficiencies can be very different, not only among but also within species as well as among laboratories. The examination of semen quality before and after freezing is necessary to achieve adequate fertility. As a result of prefreezing sperm cell quality control, individuals producing poor semen should be excluded from production.

Steps of sperm cryopreservation:

1. sperm dilution
2. precooling
3. addition of cryoprotectant and equilibration
4. packaging systems
5. freezing
6. thawing

13.1.4.1 Diluents and precooling

Many protocols for preparing diluents for avian semen are nowadays available. Semen dilution is generally carried out at room temperature in two steps. As a first step, a diluent without cryoprotectant is always added to the samples. The dilution may be performed in fixed dilution volume or fixed sperm cell concentration. The dilution rate depends on the breed, and among poultry species, turkey and goose have the most and the least concentrated semen, respectively. In Table 13.3, the most common diluents and dilution ratios in poultry species have been summarized. As a second step, diluted semen is placed in a cold cabinet at 3°C−5°C for precooling and to restore the homeostasis of spermatozoa. The duration of this step depends largely on the amount of samples. According to the literature, it can be from 10 minutes to 2 hours. From this point of the cryopreservation protocol, all manipulations should be performed in a cold room or cold cabinet with chilled devices (Table 13.2).

The cryoprotectant used in the freezing protocol is very important for a successful avian semen cryopreservation [131]. Among the penetrating cryoprotectants, the most commonly used is glycerol (Table 13.3), although its use is more difficult in birds because on the one hand, it is the most effective and least toxic cryoprotectant [57,58], while on the other hand it has a contraceptive effect in the avian vagina. Therefore, in thawed semen samples the amount of glycerol should be reduced to less than 2% by centrifugation (700 g for 15 minutes at 5°C). Several cryoprotective agents are used to replace glycerol during avian sperm freezing [(dimethyl sulfoxide (DMSO), dimethyl acetamide (DMA), dimethyl formamide (DMF), ethylene glycol (EG), methyl acetamide (MA)]. If a comparison has to be made with regard to the efficacy of different compounds for their cryoprotective activity, or for their toxicity toward spermatozoa, several factors should be taken into consideration: species of birds, cryoprotectant concentration, equilibration temperature and time, freezing rate, freezing methods, and postthaw treatment [132]. In the case of many nondomestic species, a small amount of ejaculate is usually available, and therefore the use of DMSO and DMA is the most obvious choice. Cryoprotectant agents are usually added to precooled semen, directly in the sample, or as a second step, samples can be diluted with an extender containing cryoprotectant. In Table 13.3, the most commonly used cryoprotectant agents in poultry are shown.

page 200 Reproductive Technologies in Animals

TABLE 13.3 Cryoprotective agents and equilibration times for different poultry species.

Species	Cryoprotective agent	Amount of cryoprotectant (%)	Equilibration time with cryoprotectant (min)	References
Domestic Fowl	Glycerol	7 – 20	10 – 60	[56–58,61–63,65,77–79,120,121]
	DMA	4 – 9	0 – 30	[35,55,57,59,60,62,64,66,67,69,72,74,121,122]
	DMSO	3 – 8	0 – 60	[65,68,70,71,74,76,123]
	DMF	6.5 – 7.5	0 – 25	[57,73,74,124]
	EG	5.6 – 8	30 – 45	[69,125]
	MA	7.5 – 12	–	[74,126,127]
	MA + DMF	6.5 + 6.5	–	[75]
Turkey	Glycerol	9 – 11	5 – 30	[34,83,128]
	DMA	3.3 – 26	0 – 40	[34,81–83,85–90]
	DMSO	4 – 10	0 – 120	[80,90,114]
	EG	4.8 – 8	6 – 30	[84,85,114,128]
Guinea fowl	DMA	6	0 – 2	[91,92]
	DMF	6	2 – 10	[93]
	EG	10	0 – 25	[92]
Goose	Glycerol	11	8 – 10	[99,129]
	DMF	6 – 9	5	[36,95–98]
	DMA	4 – 9	0 – 5	[35,99,129]
Pekin drakes	Glycerol	2 – 8	15	[101]
	DMA	5 – 6	0 – 40	[35,103]
	DMSO	1 – 10	15 – 120	[101,102]
Cairina moschata	Glycerol	3 – 7	30 – 60	[105,130]
	DMSO	3 – 10	60	[105]
	EG	6 – 7	30	[104]

13.1.4.2 Packaging systems

Before freezing, semen must be packed in 0.25 or 0.5 mL straws. In our experience, the use of 0.5 mL French medium straw for goose and guinea fowl semen, and 0.25 mL French mini straw for other species, is the preferred choice. Earlier, polyvinyl alcohol or carbon was commonly used, whereas more recently, a manual sealing unit for straws is widely adopted. Straws allow a more complete delivery of semen during insemination, and probably the most important advantage is given by a more uniform control of the freezing and thawing process, leading to improved sperm cell recovery. The major disadvantage of the straw system is the vulnerability related to mishandling. Straws as packaging system have many benefits such as sperm traceability and the safe transport of semen for breeding or storage in gene banks [90]. In addition, cryovials for freezing can also be used.

13.1.4.3 Freezing methods

13.1.4.3.1 Dynamic programmable freezing

Programmable freezing units are nowadays amply employed and sold by a number of companies around the world, characterized by the automatic control in the rate of cooling. Within programmable freezing, one-step and two-step programmed slow and fast freezing protocols can be distinguished (Table 13.4). During programmable freezing, the

TABLE 13.4 Freezing methods for different poultry species.

Species	Freezing method	Packaging system	Motility (%)	Viability (%)	Fertility (%)	References
Domestic Fowl	Dynamic programmable freezing	0.25 mL straw	17.6−43		10.8−85	[59,62,77,123,124]
		0.5 mL straw			2.8−92	[57,61,70,121]
		Cryovial			41−82	[55−57,64,69,72,73]
	Static freezing in nitrogen vapor	0.25 mL straw	31.1−46.6		1.9−87.6	[60,67,76,78,79,125,133]
		0.5 mL straw			5.38−88	[65,74,126,127]
	Pellet		1.2−24.7	33.6	4.2−94	[35,60,62,66,71,123,133−135]
Turkey	Dynamic programmable freezing	0.25 mL straw			2.4−32.6	[34,83,128]
		0.5 mL straw			26.7−63.9	[58]
		Cryovial	40	48−80	0−55.7	[80,81,84]
	Static freezing in nitrogen vapor	0.25 mL straw		22.8−47		[82,90]
	Pellet			32.4	0−95.6	[58,86,88,89,114]
Guinea fowl		0.5 mL straw		38	71	[91,93]
		Cryovial			29.1	[92]
	Pellet				63.6	[92]
Goose	Dynamic programmable freezing	0.25 mL straw		10.53−74.4	7.8−92.3	[36,95,96,99,129,136]
		0.5 mL straw		40		[97]
		Cryovial		35.5−41.1	67.5−85.1	[94,97,98]
	Static freezing in nitrogen vapor	0.25 mL straw		14.3		[129]
		Cryovial		36.9−37.8		[97,98]
	Pellet				81−91	[35]
Pekin drakes	Dynamic programmable freezing	0.25 mL straw	21.7−26.04	1−35		[101,103]
	Static freezing in nitrogen vapor	0.25 mL straw		5−32.5		[101,103]
		Cryovial	58		75.1−83.6	[35,102]
	Pellet			25.8		[103]
Cairina moschata	Dynamic programmable freezing	Cryovial			56.2−80.3	[104]
	Pellet		8.25−41.4			[105,130]

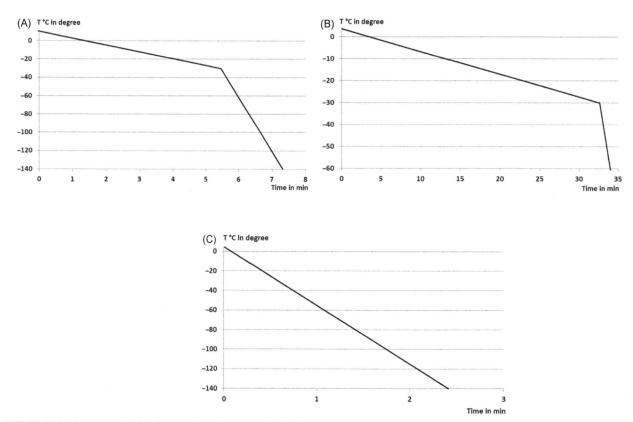

FIGURE 13.2 Programmable freezing rates for (A) domestic fowl [61], (B) guinea fowl [92], and (C) goose [36].

cooling rate and the final freezing temperature can also be specified. In Fig. 13.2 examples of different programmable cooling rates are shown. The optimum cooling rate varies from species to species, and in Table 13.5 the cooling rates used for different species are listed. The freezing methods were first developed for the chicken and then adapted to other bird species [138]. Lake and Stewart [56] reported the first successful slow freezing-thawing method. Since then, many studies have been done to improve the efficiency of freezing protocols that can be adjusted to the nondomestic avian species as well. In nondomestic birds, rapid cooling was detrimental to the sperm of *Bonelli*'s eagle, golden eagle, and peregrine falcon, but it was advantageous though when cryopreserving imperial eagle (*Aquila adalberti*) and chicken sperm [139].

13.1.4.3.2 Static freezing in nitrogen vapor

Samples, usually in plastic straws, are placed in nitrogen vapor at a given distance above the liquid nitrogen and held there for different times before plunging into liquid nitrogen. Because the cooling rate depends on the distance from the nitrogen level, it is more difficult to control the rate of freezing. With this method (Fig. 13.3), liquid nitrogen is filled into a polystyrene box and the straws can be placed even on a tube rack. The great advantage of this method is that it does not require expensive equipment and needs less liquid nitrogen. More comparative studies have been carried out with sperm samples at different liquid nitrogen levels for each poultry species during static freezing method [60,67,76,78,79,82,90,101,103,125,129,133].

13.1.4.3.3 Pellet method

In the pellet method, the diluted and cryoprotected semen is pipetted directly into liquid nitrogen with a droplet size of $10-200\ \mu L$. The globules are then collected from liquid nitrogen with a tweezer and placed in a cryovial (Fig. 13.4).

Also in this case, expensive freezing equipment is not needed and it is easily adaptable to field conditions. For thawing sperm it is required to have an apparatus with a temperature of $60°C-70°C$ (see Section 13.1.4.4). The disadvantage of this method is that it is difficult to keep track of the amount of insemination dose in a vial and that up to 25% of a given sample may be lost during the freezing and thawing process. According to the experience of other researchers,

FIGURE 13.3   Static freezing in polystyrene box.

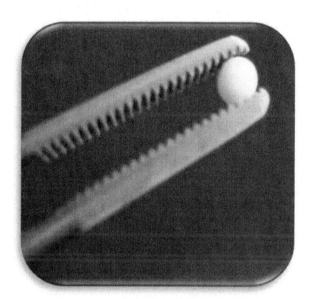

FIGURE 13.4   0.25 µL droplet made by pellet method.

storage in the pellet form is not recommended for long-term gene preservation, due to pellet shrinkage (personal communication). In contrast, for short-term storage, it can be easily applied to some wild bird species under field conditions. In Table 13.4, the results of various freezing methods in different poultry species are shown. With different freezing/thawing methods, the duration of thawing depends on the type of cryo container and the thawing temperature.

13.1.4.4   Thawing procedure

During sperm preservation, cell survival is affected by thawing at least as much as deep-freezing techniques. Choosing the right thawing rate avoids the re-formation of ice within the cells and the osmotic conditions in the solution are resolved without damaging the cells. It has been observed in semen of several species that the combination of fast cooling and slow thawing is particularly damaging to the cells [140−143]. It is generally accepted that slow freezing requires slow thawing, whereas fast freezing of cells requires faster thawing [144]. Frozen semen samples may be thawed in alcohol [110] or water bath [78,145], but in the pellet method a special thawing unit can be used [58,98]. In fact, pellets need to be filled into a warmed funnel placed within the apparatus, which at the same time thaws and flows out of the funnel. Originally, [58], the funnel was heated by surrounded water at 60°C. In Fig. 13.5 a thawing apparatus in which the funnel is not heated by the water but by the funnel itself is shown.

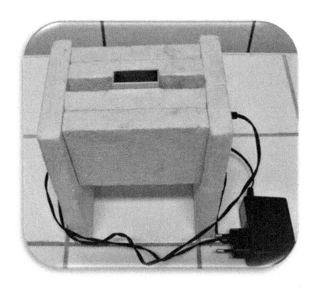

FIGURE 13.5 Special thawing unit for pellet method.

13.1.5 Insemination

For an optimal insemination technique, it is necessary to have a good knowledge of the reproductive anatomy and physiology of the female bird, as reviewed by Bakst et al. [146]. It is well known that birds commonly have a single functional ovary and oviduct on the left side of the body; therefore the insemination procedure needs to be directed to the left side of the cloaca. For a successful AI, it is important to rely on trained personnel, familiar with the specific anatomical features of the given avian species and with the skill to locate the vaginal orifice, since in most avian species the insemination is carried out by intravaginal route.

There are different ways of insemination in nonconditioned and conditioned birds. The steps of inseminations in the case of noncooperative species involve female restraint (legs and wings) and keeping it in such position so that vaginal eversion is possible. In cases where the eversion is not possible, the use of a vaginal speculum is needed (Fig. 13.6).

There are three ways to perform the insemination: cloacal, intravaginal, and intramagnal routes. (1) Cloacal insemination is the easiest and less stressful method, but fertility is lower compared to the intravaginal route. Nevertheless, with increased sperm cell dose and insemination frequency the fertility can be improved in some species [147]. (2) Certainly, an intravaginal approach is used more often in birds by manually everting the cloaca and by applying a stimulation massage on the back with gentle and continuous pressure to the lower abdomen, on the left side, to expose the oviduct. If the vaginal orifice is too deep in the cloaca, such is the case in larger birds (goose, ostrich, emu, and crane), then the palpation method is required with a sterile pipette or finger. (3) The intramagnal insemination is the most invasive method requiring anesthesia, with some development by the use of an endoscope [139]. The disadvantage of the latter method is the limitation of in vivo selection of unfit spermatozoa by the female and the higher risk of contamination, which increases embryonic mortality [139,148].

Inseminations of properly trained, imprinted females which voluntary explicit copulatory postures and oviductal exposure are free of stress, easier, and more comfortable for both the birds and the handlers. The only limitation of the method is the long time for training birds. A detailed description of the training of imprinted birds such as raptors, storks, cranes, some waterfowls, and ratites is available in literature [8,149−151].

The ideal time of AI depends on the season when the hens are sexually active just before and/or during egg production. The abdominal distension and the distances between the pubic bones are common indicators for the timing of AI. The frequency of inseminations depends on the sperm storage capacity, that is, the length of the fertile period of the female, which in most species is not known. Generally, the smaller the body mass, the shorter the duration of fertility. Typically, 5−10 days intervals between inseminations are used routinely in poultry industry; however, it is highly variable in nondomestic birds. Birds with a short duration of fertility require more frequent inseminations [137,152,153]. Recommended insemination frequency varies from once every other day in Japanese quail [154] to once every other week in turkeys [155]. Chinese pheasants inseminated with fresh semen 2−3 days per week during the season produced excellent fertility [156]. Some birds need the first inseminations before the onset of egg production, turkey for example, among the domesticated species. Generally, among nondomestic birds larger species are inseminated twice a week during the 2 weeks prior to the onset of egg-laying, while smaller ones are inseminated three times during the week before oviposition begins. During the laying

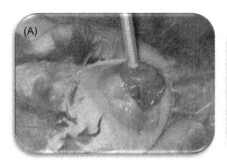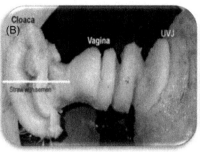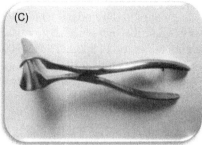

FIGURE 13.6 Intravaginal insemination. (A) Everted vagina with insemination catheter (*Gallus Gallus domesticus*). (B) Preparation of the hen's cloaca, vagina, and uterovaginal junction where the sperm storage tubules are located. (C) Vaginal speculum helps to find the orifice of the vagina in cases where the eversion of vagina is not possible. *(B) Photo by M. Bakst.*

TABLE 13.5 Frequency of AI and sperm concentration in poultry species (based on table of Bakst and Brillard [137]).

Species	Inseminations	Sperm number ($\times 10^6$)		AI frequencies	
		First half of laying season	Second half of laying season	At onset of reproductive season	During season
Turkey	14 days postphotostimulation	150–200	200–300	3 AI within 10 days	1/week
Domestic fowl	15%–20% lay				
Guinea fowl	15%–20% lay	80–100	100–120		
Duck (Muscovy male × Pekin female)	25%–30% lay	70–100	100–120	2/week	
Goose	30% lay	70		2/week	

season, AI should be performed as soon as possible after each oviposition, while AI just prior to oviposition has been related to a decreased fertility, since most of the spermatozoa are expelled during egg-laying and fewer sperm cells reach the sperm storage tubules (SSTs) in the uterovaginal junction of the oviduct [8,157].

As mentioned earlier, semen volume and sperm cell concentrations can vary between species but most birds produce a sperm cell concentration that ranges from 2 to 10×10^9/mL. Most nondomestic bird ejaculates contain more spermatozoa than what needed for one insemination [158] and the ejaculate can therefore be divided for multiple inseminations. The minimum amount of sperm cells appears to be about 1 million for the American kestrel [17], 16 million for cranes [158], 150 million for chickens and turkeys, 80 million for guinea fowl, and 70 million for ducks [137]. In geese the inseminated sperm cell doses vary between 9 and 40 million of motile spermatozoa weekly with good fertility results [159]. Usually, in the poultry industry and presumably in wild birds and ratites the inseminated sperm cell concentration has to be increased following the peak production period since the sperm cell storage capacity of SSTs decreases, while the sperm cell release from the SSTs increases in the second half of the production period [152,160,161]. Repeated ejaculations decrease the sperm cells availability and can exceed the rate at which the bird replaces cells (Table 13.5) [162].

13.2 Other assisted reproductive techniques in birds

13.2.1 In vitro fertilization and embryo transfer

Other reproductive technologies in avian species are not used in the everyday practice. The first attempt to carry out in vitro fertilization (IVF) in domestic fowl was in 1970 by Howarth [163] in order to find direct evidence that avian spermatozoa do not need a period of capacitation within the hen oviduct for fertilization. His method was simple: for the successful collection of ova, the knowledge of the exact time of oviposition is important since ovulation of the succeeding ovum usually occurs within 30 minutes after oviposition [164]. The hens were sacrificed 30 minutes after oviposition and the ova obtained from the infundibulum, cultured in modified Ringer solution individually in a plastic shell container for 24 hours. Then, 100 μL of freshly collected and diluted semen was placed directly over the germinal disc (GD). The plastic shell with the ova was then placed in 150 mL beaker and incubated at 41°C for 15 minutes (fertilization time). Afterward, the Ringer solution was removed and freshly collected albumen was added to the beakers to cover the ova. An air space of 15 mm was left between the albumen and the top of the beaker. A Petri dish was used to cover the beaker and the fertilized ova were incubated again at 41°C for 24 hours in order to check the fertilization by examination of blastodiscs. By this simple in vitro system, an embryonic development until the blastula stage could be achieved. Later, Nakanishi et al. [165] demonstrated the formation of female and male pronuclei by IVF of chicken oocytes and found that pronuclear formation during the first 1−4 hours does not differ between in vivo fertilization and IVF. Tanaka et al. [166] were the first to obtain viable chicks by surgically transplanting IVF ova into the oviduct of recipient hens followed by incubation of the eggs. The disadvantage of the above method is the inevitability of surgical intervention. The oocytes used for IVF were ovulated and removed from the infundibulum of the donor oviduct.

Perry [167] was the first who elaborated a complete culture system for the chick (*Gallus domesticus*) embryo. He used in vivo fertilized eggs and cultured them in three different culture media from 0 to 22 days of hatching, according to the different developmental stages (first day—blastoderm formation; 2−4 days—embryogenesis; 5−22 days—embryonic growth). Later Li et al. [168] tested the IVF techniques using various semen volumes, different times of ova collection and reproductive stages of recipient hens. These developed methods are nowadays used for gene transfer or other gene manipulations in poultry.

Various successful investigations were made with regard to culturing chicken embryos from freshly laid eggs in surrogate eggshells [169]. They investigated the effect that a moving apparatus has over embryo cultures and compared shell-less culture to surrogate eggshell. The latter was found more efficient and it was proved that the shape of the eggshell plays an important role in the normal development of the chorioallantois of the embryo, while at the same time the surrogate eggshell does not have to be species-specific. Later Li and Qi [170] successfully cultured embryos of the Peking duck in surrogate eggshell.

13.2.2   Intracytoplasmic sperm injection

Establishment of the intracytoplasmic sperm injection (ICSI) method in birds is advantageous for studying the mechanism of fertilization and the role of gametes, and also for protection of endangered species. The use of sperm cell as a vector to carry foreign DNA to the oocyte may help the production of transgenic birds as well. Despite the fact that the avian fertilization is polyspermic, the technique of ICSI proved first that intracytoplasmic injection of a single sperm into a quail ovulated ovum could activate the GD and initiate embryonic development.

A short description of the steps of the basic technique according to Hrabia et al. [171] is as follows:

1) To obtain ova, in the case of spontaneous ovulation, the females have to be killed 30−60 minutes after oviposition, and oocytes taken from the infundibulum or the upper part of the magnum. Whereas, in case of in vitro ovulation quail have to be subcutaneously injected with a LHRH 12 hours before expected ovulation and killed 9−11 hours after the treatment.
2) Washed sperm cells can be diluted with 12% PVP in Dulbecco's modified Eagle medium (DMEM) and kept until injection at room temperature.
3) To prepare pipettes for ICSI, glass capillary tubing is drawn with a pipette puller and the tip of the pipette is broken with a microforge such that the inner diameter at and near the tip is approximately 10 mm. Next, the tip of the pipette is beveled at a 30-degree angle with a micropipette beveler, wetted in 70% ethanol, and finally dried. Then the injection pipette has to be connected to a micromanipulator and half-filled with silicon oil, and a few microliters of 0.9% NaCl sucked into the pipette.
4) Immediately before injection, a single sperm cell has to be isolated into the injection pipette under the microscope. The ovum is placed into DMEM in a plastic dish, and the sperm or spermatid is injected into the central area of the GD using a micromanipulator under a stereomicroscope.
5) The injected ova can be cultured in plastic dishes in DMEM containing 3 mg/mL bovine serum albumin (BSA) and antibiotic-antimycotic solution (with 10,000 units penicillin, 10 mg streptomycin, and 25 mg amphotericin B/mL) for 24 hours at 41°C under 5% CO_2, in air. The blastoderms after ICSI and 24 hours culture can reach stages II−VII (see Ref. [172]).

Later, a slightly modified method was used successfully, confirming that a single sperm cell is capable of activating the oocyte even when quail oocyte is injected with chicken spermatozoa [173]. To improve fertility rate, Shimada et al. [174] used phospholipase Czeta and inositol triphosphate, reaching an increase in fertility up to 85%, and to extend embryonic development, chicken eggshell was used as a surrogate culture at 37°C after 24 hours of incubation at 41°C under 5% CO_2, in air. Kang et al. [175] tested the fertilizing capacity of cryopreserved spermatozoa using ICSI and proved that quail embryos could develop in surrogate eggshell culture even up to 11 days of incubation.

13.3 Gonadal tissue transplantation in domestic chicken

Most of the reproductive biology techniques developed so far in birds to preserve genetic material have focused on sperm and embryonic cells. The manipulation of the female gametes, mainly because of the huge amount of vitellus as well as due to the special physicochemical traits of the egg, was not at the center of attention. The structure of the immature ovary, which is removed at the age of one day, is suitable for cryopreservation, it can be cut, and the stored tissue pieces can be transplanted. Freezing and transplanting of testis of 1-day-old chicken can be useful in those special cases, where semen cannot be obtained, so that sperm cells cannot be preserved in the frozen state. Transplantation of gonadal tissues is not yet widespread, and the technique is being developed and improved. Not all genotypes are suitable recipients among the domestic chicken breeds [176,177]. Methods for developing appropriate donor/recipient combinations are still in progress. So far, the ratio of adherence of gonadal tissues in successful pairs is 70%−80% and the proportion of donor-derived progeny from them is 33%−50% [176,178]. A slightly different method was used earlier successfully in Japanese quail [179−181] and duck [182]. The ratio of donor-derived offspring was 71% and 25%, respectively, in these species.

The following is a detailed description of the gonadal tissue grafting of domestic chicken. Certainly, the presented method can be applied in the case of nondomestic birds as well, with modification according to the requirements of the given species. Unless otherwise stated, all chemical materials used were acquired from Sigma.

13.3.1 Preparation and vitrification of donor tissues

The 1-day-old chicks are euthanized by cervical dislocation, feathers are removed, and the abdominal region is cleansed with 70% ethanol. Then, a transversal, 2−3 cm long incision should be done on the left side, and through that, the yolk sac can be carefully removed. The next steps should be done under sterile conditions and in a laminar flow box. The lengthening of the incision is followed by pushing the gastrointestinal tract aside, so the gonads become accessible. In the first step, the air sacs and serous membranes are opened, so that the ovary or the testes can be removed in caudocranial direction with fine forceps or microsurgical scissors (Fig. 13.7).

The donor organs are placed on ice, in antiseptic Petri dishes with Dulbecco's phosphate-buffered saline (DPBS) and 20% fetal bovine serum (FBS), for a maximum of 25−30 minutes. Depending on the size, organs can be halved, and thereafter three of these uncut or halved gonads are put into each human hand acupuncture needle (Fig. 13.8) (0.18 × 8 mm—Dongbang Acupuncture Inc, Korea).

The needles are placed into two different vitrification solutions in sterile Petri dishes for 5 minutes each at room temperature [183]. Wang et al. [184] treated the donor organs in two solutions for 10 and 2 minutes, respectively. The content of the two vitrification solutions is as follows: the first one is DPBS and 20% FBS with 7.5% DMSO and 7.5% EG and the second one is DPBS and 20% FBS with 15% DMSO and 15% ethylene glycol (EG) plus 0.5 M sucrose. The equilibration is achieved by plunging the needles into liquid nitrogen, and then putting them in labeled 1-mL volume cryovials (Labsystem Comp., Hungary). The cryovials are closed under liquid nitrogen with a long forceps and then placed into nitrogen tanks for long-term storage.

13.3.2 Thawing of donor organs

Immediately prior to implantation, needles containing organs are placed following three steps in three different thawing solutions under sterile conditions for 5 minutes each at 38.5°C [183], whereas Wang et al. [184] reduced the temperature at 37°C. It is important that thawing is done as quickly as possible to maintain the integrity of the cells. At least 3 mL of sterile-filtered solution at stable temperature is required. Filtration is performed in a laminar flow box by using a microfilter and a syringe, and the solutions are placed on a fixed-temperature heating plate in labeled vials, which are closed with sterile caps until use (Fig. 13.9). The composition of the solutions, in the order of application, is the

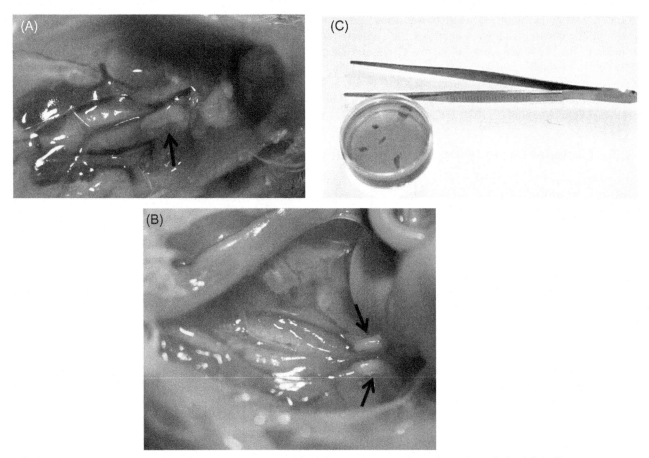

FIGURE 13.7 Location of ovary (A) and testes (B) in donor day-old chicks as well as the collected organs in sterile Petri dish (C).

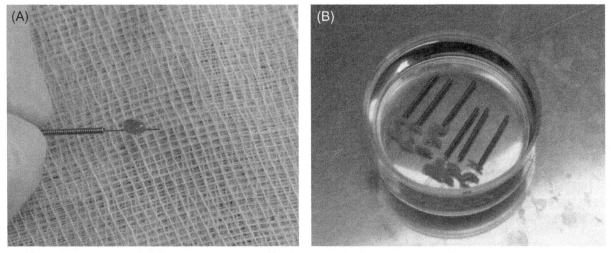

FIGURE 13.8 Gonads are put into acupuncture needle (A), and then they are transferred in vitrification solution (B).

following: DPBS and 20% FBS with 1M sucrose, DPBS and 20% FBS with 0.5 M sucrose, and finally DPBS and 20% FBS with 0.25 M sucrose. Storage until intervention is done in DPBS and 20% FBS for up to 1 hour at 0°C. Each solution should be freshly prepared and sterile filtered before use. Following thawing, the organs are cut into 2−4 pieces under a stereomicroscope (20 ×) using an iris scissor.

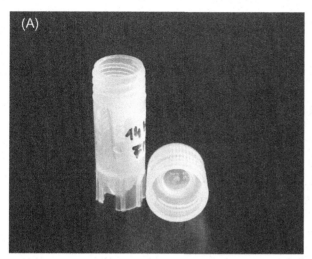

FIGURE 13.9   Gonads on needle in a cryovial (A). Vitrification solution on heating plate in sterile box (B).

13.3.3   Surgical transfer of gonads

13.3.3.1   Preparation of the recipient animals

Anesthesia of recipients is induced by administration of 0.1 mg xylazine (Narcoxyl 2) and 0.5 mg ketamine (CP-Ketamin 10%)/chick intramuscularly. Once the feathers are removed, the abdominal area is cleansed with 70% ethanol. The chick is placed in recumbent position on a heating pad. During the operation, anesthesia is maintained by a mask containing isoflurane (Forane). The opening of abdominal area is accomplished by a 2–3 cm long transversal incision, which is continued in 1 cm length to the last rib. The yolk sac can be seen through the peritoneum and removed. The gastrointestinal tract is gently pushed aside, so the genital organs can be accessed as shown in Fig. 13.10. In 1-day-old chicks, the ovary is located in the left part of the abdominal cavity, enclosed by air sacs, the left kidney, and the mesentery of colon. The ovary is nearby the abdominal aorta and vena cava and therefore, to avoid excessive bleeding, the ovariectomy should be done precisely, in smaller pieces, in caudo-cranial direction [185]. Then, the place of the removed ovary is cleansed with sterile cotton. Removal of testes is carried out similarly as done for the ovary [176,178,181,186,187]. The chicks should be kept in a heated room, under infrared heat lamps during the whole pre-, peri-, and postoperative care.

13.3.3.2   Transplantation

The prepared donor gonadal tissue is grafted as close as possible to the anatomical location (Fig. 13.10). For fixation, the transplanted organ is covered by the parietal layer of the abdominal air sac. The abdominal incision is closed with two layers of polyglactin suture (Safil 3.0). In case of transplanting testes, the procedure is the same, with the exception that the donor tissue is positioned under the mesenterium.

13.3.4   Pre- and postoperation treatments

13.3.4.1   Busulfan treatment of eggs

If the recipient's own gonad is not removed surgically, there is a possibility to hinder its development with busulfan (1,4-butanediol dimethanesulfonate) during the incubation [188]. Eggs should be incubated horizontally for 24 hours and injected with the mixture of sesame oil, DMF, and busulfan into the yolk through the blunt end of the egg. To avoid the fractioning of the solution in the syringe, Realsonic Cleaner (AA Labor Ltd., Hungary) ultrasound sonar can be applied for 20 minutes, which renders the oil droplets smaller and the solution more stable. To allow the transmission of the active agent to the embryo, the eggs should be incubated for further 24 hours without turning [176,189].

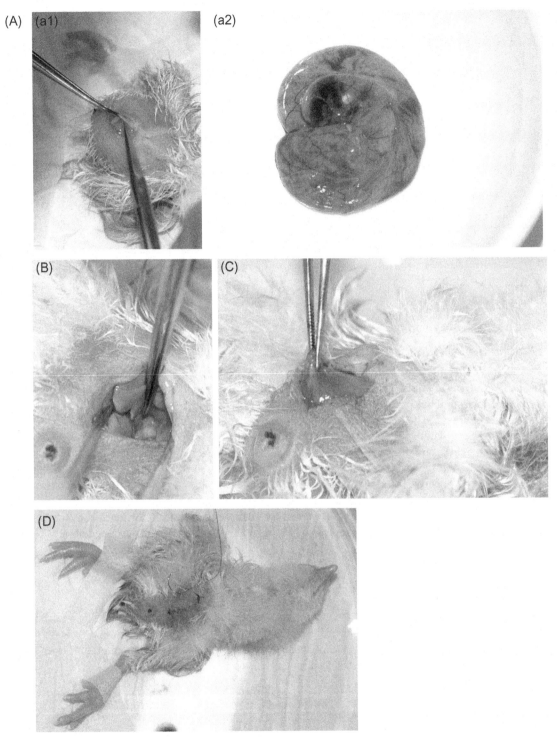

FIGURE 13.10 Removal of the yolk sack (A) and recipient's own gonads is followed by the transplantation of donor tissue (B). The incision should be closed by two-layer suture (C and D).

13.3.4.2 Antibiotics treatment

Postoperative administration of antibiotics (2.5−5 mg ceftiofur, Excenel sc.) is possible [176,178,182,186], while other authors consider it unnecessary since it does not influence either the mortality or the adherence of the grafted organs [182,183,189].

13.3.4.3 Immunosuppression

To prevent acute immune response and edema, operated chicks should be administered immediately after surgery an intramuscular steroid injection (0.05 mg dexamethasone). Although transplanted gonads can adhere without any further treatment, mycophenolate mofetil (CellCept) can be used for supporting later gametogenesis, because due to the inhibition of B and T cells, it facilitates the implantation of donor tissue [190]. The animals then receive 4 mg/kg body weight mycophenolic acid *per os* (individually) for 2 weeks daily, then once a week for further 6 weeks [176,178,182,186,187].

13.4 Manipulation of avian embryonic cells—blastodermal cells and primordial germ cells

13.4.1 Introduction

In oviparous animals, including avian species, the cryopreservation of intact embryos would be the most obvious solution for ex situ in vitro preservation, but this is currently not possible due to the large amount of yolk and albumin in the egg. In case of birds, a further limiting factor is that female represents the heterogamous sex (with ZW sex chromosomes), so methods based on sperm-freezing only are not able to preserve the information stored in the W chromosome or the mitochondrial DNA. The appropriate method to preserve the genetic content of birds is by cryopreserving the embryonic cells and reintegrating them into a recipient embryo [191–193], thereby preserving the entire genetic material. Currently, two methods are known to produce germline chimera depending on the embryonic cell type (Fig. 13.11), blastodermal cell (BC) and primordial germ stem cell (PGC) transfer.

Both BCs [194–196] and primordial germ cells (PGCs have been successfully used [197–199] with this technology, but most of the experiments have been directed to domestic fowl [200–202] or the quail [203,204] species. Maya duck chimera production has also been reported [205] with embryonic BCs injected into stage X recipient White Leghorn embryos. PGCs enter the bloodstream with active movement in a specific phase of embryonic development and reach the developing gonads through the vascular system. During this period, Hamburger–Hamilton (HH) stages 13–17 are easily accessible from the embryonic blood and can be reintegrated in the same way [206,207]. PGCs are the precursors of adult germ cells and are suitable for genome preservation. Furthermore, the genetic modification of PGCs (via transcription activator-like effector nuclease or CRISPR/Cas9) is the most efficient way to create genetically engineered bird models for basic and applied purposes [208–210].

The cryopreservation of different embryonic cell types is an essential aspect to perform efficiently these methods.

13.4.2 Genome preservation using cryopreserved embryonic cells of different poultry species

13.4.2.1 Cryopreservation and manipulation of avian blastodermal cells

Successful reimplantation of BCs after freezing and thawing was first reported in 1992 by Naito et al. [211]. Japanese quail BCs were isolated from the embryo and after adding cryoprotectant (10% DMSO), the cell suspension was stored in cryotubes for 3 hours in a deep freezer at $-80°C$. Petitte et al. [212] also reported successful cryopreservation of chicken BC.

In these first studies, 2%–3% chimera production was achieved after microinjection of frozen/thawed cells. The poor efficiency of chimera production was explained by Reedy et al. [213], being that no more than 500 cells could be injected into the sub germinal cavity and that the viability of the frozen/thawed cells was only around 33%. Thus the amount of living cells is much lower than with the injection of fresh cells. In their investigations, 12.3% of the embryos hatched, and among them 22.4% proved to be chimeras based on the feather color. Kino et al. [214] were the first to produced germline chimera by injecting frozen/thawed chicken BCs.

13.4.2.1.1 Isolation of blastodermal cells

The workflow is carried out in a sterile laminar box. Prior to use, the eggs are disinfected with 70% ethanol. The eggs are broken individually, and the albumin is separated from the yolk (Fig. 13.12A). Filter paper ring is placed on the GD to isolate the BCs. The perivitelline membrane is cut around the paper ring (Fig. 13.12B), and the GD is removed [215]. Most of the adhering yolk is removed with sterile tissue paper, and the cells are washed by syringe into 10 mL DMEM High glucose medium containing 10% FBS (Fig. 13.12C). BCs are dispersed mechanically in DMEM High glucose medium by gentle pipetting. The whole process should be done on ice ($+4°C$) for better cell survival. Then the

1. Production of avian chimaeras using blastodermal cells (BCs)

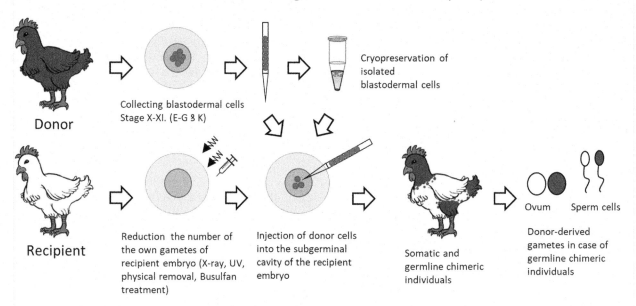

2. Production of avian chimaeras using primordial germ cells (PGCs)

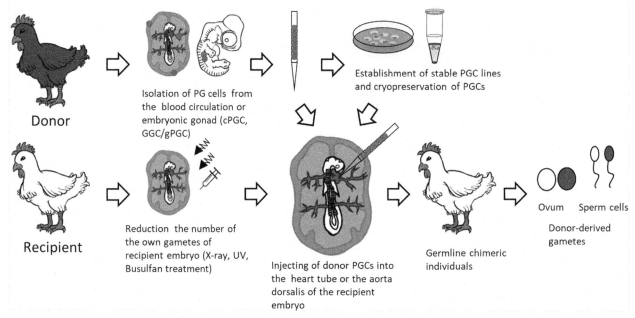

FIGURE 13.11   Diagram showing the workflow of the production of avian chimeras using different embryonic cell types: (1) production of avian chimeras using blastodermal cells (BCs); (2) production of avian chimeras using primordial germ cells (PGCs). *Designed by Luca Pataki and Bence Lázár.*

mixture is centrifuged for 3 minutes at 2300 rpm at +4°C (see Fig. 13.12D). Then 7.5 mL of the supernatant is removed; the cells are aspirated carefully from the top layer of the precipitate and resuspended in approximately 2.5 mL DMEM High glucose. This step is repeated to maximize removal of debris and yolk. After the second centrifugation, the supernatant is removed, leaving only 0.5 mL medium, and then the cells are ready for injection or freezing [216–218].

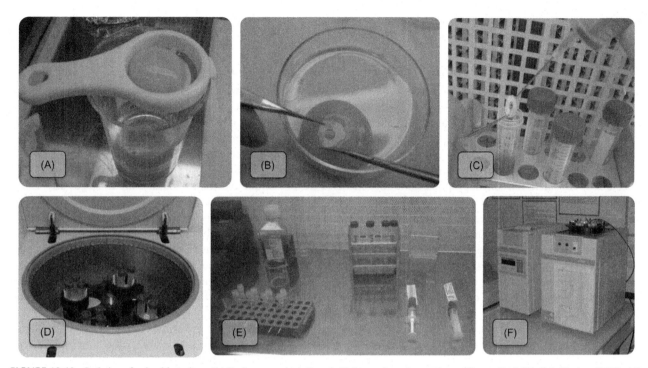

FIGURE 13.12 Isolation of avian blastodermal cells from germinal disc of chicken embryo (stage X according to Eyal-Giladi & Kochav [172]). (A) Separate albumin from the yolk. (B) Filter paper ring is placed on the germinal disc to isolate the blastodermal cells. The perivitelline membrane is cut around. (C) The cells are washed by syringe into 10 mL DMEM high glucose medium containing 10% FBS. (D) The mixture is centrifuged for 3 minutes at 2300 rpm at +4°C. (E) The cell suspension is loaded into straws or cryotubes. (F) The freezing is carried out in a programmable freezing machine. *Photos by Eszter Patakiné Várkonyi.*

13.4.2.1.2 Efficient slow freezing method for blastodermal cells

Cell suspension is centrifuged at 2300 rpm at +4°C for 5 minutes, and the supernatant is replaced by 2.5 mL cryoprotectant solution containing DMEM High glucose medium supplemented with 10% DMSO [219] or 5% DMSO and 5% ethylene glycol. Then the cell suspension is loaded into straws or cryotubes. The freezing is carried out in a programmable freezing machine (see Fig. 13.12F). The cryopreservation protocol according to Sawicka et al. [220] starts with cooling at +18°C. The temperature decreases at a rate of 4°C/minute until has reached 0°C, and the equilibration time is carried out for 5 minutes at 0°C. The cooling is continued with a rate of 1°C/minute until −7°C and a rate of 0.3°C/ minute until −37°C. Finally, the temperature is reduced with a rate of 30°C/minute until −130°C. After the program has ended, the straws or cryovials are removed from the freezing machine into a polystyrene box containing liquid nitrogen. The samples are then transferred to the cryobank or liquid N tank [221].

13.4.2.1.3 Thawing of the frozen embryonic blastodermal cells

From the liquid nitrogen storage tank the samples are transferred to a smaller polystyrene box containing liquid nitrogen. Thereafter, they are placed into the water bath and by continuously moving, the ampoules are thawed in 2−3 minutes at +27°C, whereas straws should be warmed for 20−30 seconds only.

The thawed cell suspension is placed in 2 mL of DMEM High glucose medium containing 10% FBS and then centrifuged at +20°C for 3 minutes at 2300 rpm. The supernatant, cryoprotectant-containing medium is removed and the remaining cell suspension is refilled with 300 μL of DMEM medium [221].

13.4.2.1.4 Management of recipient eggs before injection

Fertile eggs are used as recipients and are stored at 13°C−15°C for 6−10 days. According to our experience, prestorage is required, since it does not reduce viability, while providing significantly better conditions for the manipulations, because of the increased air chamber after natural water loss. Moreover, resting is necessary for the low-density GD to reach the top of the more dense yolk, thus becoming easily visible. The chicken and turkey eggs are stored with their blunt ends up for 6−10 days. Goose eggs are stored horizontally for 7−10 days, because the GD of this species does not rise below the air chamber, but usually stays sideways [217,218].

13.4.2.2 Cryopreservation and manipulation of avian primordial germ cells

Nowadays, the establishment of PGC cultures in many domestic fowl breeds, including indigenous lines, has been proved to be successful [202,222−224]. PGCs can be collected from multiple sites within the embryo. The entrance of PGCs from the anterior part of the extra-embryonic region into the vascular network starts to evolve at stage 10 [225] and is completed at stage 13. The migration of PGCs to the gonadal ridge begins at stage 15 and is completed at stage 17 [226]. Both male and female PGCs are collected from 5- to 7-day-old chicken embryos (HH stages 27−31), and then the cells have the ability to differentiate into functional gametes following transplantation [227]. Taking these aspects into consideration, the appropriate time for chicken PGC collection and transplantation are at stages 13−17 as well as 27−31. PGCs circulating in the bloodstream can be collected the earliest (circulating PG cells, cPGCs). The time of the migration period depends on the hatching time. This period occurs 2.5−3 days after the start of the incubation (HH14−17) in domestic fowl. The migration period of the PGCs in goose embryos is 69−84 hours (2.8−3.5 days) of development [218]. The benefit of collecting cPGCs is that the donor embryo can be brought to the hatching machine for further development and for hatching eventually [228]. Using the gonad-derived PG cells (gPGCs or GGCs) is simple and results in large amount of pure cell population [229,230]. However, the death of donor embryos is unavoidable and therefore not applicable in cases where the number of expendable embryos is limited, or because it belongs to rare breeds or endangered species.

To apply this method, it is essential to use recipient host embryos. Since the gonad of the recipient embryo contains its own germ cells, the offspring will have dual origin. Therefore, to improve the efficiency of the colonization and transmission of the exogenous PGCs, it is desirable to reduce the number of endogenous PGCs in the recipient [197,198,231,232].

Recently, xenotransplantation experiments have already been performed between distant phylogenetic groups (pheasant, duck, and Houbara PG cells into chicken recipient), but only male PG cells have been proved successful until now [205,233−236].

13.4.2.2.1 Isolation of circulating primordial germ cells

cPGCs were isolated from the blood of HH 14−17 (2.5−3 days) embryos [225]. Eggs are disinfected with 70% alcohol and broken into a glass Petri dish, making the entire vascular network of the embryo accessible. As a next step, with a sterile, finely drawn glass capillary (∼20 μm end diameter) and a pipette, 1−3 μL of blood is recovered from the dorsal aorta of the embryos under inverse stereomicroscope (Fig. 13.13) and is immediately placed into a specific culture medium for PG cells [237] in tissue culture plates (48 well plates). If it is possible, a tissue sample from the isolated embryos is kept for subsequent sex determination. The samples need to be stored at −20°C until further processing. The time of isolation and the exact age of the embryos (HH stages) are recorded.

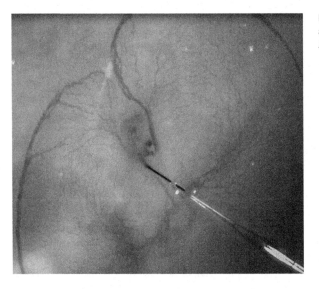

FIGURE 13.13 Isolation of circulating PG cells (cPGCs) from the dorsal aorta of 2.5−3-day-old chicken embryo (stage HH14−17). *Photo by Bence Lázár.*

13.4.2.2.2 Isolation of gonadal primordial germ cells (GGCs or gPGCs)

After incubation, embryonic gonads are isolated from 7-day-old embryos at stages 27–31 of development [225]. Isolated gonads are placed in 1.5 mL centrifuge tubes containing 200 μL of Dulbecco's PBS[−] (calcium- and magnesium-free). The centrifuge tubes are placed in an incubator maintained at 37.8°C for approximately 1–1.5 hours. Incubated gonads are pipetted gently, and 100 μL of cell suspension is placed on 48-well, nonadhesive tissue culture plate [230,238].

The samples need to be stored at − 20°C until further processing. The time of isolation and the exact age of the embryos (HH stages) are recorded.

13.4.2.2.3 Establishment of stable primordial germ cell lines

Isolated embryonic blood, which contains the cellular elements and also the cPGCs, or the GGCs suspension, is cultured for 3–4 weeks in an in vitro PG specific culture medium [237]. This period is long enough to eliminate all other cell types from the tissue culture plate except for the PG cells. Due to this specific medium, PG cells start to multiply and optimally create a homogeneous cell population. If the cell number of a single embryo-based culture reaches 5.0×10^4 PG cells after 3 weeks, then the establishment of the cell line can be considered successful. The PG cells will be floating freely in the medium at the bottom of the culture vessel, but they will not adhere; therefore the media should be changed every 2 days by carefully removing one-third of the medium and replacing it with freshly prepared solution (Fig. 13.14).

Once a week, the entire cell volume is suspended and centrifuged at $800 \times g$ for 4 minutes in a sterile 1.5 mL Eppendorf tube. Subsequently, the entire amount of medium is replaced with a fresh one, and finally the cells are placed in a new nonadhesive culture plate.

13.4.2.2.4 Freezing and thawing of primordial germ cell lines

PGCs can be frozen most effectively in media containing more than 10% serum and 5%–10% DMSO or 10% ethylene glycol at cooling rate of − 2°C/minute [239]. There are also commercially available cryo-solutions, such as CELLBANKER 1 (Nippon Zenyaku Kogyo, Japan) that is frequently used for gene preservation of species and breeds due to their high efficiency and easy usability [232,240,241]. In addition, there are efforts to replace animal sera from the freezing medium.

Freezing of the established PG cell lines is accomplished with freshly mixed PG cryo-solution containing the following components: DMEM and sterile water in a ratio of 2:1, 8% DMSO, 10% chicken serum, and 0.75% 20 mM $CaCl_2$. PG cells are gently resuspended from the bottom of the culture vessel and then pipetted into a 1.5 mL Eppendorf tube. After centrifugation ($1000 \times g$, 3 minutes), the supernatant is removed and the cells are resuspended in 250 μL DMSO-

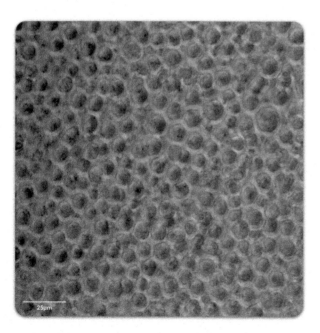

FIGURE 13.14 Transylvanian naked neck chicken–derived ZW (female) PGC line. *Photo by Elen Gócza.*

free freezing medium and transposed into labeled cryotubes. 250 μL of cryo-solution for PGCs is added slowly, drop by drop, and the cryotubes are placed in a deep freezer at −70°C. For long-term storage, the samples are transferred to −150°C or liquid nitrogen after one night.

For thawing of PGCs, a 37°C water bath is used and the entire content of the cryovial is pipetted into 2 mL of PGC culture medium. After centrifugation ($1000 \times g$, 3 minutes), the supernatant is removed, and after suspension in fresh culture medium the cells are placed in a nonadhesive 24-well plate and cultured further.

13.4.2.2.5 In vitro validation of the primordial germ cell lines

The established PG cell lines can be subjected to in vitro analysis to determine basic characteristics and to ensure that they belong to homogeneous cell populations, showing the characteristics of the cell type.

Sex determination of the primordial germ cell lines For sex determination of PGC lines, embryonic tissue samples are used, usually collected during the isolation. After digestion, the DNA is isolated from the tissues and the samples are diluted to 25 ng/μL concentration.

Subsequently, P2−P8 [242] or CHD1 primer pair [243] is used to carry out sex polymerase chain reaction (PCR). The PCR product undergoes gel electrophoresis on 3% gel for P2−P8 primer pairs and on 1.5% gel in case of CHD1 primer pair for 90 and 30 minutes at 90 V. The results are recorded in gel documentation system.

In vitro characterization of primordial germ cell lines based on immunohistochemistry Samples collected from the established PGC lines need to be fixed in 4% PFA for 10−15 minutes. After washing with PBS (three times), cells are permeabilized with 0.5% Triton X-100 (Merck Millipore, United States) for 5 minutes, so that the antibodies are allowed to enter the cytoplasm. After washing with PBS, to minimize nonspecific binding of antibodies, the fixed cells need to be blocked for 45 minutes with a blocking buffer containing PBS with 1% (v/v) BSA. Then, three more rounds of PBS washing follow, until cells are incubated with the primary antibodies (overnight at 4°C in a humid chamber). Remnants of the primary antibodies are being washed with PBS (three times); then cells are incubated with the secondary antibodies, in a dark humid chamber for 1 hour at room temperature. After washing with PBS again, TO-PRO-3 (1:500, Molecular Probes Inc., United States) can be used, which is a far-red fluorescent nuclear and chromosome counterstain. It is advised to use a specific solution for mounting the coverslips to the slides such as the VECTASHIELD Mounting Media (Vector Laboratories Inc., United States). Confocal microscopy is an ideal way to analyze the slides, because of its capability to show not just the cell membrane but the inside of the cells as well (Fig. 13.15).

Negative controls, where only the secondary antibody is used, are necessary for the appropriate evaluation of the samples. One of the best commercially available germ cell−specific antibodies nowadays is DAZL, whereas SSEA-1 is a stem cell−specific antibody wildly used in PGC studies [244] (Table 13.6).

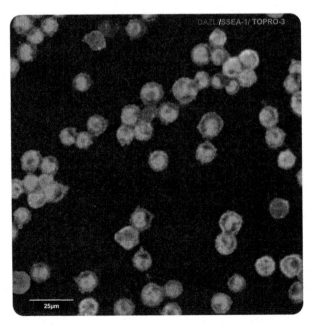

FIGURE 13.15 Immunostaining of chicken PGC—Hungarian Yellow chicken embryo−derived PGC line (#815): DAZL (green), SSEA-1 (red), and nuclear (blue) staining. *Photo by Elen Gócza.*

TABLE 13.6 Primary antibodies required for PGCs immunostaining method.

Name of primary antibody	Company	Catalog number	Usage	Type
DAZL	Abcam	ab34139	1:500	Rabbit IgG
SSEA-1	Millipore	A21432−1750277, MC480	1:100	Mouse IgM

TABLE 13.7 Most commonly used PGC-specific primers.

Symbol	Gene	NCBI ID	Primer sequences		Product length (bp)
cGAPDH	Glyceraldehyde-3-phosphate dehydrogenase (*Gallus gallus*)	NM_204305.1	FW	GACGTGCAGCAGGAACACTA	112
			RV	CTTGGACTTTGCCAGAGAGG	
cNANOG	Nanog homeobox (*Gallus gallus*)	NM_001146142.1	FW	ATACCCCAGACTCTGCCACT	100
			RV	GCCTTCCTTGTCCCACTCTC	
cPOUV	POU domain class 5 transcription factor 3 (Pou5f3) (*Gallus gallus*)	NM_001110178.1	FW	GAGGCAGAGAACACGGACAA	109
			RV	TTCCCTTCACGTTGGTCTCG	
CVH	DEAD-box helicase 4 (DDX4) (*Gallus gallus*)	NM_204708.1	FW	GAACCTACCATCCACCAGCA	113
			RV	ATGCTACCGAAGTTGCCACA	
cDAZL	Deleted in azoospermia like (*Gallus gallus*)	NM_204218.1	FW	TGGTACTGTGAAGGAGGTGA	148
			RV	TGGTCCCAGTTTCAGCCTTT	

In vitro characterization of primordial germ cell lines based on the expression of stem- and germ cell−specific marker genes The process follows the standard RNA isolation, cDNA synthesis, and quantitative real-time PCR workflow:

Total RNA from the established PGC lines is isolated after washing in $1 \times$ PBS for removing the medium components. Good-quality RNA can be obtained by using, for example, TRI Reagent (MRC, United Kingdom) [245] following the instructions of the manufacturer. The isolated RNA can be stored at $-70°C$ until later use. As a next step, the extracted RNA samples need to be reverse transcribed into cDNA (e.g., High-Capacity cDNA Reverse Transcription Kit from Applied Biosystems, Life Technologies is an appropriate method for PG cells). RT Master Mix is used for cDNA writing. cDNA can be stored at $-20°C$. The SYBR Green PCR Master Mix is a good choice for the qPCR as a double-stranded fluorescent DNA-specific dye (Applied Biosystems, Life Technologies). For each gene examined, it is worth to measure and analyze at least three in parallel so as to get the most accurate results [244].

Probably the most important and widely used germ cell−specific marker genes are CVH and cDAZL, whereas cPOUV and cNANOG are good examples for stem cell−specific genes used in the PGC validation process. cGAPDH is common as a housekeeping gene. The following table contains all the details for the abovementioned genes including the primer sequences and the length of the products (Table 13.7).

13.4.3 Producing presumptive germline chimeras

13.4.3.1 Injection of blastodermal cells

The basic workflows, regardless of the species involved, can be written as: (1) isolation of BCs from the donor species (see Fig. 13.13) and (2) injection of donor BCs into the recipient embryo: chimera production.

The recipient egg is prepared and a window is opened in air chamber by use of forceps, and 3 μL of the BC suspension (prepared as described below) is injected directly into the subgerminal cavity of the recipient embryo GD with a

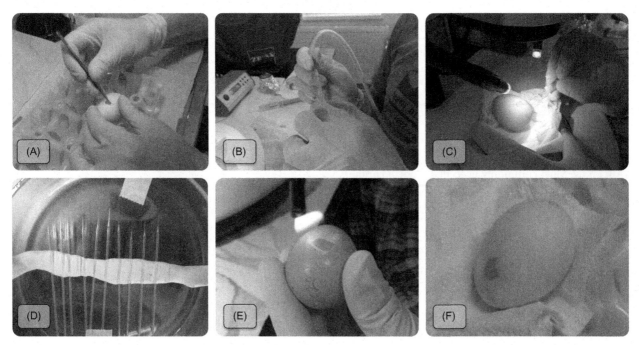

FIGURE 13.16 Injection of avian primordial germ cells into recipient embryo. (A) 10 mm in diameter "window" is opened for approximately one-third of the egg closer to the pointed end. (B) Wicking of 300–500 PGCs into the finely drawn glass capillary. (C) Injection of PGCs into the heart or the dorsal aorta of the 2.5–3-day recipient embryos. (D) Finely drawn glass capillary with a diameter of ~20 μm at the pointed end. (E) Covering the egg with two-layer laboratory parafilm. (F) Successfully injected recipient domestic fowl embryo. *Photos by Eszter Patakiné Várkonyi.*

finely drawn glass capillary under sterile conditions, as in the case of domestic fowl and turkey. In most cases, the GD is clearly visible through the shell membrane, so that the injection can be performed without any major injury for which the egg is very sensitive. If the injection is not a feasible approach, the shell membrane can be removed. In this case, after the injection, the injury has to be closed with a piece of prepared shell membrane. In goose, the germinal disk does not rise up during the storage, so the embryo is injected through a prepared window in the upper side of the egg. After injection, the window is covered with a two-layer laboratory parafilm. During the incubation, the embryonic development is controlled with periodic candling.

13.4.3.2 Injection of primordial germ cells

PG cells are collected and counted (Arthur Image Based Cell Analyzer, NanoEnTek). After centrifugation in a mixture of DMEM and sterile water (ratio 2:1) the cell pellet is resuspended. The alcoholic disinfection of the recipient eggs is pursued by opening a 10-mm diameter "window" at approximately one-third of the egg (Fig. 13.16A), closer to the pointed end. Then 1 μL (~3–5,000 PG cells) of the prepared cell suspension is injected into the heart or into the dorsal aorta of the 2.5–3-day recipient embryos (Fig. 13.16B and C). Injection takes place at the pointed end, by a finely drawn glass capillary with a diameter of ~20 μm (Fig. 13.16D). Following the injection, ~50 μL of sterile prewarmed 1 × PBS is dripped onto the embryo and then covered with two-layer laboratory parafilm, until finally the egg is brought to the hatching machine after individual marking (Fig. 13.16F).

13.4.4 Incubation of the manipulated eggs and investigation of embryo survival

The injected eggs are returned to the hatchery. They need to be candled on the 8th, 14th days (chicken) and 12th, 21st (turkey, goose, guinea fowl) days of incubation. The success of the injection can also be confirmed by the injection of fluorescently labeled PG cells (Fig. 13.17).

Examination of dead embryos, together with recording date, possible cause of death, possible malformations, and the developmental stages should be recorded. In case of goose eggs, holes have to be made in the parafilm 2 days before the expected hatching, in order to provide adequate oxygen supply for the embryos and eggs.

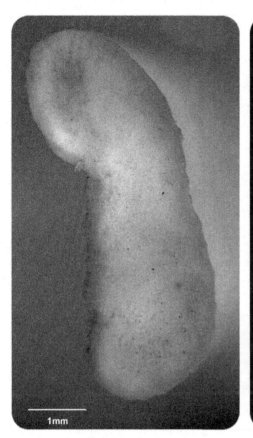

FIGURE 13.17 Successful integration of GFP expressing PG cells into the gonad of 14-day-old chicken embryo. 4ZP PGC line, Hungarian Partridge color recipient #4 embryo, left gonad. *Photo by Elen Gócza.*

13.5 Fertility determination

The fertility of eggs can be assessed nondestructively by examining eggs that have failed to hatch or by candling during incubation or destructively by opening a sample of eggs to look for the presence of an embryo, either before or during incubation.

Fresh, unincubated egg breakouts are performed when an estimate of true fertility needs to be determined as quickly as possible, and it is made by simple visual inspection [246]. The unfertilized GD appears as a single small white mass while the fertilized GD appears as a double ring and larger white disk on the surface of the egg yolk (Fig. 13.18).

The ability to correctly differentiate fertilized from unfertilized GD and fertilized but dead preincubation embryos could help in determining the basis of decreased hatchability [247].

If differentiating an unfertilized GD from an early dead is difficult, it can be made clear with the help of nuclear fluorescent stains such as either propidium iodide [248] or bisbenzimide (Hoechst 33342) [247], which can dye the dividing BC nuclei (live or dead), while the infertile GD show a uniform, light red or blue background [247,248] (Fig. 13.19).

13.6 Methods for predicting the fertility—sperm in eggs

The above assessments of fertility only tell us what happened to a single spermatozoon that fertilized the egg; however, we know nothing about the hundreds of millions of spermatozoa that are transferred into the oviduct by natural mating or AI. It is well known that polyspermic fertilization occurs in avian species, that is, presence of spermatozoa that associate with the perivitelline layers of the ovum can be determined in freshly laid eggs. With the help of these examinations, the efficiency of the mating or AI as well as the expected lengths of the duration of fertility can be predicted both in domestic and nondomestic species.

In the infundibulum of the avian oviduct, the ovulated ovum is surrounded with the IPVL derived from the ovary. Penetrations of spermatozoa can be detected as small holes most frequently on and around the GD. After fertilization, spermatozoa that have not penetrated are trapped in the OPVL) derived from the distal part of the infundibulum and/or

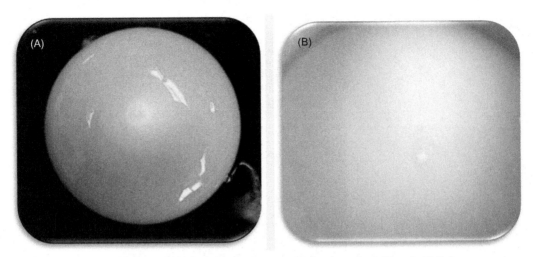

FIGURE 13.18 Fertilized (A) and unfertilized (B) germinal disc of avian ovum (*Gallus domesticus*). *Photo by M. Bakst.*

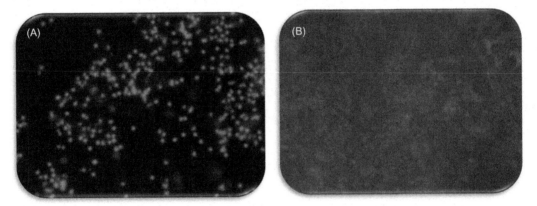

FIGURE 13.19 Propidium iodide fluorescent staining of germinal discs for differentiating of fertilized ova with early dead embryo and unfertilized germinal disc: (A) stained dividing blastodermal cell nuclei and (B) infertile germinal disc with uniform reddish background.

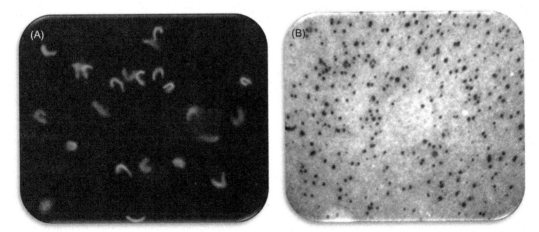

FIGURE 13.20 Predicting of fertility of avian eggs. (A) Sperm nuclei from the outer perivitelline layer of goose egg stained with DNA-specific fluorochrome Hoechst 33342 (bis-benzimide). (B) Penetration holes in inner perivitelline layer of domestic hen's egg under dark ground optics.

the cranial part of the magnum. The number of sperm cells interacting with the egg is proportional to the numbers in the SSTs, and therefore these examinations are useful reporters of successful sperm transfer.

For measurement, the whole yolk from the unincubated egg should be removed, washed thoroughly in a simple salt solution, and then a 2 cm squared piece of whole perivitelline layer from over the GD (IPVL-holes) or the vegetative pole (OPVL spermatozoa) should be cut. The membrane free of egg white and yolk should be washed and placed on a microscope slide. For viewing OPVL-sperm, a DNA-specific fluorochrome should be used for staining, while for viewing IPVL-holes, either a stain for glycoproteins [249] or left unstained for viewing under dark ground optics can be successfully used [44,250] (Fig. 13.20).

References

[1] Blanco JM, Wildt DE, Höfle U, Voelker W, Donoghue AM. Implementing artificial insemination as an effective tool for ex situ conservation of endangered avian species. Theriogenology 2009;71:200−13.

[2] Ivanov E. Expériences sur la fécondation artificielle des oiseaux. C R Soc Biol Paris 1913;371−4.

[3] Quinn JP, Burrows WH. Artificial insemination in fowls. J Hered 1936;27.1:7−31.

[4] Gee GF. Artificial insemination and cryopreservation of semen from non-domestic birds. In: Bakst, MR and Wishart, GJ, editors. First international symposium on the artificial insemination of poultry. Poultry Science Association; 1995. p. 262−79.

[5] Gee GF, Bertschinger H, Donoghue AM, Blanco J, Soley J. Reproduction in nondomestic birds: physiology, semen collection, artificial insemination and cryopreservation. Avian Poult Biol Rev 2004;15.2:47−101.

[6] King AS. Phallus. In: King AS, McLelland J, editors. Form and function in birds, 2. New York, NY: Academic Press; 1981. p. 107.

[7] Hammerstrom F. An eagle to the sky. Ames, IA: Iowa University Press; 1970. p. 142.

[8] Malecki IA, Rybnik PK, Martin GB. Artificial insemination technology for ratites: a review. Aust J Exp Agric 2008;48:1284−92.

[9] Samour HJ, Sprat DMJ, Hutton RE, Jones DM. Studies on semen collection in waterfowl by electrical stimulation. Br Vet J 1985;141:265.

[10] Betzen KM. Techniques for electrical semen collection of birds. MSc thesis. Stillwater, OK: Oklahoma State University; 1985. p. 85.

[11] Harrison GJ, Wasmund D. Preliminary studies of electroejaculation to facilitate manual semen collection in psittacines. In: Proceedings of the annual meeting of association of avian veterinarians; 1983. p. 207.

[12] Burrows WH, Quinn JP. The method of obtaining spermatozoa from the domestic fowl. Poult Sci 1935;14(4):251−4.

[13] Cooper DM. Artificial insemination. In: Gordon RF, editor. Poultry diseases. London: Bailliere Tindall; 1977. p. 302−7.

[14] Gee GF, Sexton TJ. Cryopreservation of semen in the Aleutian Canada goose (*Branta canadensis leucopareia*). Zoo Biol 1990;9:361.

[15] Gee GF, Mirande CM. Artificial insemination. In: Ellis DH, Gee GF, Mirande CM, editors. Cranes: their biology, husbandry and conservation. Washington, DC/Baraboo, WI: National Biological Service/International Crane Foundation; 1996. p. 205.

[16] Barna J, Végi B, Váradi É, Ferencziné Szőke Zs, Liptói K. Comparative study on semen traits of two genotypes of goose at various frequencies of sperm collection. Proceedings of the XXIII world's poultry congress, 64. Brisbane: World's Poultry Science Journal; 2008. p. 565. Suppl.2.

[17] Gee GF, Morrel CA, Franson JC, Pattee OH. Cryopreservation of American kestrel semen with dimethyl sulfoxide. J Raptor Res 1993;27:21.

[18] Gee GF, Blanco JM, Höfle U. Evaluation of semen of non-domestic birds Chapter V Techniques for semen evaluation, semen storage and fertility determination. 2nd Edition St. Paul, MN: The Midwest Poultry Federation; 2014. p. 55−61.

[19] Barna J, Boldizsár H. Motility and agglutination of fowl spermatozoa in media of different amino acid content and pH value *in vitro*. Acta Vet Hung 1996;44(2):221−32.

[20] McFarlane RW. The taxonomic significance of avian sperm. MSc dissertation. University of Florida; 1962. p. 32.

[21] Brock MK, Bird DM. Pre-freeze and post-thaw effects of glycerol and dimethylacedamide on motility and fertilizing ability of American kestrel (*Falco sparverius*) spermatozoa. J Zoo Wildl Med 1991;22(4):453.

[22] Barna J, Wishart GJ. Excess nuclear DNA in spermatozoa of guinea fowl. Theriogenology 2003;59:1685−91.

[23] Soley JT, Bertschinger HJ, Hels HJ, Burger WP. The morphology and incidence of retained cytoplasmic droplets in ostrich spermatozoa. In: Deeming DC, editor. Improving our understanding of ratites in a farming environment. Ratite Conference, Oxfordshire; 1996. p. 16.

[24] Bakst MR, Long JA. Techniques for semen evaluation, semen storage, and fertility determination. St. Paul, MN: The Midwest Poultry Federation; 2010. p. 1−113.

[25] Hargrove TL. Cryogenic preservation of budgerigar, *Melopsittacus undulatus* semen. Master of Science thesis. Florida Atlantic University 1986;40.

[26] Donoghue AM, Garner DL, Donoghue DJ, Johnson LA. Viability assessment of turkey sperm using fluorescent staining and flow cytometry. Poult Sci 1995;74:1191−200.

[27] Miranda M, Kuliková B, Vasicek J, Olexiková L, Iffaldano N, Chrenek P. Effect of cryoprotectants and thawing temperatures on chicken sperm quality. Reprod Dom Anim 2017;1−8.

[28] Santiago-Moreno J, Castano C, Coloma MA, Gomez-Brunet A, Toledano-Diaz A, Lopez-Sebastian A, et al. Use of the hypoosmotic swelling test and aniline blue staining to improve the evaluation of seasonal sperm variation in native Spanish free-range poultry. Poult Sci 2009;88:2661−9.

[29] Ramu S, Jeyendran RS. The hypo-osmotic swelling test for evaluation of sperm membrane integrity. Methods Mol Biol 2013;927:21−5.

[30] Wishart GJ, Wilson YI. Sperm motility and metabolism I. Visual scoring of motility using the hanging drop method. In: Bakst MR, Cecil HC, editors. Techniques for semen evaluation, semen storage, and fertility determination. Savoy, IL: The Poultry Science Association, Inc.; 1997. p. 46–7.

[31] Lake PE, Stewart JM. Artificial insemination in poultry. London: Ministry of Agriculture, Fish and Food, Bulletin 213. Her Majestry's Stationery Office; 1978.

[32] Sexton TJ. Studies the fertility of frozen fowl semen. In: Proceedings of the eighth international congress on animal reproduction and artificial insemination, Cracow; 1976. pp. 1079–82.

[33] Schramm GP. Ein neues Medium zur Verdünnung und Flüssigkonservierung von Hahnensperma. Mh Vet Med 1982;900–2.

[34] Long JA, Purdy PA, Zuidberg K, Hiemstra SJ, Velleman SG, Woelders H. Cryopreservation of turkey semen: effect of breeding line and freezing method on post-thaw sperm quality, fertilization and hatching. Crybiology 2014;68:371–8.

[35] Tselutin K, Narubina L, Mavrodina D, Tur B. Cryopreservation of poultry semen. Br Poult Sci 1995;36:805–11.

[36] Łukaszewicz E. DMF effects on frozen gander semen. Br Poult Sci 2001;42:308–14.

[37] Bakst MR, Cecil HC, editors. Techniques for semen evaluation, semen storage and fertility determination. Savoy, IL: Poultry Science Association. Inc; 1997.

[38] Froman DP, McLain DJ. Objective measurement of sperm motility based upon sperm penetration of Accudenz®. Poult Sci 1996;75(6):776–84.

[39] Froman DP. Application of the sperm mobility assay to primary broiler breeder stock. J Appl Poult Res 2006;15(2):280–6.

[40] Gravance CG, Garner DL, Miller MG, Berger T. Fluorescent probes and flow cytometry to assess rat sperm integrity and mitochondrial function. Reprod Toxicol 2001;15:5–10.

[41] Gadella BM, Harrison RA. Capacitation induces cyclic adenosine 3',5'-monophosphate-dependent, but apoptosis-unrelated, exposure of aminophospholipids at the apical head plasma membrane of boar sperm cells. Biol Reprod 2002;67:340–50.

[42] Partyka A, Nizanski W, Lukaszewicz E. Evaulation of fresh and frozen-thawed fowl semen by flow cytometry. Theriogenology 2010;74:1019–27.

[43] Robertson L, Brown HL, Staines HJ, Wishart GJ. Characterization and application of an avian in vitro spermatozoa-egg interaction assay using the inner perivitelline layer from laid chicken eggs. J Reprod Fertil 1997;110:205–11.

[44] Wishart GJ, Robertson LR, Wood L. Sperm: egg interaction in turkeys: mechanisms and practical applications. Turkeys 1995;43:19–22.

[45] Sood S, Tawang A, Malecki IA, Martin GB. Artificial insemination technology for the emu – improving sperm survival. Reprod Biol 2011;(Suppl 3):43–9.

[46] Partyka A, Nizanski W, Ochota M. Methods of assessment of cryopreserved semen Chapter 20 In: Katkov I, editor. Current frontiers in cryobiology. 2012. p. 547–74.

[47] Verstegen J, Iguer-ouada M, Onclin K. Computer assisted sperm analyzers in andrology research and veterinary practice. Theriogenology, 2002;57:149–79.

[48] Blesbois E, Grasseau I, Seigneurin F, Mignon-Grasteau S, Saint Jalme M, Mialon-Richard MM. Predictors of success of semen cryopreservation in chickens. Theriogenology 2008;69:252–61.

[49] Agarval A, Ozturk E, Loughlin KR. Comparison of semen analysis between the two Hamilton-Thorn semen analyzers. Andrologia. 1992;24:327–9.

[50] Sexton TJ. Research note: influence of damaged spermatozoa on the fertility of turkey semen stored for 24 hours at 5°C. Poult Sci 1988;67:1483–5.

[51] Bakst MR. Preservation of avian cells. In: Crawford RD, editor. Poultry breeding. Amsterdam: Elsevier; 1990. p. 91–108.

[52] Bootswala SM, Miles RD. Development of diluents for domestic fowl semen. Worlds Poult Sci J 1992;48:121–8.

[53] Sexton TJ. Oxidative and glycolytic activity of chicken and turkey spermatozoa. Comp Biochem Physiol 1974;48B:38–65.

[54] Wishart GJ. The effect of continuous aeration on the fertility of fowl and turkey semen stored above 0°C. Br Poult Sci 1981;22:445–50.

[55] Lake PE, Ravie O. An expoloration of cryoprotective compounds for fowl spermatozoa. Br Poult Sci 1986;25:145–50.

[56] Lake PE, Stewart JM. Preservation of fowl semen in liquid nitrogen – an improved method. Br Poult Sci 1978;19(2):187–94.

[57] Chalah T, Seigneurin F, Blesbois E, Brillard JP. In vitro comparison of fowl sperm viability in ejaculates frozen by three different techniques and relationship with Subsequent fertility in vivo. Cryobilogy 1999;39:185–91.

[58] Tselutin K, Seigneurin F, Blesbois E. Comparison of cryoprotectants and methods of cryopreservation of fowl spermatozoa. Poult Sci 1999;78:586–90.

[59] Santiago-Moreno J, Castaño C, Toledano-Díaz A, Coloma MA, López-Sebastián A, Prieto MT, et al. Semen cryopreservation for the creation of Spanish poultry breeds cryobank: optimization of freezing rate and equlibration time. Poult Sci 2011;90:2047–53.

[60] Mosca F, Madeddu M, Sayed AA, Zaniboni L, Iaffaldano N, Cerolini S. Combined effect of permeant and non-permeant cryoprotectants on the quality of frozen/thawed chicken sperm. Cryobiology 2016;76:343–7. Available from: https://doi.org/10.1016/j.cryobiol.2016.10.001.

[61] Seigneurin F, Blesbois E. Effects of the freezing rate on viability and fertility of frozen-thawed fowl spermatozoa. Theryogenology 1995;43:1351–8.

[62] Abouelezz FMK, Castaño C, Toledano-Díaz A, Estesi MC, López-Sebastián A, Campo JL, et al. Effect of the interaction between cryoprotectant concentration and cryopreservation method on frozen/thawed chicken sperm variables. Reprod Domest Anim 2015;50:135–41.

[63] FAO Cryoconservation of animal genetic resources. FAO Animal Production and Health Guidelines, No. 12 Rome. 2012.

[64] Váradi É. Hímivarsejtek és korai ivarszerv-szövetek mélyhűtéses tartósításának fejlesztése baromfifajokban génmegőrzési célokból. PhD thesis 2006; p. 48–50.

[65] Rakha BA, Ansari MS, Akhter S, Hussain I, Blesbois E. Cryopreservation of indian red jungle fowl (*Gallus gallus murghi*) semen. Anim Reprod Sci 2016;174:45−55.

[66] Zaniboni L, Cassinelli C, Mangiagalli MG, Gliozzi TM, Cerolini S. Pellet cryopreservation for chicken semen: effects of sperm working concentration, cryoprotectant concentration and equilibration time during in vitro processing. Theriogenology 2014;82:251−8.

[67] Madeddu M, Mosca F, Abdel Sayed A, Zaniboni L, Mangiagalli MG, Colombo E, et al. Effect of cooling rate on the survival of cryopreserved rooster sperm: comparison of different distances in the vapor above the surface of the liquid nitrogen. Anim Reprod Sci 2016. Available from: https://doi.org/10.1016/j.anireprosci.2016.05.014.

[68] Williamson RG, Etches RJ, Reinhart BS, MacPherson JW. The effect of cooling rate before of freezing and the temperature of the semen upon addition of DMSO on the fertilizing capacity of chicken semen stored at − 196°C. Reprod Nutr Dev 1981;21:1033−42.

[69] Hübner R, Schramm GP. Untersuchungen über die kryoprotektive Eignung von Äthylenglykol und Dimethylacetamid zur Tiefgefrierkonservierung von Hahnensperma. Monatshefte für VetMed 1988;43(8):279−82.

[70] Van Voorst A, Leestra FR. Fertility rate of daily collected and cryopreserved fowl semen. Poult Sci 1995;74:136−40.

[71] Herrera JA, Quintana JA, Lopez MA, Betancourt M, Fierro R. Individual cryopreservation with dimethyl sulfoxide and polyvinylpyrrolidone of ejaculates and pooled semen of three avian species. Arch Andr 2005;51:353−60. Available from: https://doi.org/10.1080/014850190944401.

[72] Schramm GP, Hübner R. Konservierung von Geflügelsperma. Archiv Tierzucht 1989;32:51−61.

[73] Schramm GP. Untersuchungen zur genotypspezifischen Modifizierung des Verfahrens der Tiefgefrierkonservierung von Hahnensperma. Züchtungskunde 2008;80(2):137−45 ISSN 0044-5401.

[74] Hanzawa S, Niinomi T, Takahashi R, Yamaguchi K, Miyata T, Tajima A. New method of freezing chicken semen using N-methyl-acetamide as cryoprotective agent. Proceedings of the twelfth European poultry conference, Verona; September 10−14, 2006. p. 519.

[75] Ehling C, Taylor U, Baulain U, Weigend S, Henning M, Rath D. Cryopreservation of semen from genetic resource chicken lines. Agric For Res 2012;62:151−8.

[76] Fujihara N, Ohboshi S. Simple and rapid cryopreservation of rooster spermatozoa. Low Temp Med 1991;17:128−31.

[77] Tajima A, Graham EF, Shoffner RN, Otis JS, Hawkins DM. Research note: cryopreservation of semen from unique lines of chicken germ plasma. Poult Sci 1990;69:999−1002.

[78] Phillips JJ, Bramwell RK, Graham JK. Cryopreservation of rooster sperm using methyl cellulose. Poult Sci 1996;75:915−23.

[79] Nabi MM, Kohram H, Yegane HM, Shahne AZ, Sharideh H, Esmaili V. Comparative evaluation of Nabi and Beltsville extenders for cryopreservation of rooster semen. Cryobiology 2015. Available from: https://doi.org/10.1016/j.cryobiol.2015.11.005.

[80] Bakst MR, Sexton TJ. Fertilizing capacity and ultrastructure of fowl and turkey spermatozoa before and after freezing. J Reprod Fertil 1979;55:1−7. Available from: https://doi.org/10.1530/jrf.0.0550001.

[81] Blanco JM, Long JA, Gee G, Wildt DE, Donoghue AM. Comparative cryopreservation of avian spermatozoa: effects of freezing and thawing rates on turkey and sandhill crane sperm cryosurvival. Anim Reprod Sci 2012;131:1−8.

[82] Végi B., Drobnyák Á., Szabó Zs, Barna J. Development of cryopreservation protocols of Copper Turkey's semen. In: Prukner-Radovcic E, Medic H, Prukner-Radovcic E, editor. Proceedings of the fifteenth European poultry conference, conference information. Dubrovnik, Horvátország: World's Poultry Science Association (WPSA); 2018. p. 497.

[83] Oderkirk AHF, Buckland RB. A comparison of diluents and cryopreservatives for freezing turkey semen. Poult Sci 1977;56:1861−7.

[84] Schramm GP. Flüssig- und Gefrierkonservierung von Putersperma. Vet Med 1986;41:460−3.

[85] Schramm GP, Hübner R. Einfluss differenzierter Kryoprotektiva und Gefrierverfahren auf die reproduktive Leistung von langzeitgelagerteem Putersperma. Monatshefte für Veterinärmedizin 1988;43:426−7.

[86] Tur B, Mavrodina T. Turkey sperm cryopreservation. Ptitsevodstvo 1987;8:28−9.

[87] Kurbatov AD, Platov EM, Korban NV, Moroz LG, Nauk VA. Cryopreservation of farm animal sperm. Agropromizdat Leningr 1988;195−245.

[88] Cerolini S, Zaniboni L, Mangiagalli MG, Cassinelli C, Marzoni M, Castillo A, et al. Sperm cryopreservation by the pellet method in chickens, turkeys and pheasants: a comparative study. Avian Biol Res 2009;1:1758−9.

[89] Iaffaldano N, Romagnoli L, Manchisi A, Rosato MP. Cryopreservation of turkey semen by the pellet method: effects of variables such as the extender, cryoprotectant concentration, cooling time and warming temperature on sperm quality determined through principal components analysis. Tehriogenology 2011;76:794−801.

[90] Iaffaldano N, Di Ioro M, Miranda M, Zaniboni L, Manchisi A, Cerolini S. Cryopreserving turkey semen in straws and nitrogen vapour using DMSO or DMA: effects of cryoprotectant concentration, freezing rate and thawing rate on post-thaw semen quality. Br Poult Sci 2016;57 (2):264−70. Available from: https://doi.org/10.1080/00071668.2016.1148261.

[91] Seigneurin F, Blesbois E. The first method of cryopreservation of guinea fowl semen. Journées de la Recherce Avic 2006;23:1−2.

[92] Váradi É, Végi B, Liptói K, Barna J. Methods for cryopreservation of guinea fowl sperm. PLoS One 2013;8(4):e62759.

[93] Seigneurin F, Grasseau I, Chapuis H, Blesbois E. An efficient method of guinea fowl sperm cryopreservation. Poult Sci 2013;92:2988−96.

[94] Sakhatsky NI, Andreyev VI, Artemenko AB. Technology of goose sperm low temperature conservation. In: Proceedings of tenth European symposium on waterfowl, Halle; 1995. p. 283−85.

[95] Łukaszewicz E. An effective method for freezing White Italian gander semen. Theriogenology 2002;58:19−27.

[96] Łukaszewicz E. Characteristics of fresh gander semen and its susceptibility to cryopreservation in six generations derived from geese inseminated with frozen-thawed semen. CryoLetters 2006;27(1):51−8.

[97] Barna J, Végi B, Váradi É, Liptói K. Comparative study on cryopreservation procedures of gander sperm. In: Proceedings of the thirteenth European poultry conference, Tours, France; August 23−27, 2010. Worlds Poultry Sci J 2010;66:508.

[98] Váradi É, Drobnyák Á, Végi B, Liptói K, Kiss Cs, Barna J. Cryopreservation of gander semen in cryovials − comparative study. Acta Vet Hung 2019;67(2):246−55. Available from: https://doi.org/10.1556/004.2019.026.

[99] Dubos F, Seigneurin F, Mialon-Richard MM, Grasseau I, Guy G, Blesbois E. Cryopreservation of landese gander semen. In: Symposium COA/INRA scientific cooperation in agriculture, Taiwan R.O.C.; November 7−10, 2006. p. 169−72.

[100] Tai JJ, Chen JC, Wu KC, Wang SD, Tai C. Cryopreservation of gander semen. Br Poult Sci 2001;42:384−8.

[101] Penfold LM, Harnal V, Lynch W, Bird D, Derrickson SR, Wildt DE. Characterization of Northern pintail (Anas acuta) ejaculate and the effect of sperm preservation on fertility. Reproduction 2001;121:267−75.

[102] Han XF, Niu ZY, Liu FZ, Yang CS. Effects of diluents, cryoprotectants, equilibration time and thawing temperature on cryopreservation of duck semen. Int J Poult Sci 2005;4:197−201.

[103] Végi B, Drobnyák Á, Bíró E, Szabó Zs, Barna J. Különböző ondómélyhűtési protokollok in vitro összehasonlítása magyar kacsában. In: Irinyiné Oláh K, Cs Tóth, editors. Az "Őshonos- és Tájfajták - Ökotermékek - Egészséges táplálkozás - Vidékfejlesztés - Minőségi élelmiszerek - Egészséges környezet: Az agrártudományok és a vidékfejlesztés kihívásai a XXI. században Nyíregyháza, Magyarország: Nyíregyházi Egyetem, Műszaki és Agrártudományi Intézet; 2018. p. 74.

[104] Schramm GP. Kurz- und Langzeitkonservierung von Cariniasperma (Cairina moschata). Tierzucht 1987;41(9):418−19.

[105] Gerzilov V. Influence of various cryoprotectants on the sperm mobility of Muscovy semen before and after cryopreservation. Agric Sci Technol 2010;2:57−60.

[106] Sood S, Malecki IA, Tawang A, Martin GB. Sperm viability motility and morphology in emus (Dromaius novaehollandiae) are independent of the ambient collection temperature but are influenced by storage temperature. Theriogenology 2012;77:1597−604.

[107] Smith AMJ. A protocol for liquid storage and cryopreservation of ostrich (Struthio camelus) semen. Doctoral dissertation. South Africa: University of Stellenbosch; 2016. p. 51. Chapter III.

[108] Fujihara N, Nishiyama H, Koga O. In vitro viability and fertilizing capacity of Guinea fowl spermatozoa. Zool Sci 1989;6:731.

[109] Gale C, Brown KI. The identification of bacteria contaminating collected semen and the use of antibiotics in their control. Poult Sci 1960;39:50.

[110] Sexton TJ, Jacobs LA, McDaniel GR. A new poultry semen extender. 4. Effect of antibacterials in control of bacterial contamination in chicken semen. Poult Sci 1980;59:274.

[111] Shaffner CS, Henderson EW, Card CG. Viability of spermatozoa of the chicken under various environmental conditions. Poult Sci 1942;20:259−65.

[112] Polge C, Smith AU, Parkes AS. Revival of spermatozoa after vitrification and dehydration at low temperatures. Nature 1949;164:666.

[113] Polge C. Functional survival of fowl spermatozoa after freezing at −70°C. Nature 1951;167:949−50.

[114] Graham EF, Nelson DS, Schmehl MKL. Development of extender and techniques for frozen turkey semen. 2. Fertil Trials Poult Sci 1982;61:558−63.

[115] Bellagamba F, Cerolini S, Cavalchini LG. Cryopreservation of poultry semen: a review. Worlds Poult Sci J 1993;49:157−66.

[116] Hammerstedt RH. Cryopreservation of poultry semen − current status and economics. In: Proceedings of the first international symposium on the artificial insemination of poultry. Poultry Science Association, Savoy, IL; 1995. p. 229−50.

[117] Blesbois E. Advances in avian semen cryopreservation. In: Proceedings of the twelfth European poultry conference, Verona; September 10−14, 2006. Worlds Poultry Sci J 2006;62:519.

[118] Long JA. Avian semen cryopreservation: what are the biological challenges. Poult Sci 2006;85:232−6.

[119] Holt VW. Basic aspects of frozen storage of semen. Anim Reprod Sci 2000;62:3−22.

[120] Buss EG. Cryopreservation of rooster sperm. Poult Sci 1993;72:944−54.

[121] Blesbois E. Current status in avian semen cryopreservation. Worlds Poult Sci J 2007;63:213−22.

[122] Lake PE. The history and future of the cryopreservation of avian germ plasma. Poult Sci 1986;65:1−15.

[123] Mphaphati ML, Luseba D, Sutherland B, Nedabale TL. Comparison of slow freezing and vitrification methods for Venda Cockere's spermatozoa. Open J Anim Sci 2012;2(No.3):204−10. Available from: https://doi.org/10.4236/ojas.2012.23028.

[124] Tereshchenko A.V., Artemenko A.B., Sakhatsky N.I. Cryopreservation of chicken semen. In: Proceedings of the twelfth international congress of animal reproduction, Hague; August 23−27, 1992. p. 1602−4.

[125] Miranda M, Kuliková B, Vašíček J, Olexiková L, Iaffaldano N, Chrenek P. Effect of cryoprotectants and thawing temperature on chicken sperm quality. Reprod Domest Anim 2017;1−8. Available from: https://doi.org/10.1111/rda.13070.

[126] Sasaki K, Tatsumi T, Tsutsui M, Niinomi T, Imai T, Naito M, et al. A method for cryopreserving semen from Yakido roosters using N-methylacetamide as a cryoprotective agent. J Poult Sci 2010;47:297−301.

[127] Pranay Kumar K, Swathi B, Shanmugan M. Cryopreservation of rooster semen using N-methylacetamide as cryoprotective agent J Agric Sci 2018(10):5123−6.

[128] Pandian C, Prabakaran R, Venukopalan K, Kalatharan J. Effect of genetic groups and cryoprotectants on preservation of turkey semen. Indian Vet J 2011;88(2):16−17.

[129] Dubos F., Lemoine M., Seigneurin F., Mialon-Richard M.M., Grasseau I., Guy G. et al. Cryopreservation of landese gander semen. In: Proceedings of the XXIII world's poultry congress, Brisbane; June 30−July 4. Worlds Poultry Sci J 2008;64(2):147.

[130] Gerzilov V. Sperm motion of in vitro stored Muscovy drake semen. 2018; <https://www.researchgate.net/publication/324485351>.

[131] Çiftci HB, Aygün A. Poultry semen cryopreservation technologies. Worlds Poult Sci J 2018;74. Available from: https://doi.org/10.1017/S0043933918000673.

[132] Donoghue AM, Wishart GJ. Storage of poultry semen. Anim Reprod Sci 2000;62:213–32.

[133] Woelders H, Zuidberg CA, Hiemstra SJ. Animal genetic resources conservation in The Netherlands and Europe: poultry perspective. Poult Sci 2006;85:216–22.

[134] Voronina MS, Komarova VV, Moskalenko LI. Effect of different diluents and cock sperm cryopreservation methods on results of artificial insemination (Russian). Sb Nauchnikh Trudov VNIRGJ, Leningr 1986;71–9.

[135] Matskova LI, Romanov VS, Narubina LE. The effect of diluents used for cock sperm cryopreservation on fertility and hatchability. (Russian) Sb Nauchnikh Trudov VNIRGJ Leningr 1986;68–70.

[136] Kowalczyk A, Łukaszewicz E. The possibility of obtaining intergeneric hybrids via White Koluda (*Anser anser* L.) goose insemination with fresh and frozen-thawed Canada goose (*Branta canadensis* L.) gander semen. Theriogenology 2012;77:507–13.

[137] Bakst MR., Brillard JP. Mating and fertility. In: P. Hunton, editor. World animal science. Vol. C9. Poultry production. Amsterdam: Elsevier Science Publishers; 1995. [Chapter 12].

[138] Blesbois E. Freezing avian semen. Avian Biol Res 2011;4:52–8.

[139] Blanco JM, Gee GF, Wildt DE, Donoghue AM. Producing progeny from endangered birds of prey: treatment of urine contaminated semen and a novel intramagnal insemination approach. J Zoo Wildl Med 2002;33:1–7.

[140] Rodriguez OL, Berndtson WE, Ennen BD, Pickett BW. Effects of rates of freezing, thawing and level of glycerol on the survival of bovine spermatozoa in straws. J Anim Sci 1975;41:129–36.

[141] Fiser PS. Interactions of cooling velocity, warming velocity and glycerol concentrations on the survival of frozen-thawed boar sperm. In: Johnso LA, Rath D, editors. Boar semen preservation II: reproduction in domestic animals. Berlin: Paul Parey Publishers; 1991. p. 123–37. Suppl.1.

[142] Henry MA, Noiles EE, Gao D, Mazur P, Critser JK. Cryopreservation of human spermatozoa. IV. The effects of cooling rate and warming rate on the maintenance of motility, plasma membrane integrity, and mitochondrial function. Fertil Steril, 1993;360:911–18.

[143] Woelders H, Malva AP. How important is the cooling rate in cryopreservation of (bull) semen, and what is its relation to thawing rate and glycerol concentration. Reprod Domest Anim 1998;33:299–305.

[144] Mazur P. Principles of cryobiology. In: Benson E, Fuller B, Lane N, editors. Life in the frozen state, 1. London: Taylor and Francis Group; 2004. p. 4–65.

[145] Terada T, Ashizawa K, Maeda T, Tsutsumi Y. Efficiacy of trehalose in cryopreservation of chicken spermatozoa. Jpn J Anim Reprod 1989;35:20–5.

[146] Bakst MR, Wishart GJ, Brillard JP. Oviductal sperm selection, transport and storage in poultry. Poult Sci Rev 1994;5:117.

[147] Gee GF. Avian artificial insemination and semen preservation. Jean Delacour/IFCB symposium on breeding birds in captivity. North Hollywood, CA; 1983. p. 375.

[148] Lorenz FW, Ogasawara FX. Distribution of spermatozoa in the oviduct and fertility. VI. The relations of fertility and embryo mortality with the site of experimental insemination. J Reprod Fertil 1968;16:445–55.

[149] Temple SA. Artificial insemination with imprinted birds of prey. Nat Lophophorusihu (Lond) 1972;237:287–8.

[150] Grier JW. Techniques and results of artificial insemination with golden eagles. Raptor Res 1973;7:1–2.

[151] Boyd LL, Boyd NS, Dobler FC. Reproduction of prairie falcons by artificial insemination. J Wildl Mgmt 1977;41:266–71.

[152] Brillard JP. Sperm storage and transport following natural mating and artificial insemination. Poult Sci 1993;72:923.

[153] Brillard JP, McDaniel GR. Influence of spermatozoa numbers and insemination frequency on fertility in dwarf broiler breeder hens. Poult Sci 1986;65:2330–4.

[154] Lepore PD, Marks AL. Intravaginal insemination of Japanese quail: factors influencing the basic techniques. Poult Sci 1966;45:888.

[155] Smyth Jr. JR. Poultry. In: Perry EJ, editor. The artificial insemination of farm animals. New Brunswick, NJ: Rutgers University Press; 1968. p. 258.

[156] Durrant BS, Burch CD, Yamada JK, Good J. Seminal characteristics and artificial insemination of Chinese pheasants, *Tragopan temminckii*, *Lophophorus impeyanus, and Lophophorus ihuysii*. Zoo Biol 1995;14:523.

[157] Christensen VL, Johnston NP. Effect of time of day of insemination and the position of the egg in the oviduct on the fertility of turkeys. Poult Sci 1977;56:458–62.

[158] Gee GF, Sexton TJ. A comparative study of the cryogenic preservation of semen from the sandhill crane and the domestic fowl. Symp Zool Soc Lond 1978;43:89.

[159] Lukaszewicz E. Artificial insemination in geese. Worlds Poult Sci J 2010;66:647–58.

[160] Barna J, Végi B, Váradi É, Ferencziné Szőke Z, Péczely P. Studies related to fertility in broiler breeders. In: XXI International Poultry Symposium PB WPSA, Szklarska Poręba; 2009. p. 18–23.

[161] Török J, Michl G, Garamszegi L, Barna J. Repeated inseminations required for natural fertility in a wild bird population. Proc R Soc Lond 2003;B 270(1515):641–8.

[162] Birkhead TR, Fletcher F, Pellatt EJ, Staples A. Ejaculate quality and the success of extra-pair copulations in the zebra finch. Nature 1995;377:422.

[163] Howarth Jr. B. An examination for sperm capacitation in the fowl. Biol Reprod 1971;3:338–41.

[164] Olsen MW, Neher BH. The site of fertilization in the domestic fowl. J Exp Zool 1948;109:355–66.

[165] Nakanishi A, Utsumi K, Iritani A. Early nuclear events of *in vitro* fertilization in the domestic fowl (*Gallus domesticus*). Mol Reprod Dev 1990;26:217–21.

[166] Tanaka K, Wada T, Koga O, Nishio Y, Hertelendy F. Chick production by in vitro fertilization of the fowl ovum. J Reprod Fertil 1994;100:447–9.

[167] Perry MM. A complete culture system for the chick embryo. Nature. 1988;331:70–2.

[168] Li BC, Li W, Chen H, Zhang YL, Zhang ZT, Wang XY, et al. The influencing factor of in vitro fertilization and embryonic transfer in the domestic fowl (Gallus domesticus). Reprod Dom Anim 2013;48:368–72.

[169] Rowlett K, Simkiss K. Explanted embryo culture: in vitro and in ovo techniques for domestic fowl. Br Poult Sci 1987;28:91–101.

[170] Li ZD, Qi SZ. Hatching of cultured embryos of the Peking duck. AJAS 1996;9(2):195–7.

[171] Hrabia A, Takagi S, Ono T, Shimada K. Fertilization and development of quail oocytes after intracytoplasmic sperm injection. Biol Reprod 2003;69:1651–7.

[172] Eyal-Giladi H, Kochav S. From cleavage to primitive streak formation: a complementary normal tables and a new look at the first stages of the development of the chick. I. General morphology. Dev Biol 1976;49:321–37.

[173] Takagi S, Ono T, Tsukada A, Atsumi Y, Mizushima S, Saito N, et al. Fertilization and blastoderm development of quail oocytes after intracytoplasmic injection of chicken sperm bearing the W chromosome. Poult Sci 2007;86:937–43.

[174] Shimada K, Ono T, Mizushima S. Application of intracytoplasmic sperm injection (ICSI) for fertilization and development in birds. Gen Comp Endocrinol 2014;196:100–5.

[175] Kang KS, Park TS, Rengaraj D, Lee HC, Lee HJ, Choi HJ, et al. Fertilization of cryopreserved sperm and unfertilized quail ovum by intracytoplasmic sperm injection. Reprod Fertil Dev 2016;28:1974–81.

[176] Liptoi K, Horvath G, Gal J, Varadi E, Barna J. Preliminary results of the application of gonadal tissue transfer in various chicken breeds in the poultry gene conservation. Anim Reprod 2013;141(1-2):86–9.

[177] Lessard C., Svendsen E., Auckland C., Hind P. Genetic Preservation through Cryopreservation, In: Proceedings of genetic preservation summit, Alberta, Canada, May 24–25, 2017. p. 42.

[178] Song Y, Silversides FG. The technique of orthotopic ovarian transplantation in the chicken. Poult Sci 2006;85:1104–6.

[179] Song Y, Silversides FG. Transplantation of ovaries in Japanese quail (Coturnix japonica). Anim Reprod Sci 2008;105:430–7.

[180] Liu J, Song Y, Cheng KM, Silversides FG. Production of donor-derived offspring from cryopreserved ovarian tissue in Japanese quail (Coturnix japonica). Biol Reprod 2010;83:15–19.

[181] Liu J. Cryopreservation and transplantation of gonadal tissue for genetic conservation and biological research in avian species. Doctoral thesis, The University of British Columbia, Vancouver. 2013. p. 178.

[182] Song Y, Cheng KM, Robertson MC, Silversides FG. Production of donor-derived offspring after ovarian transplantation between Muscovy (Cairina moschata) and Pekin (Anas platyrhynchos) ducks. Poult Sci 2012;91:197–200.

[183] Liptoi K, Buda K, Rohn E, Drobnyak A, Edvine Meleg E, Palinkas-Bodzsár N, et al. Improvement of the application of gonadal tissue allotransplantation in the in vitro conservation of chicken genetic lines. Anim Reprod Sci 2020;213:106280. Available from: https://doi.org/10.1016/j.anireprosci.2020.106280.

[184] Wang Y, Xiao Z, Li L, Fan W, Li SW. Novel needle immersed vitrification: a practical and convenient method with potential advantages in mouse and human ovarian tissue cryopreservation. Hum Reprod 2008;Vol. 23(10):2256–65.

[185] Buda K, Rohn E, Barna J, Liptói K. Technical difficulties in gonadal tissue removal of newly hatched chicken in gene preservation practice. MÁL 2019;141:355–62.

[186] Song Y, Silversides FG. Heterotopic transplantation of testes in newly hatched chickens and subsequent production of offspring via intramagnal insemination. Biol Reprod 2007;76:598–603.

[187] Song Y, Silversides FG. Offspring produced from orthotopic transplantation of chicken ovaries. Poult Sci 2007;86:107–11.

[188] Song Y, D'Costa S, Pardue SL, Petitte JN. Production of germline chimeric chickens following the administration of a busulfan emulsion. Mol Reprod Dev 2005;70(4):438–44.

[189] Song Y, Silversides FG. Long-term production of donor-derived offspring from chicken ovarian transplants. Poult Sci 2008;87:1818–22.

[190] Morris RE, Hoyt EG, Murphy MP, Eugui EM, Allison AC. Mycophenolic acid morpholinoethylester (RS-61443) is a new immunosuprestant that prevents and halts heart allograft rejection by selective inhibition of T- and B-cell purine synthesis. Transpl Proc 1990;22:1659–62.

[191] Naito M, Sano A, Matsubara Y, Harumi T, Tagami T, Sakurai M, et al. Localization of primordial germ cells or their precursors in stage X blastoderm of chickens and their ability to differentiate into functional gametes in opposite-sex recipient gonads. Reproduction 2001;121 547–52.

[192] Naito M. Cryopreservation of avian germline cells and subsequent production of viable offspring. J Poult Sci 2003;40:1–12.

[193] Tajima A. Production of germ-line chimeras and their application in domestic chicken. Avian Poult Biol Rev 2002;13:15–30.

[194] Petitte JN, Clark ME. Production of somatic and germline chimeras in the chicken by transfer of early blastodermal cells. Development 1990;108:185–9.

[195] Fraser RA, Carsience RS, Clark ME, Etches RJ, Gibbins AM. Efficient incorporation of transfected blastodermal cells into chimeric chicken embryos. Int J Dev Biol 1993;37:381–5.

[196] Pain B, Clark ME, Shen M, Nakazawa H, Sakurai M, Samarut J, et al. Long-term in vitro culture and characterisation of avian embryonic stem cells with multiple morphogenetic potentialities. Development 1996;122:2339–48.

[197] Aige-Gil V, Simkiss K. Sterilising embryos for transgenic chimaeras. Br Poult Sci 1991;32:427–38.

[198] Aige-Gil V, Simkiss K. Sterilisation of avian embryos with busulphan. Res Vet Sci 1991;50:139–44.

[199] Naito M, Tajima A, Yasuda Y, Kuwana T. Production of germline chimeric chickens, with high transmission rate of donor-derived gametes, produced by transfer of primordial germ cells. Mol Reprod Dev 1994;39:153−61.

[200] Carsience RS, Clark ME, Verrinder Gibbin AM, Etches RJ. Germline chimeric chickens from dispersed donor blastodermal cells and compromised recipient embryos. Development 1993;117:669−75.

[201] Bednarczyk M, Łakota P, Siwek M. Improvement of hatchability of chicken eggs injected by blastoderm cells. Poult Sci 2000;79:1823−8.

[202] van de Lavoir M-C, Diamond JH, Leighton PA, Mather-Love C, Heyer BS, Bradshaw R, et al. Germline transmission of genetically modified primordial germ cells. Nature 2006;441:766−9.

[203] Naito M, Watanabe M, Kinutani M, Nirasawa K, Oishi T. Production of quail-chick chimaeras by blastoderm cell transfer. Br Poult Sci 1991;32:79−86.

[204] Li HC, Kagami H, Matsui K, Ono T. Restriction of proliferation of primordial germ cells by the irradiation of Japanese quail embryos with soft X-rays. Comp Biochem Phys A 2001;130:133−40.

[205] Li ZD, Deng H, Liu CH, Song YH, Sha J, Wang N, et al. Production of duck-chicken chimeras by transferring early blastodermal cells. Poult Sci 2002;81:1360−4.

[206] Naito M, Matsubara Y, Harumi T, Tagami T, Kagami H, Sakurai M, et al. Differentiation of donor primordial germ cells into functional gametes in the gonads of mixed-sex germline chimaeric chickens produced by transfer of primordial germ cells isolated from embryonic blood. J Reprod Fertil 1999;117:291−8.

[207] Tagami T, Kagami H, Matsubara Y, Harumi T, Naito M, Takeda K, et al. Differentiation of female primordial germ cells in the male testes of chicken (Gallus gallus domesticus). Mol Reprod Dev 2007;74:68−75.

[208] Doran TJ, Cooper CA, Jenkins KA, Tizard MLV. Advances in genetic engineering of the avian genome: realising the promise. Transgenic Res 2016;25(3):307−19.

[209] Woodcock ME, Idoko-Akoh A, McGrew MJ. Gene editing in birds takes flight. Mammalian Genome 2017;28(7-8):315−23.

[210] Lee HJ, Lee HC, Han JY. Germline modification and engineering in Avian species. Mol Cell 2015;38(9):743−9. Available from: https://doi.org/10.14348/molcells.2015.0225.

[211] Naito M, Nirasawa K, Oishi T. Preservation of quail blastoderm cells in liquid nitrogen. Br Poult Sci 1992;33:449−53.

[212] Petitte JN, Brazolot CL, Clark ME, Liu G, Verrinder Gibbins AM, Etches RJ. Accessing the genome of the chicken using germline chimeras. In. In: Etches RJ, Verrinder Gibbins AM, editors. Manipulation of the avian genome. Boca Raton: CRC Press; 1993. p. 81−101.

[213] Reedy S.E., Leibo S.P., Etches R.J. Cryopreservation of chick blastodermal cells to produce chimeric chickens. In: 31st annual meeting of society for cryobiology, Kyoto; August 21−26, 1994.

[214] Kino K, Pain B, Leibo SP, Cochran M, Clark ME, Etches RJ. Production of chicken chimeras from injection of frozen-thawed blastodermal cells. Poult Sci 1997;76:753−60.

[215] Petitte JN, Clark ME, Liu G, Verrinder Gibbins AM, Etches RJ. Production of somatic and germline chimeras in the chicken by transfer of early blastodermal cells. Development 1990;108:185−9.

[216] Patakiné Várkonyi E, Végi B, Váradi É, Barna J. Comparison of cryopreservation methods of blastodermal cells of various indigenous Hungarian poultry breeds. Worlds Poult Sci J 2010;66:485.

[217] Sztán N, Patakiné Várkonyi E, Liptói K, Barna J. Observations of embryonic cell manipulations in different poultry species. Hung Vet J/ Magy Allatorvosok 2012;8(134):475−81.

[218] Sztán N, Lázár B, Bodzsár N, Végi B, Liptói K, Pain B, et al. Successful chimera production in the Hungarian goose (Anser anser domestica) by intracardiac injection of blastodermal cells in 3-day-old embryos. Reprod Fertil Dev 2017;29(11):2206−16. Available from: https://doi.org/10.1071/RD16289.

[219] Svoradová A, Kuzelova L, Vasicek J, Olexikova L, Chrenek P. Cryopreservation of chicken blastodermal cells and their quality assessment by flow cytometry and transmission electron microscopy. Biotechnol Prog 2018;34(3). Available from: https://doi.org/10.1002/btpr.2615.

[220] Sawicka D, Chojnacka-Puchta L, Brzezinska J, Lakota P, Bednarczyk M. Cryoconservation of chicken blastodermal cells: effects of slow freezing, vitrification, cryoprotectant type and thawing method during in vitro processing. Folia Biologica (Krakow) 2015;63(2):129−34. Available from: https://doi.org/10.3409/fb63_2.129.

[221] Patakiné Várkonyi E, Molnár M, Sztán N, Váradi É, Végi B, Pusztai P. Cryopreservation of embryonic blastodermal cells of a valuable domestic poultry breed, the Hungarian landrace guinea fowl (Numida meleagris) as a biodiversity preservation method. Magy Állatorvosok Lapja 2016;138:673−80.

[222] Macdonald J, Glover JD, Taylor L, Sang HM, McGrew MJ. Characterisation and germline transmission of cultured avian primordial germ cells. PLoS One 2010;5(11):e15518.

[223] Miyahara D, Oishi I, Makino R, Kurumisawa N, Nakaya R, Ono T, et al. Chicken stem cell factor enhances primordial germ cell proliferation cooperatively with fibroblast growth factor 2. J Reprod Dev 2016;62:143−9.

[224] Tonus C, Cloquette K, Ectors F, Piret J, Gillet L, Antoine N, et al. Long term-cultured and cryopreserved primordial germ cells from various chicken breeds retain high proliferative potential and gonadal colonisation competency. Reprod Fertil Dev 2016;28:628−39.

[225] Hamburger V, Hamilton HL. A series of normal stages in the development of the chick embryo. J Morphol 1951;88:49−92.

[226] Nakamura Y. Poultry genetic resource conservation using primordial germ cells. J Reprod Dev 2015;62(5):431−7.

[227] Tajima A, Naito M, Yasuda Y, Kuwana T. Production of germ-line chimeras by transfer of cryopreserved gonadal primordial germ cells (gPGCs) in chicken. J Exp Zool 1998;280(3):265−7.

[228] Nakamura Y, Usui F, Miyahara D, Mori T, Ono T, Takeda K, et al. Efficient system for preservation and regeneration of genetic resources in chicken: concurrent storage of primordial germ cells and live animals from early embryos of a rare indigenous fowl (Gifujidori). Reprod Fertil Dev 2010;22:1237−46.

[229] Park TS, Jeong DK, Kim JN, Song GH, Hong YH, Lim JM, et al. Improved germline transmission in chicken chimeras produced by transplantation of gonadal primordial germ cells into recipient embryos. Biol Reprod 2003;68:1657−62. Available from: https://doi.org/10.1095/biolreprod.102.006825.

[230] Nakajima Y, Minematsu T, Naito M, Tajima A. A new method for isolating viable gonadal germ cells from 7-day-old chick embryos. J Poult Sci 2011;48:106−11.

[231] Nakamura Y, et al. Increased proportion of donor primordial germ cells in chimeric gonads by sterilisation of recipient embryos using busulfan sustained-release emulsion in chickens. Reprod Fertil Dev 2008;20:900−7.

[232] Nakamura Y, et al. Germline replacement by transfer of primordial germ cells into partially sterilized embryos in the chicken. Biol Reprod 2010;83:130−7.

[233] Kang SJ, Choi JW, Kim SY, Park KJ, Kim TM, Lee YM, et al. Reproduction of wild birds via interspecies germ cell transplantation. Biol Reprod 2008;79(5):931−7. Available from: https://doi.org/10.1095/biolreprod.108.069989.

[234] van de Lavoir MC, Collarini EJ, Leighton PA, Fesler J, Lu DR, Harriman WD, et al. Interspecific germline transmission of cultured primordial germ cells. PLoS One 2012;7(5):e35664. Available from: https://doi.org/10.1371/journal.pone.0035664.

[235] Liu C, Khazanehdari KA, Baskar V, Saleem S, Kinne J, Wernery U, et al. Production of chicken progeny (*Gallus gallus domesticus*) from interspecies germline chimeric duck (*Anas domesticus*) by primordial germ cell transfer. Biol Reprod 2012;86:1−8.

[236] Wernery U, Liu C, Baskar V, Guerineche Z, Khazanehdari KA, Saleem S, et al. Primordial germ cell-mediated chimera technology produces viable pure-line Houbara Bustard offspring: potential for repopulating an endangered species. PLoS One 2010;5(12):e15824.

[237] Whyte J, Glover JD, Woodcock M, Brzeszczynska J, Taylor L, Sherman A, et al. FGF, insulin, and SMAD signaling cooperate for Avian primordial germ cell self-renewal. Stem Cell Rep 2015;5(6):1171−82. Available from: https://doi.org/10.1016/j.stemcr.2015.10.008.

[238] Nakajima Y, Asano A, Ishikawa N, Tajima A. Factors involved in spontaneous discharge of gonadal germ cells from developing gonad of 7-day-old chick embryos. J Poult Sci 2014;51:416−23. Available from: https://doi.org/10.2141/jpsa.0130214.

[239] Kim H, Kim DH, Han JY, Choi SB, Ko Y-G, Do YJ, et al. The effect of modified cryopreservation method on viability of frozen-thawed primordial germ cell on the Korean native chicken (Ogye). J Anim Sci Technol 2013;55(5):427−34. Available from: https://doi.org/10.5187/JAST.2013.55.5.427.

[240] Nakamura Y, Usui F, Miyahara D, Mori T, Watanabe H, Ono T, et al. Viability and functionality of primordial germ cells after freeze-thaw in chickens. J Poult Sci 2011;48:57−63.

[241] Nakamura Y, Tasai M, Takeda K, Nirasawa K, Tagami T. Production of functional gametes from cryopreserved primordial germ cells of the Japanese quail. J Reprod Dev 2013;59:580−7.

[242] Griffiths PR. Sex identification in birds. J Exotic Pet Med 2000;9:14−26.

[243] Cakmak E, Peksen CA, Bilgin CC. Comparison of three different primer sets for sexing birds. J Vet Diagn Invest 2017;29(1):59−63.

[244] Lázár B, Anand M, Tóth R, Várkonyi EP, Liptói K, Gócza E. Comparison of the microRNA expression profiles of male and female avian primordial germ cell lines. Stem Cell Int 2018;1−17. Available from: https://doi.org/10.1155/2018/1780679 eCollection 2018.

[245] Rio DC, Ares M, Hannon GJ, Nilsen TW. Purification of RNA using trizol (TRI reagent). Cold Spring Harb Protoc 2010;(6).

[246] Kosin IL. Macro- and microscopic methods of detecting fertility in unincubated hens' eggs. Poult Sci 1944;23:266.

[247] Bakst MR. Predicting fertility. Techniques for semen evaluation, semen storage and fertility determination. Chapter VII. Section 2. 2nd Edition St. Paul, MN: The Midwest Poultry Federation; 2014. p. 79−85.

[248] Liptói K, Varga Á, Hidas A, Barna J. Determination of the rate of true fertility in duck breeds by the combination of two *in vitro* methods. Acta Vet Hung 2004;52:227−33.

[249] Bramwell RK, Marks HL, Howarth Jr. B. Quantitative determination of spermatozoa penetration of the perivitelline layer of the hen's ovum as assessed on oviposited eggs. Poult Sci 1995;74:1875−83.

[250] Wishart GJ, Bakst MR. Predicting fertility Chapter VII. Section 3-4 Techniques for semen evaluation, semen storage and fertility determination. 2nd Edition St. Paul, MN: The Midwest Poultry Federation; 2014. p. 86−94.

Chapter 14

Reproductive technologies in the honeybee (*Apis mellifera*)

Ajda Moškrič[1], Giovanni Formato[2], Maja Ivana Smodiš Škerl[1] and Janez Prešern[1]

[1]*Agricultural Institute of Slovenia, Ljubljana, Slovenia,* [2]*Istituto Zooprofilattico Sperimentale del Lazio e della Toscana, Roma, Italy*

14.1 Introduction

Beekeeping is often related to plant production due to its role in pollination of many economically important crops, such as almonds, oil rape, and citruses. However, in its basics, beekeeping itself is a subsector of the animal production field: in most countries people keep bees to harvest their products, namely honey, pollen, royal jelly, wax, and others, for western markets at least, less important products such as honeybee venom. Beekeepers also make profit by multiplying and selling colonies as well as rearing and selling queen bees. The knowledge of basic biology of honeybees is the prerequisite to understand the reproduction and technology of beekeeping.

Phylogenetically, vertebrates reared and bred for human needs are much closely related to each other than to honeybees. Yet honeybees and their products have been part of human culture for very long time. The paintings in the Araña Cave (Spain) are dated at least 8000 years ago and a picture shows a honey hunter, collecting honey from a tree cavity while bees swarm around it. There are several proofs of the usage of honeybee products from the Neolithic period, such as clay jar fragments containing bee wax [1] or a wax filling of the teeth [2]. The first direct proof of humans keeping bees comes from Tel Rehov in Jordan valley (Israel) where clay cylinders dated to the Iron Age (1000−900 BCE) with remnants of combs and honeybees were found. These cylinders acted as hives and were equipped with opening for bees on one side and lid on the opposite side to access the combs. Honeybee remnants belong to Anatolian honeybee, *Apis mellifera anatoliaca* [3], which is currently not identified as native species for the region [4]. This likely means that at the time there was already both (1) trade with bees in the Middle East and (2) knowledge about keeping bees and perhaps—this is speculation—even rudimentary knowledge of such simple breeding approaches such as culling less productive colonies and keeping swarms of colonies that performed well in economic sense.

14.2 Differences with other managed animals

From the perspective of animal keeping and rearing, there are important differences between honeybees and vertebrates under human care and management. Honeybees are true social animals that live in colonies. Consequently, the standard unit of work is a colony and not an individual—with some exceptions. The honeybee colony is composed of the incumbent queen bee and several thousand —depending on the time within season—worker bees. While the majority of individuals in the colony are females, the queen bee is the only reproductive individual of the female sex at the time. When present, the males—termed drones—can represent a large but never-dominating fraction of the population in the colony. Their presence in the colony is dependent on the status of the colony and availability of food resources. In feral colonies, the percentage of drone comb was established at around 17% of the total comb surface [5]. Honeybee colony is often described as a superorganism due to the multitude of individuals within. The parallels may be drawn between worker bees that "gave up" sexual reproduction for the greater good of the colony and somatic cells in a body of a vertebrate. Though the specialization in workers and reproductive individuals is not as diverse as in ants, there is definite specialization of workers for various tasks [6]. Drones, however, are all reproductive individuals, yet they are offspring of the incumbent queen bee. This peculiarity is important in both conservation biology and breeding. The latter is explained below.

Reproductive Technologies in Animals. DOI: https://doi.org/10.1016/B978-0-12-817107-3.00014-X

Domestication of honeybees is considered incomplete, which is another important difference between honeybees and other production animal species. The beekeeper has limited control of the colonies' lifecycle. Attentive beekeepers know which sources are available in the environment at any moment, yet they have no or very limited direct control over diet selection of the individual colony. The same is true for mating: the attentive beekeepers know the time interval in which virgin queen bee goes for her nuptial flights, yet they have no control over the males she mates with, rare exceptions being isolated mating stations where control can be provided to some degree. Furthermore, each year many swarms "escape" their hive of origin and find their new home somewhere else outside of beekeepers reach. A notion that feral colonies are extinct in the era of *Varroa* has been proven wrong [7—9], and constant gene flow between managed and feral colonies must also be considered.

Furthermore, there are more than 30 different subspecies of western honeybee *Apis mellifera* described so far. From an economic point of view, the most important are the Italian honeybee (*Apis mellifera ligustica*) and the Carniolan honeybee (*Apis mellifera carnica*). Other local subspecies, such as dark honeybee (*Apis mellifera mellifera*), Caucasian honeybee (*Apis mellifera caucasica*), Anatolian honeybee (*A. m. anatoliaca*), Iberian honeybee (*Apis mellifera iberiensis*), Macedonian honeybee (*Apis mellifera macedonica*), or even often overlooked island subspecies (*Apis mellifera siciliana*, *Apis mellifera ruttneri*, *Apis mellifera adami*) are steadily gaining importance through conservation efforts and specialized breeding programs.

14.3   Mating biology

Honeybee queens are polyandrous. Mating occurs about a week after the queen hatches, with several drones during one or several mating (nuptial) flights. This is the only mating of the queen bee during her lifespan: drone sperm is stored for the duration of her life (3 years on average) in a specialized organ called spermatheca, which keeps the sperm viable. Number of males that the queen bee mates with was evaluated with microsatellite molecular markers and was set to be between 10 and 20 with an average value of 16 [10] or 17 [11].

Mating with several drones occurs in rapid succession: the flying queen bee has no behavioral control over the number of drones she mates with and she does not extend or shorten mating flight time to achieve "optimum" number of "partners." It is estimated that two to three nuptial flights are needed to achieve the average number of drones the queen mates with [12]. Indeed, older estimations had queens engaged in up to five mating flights [13]. The queens that take more than one nuptial flight seem to have significantly fewer matings during their first flight compared to those who flew only once. It was also suggested that queens take additional nuptial flights if they fail to achieve mating success threshold [14]. Successful copulation is essentially required to initiate egg laying [15].

A single drone produces between 0.75 and 3.00 (1.73 ± 0.06) mm^3 of sperm while the volume of spermatheca of the newly mated queen (*A. m. carnica*) is between 0.5 and 1.7 mm^3 [16]. Thus the volume of the single drone's sperm should be great enough to fill the spermatheca. Yet the sperm of a single drone fills up only 20% of the spermathecal volume after instrumental insemination (IIS) [17]. The volume of semen required to fill the sprematheca during IIS is about 8 μL.

The volume of semen received by the queen during her multiple mating flights is much greater than the capacity of spermatheca. The semen is initially stored cranially in the lateral oviducts, away from the opening of spermatheca. The transfer of sperm into the spermatheca is a slow process, which can take almost 40 hours [18]. During the process, the sperm is pumped in the reverse direction, passing the spermathecal opening again, where only a small quantity ($\sim 5\%$; 5×10^6 sperm cells) is pumped in the spermatheca and the rest is excreted in small packages [17,19]. The transfer is an active process that involves muscle activity [20], but is also uneven—it seems that sperm from different drones do not enter the spermatheca in equal amounts [21]. Furthermore, sperm from different drones gets mixed in the spermatheca. This was proven experimentally for both IIS, where the semen from different drones was arranged in layers, and in the course of natural matings. The last male does not have precedence over its competitors in the share of the offspring. Paternity frequencies of offspring of different drones are not equal but usually their rank of precedence is kept over time [22]. This cannot be explained by the different amount of sperm they deliver, since similar results were also observed using IIS [21,23].

14.4   Genetics in honeybees

Unlike other managed animal species, honeybees (like all other members of insect order Hymenoptera) maintain haplodiploid sex determination system. There are 16 chromosome pairs in female honeybees (diploid) and 16 single chromosomes in drones (haploid). In the haplodiploid system, males develop from unfertilized eggs while the females—both reproductive and nonreproductive—develop from fertilized eggs. The practical consequences are the following: (1) drones do not have a father—all their genetic information stems from the genetic material of the queen, (2) there is no recombination in development of male gametes due to the haploid nature of the individuals, (3) the genetic material of drone's own gametes is identical to the material of drone itself [24,25] (Fig. 14.1), and this too, has a consequence for breeding, (4) recessive lethal and deleterious alleles are expressed in drones as there is no backup or alternative, and such drones are quickly

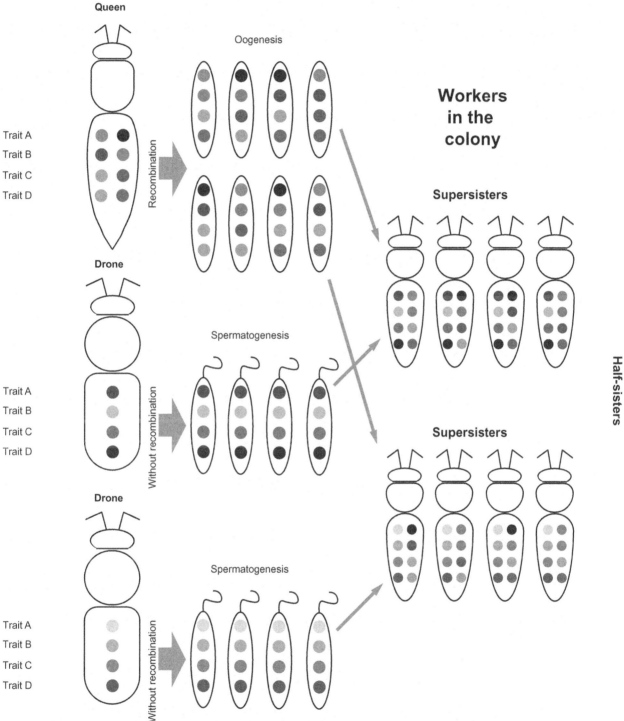

FIGURE 14.1   Haplodiploid system in honeybees. Queen bee (top left) and workers (right) are diploid. Drones (middle and bottom left) are haploid. Oogenesis includes recombination, spermatogenesis does not—all the spermatozoids are exact copy of the drone. Due to the polyandric behavior of the queen, there are several patrilines within the colony: workers belonging to the same patriline are termed "supersisters" and are more related to each other than to the workers belonging to other patrilines ("half-sisters").

removed from the population before mating. The differentiation of larvae to queens and worker bees occurs in early larval stages and depends on nutrition-based caste-specific transcriptional programs that are established and maintained by chromatin-based epigenetic modifications [26]. Differential feeding results in distinct phenotypes that are genetically indistinguishable. There is also an important *caveat* in sex determination in honeybees. For a successful development of a female individual, a sex determination allele called *complementary sex determiner* (*csd*) should be considered. The *csd* is a single locus allele that encodes an SR-type protein *Csd*. In the heterozygous condition, this protein triggers the female sex pathway by female splicing of the transcripts of the *fem* gene [27,28], resulting in functional Fem protein, important in ontogenetic development of female individuals. The hemizygous condition of nonfertilized eggs does result in male splicing, which terminates translation of *fem* mRNA prematurely, thus leading the development into haploid drones [28]. Similar result is achieved in homozygous diploid condition. The result is not a female, but a diploid male. Such drones are removed early after eclosion. Numerous diploid drone larvae are often consequence of mating between closely related individuals, such as brother and sister. High occurrence of diploid drones can weaken the colony by reducing the number of worker bees [29,30]. The initial estimation of around 20 different *csd* alleles has been exceeded; recent papers show much higher numbers: 116−145 worldwide [31], claiming also uneven distribution [32] as a consequence of frequent generation of new alleles. The possibility of homozygous males and consequential weakening of the colony is important when designing a breeding scheme.

14.5 Relationship between individuals in the honeybee colony

The colony itself acts as a superorganism; the tasks are normally divided among worker bees, and the queen is not an absolute monarch but rather an elected official who gets removed from the office when her performance is inadequate [33]. A queen is also the mother of all the individuals in the colony, yet she is not the foundress of the colony like in bumblebees, ants, or wasps. Newly mated queen in natural situation inherits the colony from her predecessor, either through swarming, through supersedure (replacement) or through emergency queen rearing.

The consequence of both the polyandry of the queen bee and the haplodiploid system is a genetically composite colony. As noted above, there is no recombination involved in sperm development. The only recombination involved is the one in queen bee at the time of oogenesis. This means that although drones do differ from each other, all the spermatozoids of a single drone are identical. Taking into account the variability due to the recombination on the mother's side, the coefficient of relatedness between honeybees of the same patrilines R is 0.75. The coefficient R between honeybees of two different patrilines is 0.25. Yet the expected coefficient of the relationship between the individual workers within the colony is inverse to the number of drones that inseminated the queen under condition that frequencies of patriline representations are equal [34,35]. The workers of the subgroup sharing the same father are termed supersisters (Fig. 14.1). The question that needs to be addressed is whether the supersisters favor their own patriline? Would it be better if a queen breeder would select larvae belonging to the same patriline as the majority of nurse bees? Experimentally it was shown that nurse bees select larvae on other parameters such as the nutrition status [36]. However, bees belonging to the same patriline might have a preference for a certain task within the colony. Worker bees belonging to some of the subfamilies have more affinity to carry out certain subset of tasks, such as pollen foraging, scenting, and water foraging [10]. Tasks that honeybees perform within the colony are changing in relationship with the individual's age—so-called age polyethism [37,38]. Certain behaviors, such as foraging, can be selected for and have clear genetic determinants [39]. This is also true for grooming behavior [40], guarding, undertaking, fanning [41,42], hygienic behavior [43], and varroa sensitive hygiene [44]. Especially the latter has received a lot of attention in recent years since differential gene expression studies [45−47], quantitative trait loci mapping of genotypes and determination of single nucleotide polymorphisms [48−52] have confirmed the relationship between hygienic behavior and Varroa-sensitive hygiene in connection to resistance of the colony to varroa mites and other diseases. All these discoveries give possible foundation for using genomic selection within breeding programs.

14.6 Life cycle and reproduction in the honeybee colonies

Two reproductive targets can be distinguished in the colony. The first is reproduction for the colony growth: the queen bee—as the only female sexual individual—lays eggs to increase the number of workers at the time when resources are in abundance. The normal life cycle of the colony in central European geographical latitudes is related to the seasons of the year with peak population in late May or early June (Fig. 14.2). At the time when population approaches its peak, the worker bees rear some of the female larvae as future queens, which is a precondition for the second type of

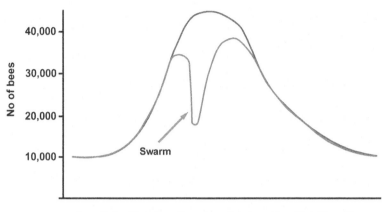

FIGURE 14.2 Rough schematics of seasonal dynamics of the colony in Central Europe. Modern beekeeping practice suggests technological prevention and selection against swarming (dark blue line). Population of colonies that swarm follows the light blue line. Gray arrow marks departure of swarm. New queen has high fecundity and the population often surpasses the one before the swarm.

reproduction: the reproduction of the colony as the unit, termed swarming. Incidentally, this is also the time when drone population reaches its peak as well [53]. Both types of reproduction are used in breeding and rearing bees.

Queen rearing in the colony starts by worker bees designed to prepare special queen cells. These cells are quite often at the edges of the brood nest, resembling cups pointing downwards. If the queen bee lays eggs into those cells, they get special dietary treatment after hatching. This dietary treatment, consisting of royal jelly only, ensures development of ovaries [54]. A colony normally seals between 15 and 25 queen cells [55]. Just before hatching of the virgin queens, the old incumbent queen takes flight with up to two-thirds of individuals [56], whose age-structure analysis shows domination of younger bees over the composition of the original colony [57]. Bees in such swarm take supplies which should suffice for 2−3 days and land at an appropriate natural site—such as a tree branch, normally not far from the hive of origin. Natural swarm can be collected in a short time window, before the honeybees in the swarm reach consensus where to go and which place to colonize next [33]. There is a natural brood interruption starting before swarming. The workers decrease queen feeding and the incumbent queen starts losing weight to be able to fly. Consequently, egg laying almost halts several days before the swarm. The egg laying by the new queen can take up to 11 days, making the pre- and postmating gap up to 16 days long [58−60]. Meanwhile in the hive of origin, the first of the virgin queens that hatches kills other competitors that are still in their cells and prepares for her nuptial flight in 6−7 days. After successful nuptial flights, the new queen starts laying eggs.

The old queen takes residence at some other location, either by choice of her minions or by beekeepers'. With aging, her performance falters. Decreased performance shows in her diminished pheromone profile and lower number of eggs laid. In that case worker bees perform a supersedure of the queen (queen replacement). The signs are similar as in preparation for swarming, the queen cells appear, albeit in lesser numbers. There is also a possibility that queen dies before she is able to lay eggs in queen cell cups. As long as there is any young brood present (larvae less than 3 days old), the colony can rear so-called emergency queen. Both supersedure queen or emergency queen takes nuptial flights and starts laying eggs several days after the flights are completed [55].

14.7   Reproduction of colonies

Honeybee colony requires a queen bee to sustain itself. Despite longer life span, many apiculturists choose to replace their queens every 2 years. Modern beekeeping tends to prevent swarming: swarming halves the population of worker bees in the colony and decreases the honey yield. Beekeepers prefer to prevent this by removing swarm queen cells at the early stage and/or by adding more volume with additional brood chambers to reduce overcrowding. This technique, however, prevents installment of fresh "management" of the hive, which would occur in natural circumstances periodically, as often as once a year.

Queen rearing is a very important activity within the beekeeping field: it provides fresh queens to the industry. For example, official production of queens by the queen breeders within the breeding program for Carniolan honey bee for Slovenian domestic market is about 30,000 [61]. This means that one out of seven of the registered colonies in Slovenia gets new queens. The unrecorded production most likely doubles or increases this number threefold.

Queens reared by the queen breeders are important also for multiplication of the colonies. The apiculturists often multiply their colonies to replace lost ones or to sell so-called nucleus colonies (nucs) to other beekeepers. The queens or queen cells are inserted into freshly formed queenless colonies for sustainment and development. Below we discuss both types of reproduction techniques: those that do and those that do not require extra rearing of the queens.

14.8 Reproduction of colonies without extra queen source

14.8.1 Colony split

The split colony technique uses the honeybee instinct to rear emergency queens. Nucs are formed by removing one or more combs (thus splitting the colony of origin), with both open and covered brood together with worker bees. Such combs should preferably contain some eggs as well. These combs are added in nuc box, together with several combs with empty space for brood/food or comb wax foundations. The bees will attempt to transform one (or several) of the cells with suitable larvae into queen cells. The number of brood combs required differs depending on the time in the season. In Central Europe in May it is possible to use a single brood comb to form the nucleus colony. The same technique in early July can require three, four, or even more brood combs. Presence of drones is also required to mate with the new queen. Beekeeper producing and selling nucs often uses such technique to replenish colonies for the next year: five frames are sold as the nuc and one with brood is left in the colony to develop until next year. This technique of obtaining queens is also used in countries that keep *A. mellifera* but queen breeding is not yet established, like in Bangladesh.

14.9 Reproduction of colonies with extra mated queens

It is possible to create a completely new colony by using bees and/or brood from other colonies and adding queen from extra source. Queen is carefully added in a small queen cage, normally covered with mesh on the side and with several escort bees to tend to her. The mesh allows the odor exchanges and contact between the host colony and the queen. The queen cage is equipped with small piece of sugar dough, which acts as a food source and a removable plug. When releasing queen into the colony, the beekeeper opens the cage to allow access to the sugar plug. Workers in the host colonies eat the dough and thus release the queen into the colony. The process takes some time—depending on the colony strength—which allows the odorant exchange and improves acceptance of the queen into the colony. The beekeeper should inspect the colony in a week to verify the presence of eggs, a sign that the queen was accepted and performs normally. In 2 weeks, the brood should be covered, homogenous, without spotty pattern of the brood (shot-through look), which would be an indication that one or more drones were close relatives to the queen they inseminated (see above).

14.9.1 Shook swarms

Shook swarms are created by brushing worker bees (excluding the queens) from combs of the several different colonies into a temporary vessel (e.g., bucket). Depending on the purpose of the swarm, an appropriate quantity of bees in weight is then transferred into a transport box or nucleus hive with comb foundations/combs without brood. The queen is added from extra source in a special transport cage. If the swarm is sent as the "package bees," the queen cage is kept closed until it is inserted into the hive. If the swarm is inserted directly into the nucleus hive, the queen cage is opened immediately to allow releasing of the queen through sugar dough. Alternatively, a carrier impregnated with queen pheromone can be used instead, and queen added later.

14.9.2 Colony split

It is possible to split the colony and/or assemble nucleus colony in the manner discussed above. After few days, when brood is completely covered, the beekeeper removes any emergency queen cells that might get formed and inserts the queen cage with the newly mated queen and opens it to allow access to dough. This interim period is very important: if emergency cells are present, the newly formed colony will most likely not accept the new queen.

14.10 Reproduction of colonies with extra queen cells

Shook swarms or splits can be both equipped with queen cells from external source; the latter only when "domestic" queen cells have been removed. In this case, the virgin queen will mate on the current location and brood will appear later. Nevertheless, queen cells are sufficient to keep the colony together.

14.11 Reproduction of honeybee queens

14.11.1 Queen rearing

Queen rearing is an important part of beekeeping. There are two aspects that will be highlighted: (1) rearing of queens as a method of multiplication and (2) selection of the colonies that are appropriate for multiplication and breeding. In nature, the colony starts rearing its own queens when two conditions are met: (1) colony is strong enough in terms of population, so that the space within the hive gets crowded, and (2) resources are abundant. In central Europe, the second condition is usually met in late spring, with some exceptions. The beekeeper who wishes to rear queens looks for the colony that provides the source material—newly hatched larvae and the colony, which will initiate the actual rearing—"starter" colony. Starter colony must be prepared in terms of additional brood, bees, and feed if a greater number of reared queens are planned. Starter colonies can be queenright or queenless.

A special rearing frame is prepared, carrying the plastic or wax rearing cups pointing downwards. The larvae of suitable age are then grafted—transferred from their original cells into a rearing cup. The best larval age for grafting is between 0 and 24 hours after eclosion from the egg [62,63]. Some queen breeders prefer simple starter colonies with one frame of brood, one frame of food, one rearing frame, and two feeder frames. These starters need to be often replenished with young bees and brood, as they have no queen to sustain themselves, yet they are simple to monitor and handle. Alternatively, the starter with two or more brood supers could be set up. One of them serves as rearing super. The queen is excluded from the rearing super with queen excluder. The emerging brood is periodically transferred from queenright brood chamber to rearing super to provide a steady quantity of nursing bees; frame with open-brood keeps nursing bees close to the queen cells. Such constellation is more difficult to monitor, and yet the hive is mostly self-sustainable.

After the queen cells are completed and closed by nurse bees, and it is possible to remove the breeding frame and release the space for another batch of grafted larvae. The breeding frame can be then transferred into another, queenright colony, where it can be inserted into a super, separated with queen excluder. Alternatively, the queen cells can be transferred into the incubator. In any case, the first queen that hatches attempts to obliterate the others. In order to prevent that, a special mesh cage is often added over the cells to separate them physically one from another. Cells in the incubator can be carefully inserted in their own chamber each.

14.11.2 Queen mating

Normally, the virgin queens will take mating flights 6—10 days after hatching. Even earlier, orientation flights take place: their purpose is reconnaissance. Newly hatched queen needs to be added to its mating colony as soon as possible. Mating boxes, or mating hives, are special hives used for the purpose of mating. Mating hives come in various sizes and volumes and are normally initiated as a queenless shook swarm. In most cases, such mating boxes are of much smaller volume than the normal hives, to reduce the number of bees required. For example, Zander-style mating box has a single comb with $\sim 86\,\mathrm{cm}^2$ surface and requires 0.25 kg of "dry" bees. Such shook swarms are made by brushing combs of several colonies and are probably the greatest expense in large-scale queen breeding. Brushing depletes colonies of workers and consequentially strength and yield of individual colony. After formation, the mating hives are provisioned with sugar dough, kept enclosed in the darkness for a few days to stimulate community building and prevent migrating of bees to another hive, and then transported to another location away from the hives of origin. A few days after establishment, the virgin queens are added to the colony in the transport cage, plugged with sugar dough. This prevents immediate and direct exiting of the queen into the hive, allowing odors to mix, and results in higher acceptance rate of the queen by the workers. New queen starts laying eggs a few days after the last mating flight. Queen breeder must inspect type and shape of covered brood to verify the quality of mating before offering the queen to the market. Mating colonies can be recycled: after the mated queen is removed, new virgin queen can be inserted for the next cycle of mating. It is possible to use queen cell instead of a hatched queen. Such approach would have queen hatching in the hive, thus enabling direct contact and immediate service of queen attendant bees that feed and groom the queen bee. The downside is that such approach prolongs cycles between queen generations and mating box recycling for several days.

The fact that queen mates during the flight in the air, and with more than one drone, makes the mating process difficult to control. Therefore there are two concepts queen breeders may use in queen mating: open mating and controlled mating. Open mating attempts no control over source of the drones in the area; the only requirement is a sufficient quantity of drones. The alternative concept is controlled mating with known source of drones. This may be achieved by either IIS or an isolated area—often a mountain valley or an island where there are no other drone sources except the selected colonies. In controlled mating, a special care needs to be taken at the initialization of mating boxes: all the drones need to be "filtered out" during population of the mating hives.

Drones start exiting the hive at noon and are active till 17:.00, with peak of their departures between 14:.00 and 16:.00 [64]. The majority ($>80\%$) of queens' mating flight were observed between 13:.00 and 16:.00 [65]. Weather is an important factor: proper day for mating must be as clear, warm, and windless as possible. Location where mating takes place is termed drone congregation area (DCA). DCAs are not well defined in qualitative terms, yet they seem to be persistent from year to year. Drones wait for possible queen arrival while flying in the air.

DCA properties are probably of less interest if open mating is considered: the only consideration being that there are plenty of drones available. On the contrary, knowledge of DCAs in neighborhood is very important in controlled mating. To achieve controlled mating, two conditions need to be met: (1) the area where mating takes place must be isolated in a sense that there are no other colonies than the desired ones and (2) drones at the DCAs must belong to the known source.

14.11.3 Mating control at mating stations

The lack of direct mating control is one of the reasons why honeybees are not considered as domesticated as other farm animals. Yet it is possible to achieve situation in which the genetic material involved is largely under human control. Queen breeders and managers of breeding programs often try to make use of isolated regions, namely small islands in North Sea, Baltic or Adriatic and alpine valleys, or regions away from managed population.

For the purpose of isolation testing, sealed mating boxes with virgin queens are placed at the location with simultaneous control at the nearby location where the presence of honeybee colony is known without doubt. After a given period (2–3 weeks), the boxes are inspected for the presence of eggs and brood. Additional inspection is performed later to verify the type of brood: drone laying queen could signal absence of drones at the location. Verification practice in Slovenia suggests that at least two cycles of testing are required. Verification of isolation is, however, time-consuming and therefore the tested location can rarely be used as mating station in the same season. Isolation ensures that only drones placed at the location will actually play a role in mating: thus, the pedigree is known on both maternal and paternal side, which is prerequisite for calculation of reliable breeding values, and genetic improvement evaluated and predicted.

Sufficient number of drones is required at the mating station: DCAs are stable only when several thousand drones are available [66]. Such numbers are too high to be provided by a single colony, which can manage at most around 2000 drones: eight or even more drone-producing colonies are suggested at the isolated location. These colonies should be of known genetics. Also, the number of drone-providing colonies seems to be important: it turned out that greater the number of mating hives, greater the required number of drones. For isolated mating stations, two additional drone-producing colonies per 25–50 virgin queens are suggested [67]. In case of pedigree mating station, it is important that drone-producing colonies are offspring of the same queen (see 14.13: Selection in honeybees).

Mating stations in alpine valley are often a good choice. Being fenced on three sides, there is only one direction that needs to be monitored for foreign drones. If other, nonrelated apiaries are present, it is best that queen breeders provide these nonrelated beekeepers with queens of the same line as those in the drone colonies at the mating station. Alpine valleys are quite poor in resources and there is scarce danger that mobile apiaries for honey production would be suddenly crammed in. The absence of physical isolation of the mating station requires different approach: a large surrounded area, often more than 10 km in radius, needs to be cleared of foreign drone sources. Alternatively, beekeepers in the area can be offered queens of known genetics. In geographically monotonous surroundings, the DCAs are not formed: the drones are themselves evenly disperse [66]. Such situation probably requires even higher numbers of drone-producing colonies.

14.12 Instrumental insemination

IIS is the only method that ensures that only selected genetic material is used. IIS does not require huge multitude of drones like the mating station does, yet the source of drones needs to be strictly controlled. Often, an empty drone comb is inserted in the colony at the edge of the nest. After the brood is covered, the comb is transferred to honey super separated with queen excluder from brood chamber. The excluder prevents entrance of foreign drones and exit of domestic drones before they are marked. Every few days, fresh drones are marked and transferred to brood chamber. Marks can be designed to label the age of drones, too.

The first phase drone collection at the hive. Drones are best to be allowed to defecate first, which they do during their attempted mating flights. The most comfortable approach is to collect drones when they return to the hive. Drones are very sensitive: without the worker bees they tend to wither away in about half an hour. To prevent this, beekeepers performing IIS use special cage, which features the bottom part with worker bees, and some honeycomb. This bottom part is separated with mesh from the top part, which acts as drone storage. This way, the drones are easily accessible to the person in charge of IIS and have access to worker bees who feed them.

In the second phase, sperm is collected. Individual drones are held between thumb and forefinger at the thorax. The person in charge applies gentle pressure and rolling to stimulate the contraction of abdominal musculature. Further pressure causes full eversion of endophallus and ejaculation. Sperm is then sucked into a microcapillary. The volume of the sperm collected from a single drone is between 0.8 and 1 µL. It is important to realize that sperm collection is a delicate process requiring sterile work and special equipment [68].

In order to perform the insemination, the virgin queen is anesthetized with CO_2 and inserted in a tube-like holder with the posterior part of the abdomen exposed. Abdominal plates are separated with special hooks or grip forceps to expose the vaginal orifice. The microcapillary containing semen is then carefully maneuvered to bypass the valvefold and reach median oviducts. The semen, approximately 8 µL, is slowly injected. The queen is reinserted into the colony when fully awake. CO_2 stimulates the beginning of the egg laying: some experts suggest a preinsemination short-term (2−3 minutes) of narcosis a day or two before the insemination. It is very important to prevent any uncontrolled mating flight after insemination, which may be achieved by wing clipping [68].

A comparison between naturally mated queens and instrumentally inseminated showed similar performances, although often the instrumentally inseminated queens performed better [69]. Despite that, it seems that IIS is well suited as a tool for queen breeders to assist in selection, rather than a tool for mass production.

14.13   Selection in honeybees

The honeybee queen is the single vessel of genetic material that matters in reproduction in the colony. After completing her mating flights, she is a vector of both female and male gametes. Yet in stark contrast with vertebrate species discussed in this book, the queen itself is never evaluated directly, but via the colony performance. Colony performance in any trait is regarded as composite of a queen effect and workers effect, basically covering two generations of bees [70]. The queen contributes to her colony performance in two ways, through her direct genetic effect when her genes pass to her progeny and through her maternal effect, the ability to provide a common environment for the colony [71].

There are several colony traits that are normally of interest to a beekeeper: calmness, defensive behavior, swarming, and honey yield. For each trait, the performance of a colony is graded on the scale 1−4, with 1 being the lowest and 4 being the best grade. Grading is performed several times during the season, with exception of swarming, which is evaluated only during the swarming season (Fig. 14.3). Additionally, colony development is being evaluated as a number of frames covered with brood and bees. Calmness and low aggressivity are greatly prized among beekeepers and they define work intensity requirements. Honey yield is often an important trait, the unit being absolute kilograms extracted from the colony. Nowadays, varroa-related traits are gaining importance. Colony infestation level is being measured with powdered sugar test, while natural mite mortality is being monitored and colony's hygienic behavior is being tested [72]. Some of the abovementioned economically important colony traits were shown to be heritable at a certain rate. The queen and worker effect for a certain trait were found to be negatively correlated [73]. A selection index was proposed to improve both effects simultaneously [74].

Morphometry and other physical features are often being considered, especially where the purity of subspecies (race) is important. Morphometric analysis is used to evaluate a set of external characters of the worker or drone with the help of a microscope and compare them with reference standards of the different subspecies of *A. mellifera* [4] (Fig. 14.4). The most important characters are: length and width of the right forewing, distances between nodes of right forewing venation, length of proboscis, length of femur, tibia and metatarsus, width of metatarsus, longitudinal diameter of tergite 3 (T3) and 4 (T4), measurement of sternite 3 (S3), longitudinal (L6) and transversal (T6) lengths of sternite 6, and pigmentation of tergites. Slovenian breeding program for *A. m. carnica*, for example, requires evaluation of worker coloration (the first and second rings on the abdomen should be grayish to brownish color). Grades are again 1−4, 4 being the only acceptable form.

Marker-assisted selection (MAS) is being considered, mostly in varroa-related research to contribute in breeding programs of varroa-resistant lines [47,75]. Promising results show that net beekeeper's profit gain from a MAS colony

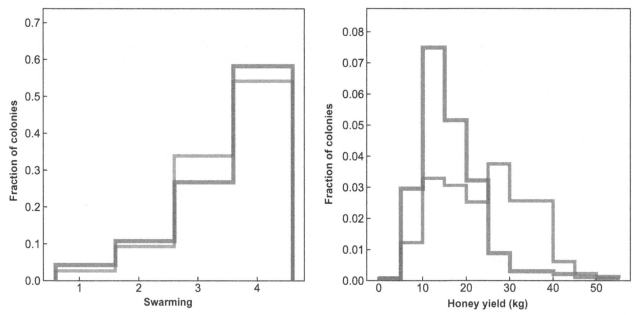

FIGURE 14.3 Distribution of grades for swarming (left) obtained in performance testing in Slovenian breeding program during years 2017 (blue) and 2018 (orange). Distribution of honey yield obtained in performance testing in Slovenian breeding program during years 2017 (blue) and 2018 (orange).

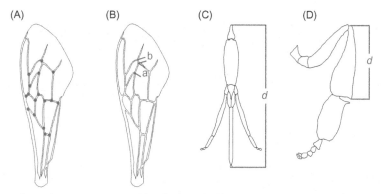

FIGURE 14.4 Some of the morphometric measurements used in selection. (A) Distances between wing venation nodes (red points) are used in morphometric analysis. (B) Cubital index: it represents ratio of distances between three nodes at the venation (a/b). (C) Length of proboscis (*d*). (D) Length of tibia (*d*).

is ranging between 9% and 96% in comparison to field-assisted selection, depending on the varroa load in case of ineffective varroa treatment [76]. MAS is not yet commercially available in honeybees at the moment (late 2019).

Controlled mating requires planning of selection activities well ahead. Both in maternal and paternal line the parents are dam of queens (DoQ). Drones do not have a father; DoQ in paternal line is therefore often considered as sire of queens (SoQ) [73]. The lack of recombination at the level of gamete production in drones adds another generation of individuals in the calculation: drones, which require approximately 40 days from egg to sexual maturity. Also, an adequate number of drones must be provided, especially in isolated mating stations. To achieve this, sister queens are reared from eggs laid by SoQ, mated, and in the following year installed to the mating station to produce drones as a single purpose (drone-producing queens, DpQ). DpQ can be open mated at the mating station, since all the genetic material for drones comes from the mother side. Due to requirement of DpQ, selection of SoQ needs to take place 1 year ahead of selection of DoQ (Fig. 14.5). On the contrary, IIS does not require such multitude of drones since DpQ are not needed as drone replicators. Using IIS, it is therefore possible to skip one generation and use drones produced directly by SoQ.

(A)

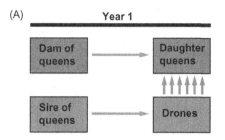

(B)

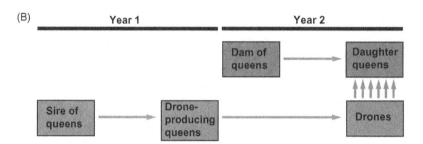

FIGURE 14.5 Time required for controlled mating after selection has been completed. Comparison between instrumental insemination (A) and natural mating at isolated mating station (B). In (B), high drone numbers are required. Drone-producing queens take an additional year to prepare. Therefore paternal line needs to be selected 1 year in advance.

14.14 Breeding programs

Breeding programs for honeybees are established for two purposes: (1) genetic progress, like in any other production animal species under human management, or (2) conservation, to sustain some local subspecies [72]. Many beekeeping organizations around the world have their own bee breeding programs, and performance tests are their essential component. The performance testing should answer, which colonies are preferable as the source of genetic material for the next generation in order to achieve the breeding objective. For the purpose of breeding objectives such as calmness and honey yield, design of performance tests reflects those used in other farm animals, with an important difference that the performance observed is performance of the colony, not of the individual.

In performance testing, like the one implemented in BEEBREED, a Germany-based breeding program, each breeder receives queens from other breeders for testing. At the same time, the breeder distributes queens of his own production among other queen breeders. Queens are introduced to nucs. These nucs develop into full sized colonies following the season when they are evaluated. Based on results, a DoQ and SoQ are selected for further multiplication. After successful mating of her daughters, some of the newly mated queens are distributed among the queen breeders in an organized queen exchange. From a practical standpoint, this means that each queen breeder receives genetic material to be evaluated from other breeders and, if acceptable, to be reared. Slovenian breeding program for native Carniolan honeybee *A. m. carnica* has a different design. Several daughters of each selected DoQ are distributed to beekeepers at various locations, who evaluate their performance following a set of prescribed objective rules. At the end of the second season, these queens are evaluated, and decisions are made about selected DoQ and SoQ for the next seasons. Alternatively, queen breeders can gain back individually tested queens, if they like their performance [61].

Results of performance testing are used to predict the breeding value of future generations and to suggest combinations of dams and sires. The evaluation of the contribution of the maternal line is very straightforward. The paternal line, however, requires some simplifications due to polyandric behavior of queen, single time mating and the number of drones required at the mating station. The set of DpQ at the mating station can be treated as single individual (sire), providing that DpQ are sisters [77].

Breeding programs have a role in conservation as well. Often beekeepers, in order to bring immediate advantages of breeding to production level, introduce bees of allochthonous origin (Fig. 14.6), at the expense of autochthony risking genetic erosion, integrity of local populations, and possible accompanying pests and pathogens. Recent research regarding genotype−environment interaction [78] shows that locally adapted honeybee races perform better [79−81]. Conservation breeding programs improve all economically important and desired traits in local races.

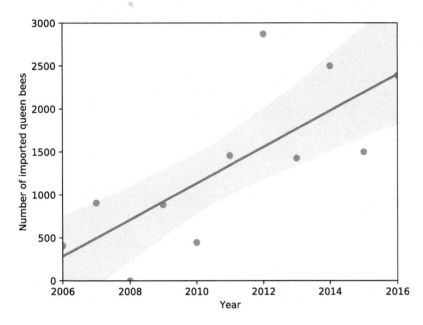

FIGURE 14.6 Increasing yearly numbers of imports of honeybee queens from non-EU countries recorded at Fiumicino International Airport, Rome, Italy.

Conclusion

Honeybees have less direct impact on the human diet in comparison to other farm animals, yet they have a profound indirect influence on global food production being the pollinators, and the global value of their ecosystem service goes into billions of euros yearly, together with being the providers of their own products (honey). It is therefore of paramount importance to safeguard this species, through both an integrated ecosystem approach and the application and improvement of breeding schemes and genetic programs.

References

[1] Roffet-Salque M, Regert M, Evershed RP, Outram AK, Cramp LJE, Decavallas O, et al. Widespread exploitation of the honeybee by early neolithic farmers. Nature 2015;527(7577):226–30 <https://doi.org/10.1038/nature15757>.

[2] Bernardini F, Tuniz C, Coppa A, Mancini L, Dreossi D, Eichert D, et al. Beeswax as dental filling on a neolithic human tooth. PLoS One 2012;7(9):e44904. Available from: https://doi.org/10.1371/journal.pone.0044904.

[3] Bloch G, Francoy TM, Wachtel I, Panitz-Cohen N, Fuchs S, Mazar A. Industrial apiculture in the Jordan valley during Biblical times with Anatolian honeybees. Proc Natl Acad Sci U S A 2010;107(25):11240–4. Available from: https://doi.org/10.1073/pnas.1003265107.

[4] Ruttner F. Biogeography and taxonomy of honeybees. Springer Verlag; 1987.

[5] Seeley TD, Morse RA. The nest of the honey bee (*Apis mellifera* L.). Insectes Soc 1976;23(4):495–512.

[6] Moritz RFA, Fuchs S. Organization of honeybee colonies: characteristics and consequences of superorganism concept. Apidologie 1998;29:7–21. Available from: https://doi.org/10.1051/apido:19980101.

[7] Seeley TD. Honey bees of the Arnot Forest: a population of feral colonies persisting with *Varroa destructor* in the northeastern United States. Apidologie 2007;38(1):19–29. Available from: https://doi.org/10.1051/apido:2006055.

[8] Oleksa A, Gawronski R, Tofilski A. Rural avenues as a refuge for feral honey bee population. J Insect Conserv 2012;17(3):465–72. Available from: https://doi.org/10.1007/s10841-012-9528-6.

[9] Kohl PL, Rutschmann B. The neglected bee trees: European beech forests as a home for feral honey bee colonies. Peer J 2018;6:e4602. Available from: https://doi.org/10.7717/peerj.4602.

[10] Kryger P, Kryger U, Moritz RFA. Genotypical Variability for the Task of Water Collecting and Scenting in a Honey Bee Colony. Ethology 2000;106(9):769–79. Available from: https://doi.org/10.1046/j.1439-0310.2000.00571.x.

[11] Jensen AB, Palmer KA, Chaline N, Raine NE, Tofilski A, Martin SJ, et al. Quantifying honey bee mating range and isolation in semi-isolated valley by DNA microsatellite paternity analysis. Conserv Genet 2005;6(4):527–37. Available from: https://doi.org/10.1007/s10592-005-9007-7.

[12] Tarpy P. No behavioral control over mating frequency in queen honey bees (*Apis mellifera* L.): implications for the evolution of extreme polyandry. Amer Nat 2000;155(6):820–7. Available from: https://doi.org/10.1086/303358.

[13] Alber M, Jordan R, Ruttner F, Ruttner H. Von der Paarung der Honigbiene. Z für Bienenforsch 1955;3:1–28.

[14] Schlüns H, Moritz RFA, Neumann P, Kryger P, Koeniger G. Multiple nuptial flights, sperm transfer and the evolution of extreme polyandry in honeybee queens. Anim Behav 2005;70(1):125–31. Available from: https://doi.org/10.1016/j.anbehav.2004.11.005.

[15] Koeniger G. In welchem Abschnitt des Paarungsverhaltens der Bienenkönigin findet die Induktion der Eiablage statt? Apidologie 1981;12 (4):329−43. Available from: https://doi.org/10.1051/apido:19810403.

[16] Prešern J, Smodiš Škerl MI. Parameters influencing queen body mass and their importance as determined by machine learning in honey bees (*Apis mellifera carnica*). Apidologie 2019;50(5):745−57. Available from: https://doi.org/10.1007/s13592-019-00683-y.

[17] Woyke J. Naturalne i sztucne unasienianie matek pszczelich. Pszczelnicze Zesz Nauk 1960;4(3−4):183−275.

[18] Woyke J. Dyanmics of entry of spermatozoa into the spermatheca of instrumentally inseminated queen honeybees. J Apic Res 1983;22 (3):150−4.

[19] Gessner B, Ruttner F. Transfer der Spermatozoen in die Spermatheka der Bienenköningin. Apidologie 1977;8(1):1−18.

[20] Ruttner F, Koeniger G. Die Füllung der Spermatheka der Bienenkönigin. Zeitschrift für Vergleichende. Physiologie 1971;72(4):411−22.

[21] Moritz RFA. Homogeneous mixing of honeybee semen by centrifugation. J Apic Res 1983;22(4):249−55. Available from: https://doi.org/10.1080/00218839.1983.11100595.

[22] Franck P, Solignac M, Vautrin D, Cornuet J-M, Koeniger G, Koeniger N. Sperm competition and last-male precedence in the honeybee. Anim Behav 2002;64(3):503−9. Available from: https://doi.org/10.1006/anbe.2002.3078.

[23] Franck P, Coussy H, le Conte Y, Solignac M, Garnery L, Cornuet J-M. Microsatellite analysis of sperm admixture in honeybee. Insect Mol Biol 1999;8(3):419−21. Available from: https://doi.org/10.1046/j.1365-2583.1999.83131.x.

[24] Kerr WE, Laidlaw HH. General genetics of bees. Adv Genet 1956;8:109−53. Available from: https://doi.org/10.1016/s0065-2660(08)60501-5.

[25] Hachinohe Y, Onishi N. On the meiosis of the drone honey bee (*Apis mellifica*). Bull Natl Inst Agric Sci Ser 1952;3:83−7.

[26] Wojciechowski M, Lowe R, Maleszka J, Conn D, Maleszka R, Hurd PJ. Phenotypically distinct female castes in honey bees are defined by alternative chromatin states during larval development. Genome Res 2018;28:1532−42. Available from: https://doi.org/10.1101/gr.236497.118.

[27] Hasselmann M, Gempe T, Schiott M, Nunes-Silva CG, Otte M, et al. Evidence for the evolutionary nascence of a novel sex determination pathway in honeybees. Nature 2008;454:519−22. Available from: https://doi.org/10.1038/nature07052.

[28] Gempe T, Hasselmann M, Schiott M, Hause G, Otte M, et al. Sex determination in honeybees: two separate mechanisms induce and maintain the female pathway. PLoS Biol 2009;7:e1000222. Available from: https://doi.org/10.1371/journal.pbio.1000222.

[29] Woyke J. What happens to diploid drone larvae in a honeybee colony. J Apic Res 1963;2:73−5. Available from: https://doi.org/10.1080/00218839.1963.11100063.

[30] Woyke J. Drone larvae from fertilized eggs of the honey bee. J Apic Res 1963;2:19−24. Available from: https://doi.org/10.1080/00218839.1963.11100052.

[31] Lechner S, Ferretti L, Schöning C, Kinuthia W, Willemsen D, Hasselmann M. Nucleotide variability at its limit? Insights into the number and evolutionary dynamics of the sex-determining specificities of the honey bee *Apis mellifera*. Mol Biol Evol 2013;31(2):272−87. Available from: https://doi.org/10.1093/molbev/mst207.

[32] Zareba J, Blazej P, Laszkiewicz A, Sniezewski L, Majkowski M, Janik S, et al. Uneven distribution of complementary sex determiner (csd) alleles in *Apis mellifera* population. Sci Rep 2017;7:2317. Available from: https://doi.org/10.1038/s41598-017-02629-9.

[33] Seeley TD. Honeybee democracy. Princeton University Press; 2010. p. 273.

[34] Laidlaw HH, Page RE. Polyandry in honey bees (*Apis melliefera* L.): sperm utilization and intracolony genetic relationships. Genetics 1984;108 (4):985−97.

[35] Estoup A, Solignac M, Cornuet J-M. Precise assessment of the number of patrilines and of genetic relatedness in honeybee colonies. Proc Roy Soc B 1994;258(1351):1−7. Available from: https://doi.org/10.1098/rspb.1994.0133.

[36] Sagili RR, Metz BN, Lucas HM, Chakrabarti P, Breece CR. Honey bees consider larval nutritional status rather than genetic relatedness when selecting larvae for emergency queen rearing. Sci Rep 2018;8(1). Available from: https://doi.org/10.1038/s41598-018-25976-7.

[37] Rösch GA. Untersuchungen über die Arbeitsteilung im Bienenstaat. J Comp Phys A 1925;2(6):571−631.

[38] Rösch GA. Untersuchungen über die Arbeitsteilung im Bienenstaat. II. Die Tätigkeiten der Arbeitsbienen unter experimentell veränderten Bedingungen. Z. Vgl Pysiol 1930;12:1−17. Available from: https://doi.org/10.1007/978-3-662-41225-1_1.

[39] Page RE, Waddington KD, Hunt GJ, Fondrk MK. Genetic determinants of honey bee foraging behavior. Anim Behav 1995;50 1617-25.

[40] Frumhoff PC, Baker J. A genetic component to division of labour within honey bee colonies. Nature 1988;333(6171):358−61. Available from: https://doi.org/10.1038/333358a0.

[41] Robinson GE, Page RE. Genetic determination of guarding and undertaking in honey-bee colonies. Nature 1988;333(6171):356−8.

[42] Su S, Albert S, Zhang S, Maier S, Chen S, Du H, et al. Non-destructive genotyping and genetic variation of fanning in a honey bee colony. J Insect Phys 2007;53(5):411−17. Available from: https://doi.org/10.1016/j.jinsphys.2007.01.002.

[43] Rothenbuhler WC. Behavior genetics of nest cleaning in honey bees. IV. Responses of F1 and backcross generations to disease-killed brood. Am Zool 1964;4(2):111−23.

[44] Boecking O, Bienefeld K, Drescher W. Heritability of the Varroa-specific hygienic behaaviour in honey bees (Hymenoptera: Apidae). J Anim Breed Genet 2000;117(2000):417−24. Available from: https://doi.org/10.1046/j.1439-0388.2000.00271.x.

[45] Le Conte Y, Alaux C, Martin JF, Harbo JR, Harris JW, Dantec C, et al. Social immunity in honeybees (*Apis mellifera*): transcriptome analysis of *Varroa*-hygienic behavior. Insect Mol Biol 2011;20(3):399−408. Available from: https://doi.org/10.1111/j.1365-2583.2011.01074.x.

[46] Navajas M, Migeon A, Alaux C, Martin-Magniette M, Robinson G, Evans J, et al. Differential gene expression of the honey bee *Apis mellifera* associated with *Varroa destructor* infection. BMC Genomics 2008;9:301. Available from: https://doi.org/10.1186/1471-2164-9-301.

[47] Mondet F, Alaux C, Severac D, Rohmer M, Mercer AR, Le Conte Y. Antennae hold a key to Varroa-sensitive hygiene behaviour in honey bees. Sci Rep 2015;5:10454. Available from: https://doi.org/10.1038/srep10454.

[48] Tsuruda JM, Harris JW, Bourgeois L, Danka RG, Hunt GJ. High-resolution linkage analyses to identify genes that influence Varroa sensitive hygiene behavior in honey bees. PLoS One 2012;7(11):e48276. Available from: https://doi.org/10.1371/journal.pone.0048276.

[49] Holloway B, Tarver MR, Rinderer TE. Fine mapping identifies significantly associating markers for resistance to the honey bee brood fungal disease, Chalkbrood. J Apic Res 2013;52(3):134−40. Available from: https://doi.org/10.3896/IBRA.1.52.3.04.

[50] Spötter A, Gupta P, Mayer M, Reinsch N, Bienefeld K. Genome-wide association study of a Varroa-specific defense behavior in honeybees (*Apis mellifera*). J Hered 2016;107(3):220−7. Available from: https://doi.org/10.1093/jhered/esw005.

[51] Harpur BA, Guarna MM, Huxter E, Higo H, Moon K-M, Hoover SE, et al. Integrative genomics reveals the genetics and evolution of the honey bee's social immune system. Genome Biol Evol 2019;11:937−48. Available from: https://doi.org/10.1093/gbe/evz018.

[52] Conlon BH, Aurori A, Giurgiu AI, et al. A gene for resistance to the *Varroa* mite (Acari) in honey bee (*Apis mellifera*) pupae. Mol Ecol 2019;28:2958−66. Available from: https://doi.org/10.1111/mec.15080.

[53] Atkins EL, Grout RA. The hive and the honey bee: a new book on beekeeping which continues the tradition of "Langstroth on the hive and the honeybee. Dadant & Sons; 1975. p. 740.

[54] Page RE, Peng CY-S. Aging and development in social insects with emphasis on the honey be*e*, Apis mellifera L. Exp Gerontol 2001;36 (4−6):695−711. Available from: https://doi.org/10.1016/S0531-5565(00)00236-9.

[55] Winston ML. The biology of the honey bee. Harvard University Press; 1995. p. 281.

[56] Martin P. Die Steuerung der Volksteilung beim Schwärmen der Bienen. Zugleich ein Beitrag zum Problem der Wanderschwärme. Insect Soc 1963;10(1)):13−42.

[57] Gilley DC. The identity of nest-site scouts in honey bee swarms. Apidologie 1998;29(3):229−40. Available from: https://doi.org/10.1051/apido:19770101.

[58] Allen MD. The behavior of honeybees preparing to swarm. Brit J Anim Behav 1956;4(1):14−22. Available from: https://doi.org/10.1016/s0950-5601(56)80011-7.

[59] Morse RA, Dyce EJ, Young RG. Weight Changes by the Queen Honey Bee During Swarming. Ann Entomol Soc Am 1966;59(4):772−4.

[60] Oertel E, Drones IN. In: Root AI, editor. ABC and XYZ of Bee Culture. 1950. p. 217−9. Available from: http://dx.doi.org/10.1093/aesa/49.5.497.

[61] Kozmus P, Podgoršek P, Smodiš Škerl MI, Prešern J, Metelko M, Hrastelj M. Rejski program za kranjsko čebelo (2018-2023): (*Apis mellifera carnica*, Pollmann 1879). Lukovica: Čebelarska zveza Slovenije; Ljubljana: Kmetijski inštitut Slovenije; 2018. 47 p.

[62] Woyke J. Correlations between the age at which honeybee brood was grafted, characteristics of the resultant queens, and results of insemination. J Apic Res 1971;10(1):45−55.

[63] Nelson DL, Gary NE. Honey productivity of honeybee colonies in relation to body weight, attractiveness and fecundity of the queen. J Apic Res 1983;22(4) 209-13.

[64] Ruttner F. The life and flight activity of drones. Bee World 1966;47(3):93−100. Available from: https://doi.org/10.1080/0005772X.1966.11097111.

[65] Heidinger IMM, Meixner MD, Berg S, Büchler R. Observation of the mating behavior of honey bee (*Apis mellifera* L.) queens using radio-frequency identification (RFID): factors influencing the duration and frequency of nuptial flights. Insects 2014;5:513−27. Available from: https://doi.org/10.3390/insects5030513.

[66] Koeniger G, Koeniger N, Tiesler F-K. Paarungsbiologie und Paarungskontrolle bei der Honigbiene. Druck- und Verlagshaus Buschhausen; 2014. p. 384.

[67] Tissler F-K, Englert E. Aufzucht, paarung und verwertung von königinnen. Franckh-Kosmos Verlag; 1989. p. 251.

[68] Cobey SW, Tarpy DR, Woyke J. Standard methods for instrumental insemination of *Apis mellifera* queens. J Apic Res 2013;52(4)):1.18. Available from: https://doi.org/10.3896/ibra.1.52.4.09.

[69] Cobey SW. Comparison studies of instrumentally inseminated and naturally mated honey bee queens and factors affecting their performance. Apidologie 2007;38(4):390−410. Available from: https://doi.org/10.1051/apido:2007029.

[70] Chevalet C, Cornuet JM. Etude théorique sur la selection du caractère "production de miel" chez l'abeille. I. Modèle génétique et statistique. Apidologie 1982;13:39−65.

[71] Bienefeld K, Ehrhardt K, Reinhardt F. Genetic evaluation in honey bee considering queen and worker effects − a BLUP-animal model approach. Apidologie 2007;38:77−85. Available from: https://doi.org/10.1051/apido:2006050.

[72] Uzunov A, Büchler R, Bienefeld K. Performance testing protocol. A guide for European honey bee breeders. Smartbees, Fp7-Kbbe 2013;2015:1.3.

[73] Bienefeld K, Pirchner F. Heritabilities for several colony traits in honey bee (*Apis mellifera carnica*). Apidologie 1990;21:175−83. Available from: https://doi.org/10.1051/apido:19900302.

[74] Bienefeld K, Pirchner F. Genetic correlations among several colony characters in honey bee (Hymenoptera: Apidae) taking queen and worker effects into account. Ann Entomol Soc Am 1991;84(3):324−31. Available from: https://doi.org/10.1093/aesa/84.3.324.

[75] Behrens D, Huang Q, Geßner C, Rosenkranz P, Frey E, Locke B, et al. Three QTL in honey bee *Apis mellifera* L. suppress reproduction of the parasitic mite *Varroa destructor*. Ecol Evol 2011;1(4):451−8. Available from: https://doi.org/10.1002/ece3.17.

[76] Bixby M, Baylis K, Hoover SE, Currie RW, Melathopoulos AP, Pernal SF, et al. A bio-economic case study of Canadian honey bee (Hymenoptera: Apidae) colonies: marker-assisted selection (MAS) in queen breeding affects beekeeper profits. J Econ Entomol 2017;110 (3):816−25. Available from: https://doi.org/10.1093/jee/tox077.

[77] Bienefeld K, Reinhardt F, Pirchner F. Inbreeding effects of queen and workers on colony traits in the honey bee. Apidologie 1989;20:439−50. Available from: https://doi.org/10.1051/apido:19890509.

[78] Ottman R. Gene-environment interaction: definitions and sturdy designs. Prev Med 1996;25(6):764−70. Available from: https://doi.org/10.1006/pmed.1996.0117.

[79] Büchler R, Costa C, Hatjina F, Andonov S, Meixner MD, Uzunov A, et al. The influence of genetic origin and its interaction with environmental effects on the survival of *Apis mellifera L.* colonies in Europe. J Apic Res 2014;53(2):205−14. Available from: https://doi.org/10.3896/IBRA.1.53.2.03.

[80] Dražić MM, Filipi J, Prðun S, Bubalo D, Špehar M, Cvitković D, et al. Colony development of two Carniolan genotypes (*Apis mellifera carnica*) in relation to environment. J Apic Res 2014;53(2):261−8. Available from: https://doi.org/10.3896/IBRA.1.53.2.07.

[81] Meixner MD, Büchler R, Costa C, Francis RM, Hatjina F, Kryger P, et al. Honey bee genotypes and the environment. J Apic Res 2014 2014;53 (2):183−7. Available from: https://doi.org/10.3896/IBRA.1.53.2.01.

Chapter 15

Spermatogonial stem cells: from mouse to dairy goats

Yudong Wei[1], Daguia Zambe John Clotaire[1,2] and Jinlian Hua[1]

[1]College of Veterinary Medicine, Shaanxi Centre of Stem Cells Engineering & Technology, Northwest A&F University, Yangling, China,

[2]Laboratory of Biological and Agronomic Sciences for Development, Faculty of Science, University of Bangui, Bangui, Central African Republic

15.1   Spermatogonial stem cells

Spermatogonial stem cells (SSCs), also called male germline stem cells (mGSCs), are located at the testicular seminiferous tubule basement membrane, which exhibit the ability of self-renewal or differentiation at each division. This ability has been well demonstrated in rodents where SSCs undergo several stages to differentiate into sperm cells. Similar to other adult stem cells, the number of SSCs is quite small, accounting for only 0.03% of all germ cells in testis [1]. Similarly to embryonic stem cells, some studies have shown that SSCs possessed the ability to differentiate into all the cell types with the characteristics of the three germ layers under suitable culture conditions [2,3]. SSCs retain the characteristics of tissue stem cells and the potential of embryonic stem cells, which can be used as an in vitro model to study spermatogenesis and its molecular regulation.

15.1.1   The initiation of spermatogonial stem cells

In the embryonic period, SSCs originate from primordial germ cells (PGCs). After birth, they can be found near the basement membrane of the seminiferous tubule in the testes of animals.

15.1.1.1   Generation of primordial germ cells

At 4.5 days of development (E4.5), only three types of cells exist in the mouse embryos: trophectoderm (TE), epiblast pluripotent cells (Epi), and primitive endoderm (PE). When developed to E5.5, TE, and Epi cells become extraembryonic ectoderm (EXE) and secrete BMP4 and BMP8B. At the same time, visceral endoderm (VE) from EXE and PE begins to secrete BMP2. By the time the embryo develops to E6.0−6.25, the level of BMP4 reaches its peak, together with the appearance of PGC precursor cells. Subsequently, PGC precursor cells are surrounded by ectoderm cells, preparing for the development of reproductive organs. When embryos get to E7.25, PGCs become positive for alkaline phosphatase, differentiating then into germ cells. In mice, the fate of PGCs is mainly determined by *Prdm14*, whereas *SOX17* determines the developmental fate of the human PGC [4].

15.1.1.2   Differentiation of primordial germ cells into spermatogonial stem cells

When mouse embryos develop to E8.5, the level of H2A/H4R3 methylation in PGCs is obviously highlighted. At E10.5, PGCs migrate to the gonadal ridges and begin to differentiate into the hermaphroditic germ cells, while the methylation level of H2A/H4R3 in PGCs decreases sharply when the embryo develops at E11.5. At E12.5, the expression level of the meiosis marker *Ddh38* in PGCs is upregulated, and PGCs differentiate into the amphoteric germ cells [5], deciding thus the sex of the embryo. Before differentiation into SSCs, PGCs migrate firstly into the gonadal ridges. After sex differentiation, PGCs are surrounded by the Sertoli cells and then the peritubular myoid cells, to form the spermatic cord, with further differentiation into gonocytes. Gonocyte development stops at G0/G1 phase before the individual is born. The gonocytes continue to proliferate and form A type spermatogonial cells, also called SSCs after birth. In the process of sexual maturation, SSCs are widely differentiated into sperm cells (Fig. 15.1) [5].

Reproductive Technologies in Animals. DOI: https://doi.org/10.1016/B978-0-12-817107-3.00015-1

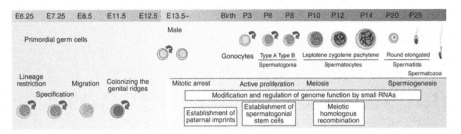

FIGURE 15.1 Pattern of germ cell differentiation [5].

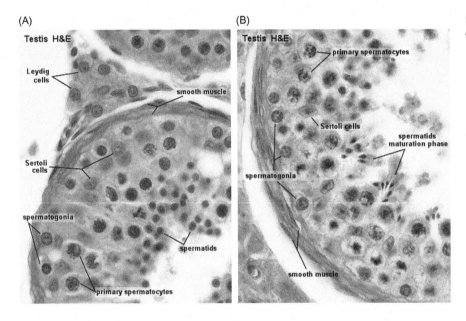

FIGURE 15.2 H&E staining of seminiferous tubes.

15.1.2 Characteristics of spermatogonial stem cells

There are two types of germ cells in the testes of animals: undifferentiated and differentiated spermatogonial cells. Undifferentiated spermatogonial cells can give rise to the differentiated spermatogonial cells, which include A1−A4 types, intermediate type (Int), and type B spermatogonial cells. The undifferentiated spermatogonial cells comprise three cell types: A_s, A_{pr}, and A_{al}. A_s cells are the SSCs in the testicles. The chromatin of A_s cells is positive for alkaline, lacking of heterochromatin in the nucleus and having some nucleolus, whereas heterochromatin is enriched in the nucleus of type B spermatogonial cells (Fig. 15.2).

A_s cells are characterized by the capability of self-renewal. By proliferation, they can form two independent A_s cells, or two A_{pr} cells connected by cytoplasmic bridge. This proliferation process of A_s cells is considered as the self-renewal of SSCs, whereas the later proliferation of Apr cells leads to differentiation into sperm cells (Fig. 15.3) [6]. Studies have confirmed that self-renewal and differentiation of SSCs are regulated by a variety of endogenous and exogenous cytokines (Fig. 15.3) [7]. The two paired Apr cells whose cytoplasms are connected can further differentiate into 4-, 8-, or 16-cell series of Aal cells, then becoming mature sperm cells through an orderly process: A1 (16-cell) → A2 (32-cell) → A3 (64-cell) → A4 (128-cell) → In (256-cell) → B type (512-cell) spermatogonial cells, which collectively can be defined as a complete spermatogenic cycle (Fig. 15.3) [3,7].

15.2 Molecular markers and purification of spermatogonial stem cells

Understanding of SSCs markers, on the one hand, helps in identifying SSCs, and on the other hand provides guidance for a more efficient separation and enrichment of SSCs. Currently, the studies of molecular markers and purification methods for SSCs are as follows:

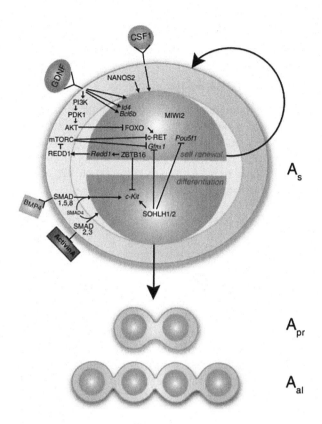

FIGURE 15.3 A schematic diagram of self-renewal and differentiation of spermatogonial stem cells [3,7].

15.2.1 Fluorescence-activated cell sorting

Cell sorting is accomplished by fluorescence activation of cell surface molecules (FACS) and flow cytometry. FACS together with cell transplantation is the commonly used method for the analysis of SSCs markers. So far, there are a number of identified SSCs markers such as α6-integrin (CD49f), β1-integrin (CD29), THY-1 (CD90), CD9, GFRA1, CDH1, av-integrin (CD51), and c-KIT (CD117). The combined use of negative and positive markers can increase the efficiency of SSCs sorting from100 to 200 times [8,9], but 100% purity of SSCs still cannot be achieved. Further efforts need to be pursued to find a more specific SSCs marker.

15.2.2 Genetic model along with cell transplantation

Cell sorting through magnetic-activated cell sorting (MACS) is usually carried out in order to observe the behavior of the selected cells. Although the combined use of the two described techniques is helpful to display the surface molecular markers of SSCs, the study of SSCs cytoplasm and nuclear markers is still hindered.

GFP-expressing transgenic mouse model driven by the promoter of SSCs candidate marker is an effective method for identifying SSCs. Scholer et al. [10] reported that the transcription factor OCT4 is expressed in type A spermatogonia and telogonia of neonatal, pubertal, and adult rats. FACS technology was used to sort OCT4/EGFP-positive spermatogonia in the testis, and cell transplantation studies demonstrated that OCT4/EGFP-positive cells had stronger stem cell characteristics than negative cells [11,12]. It has been revealed by transgenic and conditional knockout models that *Neurogenin 3* (*Ngn3*) is expressed in the earliest spermatogonia [13], including SSCs with stemness [14]. Nevertheless, *Ngn3* was not expressed in all the transplanted cells, which indicated that SSCs population are heterogeneous. Conditional knock-in technique confirms that SSCs also express *Nanos2* gene [15].

Transgenic model confirms retinoid-stimulated gene (stimulated by retinoic acid-8, Stra8) expression in undifferentiated spermatogonia, among which are SSCs. Male mice carrying the *luxoid* (*lu*) mutation gradually become infertile due to the gradual loss of SSCs and abnormal sperm development [16]. This mutation affects the *Zfp145* site that encodes the transcriptional repressor promyelocytic leukemia zinc-finger (PLZF). PLZF is expressed along with the embryonic development and plays a decisive role in limb and axial bone formation. The targeted mutated *Zfp145* site is consistent with the testicular phenotypes caused by mutation in the *luxoid* site, and PLZF is only expressed in undifferentiated spermatogonia such as A_s, A_{pr}, and A_{al} [17]. Knockout and overexpression suggested an important role of glial cell line-derived neurotrophic factor (GDNF) and its receptor in stem cell self-renewal. GDNF is essential for the expansion and the self-renewal of SSCs in vitro, by binding to the soluble GFRA1 receptor [18,19].

15.3 Overall immunofluorescence staining of seminiferous tubules

In addition to flow cytometric sorting, genetic models and cell transplantation, immunofluorescence staining of the seminiferous tubules is also used to analyze the expression profile of related proteins in the male germline. With this method, the SSCs marker is thought to be expressed only in the cells located at the basement membrane of the seminiferous tubules and has coexpression pattern with other known markers. This histological method is more persuasive in the overall staining of the seminiferous tubules, because the size of the clones (e.g., A_s, A_{pr}, A_{al}) can be clearly distinguished. In Table 15.1 the analysis of SSCs markers for each species is described.

15.4 Spermatogonial stem cells microenvironment

SSCs are surrounded in a special environment called "the stem cell niche." This microenvironment regulates the self-renewal and differentiation of SSCs. A stem cell niche contains cells including Sertoli cells and peritubular myoid cells, extracellular matrix, and soluble cytokines that regulate cell fate (Fig. 15.4) [57]. Sertoli cells and peritubular myoid cells secrete basement membrane components. By binding to SSCs through cell surface adhesion molecules, Sertoli cells and peritubular myoid cells provide nutrients and mediate extracellular signals to support SSCs and differentiated germ cells. When normal Sertoli cells are transplanted into the defective reproductive mouse testis, they are capable to initiate spermatogenesis in the testis of the recipient, demonstrating that Sertoli cells play an important role in the differentiation of germ cells.

Undifferentiated spermatogonia are mainly located at the basement membrane of the seminiferous tubules. The SSCs microenvironment mediates endocrine and paracrine signals that regulate cell self-renewal and differentiation. GDNF secreted by Sertoli cells is a key factor in the regulation of SSCs microenvironment. GDNF forms complexes

TABLE 15.1 Analysis of SSCs markers in different species.

Cell markers	Animal species						References
	Mice	Rats	Human	Pork	Cow	Goat	
c-Kit	X	X	X		X	X	[9,20–23]
α6-Integrin (CD49f)	X		X				[9,24,25]
β1-Integrin (CD29)	X	X					[2,24]
Epcam	X						[26]
Pou5f1 (OCT4)	X		X	X			[26–30]
GFRA1	X	X	X	X			[27,31–34]
Thy1 (CD90)	X	X	X	X	X	X	[8,35–39]
Nanos2	X	X					[15,40–43]
CD9	X	X					[44]
Ngn3	X	X	X				[13,45]
PLZF	X	X	X	X	X	X	[16,17,28,37,46,47]
Bcl6b	X	X	X	X			[48,49]
Ret	X	X	X				[50,51]
CDH1	X	X	X	X			[49,52]
GPR125	X		X				[53,54]
UTF1	X						[55]
Lin28	X		X				[56]

Note: X indicates that this marker is expressed in the SSCs of the mentioned species.

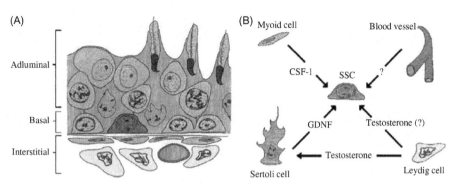

FIGURE 15.4 SSCs microenvironment. (A) Physical microenvironment of SSCs. (B) Growth factor regulation in SSCs microenvironment [57].

with tyrosine protein kinase receptors Ret and GFRA1 on the surface of A_s, A_{pr}, and A_{al} spermatogonia, which in turn activate the phosphoinositide 3-kinase (PI3K)/Akt pathway, the members of the Src kinase superfamily and the Ras/Erk1/2 pathway [58−61]. Colony-stimulating factor 1 (CSF1) produced in peritubular myoid cells enhances SSCs proliferation [62]. CSF1 was originally found in Leydig cells and peritubular myoid cells, whereas its receptor was highly expressed in THY1-positive SSCs of adolescent and adult testis.

15.5 Spermatogonial stem cells isolation, culture, and identification

Although the technology of culturing and amplifying ES cells in vitro has become more and more routine, it is still difficult to establish an SSCs culture system. Currently, only rodent SSCs can be cultured in vitro for a longer period of time. The population doubling times of mouse and rat SSCs are approximately 5−6 days [19] and 3−4 days [63], respectively. Cell transplantation experiments also confirm that these cells still maintain the characteristics of stem cells.

Cytokines are necessary for the culture and amplification of SSCs. The methods of enriching cells (FACS, MACS, or differential attachment) lead to the elimination of somatic cells from SSCs, which in turn promotes germ cell differentiation. Thus when serum-free, well-defined culture media are used, it is necessary to add some growth factors produced by the stem cell niche, such as GDNF which is essential for the maintenance and amplification of rodent SSCs [19,64]. In addition, it is usually indispensable to use STO or MEF as feeder layer. The highest SSCs content in mouse testes is from day 5 to 12 after birth; hence the success rate of SSCs isolation from the testis is higher at this stage. There are also cases of successful culture of SSCs from neonatal rats [64] and adult mice [18]. An immortalized mouse SSCs cell line can be established by overexpressing the reverse transcriptase telomerase gene [65] or the SV40 large T antigen [66].

This following section briefly introduces the isolation, culture, and identification of goat SSCs:

15.6 Animal and culture medium

15.6.1 Test animals

Select the ram from 1 to 4 months of age.

15.6.2 Culture fluid

15.6.2.1 Primary spermatogonia digestion fluid

Hundred milligrams of collagenase, 1 mg of DNase, and 100 mg of catabolic enzyme are weighed and dissolved in 100 mL of calcium-free magnesium PBS buffer. After filtration with a 0.22 μm filter, they are aliquoted and stored at −20°C.

15.6.2.2 Spermatogonia basic culture fluid

Dairy goat spermatogonia culture fluid: 10% fetal bovine serum (FBS), 0.1 mmol/L β-mercaptoethanol, 2 mmol/L glutamine, and 0.1 mmol/L nonessential amino acid are added in the high glucose DMEM/F12 basic medium and filtered with 0.22 μm filter membrane.

15.6.2.3 Red blood cell lysate

1.8675 g of NH_4Cl and 0.65 g of Tris are weighed and dissolved in 250 mL of ultrapure water, and then pH adjusted to 7.2 with hydrochloric acid.

15.7 Separation of spermatogonia

15.7.1 Testicular cell digestion

The testicular tissue is washed with PBS containing a high concentration of penicillin-streptomycin antibiotics and transferred to a cell culture chamber, which is briefly soaked with 75% medical alcohol for 1–2 minutes. After removing the tunica albuginea, the testicular tissue is cut into cubical (1 cm^3) pieces and soaked in PBS and washed 3–4 times. Then each piece of tissue is further cut into tiny pieces in a Petri dish with digestion fluid. Ten volumes of digestion fluid (CDD, 2 mg/mL Collagenase + 20 μg/mL DNase + 2 mg/mL Dispase) are added for digestion at 37°C for 30 minutes. An equal volume of basic culture fluid is added, and the supernatant is discarded after centrifugation at 2000 r/minute for 5 minutes at room temperature. Pellets are resuspended with basic culture fluid. The cells are then seeded in a 90 mm Petri dish and incubated at 37°C with 5% CO_2 and saturated humidity.

15.7.2 Differential purification of spermatogonia

After cultured in a 90 mm Petri dish at a density of 1×10^6 cells overnight, the unattached spermatogonia are collected (containing small amounts of spermatocytes and sperm cells) for subsequent screening by MACS.

15.7.3 CD49f sorting testicular cells

Goat spermatogonia are sorted using magnetic beads labeled with CD49f antibody, with a sorting efficiency of 0.867%. Most of the sorted cells are smaller than the negative and unsorted groups (Fig. 15.5). Sorted cells are assayed for mRNA and protein levels, and the SSCs-positive efficiency is analyzed on CD49f, GFRA1, OCT4, and PLZF staining. The analysis showed that the efficiency of positive and negative cell (1) for CD49f was 85% and 34.7%, respectively; (2) for GFRA1 was 79.7% and 37.3%, respectively; (3) for OCT4 was 84% and 39.7%, respectively; and (4) for PLZF was 62.3% and 40%, respectively. Overall, the rate of SSCs was significantly higher in CD49f-positive group than in CD49f-negative group (Fig. 15.5A and B). Reverse transcription-polymerase chain reaction (RT-PCR) results showed that the expression of *CD49f, CD90, Vasa,* and *Plzf* in the CD49f-positive group was significantly higher than in the CD49f-negative group (Fig. 15.5C).

15.7.4 Identification of CD49f-negative and -positive cells in vitro

Negative and positive groups obtained by CD49f-sorting were inoculated on the MEF feeder layer at the same cell density. After 14 days in a specific culture medium, the proliferation ability, morphology and gene expression of the two groups of cells were examined. The results showed that the proliferation ability of CD49f-positive group was about threefold higher than the negative group (Fig. 15.6A), and the CD49f-positive group was able to form SSCs-like typical clones (Fig. 15.6A). Gene expression tests showed that CD49f-positive group expressed CD49f and PLZF, whereas the negative group did not express them (Fig. 15.6B). RT-PCR results showed that the expression levels of *Gfra1, CD90, CD49f, Pcna,* and *Cdk2* were significantly higher in CD49f-positive group than in CD49f-negative group (Fig. 15.6C).

15.7.5 Culture and passage of goat spermatogonial stem cells

The typical cell clones obtained by sorting are disassembled into single cell or small cell clusters by mechanical discretization or TrypLE and centrifuged and inoculated on Laminin-treated culture plates or MEF feeder cells. DMEM/F12 medium is supplemented with 10% knockout serum replacement + 1% FBS + 2 mmol/L Glu + 1% NEAA + 100 μmol/Lβ-ME + 1% lipid extract + 10 ng/mL GDNF + 50 ng/mL GFRα1 + 2 ng/mL bFGF + 10 ng/mL epidermal growth factor (EGF), and the culture medium is changed every 24 or 48 hours. Immunocytochemistry or RT-PCR is used to characterize the cultured cells with the specific markers (*Gfra1, Plzf, C-kit, CD90, CD49f, PGP9.5*).

15.7.6 Self-renewal regulatory mechanism of spermatogonial stem cells

Under the action of extracellular cytokines, SSCs activate or inhibit the relevant signaling pathways and maintain self-renewal, through transcriptional level regulation or posttranscriptional level regulation.

(A)

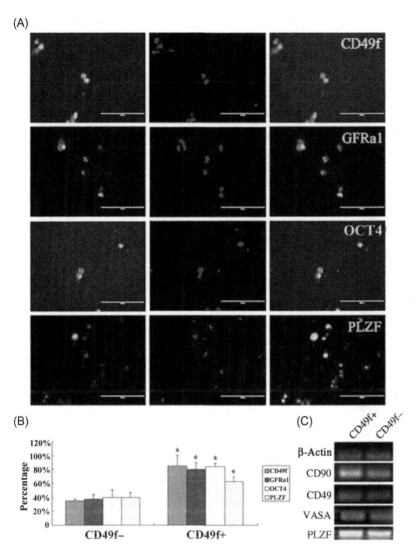

(B)

(C)

FIGURE **15.5** CD49f sorting cells. (A) Immunofluorescence staining of CD49f-positive group for GFRA1, OCT4, and PLZF. (B) Positive rates of CD49f, OCT4, GFRA1, and PLZF in both CD49f-negative and -positive groups. (C) RT-PCR detection of *CD49f*, *CD90*, *Vasa*, and *Plzf* expression levels in both CD49f-negative and -positive groups. Data are expressed as the mean SD of three independent experiments. *P $<$ 0.05

15.8   Factors affecting self-renewal of spermatogonial stem cells

The microenvironment in which SSCs are located contains many factors such as GDNF, FGF2, CSF1, EGF, WNT5A, vitamin A, and Ets-related molecule (ERM) that allow them to maintain an undifferentiated state.

GDNF was the first confirmed growth factor that can maintain the self-renewal of SSCs. GDNF secreted by Sertoli cells binds to its specific receptor GFRA1 on the surface of SSCs. Upon binding to GFRA1, GDNF activates the transmembrane tyrosine kinase RET linked to GFRA1 [61]. RET has a typical intracellular kinase domain, which has 12 auto-phosphorylation sites. RET can activate multiple signaling pathways including PI3K, mitogen-activated protein kinase (MAPK), SFK, and PLC-γ pathways, which consequently regulate the self-renewal of SSCs.

FGF2 secreted by Sertoli cells, mesenchymal cells, and germ cells affects two main aspects in the maintenance of SSCs self-renewal. In fact, FGF2 promotes the secretion of GDNF and ERM proteins from the supporting cells, two factors that play an important role in the maintenance of SSCs self-renewal. In addition, FGF2 acts as a ligand by binding to its receptor FGFR localized on the surface of SSCs. FGFR activation causes Ras protein activation, which in turn activate the MAPK pathway. Thus, FGF2 and GDNF synergistically promote the self-renewal of SSCs [67].

CSF1 is a colony-stimulating factor, which is secreted by mesenchymal cells and peritubular myoid cells. Receptors can be found on the surface of SSCs. CSF1 and GDNF cooperate to promote self-renewal of SSCs [62]. In addition, EGF is secreted by testicular Sertoli cells and cooperates with GDNF to promote the self-renewal of SSCs, despite the mechanism remains unclear [25].

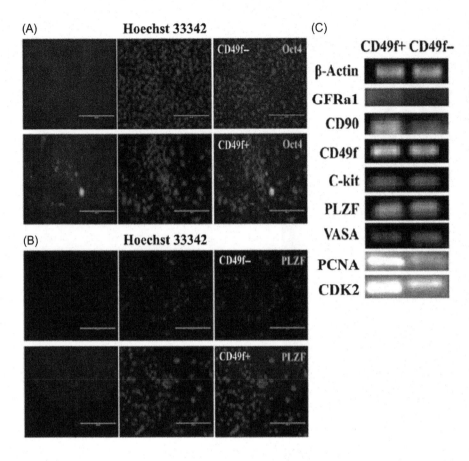

FIGURE 15.6 Characteristics of CD49f-positive cell lines. (A) Validation of CD49f-negative and -positive cell lines. (B) Immunofluorescence staining detection of CD49f and PLZF expression in CD49f-negative and -positive cell lines. (C) RT-PCR detection of germ cells—related markers *Gfra1*, *CD90*, *CD49f*, *C-kit*, *Plzf*, *Vasa*, *Pcna*, and *Cdk2* in CD49f-negative and -positive cell lines.

WNT5A is a cytokine secreted by supporting cells, and its receptors Fzd and LRP5/6 are distributed on the surface of SSCs [68]. WNT5A activates the Wnt pathway by activating its receptor. Blocking the Wnt signaling pathway leads to apoptosis of SSCs, whereas addition of WNT5A to the culture media promotes proliferation of SSCs [69].

Vitamin A promotes self-renewal of mouse male germ stem cells [70]. The active metabolite of vitamin A, retinoic acid (RA), promotes SSCs differentiation and sets them into meiosis. However, some studies also highlight the effect of vitamin A on SSCs self-renewal. Since RA receptors are present in a variety of cells within the testicular microenvironment, it has not been demonstrated whether vitamin A also affects other cells or not.

ERM protein is also known as ETV5, which is also implicated in the regulation of SSCs self-renewal. The initial studies suggested that ERM was expressed in testicular Sertoli cells, and later it was also found in spermatogonia. After the first wave of spermatogenesis, SSCs progenitor cells of ERM knockout mice disappear, and only Sertoli cells are left, which results in Sertoli cell—only syndrome. This indicates that ERM protein plays an important role in maintaining self-renewal of SSCs [61,71]. Microarray analysis showed that ERM can upregulate the expression of multiple cell chemokines that regulate the microenvironment of SSCs [72,73]. There is evidence indicating that Etv5 can directly or indirectly regulate several important self-renewal-related regulatory genes such as *Bcl6b*, *Lhx1*, *Cxcr4*, and *Brachyury*. In addition, ETV5 can directly activate the transcription of miR-21 that contributes to self-renewal [74]. *Etv5* is located at the downstream of the MAP2K/MAPK pathway, and GDNF and FGF2 can activate this pathway in different ways to promote *Etv5* expression. Upregulated chemokine CCL9 by ERM attracts undifferentiated SSCs through its receptor CCR1 on the surface of SSCs [75]. ERM regulates the expression of another chemokine CXCL12 and its receptor CXCR4, which are required for SSCs self-renewal [76,77].

15.9 Genes affecting self-renewal of spermatogonial stem cells

Many other genes including those described above, participate in the maintaining of SSCs self-renewal. They can be categorized by whether their expression depends on GDNF signaling pathway or not.

15.9.1 Genes that do not depend on GDNF regulation

PLZF was the first confirmed transcriptional factor related to SSCs self-renewal [16,17]. PLZF can interact with pathways such as GDNF and SCF. It can also regulate SSCs by inhibiting mTORC1 [78,79]. Germ cells of *Plzf* mutant mice failed to maintain spermatogenesis. In addition, Plzf regulates self-renewal of male germ stem cells in goats by targeting micro-RNA-544[80].

In SSCs, *Oct4* gene is expressed in A_s-A_{al} spermatogonia and is a marker gene for SSCs [12,81].

Foxo1 specifically marks meiocytes and part of spermatogonia. *Foxo1* plays a role in the homeostasis and spermatogenesis of SSCs. The absence of *Foxo1*, *Foxo3*, and *Foxo4* will seriously affect the self-renewal of SSCs and completely prevent the differentiation. The active expression of *Foxo1* can promote the transcription of Ret. *Foxos*, the key elements of the PI3K pathway, are downregulated by the PI3K to block the differentiating state of spermatogenesis [82].

Mili (Miwi-like) is a P element—induced wimpy testis (PIWI) protein family member. It works by forming complexes with piRNAs. It has been reported that Mili is expressed in the cytoplasm of testicular germline stem cells, spermatogonia, and early spermatocytes, and it is present in the region of enriching chromosome. Mili knockout mice first showed a decrease in mitogenic capacity of spermatogonia at all stages; then a significant decrease in the number of SSCs was recorded. It is suggested that Mili protein plays an important regulatory role in self-renewal of SSCs by working with piRNA.

NANOS2 is an RNA-binding protein that plays a key role in maintaining self-renewal of SSCs. Studies and analysis show that undifferentiated spermatogonia express *Nanos2* and maintain self-renewal. The conditional knockout of the *Nanos2* gene in postnatal mice resulted in the depletion of SSCs in the reserve. Therefore *Nanos2* is the key gene to maintain the self-renewal state of SSCs [15].

TATA-binding protein associated factor 4b (TAF4B) is a TAFs molecule. TAF4B is present in mouse A spermatogonia and forms a TFIID complex with the TATA-binding protein. This complex has a regulatory effect on the transcription of RNA polymerase II. Only the Sertoli cells were left in the seminiferous tubules of *Taf4b*-deficient mice. Transplantation of wild-type mouse SSCs into the seminiferous tubules of those mice induces spermatogenesis, and deletion of the *Taf4b* results in significant downregulation of *Plzf* expression [83].

15.9.2 Genes that depend on GDNF regulation

Bcl6b is a well-studied tumor suppressor BCL6 homolog. *Bcl6b* encodes a zinc finger transcription inhibitor that enhances the immune response [84]. *Bcl6b* was initially found to have an effect on SSCs due to the discovery that Bcl6b is enriched in rat germ cells [85]. Later, *Bcl6b* was found to be one of the series of genes that are upregulated by GDNF in Thy1$^+$ SSCs [48], and *Bcl6b* deletion mice showed the phenotype of progressive germ cell loss. When *Bcl6b* was knocked down in Thy1$^+$ SSCs, 88% SSCs were lost. The high apoptotic rate of *Bcl6b*-deficient Thy1$^+$ SSCs shows an important role of *Bcl6b* in maintaining the self-renewal and survival of SSCs. Like *Etv5*, *Bcl6b* is located at the downstream of the MAP2K/MAPK pathway and may be regulated by *Etv5* [76]. This has led to the hypothesis that *Bcl6b* promotes the maintenance of SSCs by modulating cellular interactions in the microenvironment. Surprisingly, although *Bcl6b* is a downstream target of Etv5, there is almost no overlap between *Bcl6b* and *Etv5*-regulated genes [76]. Thus although both transcriptional regulators promote self-renewal of SSCs, their mechanisms vary widely [86].

As a transcription factor, *Lhx1* (LIM homeobox 1) encodes a member of a large family containing LIM domain, a unique cysteine-enriched zinc finger domain. Oatley et al. [48] demonstrated that *Lhx1* regulates Thy1$^+$ SSCs by in vitro knockdown, and these *Lhx1*-knockdown SSCs have reduced ability to form clones upon engraftment into the body. Like *Etv5* and *Bcl6b*, *Lhx1* is regulated by the FGF2/MAP2K1 pathway [67].

Pou3f1 (POU domain, class-3 transcription factor 1), also called OCT6, is enriched in testes and brain. *Pou3f1* mutant mice die soon after birth due to various defects in the development of the nervous system [87,88]. *Pou3f1* is highly expressed in rat germ cells and enhances SSCs activity [85]. In mouse Thy1$^+$ SSCs, *Pou3f1* is regulated by the GDNF-PI3K-AKT pathway. Knockout of *Pou3f1* results in a massive decrease in the number of SSCs that leads to an increase of apoptosis rate in culture [89].

Brachyury was first discovered in Thy1$^+$ SSCs cultured in vitro and could be positively regulated by the SSCs self-renewal factor ETV5 and GDNF. *Brachyury* knockdown reduced the number of SSCs, indicating that it plays an important role in the self-renewal of SSCs. The target gene for Brachyury has not been determined and may play a role as a transcription factor regulated by *Etv5* and *Ref*.

ID4 is called the inhibitor of DNA binding (ID4). In mice, *Id4* is expressed only in some A_s spermatogonia, and A_s spermatogonia can be distinguished from Apr and Aal using *Id4*. In *Id4* knocked out mice testis, only Sertoli cells in the testicular seminiferous tubules can be found. Some studies found that reducing *Id4* significantly inhibited SSCs proliferation [90].

15.10 Effect of small RNA on spermatogonial stem cells self-renewal

Some studies have shown that miRNAs play an important role in germ cell development, germination, meiosis initiation, gametogenesis, and maturation.

In the process of spermatogenesis, in addition to coding protein genes, the noncoding part of the transcriptome also plays a positive regulatory role. One of the most studied is the micro-RNA (miRNA). miRNAs are abundantly expressed in adult mouse testes. miRNAs accumulate with Dicer and Argonaute proteins during the meiosis [91]. Knocking out Dicer leads the number of primordial germ cells and spermatogonia cells to decrease. In 8-months-old Dicer-null mice testes, the spermatogenesis was significantly depleted and the ability to reproduce was terminated prematurely [92]. This shows that miRNA plays a key role in the maintenance of spermatogenesis and fertility. In Fig. 15.7 some validated miRNAs in each stage of spermatogenesis are summarized.

15.10.1 Screening of miRNAs in CD49f cell lines

Some miRNAs can directly participate in the regulation of spermatogonial cells self-renewal and differentiation. Wu et al. [94] used immunomagnetic bead sorting and Illumina deep sequencing analysis techniques to obtain differential expression profiles of miRNAs in goat CD49f-positive and -negative spermatogonial cells. *miR-933* was found to be highly expressed in CD49f-positive group, whereas *miR-916* expression was twofold higher in CD49f-negative group (Fig. 15.8A). Comparison of differential expression profiles of miRNAs and spermatogonial marker genes in CD49f-positive and -negative spermatogonial cells revealed that some miRNAs and genes specifically expressed in mouse SSCs were also highly expressed in goat CD49f-positive spermatogonial cells, among which are *miR-221*, *miR-21*, *MiR-23a*, *miR-29a*, *miR-29b*, *miR-24*, *miR-199a*, *miR-199b*, *miR-27a*, *GFRA1*, and *PLZF* (Fig. 15.8B). Goat testis section staining confirmed CD49f mainly expressed the spermatogonia-like cells in the vicinity of seminiferous tubules basement membrane, indicating that the sorted CD49f-positive testicular cells are the SSCs population in dairy goat testes. Bioinformatics analysis revealed that 40 target genes of highly expressed miRNAs in CD49f-positive testicular cells were involved in three cellular processes: gene transcription, cell cycle, and DNA damage. Ten target genes were involved in the KEGG cell cycle pathway. Flow cytometry assay showed nearly 80% of CD49f-positive testicular cells in G1/G0 phase, and bioinformatics analysis revealed that differentially expressed miRNAs in CD49f-positive testicular cells were also involved in the regulation of cell cycle.

miR-21 is preferentially expressed in CD90-enriched SSCs. Inhibiting *miR-21* promotes SSCs apoptosis and inhibits self-renewal, suggesting that it is crucial for normal self-renewal [74]. *miR-20* and *miR-106a* directly initiate the self-renewal process of SSCs by targeting *Stat3* and *Ccnd1* [95]. *miR-221* and *miR-222* induce *Kit* expression in undifferentiated mouse spermatogonia and promote the differentiation of progenitor cells; *miR-34c* is expressed at high levels in differentiated spermatogonia, which can promote the differentiation of SSCs to spermatocytes by inhibiting retinoic acid receptor γ and the expression of pluripotency gene *Nanos2* [49,96,97].

miR-17—92 (miRc1) and *miR-106b-25 (miRc3)* clusters may play an important regulatory role in spermatogenesis. Retinoic acid—induced differentiation significantly downregulated the expression of both miRNA clusters in Thy1⁺-

FIGURE 15.7 Functional miRNA during spermatogenesis [93].

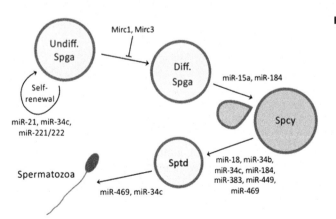

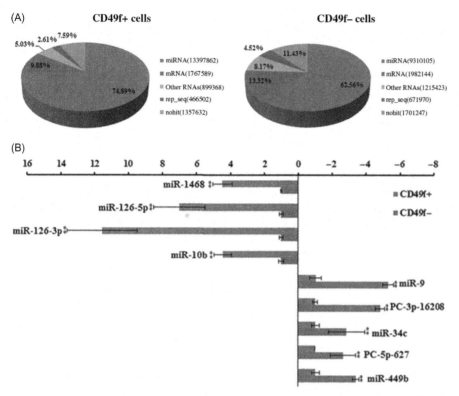

FIGURE 15.8 Screening of miRNAs in goat spermatogonial cell lines: (A) RNAs distribution of CD49f-positive and -negative spermatogonia; (B) qRT-PCR results of differentially expressed miRNAs, **$P < .01$; $n = 3$.

enriched spermatogonia. In *miRc-1* knockout mice, *miRc3* expression increased. The testicles became smaller and sperm formation decreased, suggesting that these two miRNA clusters work closely together in the regulation of spermatids [98].

miRNAs are associated with spermatogenesis: for example, *miR-34c* upregulates germline marker genes [49,99], *miR-15a* regulates early spermatogenesis by binding to *Ccnt2* [100], *miR-23b* regulates the translation of *Pten* and *Eps15* in Sertoli cells [101], and *miR-469* binds to *Tp2* and *Prm2* [102], which are both involved in the regulation of spermatogenesis, although the regulatory mechanism of these molecules is still unclear.

15.10.2 Expression level of *miR-204* and *Sirt1* in dairy goat testis

The expression and distribution of *miR-204* in dairy goat are mainly concentrated in the reproductive system, and the expression of *miR-204* reaches its highest level in 6-month-old testis of dairy goats. Fluorescence in situ hybridization and dual luciferase assay confirmed that *miR-204* also targets the 3′UTR of *SIRT1* in dairy goats. Further examination found that *SIRT1* was highly expressed in CD49f- and CD90-positive SSCs and negatively correlated with the expression of *miR-204*. These results suggest that *miR-204* may be involved in the regulation of dairy goat SSCs by targeting *SIRT1*. Further use of mGSC model found that *miR-204* regulates the self-renewal and proliferation of SSCs by targeting *SIRT1*.

15.11 Related pathways affecting self-renewal of spermatogonial stem cells

15.11.1 JAK/STAT signaling pathway

The full name of JAK/STAT is Janus kinase/signal transducer and activator of transcription signaling pathway, which was the first pathway identified to stimulate SSCs self-renewal and maintenance activity. This signal pathway is activated by the ligand Unpaired (*Upd*) in Drosophila testes. In *Upd*-mutated mice, SSCs disappeared rapidly. On the contrary, overexpressing *Upd* leads to accumulation of stem cell−like cells and hinders differentiation [103,104]. Furthermore, the blockade of JAK/STAT signaling by mutating of *JAK* or *Stat92E* results in the loss of SSCs. Conversely, activating JAK/STAT signaling pathway by *Upd* may allow SSCs to maintain self-renewal. Hindering the JAK-STAT signaling pathway leads to the differentiation of germline stem cells into interconnected spermatogonia [105].

15.11.2   Smad signaling pathway

The regulation of Smad signaling pathway on SSCs self-renewal and differentiation presents two sides. Different ligands utilize different Smad pathways to regulate the self-renewal and differentiation of SSCs. *Smad4/5* can promote spermatogonial differentiation, whereas *Nodal* activating *Smad2/3* through an autocrine pathway regulates mouse SSCs self-renewal. It has been found that *Nodal* and its receptors are expressed in SSCs but not in Sertoli cells or differentiated germ cells. *Nodal* promotes self-renewal of SSCs by activating Smad2/3 phosphorylation and the expression of Pou5f1, CyclinD1, and CyclinE [106].

15.11.3   Wnt signaling pathway

Wnts form a large family of secreted glycoproteins involved in cell proliferation, differentiation, organogenesis, and cell migration. The β-catenin-dependent Wnt signaling pathway promotes self-renewal of many cell types. Golestaneh et al. [68] reported that Wnts and their receptor Fzs are expressed in spermatogonia and C18-4 cell lines. WNT3A induces cell proliferation, morphological changes, and migration of C18-4 cells. β-Catenin is activated during testicular development, and Yeh et al. [69] reported that WNT5A is an extracellular factor that maintains the self-renewal of SSCs via β-catenin-dependent mechanism. Activating Wnt signaling pathway with GSK3 inhibitor BIO could also promote SSCs self-renewal [107]. These results indicate that Wnt/β-catenin pathway plays a key role in the self-renewal of SSCs.

15.11.4   Src signaling pathway

The Src family includes *Src*, *Yes*, *Lyn*, and *Fyn*. In nerve cells, GDNF activates Src family kinases in a *Ret*-independent way [108], but the activation of the Src family depends on the mediation of *Ret* in SSCs. It plays an important role in the proliferation of GDNF-regulated SSCs. GDNF first binds to GFRA1 to activate RET, and then the activated RET activates Src family kinases, which in turn activates the PI3K/Akt pathway and ultimately upregulates *N-myc* expression to promote SSCs proliferation [58]. Through transplantation analysis, Src family kinase pathways have been identified to be essential for GDNF-mediated self-renewal regulation in mouse SSCs [61].

15.11.5   PI3K/Akt signaling pathway

In 2007, Lee et al. [109] reported that the PI3K/Akt signaling pathway is involved in the self-renewal of GDNF-regulated SSCs. The PI3K-specific inhibitor LY294002 could hinder the proliferation of SSCs through inhibiting the PI3K pathway. In contrast, SSCs self-renewal can be stimulated by the transfection of *myr-Akt-Mer* plasmid and the addition of 4-hydroxytamoxifen to SSCs conditionally, which can also save the SSCs from apoptosis caused by the absence of GDNF, suggesting that the PI3K/Akt signaling pathway is essential for SSCs self-renewal and survival. Interestingly, the PI3K/Akt signaling pathway is also involved in the differentiation of spermatogonia. SCF binds to the receptor KIT, which in turn activates the PI3K/Akt signaling pathway. SCF promotes the proliferation of type A1−A4 spermatogonia [110,111]. In vivo experiments showed that when SCF receptor KIT failed to bind to PI3K, the reduced Akt activation lead to decreased proliferation and increased apoptosis of SSCs, and eventually spermatogenesis stagnation [112,113]. In summary, different signaling molecules regulate SSCs self-renewal, survival, proliferation, and spermatogonial differentiation through a common PI3K/Akt signaling pathway.

15.11.6   Ras/Raf/MAP2K/MAPK signaling pathway

Ras protein is a key regulator of cell proliferation and differentiation. Extracellular signal-regulated kinase (ERK) is an important member of MAPK and is involved in the regulation of many cellular functions such as proliferation, differentiation, and cell cycle [111,114]. Blocking the MAPK/ERK pathway by specific inhibitor PD098059 resulted in a slight decrease in proliferation of male germ cells in neonatal mice. Studies have shown that GDNF signals can activate the Ras/Raf/MAP2K/MAPK signaling pathway in SSCs, which is one of the two pathways by which GDNF can maintain the self-renewal of SSCs. Interestingly, the MAPK pathway is activated by SCF to stimulate the proliferation of *Kit*-expressed spermatogonia [111]. Therefore MAPK pathway is activated by different ligands to regulate the proliferation and differentiation of SSCs (Fig. 15.9).

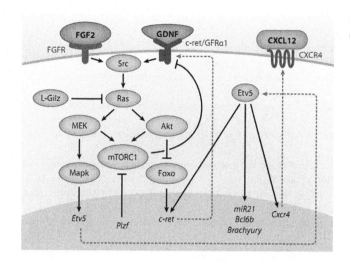

FIGURE 15.9 Signaling pathway of self-renewal of spermatogonial stem cells [115].

15.12 Effect of epigenetic modification on self-renewal of spermatogonial stem cells

There are few reports on the epigenetic modification of SSCs self-renewal and differentiation. Some histone deacetylations have been reported. It has been known that histone deacetylases (HDACs) play an important role in the self-renewal and differentiation of stem cells, and downregulation of HDACs can hinder the self-renewal of stem cells. In SSCs, Kofman et al. [116] analyzed the expression of *Hdac2*, *Hdac6*, *Hdac8*, *Hdac9*, *Sirt*, and *Sirt4* in six self-renewing, differentiating, and senescent SSCs. In the differentiated and senescent SSCs, the expression of *Sirt4* increased, and the expression of *Hdac2*, *Hdac6*, and *Sirt1* decreased. When treated with life-extending drug rapamycin, the expression of *Hdac2*, *Hdac6*, and *Sirt1* increased while the *Hdac8*, *Hdac9*, and *Sirt4* expressions declined. These results show that HDACs play an important role in SSCs self-renewal and activity maintenance.

Acknowledgment

We appreciate Dr. Furong Tang's excellent comments and revision.

List of Abbreviations

GFP (Green fluorescent protein)
SFK (Src Family Kinase)
PLC-γ, STO (SIM mouse embryo-derived thioquanine — and- quabian —resistant cells)
MEF (murine embryonic feeder)
PBS ()
NEAA ()
bFGF (Basic Fibroblast Growth Factor)
RET, SCF Stem Cell Factor (KIT ligand)
TAFs (Transcription initiation factors)
TATA (TATA)
TFIID (Transcription factor IID)
BIO (6-bromoindirubin-3'-oxime)
FGF2 (Basic Fibroblast Growth Factor).

References

[1] Tegelenbosch RA, de Rooij DG. A quantitative study of spermatogonial multiplication and stem cell renewal in the C3H/101 F1 hybrid mouse. Mutat Res 1993;290(2):193–200.

[2] Kanatsu-Shinohara M, Lee J, Inoue K, Ogonuki N, Miki H, Toyokuni S, et al. Pluripotency of a single spermatogonial stem cell in mice. Biol Reprod 2008;78(4):681–7.

[3] Oatley JA, Brinster RL. Regulation of spermatogonial stem cell self-renewal in mammals. Annu Rev Cell Dev Biol 2008;24:263–86.

[4] Irie N, Weinberger L, Tang WWC, Kobayashi T, Viukov S, Manor YS, et al. SOX17 is a critical specifier of human primordial germ cell fate. Cell 2015;160(1−2):253−68.

[5] Saitou M, Yamaji M. Primordial germ cells in mice. Cold Spring Harb Symp Quant Biol 2012;4(11).

[6] Oatley JM, Brinster RL. The germline stem cell niche unit in mammalian testes. Physiol Rev 2012;92(2):577−95.

[7] Jan SZ, Hamer G, Repping S, de Rooij DG, van Pelt AM, Vormer TL. Molecular control of rodent spermatogenesis. Biochim Biophys Acta 2012;1822(12):1838−50.

[8] Kubota H, Avarbock MR, Brinster RL. Spermatogonial stem cells share some, but not all, phenotypic and functional characteristics with other stem cells. Proc Natl Acad Sci U S A 2003;100(11):6487−92.

[9] Shinohara T, Orwig KE, Avarbock MR, Brinster RL. Spermatogonial stem cell enrichment by multiparameter selection of mouse testis cells. Proc Natl Acad Sci U S A 2000;97(15):8346−51.

[10] Scholer HR, Hatzopoulos AK, Balling R, Suzuki N, Gruss P. A family of octamer-specific proteins present during mouse embryogenesis: evidence for germline-specific expression of an Oct factor. EMBO J 1989;8(9):2543−50.

[11] Ohmura M, Yoshida S, Ide Y, Nagamatsu G, Suda T, Ohbo K. Spatial analysis of germ stem cell development in Oct-4/EGFP transgenic mice. Arch Histol Cytol 2004;67(4):285−96.

[12] Ohbo K, Yoshida S, Ohmura M, Ohneda O, Ogawa T, Tsuchiya H, et al. Identification and characterization of stem cells in prepubertal spermatogenesis in mice. Dev Biol 2003;258(1):209−25.

[13] Yoshida S, Takakura A, Ohbo K, Abe K, Wakabayashi J, Yamamoto M, et al. Neurogenin3 delineates the earliest stages of spermatogenesis in the mouse testis. Dev Biol 2004;269(2):447−58.

[14] Nakagawa T, Nabeshima Y, Yoshida S. Functional identification of the actual and potential stem cell compartments in mouse spermatogenesis. Dev Cell 2007;12(2):195−206.

[15] Sada A, Suzuki A, Suzuki H, Saga Y. The RNA-binding protein Nanos2 is required to maintain murine spermatogonial stem cells. Science 2009;325(5946):1394−8.

[16] Buaas FW, Kirsh AL, Sharma M, McLean DJ, Morris JL, Griswold MD, et al. Plzf is required in adult male germ cells for stem cell self-renewal. Nat Genet 2004;36(6):647−52.

[17] Costoya JA, Hobbs RM, Barna M, Cattoretti G, Manova K, Sukhwani M, et al. Essential role of Plzf in maintenance of spermatogonial stem cells. Nat Genet 2004;36(6):653−9.

[18] Kubota H, Avarbock MR, Brinster RL. Growth factors essential for self-renewal and expansion of mouse spermatogonial stem cells. Proc Natl Acad Sci U S A 2004;101(47):16489−94.

[19] Kubota H, Avarbock MR, Brinster RL. Culture conditions and single growth factors affect fate determination of mouse spermatogonial stem cells. Biol Reprod 2004;71(3):722−31.

[20] Tanigaki R, Sueoka K, Tajima H, Nakabayashi A, Sato K, Asada H, et al. C-kit expression in spermatogonia damaged by doxorubicin exposure in mice. J Obstet Gynaecol Res 2013;39(3):692−700.

[21] von Kopylow K, Staege H, Schulze W, Will H, Kirchhoff C. Fibroblast growth factor receptor 3 is highly expressed in rarely dividing human type A spermatogonia. Histochem Cell Biol 2012;138(5):759−72.

[22] Xie B, Qin Z, Huang B, Xie T, Yao H, Wei Y, et al. In vitro culture and differentiation of buffalo (*Bubalus bubalis*) spermatogonia. Reprod Domest Anim 2010;45(2):275−82.

[23] Bai Y, Ye Z, Zeng F. Isolation of subtype spermatogonia in juvenile rats. J Huazhong Univ Sci Technol Med Sci 2008;28(4):435−8.

[24] Shinohara T, Avarbock MR, Brinster RL. beta1- and alpha6-integrin are surface markers on mouse spermatogonial stem cells. Proc Natl Acad Sci U S A 1999;96(10):5504−9.

[25] Orwig KE, Ryu BY, Master SR, Phillips BT, Mack M, Avarbock MR, et al. Genes involved in post-transcriptional regulation are overrepresented in stem/progenitor spermatogonia of cryptorchid mouse testes. Stem Cell 2008;26(4):927−38.

[26] Choi SH, Ahn JB, Kim HJ, Im NK, Kozukue N, Levin CE, et al. Changes in free amino acid, protein, and flavonoid content in jujube (*Ziziphus jujube*) fruit during eight stages of growth and antioxidative and cancer cell inhibitory effects by extracts. J Agr Food Chem 2012;60 (41):10245−55.

[27] He Z, Kokkinaki M, Jiang J, Dobrinski I, Dym M. Isolation, characterization, and culture of human spermatogonia. Biol Reprod 2010;82(2):363−72.

[28] Luo J, Megee S, Rathi RID. Protein gene product 9.5 is a spermatogonia-specific marker in the pig testis: application to enrichment and culture of porcine spermatogonia. Mol Reprod Dev 2006;73(12):1531−40.

[29] Goel S, Fujihara M, Minami N, M Y HI. Expression of NANOG, but not POU5F1, points to the stem cell potential of primitive germ cells in neonatal pig testis. Reproduction 2008;135:785−95.

[30] Mahla RS, Reddy N, Goel S. Spermatogonial stem cells (SSCs) in buffalo (*Bubalus bubalis*) testis. PLoS One 2012;7(4):e36020.

[31] Dettin L, Ravindranath N, Hofmann MC, Dym M. Morphological characterization of the spermatogonial subtypes in the neonatal mouse testis. Biol Reprod 2003;69(5):1565−71.

[32] Gassei K, Ehmcke J, Dhir R, Schlatt S. Magnetic activated cell sorting allows isolation of spermatogonia from adult primate testes and reveals distinct GFRa1-positive subpopulations in men. J Med Primatol 2010;39(2):83−91.

[33] Nagai R, Shinomura M, Kishi K, Aiyama Y, Harikae K, Sato T, et al. Dynamics of GFRα1-positive spermatogonia at the early stages of colonization in the recipient testes of W/Wν male mice. Dev Dyn 2012;241(8):1374−84.

[34] Lee KH, Lee WY, Kim JH, Yoon MJ, Kim NH, Kim JH, et al. Characterization of GFRalpha-1-positive and GFRalpha-1-negative spermatogonia in neonatal pig testis. Reprod Domest Anim 2013;48(6):954−60.

[35] Abbasi H, Tahmoorespur M, Hosseini SM, Nasiri Z, Bahadorani M, Hajian M, et al. THY1 as a reliable marker for enrichment of undifferenti-
      ated spermatogonia in the goat. Theriogenology 2013;80(8):923—32.

[36] Eildermann K1, Gromoll J RB. Misleading and reliable markers to differentiate between primatetestis-derived multipotent stromal cells and
      spermatogonia in culture. Hum Reprod 2012;27(6):1754—67.

[37] Reding SC, Stepnoski AL, EW C JMO. THY1 is a conserved marker of undifferentiated spermatogonia in the pre-pubertal bull testis.
      Reproduction 2010;139:893—903.

[38] Wu J, Song WC, Zhu HJ, Niu ZW, Mu HL, Lei AM, et al. Enrichment and characterization of Thy1-positive male germline stem cells
      (mGSCs) from dairy goat (Capra hircus) testis using magnetic microbeads. Theriogenology 2013;80(9):1052—60.

[39] Zheng Y, He Y, An JH, Qin JZ, Wang YH, Zhang YQ, et al. THY1 is a surface marker of porcine gonocytes. Reprod Fert Dev 2014;26(4):533—9.

[40] Bellaiche J, Lareyre JJ, Cauty C, Yano A, Allemand I, Le Gac F. Spermatogonial stem cell quest: Nanos2, marker of a subpopulation of undif-
      ferentiated A spermatogonia in trout testis. Biol Reprod 2014;90(4):79.

[41] Suzuki A, Tsuda M, Saga Y. Functional redundancy among Nanos proteins and a distinct role of Nanos2 during male germ cell development.
      Development 2007;134(1):77—83.

[42] Suzuki H, Sada A, Yoshida S, Saga Y. The heterogeneity of spermatogonia is revealed by their topology and expression of marker proteins
      including the germ cell-specific proteins Nanos2 and Nanos3. Dev Biol 2009;336(2):222—31.

[43] Tsuda M, Sasaoka Y, Kiso M, Abe K, Haraguchi S, Kobayashi S, et al. Conserved role of nanos proteins in germ cell development. Science
      2003;301(5637):1239—41.

[44] Kanatsu-Shinohara M, Inoue K, Lee J, Yoshimoto M, Ogonuki N, Miki H, et al. Generation of pluripotent stem cells from neonatal mouse tes-
      tis. Cell 2004;119(7):1001—12.

[45] Raverot G, Weiss J, Park SY, Hurley L, Jameson JL. Sox3 expression in undifferentiated spermatogonia is required for the progression of sper-
      matogenesis. Dev Biol 2005;283:215—25.

[46] Song W, Zhu H, Li M, Li N, Wu J, Mu H, et al. Promyelocytic leukaemia zinc finger maintains self-renewal of male germline stem cells
      (mGSCs) and itsexpression pattern in dairy goat testis. Cell Prolif 2013;46:457—68.

[47] Grisanti L, Falciatori I, Grasso M, Dovere L, Fera S, Muciaccia B, et al. Identification of spermatogonial stem cell subsets by morphological
      analysis and prospective isolation. Stem Cell 2009;27(12):3043—52.

[48] Oatley JM, Avarbock MR, Telaranta AI, Fearon DT, Brinster RL. Identifying genes important for spermatogonial stem cell self-renewal and
      survival. Proc Natl Acad Sci U S A 2006;103(25):9524—9.

[49] Yu M, Mu H, Niu Z, Chu Z, Zhu H, Hua J. miR-34c enhances mouse spermatogonial stem cells differentiation by targeting Nanos2. J Cell
      Biochem 2014;115(2):232—42.

[50] Ebata KT, Zhang X, Nagano MC. Expression patterns of cell-surface molecules on male germ line stem cells during postnatal mouse develop-
      ment. Mol Reprod Dev 2005;72(2):171—81.

[51] Naughton CK, Jain S, Strickland AM, Gupta A, Milbrandt J. Glial cell-line derived neurotrophic factor-mediated RET signaling regulates sper-
      matogonial stem cell fate. Biol Reprod 2006;74(2):314—21.

[52] Tokuda M, Kadokawa Y, Kurahashi H, Marunouchi T. CDH1 is a specific marker for undifferentiated spermatogonia in mouse testes. Biol
      Reprod 2007;76(1):130—41.

[53] Goharbakhsh L, Mohazzab A, Salehkhou S, Heidari M, Zarnani AH, Parivar K, et al. Isolation and culture of human spermatogonial stem cells
      derived from testis biopsy. Avicenna J Med Biotechnol 2013;5(1):54—61.

[54] Seandel M, James D, Shmelkov SV, Falciatori I, Kim J, Chavala S, et al. Generation of functional multipotent adult stem cells from GPR125t
      germline progenitors. Nature 2007;449:346—50.

[55] van Bragt MPA, Roepers-Gajadien HL, Korver CM, Bogerd J, Okuda A, Eggen BJL, et al. Expression of the pluripotency marker UTF1 is
      restricted to a subpopulation of early A spermatogonia in rat testis. Reproduction 2008;136(1):33—40.

[56] Aeckerle N, Eildermann K, Drummer C, Ehmcke J, Schweyer S, Lerchl A, et al. The pluripotency factor LIN28 in monkey and human testes: a
      marker for spermatogonial stem cells? Mol Hum Reprod 2012;18(10):477—88.

[57] Dadoune JP. New insights into male gametogenesis: what about the spermatogonial stem cell niche? Folia Histochem Cyto 2007;45(3):141—7.

[58] Braydich-Stolle L, Kostereva N, Dym M, Hofmann MC. Role of Src family kinases and N-Myc in spermatogonial stem cell proliferation. Dev
      Biol 2007;304(1):34—45.

[59] He Z, Jiang J, Kokkinaki M, Golestaneh N, Hofmann M-C, Dym M. GDNF upregulates c-fos transcription via the Ras/ERK1/2 pathway to pro-
      mote mouse spermatogonial stem cell proliferation. Stem Cell 2008;26(1):266—78.

[60] Niu Z, Zheng L, Wu S, Mu H, Ma F, Song W, et al. Ras/ERK1/2 pathway regulates the self-renewal of dairy goat spermatogonia stem cells.
      Reproduction 2015;149(5):445—52.

[61] Oatley JM, Avarbock MR, Brinster RL. Glial cell line-derived neurotrophic factor regulation of genes essential for self-renewal of mouse sper-
      matogonial stem cells is dependent on src family kinase signaling. J Biol Chem 2007;282(35):25842—51.

[62] Oatley JM, Oatley MJ, Avarbock MR, Tobias JW, Brinster RL. Colony stimulating factor 1 is an extrinsic stimulator of mouse spermatogonial
      stem cell self-renewal. Development 2009;136(7):1191—9.

[63] Hamra FK, Chapman KM, Nguyen DM, Williams-Stephens AA, Hammer RE, Garbers DL. Self renewal, expansion, and transfection of rat
      spermatogonial stem cells in culture. Proc Natl Acad Sci U S A 2005;102(48):17430—5.

[64] Kanatsu-Shinohara M, Ogonuki N, Inoue K, Miki H, Ogura A, Toyokuni S, et al. Long-term proliferation in culture and germline transmission
      of mouse male germline stem cells. Biol Reprod 2003;69(2):612—16.

[65] Feng LX, Chen Y, Dettin L, Pera RA, Herr JC, Goldberg E, et al. Generation and in vitro differentiation of a spermatogonial cell line. Science 2002;297(5580):392–5.

[66] Hofmann MC, Braydich-Stolle L, Dettin L, Johnson E, Dym M. Immortalization of mouse germ line stem cells. Stem Cell 2005;23(2):200–10.

[67] Ishii K, Kanatsu-Shinohara M, Toyokuni S, Shinohara T. FGF2 mediates mouse spermatogonial stem cell self-renewal via upregulation of Etv5 and Bcl6b through MAP2K1 activation. Development 2012;139(10):1734–43.

[68] Golestaneh N, Beauchamp E, Fallen S, Kokkinaki M, Uren A, Dym M. Wnt signaling promotes proliferation and stemness regulation of spermatogonial stem/progenitor cells. Reproduction 2009;138(1):151–62.

[69] Yeh JR, Zhang XF, Nagano MC. Wnt5a is a cell-extrinsic factor that supports self-renewal of mouse spermatogonial stem cells. J Cell Sci 2011;124(14):2357–66.

[70] Zhang S, Sun J, Pan S, Zhu H, Wang L, Hu Y, et al. Retinol (vitamin A) maintains self-renewal of pluripotent male germline stem cells (mGSCs) from adult mouse testis. J Cell Biochem 2011;112(4):1009–21.

[71] Chen C, Ouyang W, Grigura V, Zhou Q, Carnes K, Lim H, et al. ERM is required for transcriptional control of the spermatogonial stem cell niche. Nature 2005;436(7053):1030–4.

[72] Choong ML, Yong YP, Tan AC, Luo B, Lodish HF. LIX: a chemokine with a role in hematopoietic stem cells maintenance. Cytokine 2004;25 (6):239–45.

[73] Christensen JL, Wright DE, Wagers AJ, Weissman IL. Circulation and chemotaxis of fetal hematopoietic stem cells. PLoS Biol 2004;2(3):E75.

[74] Niu Z, Goodyear SM, Rao S, Wu X, Tobias JW, Avarbock MR, et al. MicroRNA-21 regulates the self-renewal of mouse spermatogonial stem cells. Proc Natl Acad Sci U S A 2011;108(31):12740–5.

[75] Simon L, Ekman GC, Garcia T, Carnes K, Zhang Z, Murphy T, et al. ETV5 regulates Sertoli cell chemokines involved in mouse stem/progenitor spermatogonia maintenance. Stem Cell 2010;28(10):1882–92.

[76] Wu X, Goodyear SM, Tobias JW, Avarbock MR, Brinster RL. Spermatogonial stem cell self-renewal requires ETV5-mediated downstream activation of Brachyury in mice. Biol Reprod 2011;85(6):1114–23.

[77] Yang QE, Kim D, Kaucher A, Oatley MJ, Oatley JM. CXCL12-CXCR4 signaling is required for the maintenance of mouse spermatogonial stem cells. J Cell Sci 2013;126(4):1009–20.

[78] Filipponi D, Hobbs RM, Ottolenghi S, Rossi P, Jannini EA, Pandolfi PP, et al. Repression of kit expression by Plzf in germ cells. Mol Cell Biol 2007;27(19):6770–81.

[79] Hobbs RM, Seandel M, Falciatori I, Rafii S, Pandolfi PP. Plzf regulates germline progenitor self-renewal by opposing mTORC1. Cell 2010;142 (3):468–79.

[80] Song W, Mu H, Wu J, Liao M, Zhu H, Zheng L, et al. miR-544 regulates dairy goat male germline stem cell self-renewal via targeting PLZF. J Cell Biochem 2015;116(10):2155–65.

[81] Looijenga LH, Stoop H, de Leeuw HP, de Gouveia Brazao CA, Gillis AJ, van Roozendaal KE, et al. POU5F1 (OCT3/4) identifies cells with pluripotent potential in human germ cell tumors. Cancer Res 2003;63(9):2244–50.

[82] Goertz MJ, Wu Z, Gallardo TD, Hamra FK, Castrillon DH. Foxo1 is required in mouse spermatogonial stem cells for their maintenance and the initiation of spermatogenesis. J Clin Invest 2011;121(9):3456–66.

[83] Falender AE, Freiman RN, Geles KG, Lo KC, Hwang K, Lamb DJ, et al. Maintenance of spermatogenesis requires TAF4b, a gonad-specific subunit of TFIID. Genes Dev 2005;19(7):794–803.

[84] Manders PM, Hunter PJ, Telaranta AI, Carr JM, Marshall JL, Carrasco M, et al. BCL6b mediates the enhanced magnitude of the secondary response of memory CD8 + T lymphocytes. Proc Natl Acad Sci U S A 2005;102(21):7418–25.

[85] Hamra FK, Schultz N, Chapman KM, Grellhesl DM, Cronkite JT, Hammer RE, et al. Defining the spermatogonial stem cell. Dev Biol 2004;269(2):393–410.

[86] Song HW, Wilkinson MF. Transcriptional control of spermatogonial maintenance and differentiation. Semin Cell Dev Biol 2014;30:14–26.

[87] Bermingham JR, Scherer SS, OConnell S, Arroyo E, Kalla KA, Powell FL, et al. Tst-1/Oct-6/SCIP regulates a unique step in peripheral myelination and is required for normal respiration. Gene Dev 1996;10(14):1751–62.

[88] Ghazvini M, Mandemakers W, Jaegle M, Piirsoo M, Driegen S, Koutsourakis M, et al. A cell type-specific allele of the POU gene Oct-6 reveals Schwann cell autonomous function in nerve development and regeneration. EMBO J 2002;21(17):4612–20.

[89] Wu X, Oatley JM, Oatley MJ, Kaucher AV, Avarbock MR, Brinster RL. The POU domain transcription factor POU3F1 is an important intrinsic regulator of GDNF-induced survival and self-renewal of mouse spermatogonial stem cells. Biol Reprod 2010;82(6):1103–11.

[90] Oatley MJ, Kaucher AV, Racicot KE, Oatley JM. Inhibitor of DNA binding 4 is expressed selectively by single spermatogonia in the male germline and regulates the self-renewal of spermatogonial stem cells in mice. Biol Reprod 2011;85(2):347–56.

[91] Kotaja N, Bhattacharyya SN, Jaskiewicz L, Kimmins S, Parvinen M, Filipowicz W, et al. The chromatoid body of male germ cells: similarity with processing bodies and presence of Dicer and microRNA pathway components. Proc Natl Acad Sci U S A 2006;103 (8):2647–52.

[92] Hayashi K, Lopes SMCD, Kaneda M, Tang FC, Hajkova P, Lao KQ, et al. MicroRNA biogenesis is required for mouse primordial germ cell development and spermatogenesis. PLoS One 2008;3(3):e1738.

[93] Luk AC, Gao H, Xiao S, Liao J, Wang D, Tu J, et al. GermlncRNA: a unique catalogue of long non-coding RNAs and associated regulations in male germ cell development. Database (Oxford), 2015. 2015. p. bav044.

[94] Wu J, Liao MZ, Zhu HJ, Kang K, Mu HL, Song WC, et al. CD49f-positive testicular cells in Saanen dairy goat were identified as spermatogonia-like cells by miRNA profiling analysis. J Cell Biochem 2014;115(10):1712–23.

[95]   He Z, Jiang J, Kokkinaki M, Tang L, Zeng W, Gallicano I, et al. MiRNA-20 and mirna-106a regulate spermatogonial stem cell renewal at the post-transcriptional level via targeting STAT3 and Ccnd1. Stem Cell 2013;31(10):2205–17.

[96]   Yang QE, Racicot KE, Kaucher AV, Oatley MJ, Oatley JM. MicroRNAs 221 and 222 regulate the undifferentiated state in mammalian male germ cells. Development 2013;140(2):280–90.

[97]   Zhang S, Yu M, Liu C, Wang L, Hu Y, Bai Y, et al. MIR-34c regulates mouse embryonic stem cells differentiation into male germ-like cells through RARg. Cell Biochem Funct 2012;30(8):623–32.

[98]   Tong MH, Mitchell DA, McGowan SD, Evanoff R, Griswold MD. Two miRNA clusters, Mir-17-92 (Mirc1) and Mir-106b-25 (Mirc3), are involved in the regulation of spermatogonial differentiation in mice. Biol Reprod 2012;86(3):72.

[99]   Bouhallier F, Allioli N, Lavial F, Chalmel F, Perrard MH, Durand P, et al. Role of miR-34c microRNA in the late steps of spermatogenesis. RNA 2010;16(4):720–31.

[100]   Teng Y, Wang Y, Fu J, Cheng X, Miao S, Wang L. Cyclin T2: a novel miR-15a target gene involved in early spermatogenesis. FEBS Lett 2011;585(15):2493–500.

[101]   Nicholls PK, Harrison CA, Walton KL, McLachlan RI, O'Donnell L, Stanton PG. Hormonal regulation of Sertoli cell micro-RNAs at spermiation. Endocrinology 2011;152(4):1670–83.

[102]   Dai LS, Tsai-Morris CH, Sato H, Villar J, Kang JH, Zhang JB, et al. Testis-specific miRNA-469 up-regulated in gonadotropin-regulated testicular RNA helicase (GRTH/DDX25)-null mice silences transition protein 2 and protamine 2 messages at sites within coding region implications of its role in germ cell development. J Biol Chem 2011;286(52):44306–18.

[103]   Kiger AA, Jones DL, Schulz C, Rogers MB, Fuller MT. Stem cell self-renewal specified by JAK-STAT activation in response to a support cell cue. Science 2001;294(5551):2542–5.

[104]   Tulina N, Matunis E. Control of stem cell self-renewal in Drosophila spermatogenesis by JAK-STAT signaling. Science 2001;294 (5551):2546–9.

[105]   Brawley C, Matunis E. Regeneration of male germline stem cells by spermatogonial dedifferentiation in vivo. Science 2004;304 (5675):1331–4.

[106]   He Z, Kokkinaki M, Dym M. Signaling molecules and pathways regulating the fate of spermatogonial stem cells. Microsc Res Tech 2009;72 (8):586–95.

[107]   Zhu H, Liu C, Sun J, Li M, Hua J. Effect of GSK-3 inhibitor on the proliferation of multipotent male germ line stem cells (mGSCs) derived from goat testis. Theriogenology 2012;77(9):1939–50.

[108]   Trupp M, Scott R, Whittemore SR, Ibanez CF. Ret-dependent and -independent mechanisms of glial cell line-derived neurotrophic factor signaling in neuronal cells. J Biol Chem 1999;274(30):20885–94.

[109]   Lee J, Kanatsu-Shinohara M, Inoue K, Ogonuki N, Miki H, Toyokuni S, et al. Akt mediates self-renewal division of mouse spermatogonial stem cells. Development 2007;134(10):1853–9.

[110]   Feng LX, Ravindranath N, Dym M. Stem cell factor/c-kit up-regulates cyclin D3 and promotes cell cycle progression via the phosphoinositide 3-kinase/p70 S6 kinase pathway in spermatogonia. J Biol Chem 2000;275(33):25572–6.

[111]   S D, M P, S DA, R G, P R. Signaling through extracellular signal-regulated kinase is required for spermatogonial proliferative response to stem cell factor. J Biol Chem 2001;276(43):40225–33.

[112]   Blume-Jensen P, Jiang GQ, Hyman R, Lee KF, O'Gorman S, Hunter T. Kit/stem cell factor receptor-induced activation of phosphatidylinositol 3'-kinase is essential for male fertility. Nat Genet 2000;24(2):157–62.

[113]   Kissel H, Timokhina I, Hardy MP, Rothschild G, Tajima Y, Soares V, et al. Point mutation in kit receptor tyrosine kinase reveals essential roles for kit signaling in spermatogenesis and oogenesis without affecting other kit responses. Embo J 2000;19(6):1312–26.

[114]   Yoon S, Seger R. The extracellular signal-regulated kinase: multiple substrates regulate diverse cellular functions. Growth Factors 2006;24 (1):21–44.

[115]   Kanatsu-Shinohara M, Shinohara T. Spermatogonial stem cell self-renewal and development. Annu Rev Cell Dev Biol 2013;29:163–87 Edited by Schekman R.

[116]   Kofman AE, Huszar JM, Payne CJ. Transcriptional analysis of histone deacetylase family members reveal similarities between differentiating and aging spermatogonial stem cells. Stem Cell Rev Rep 2013;9(1):59–64.

Chapter 16

Reproductive technologies and pathogen transmission

Julie Gard

Department of Clinical Sciences, Auburn University College of Veterinary Medicine, Auburn, AL, United States

16.1 Introduction

The use of advanced reproductive technologies continues to grow. Artificial insemination (AI) is utilized worldwide to improve genetics within the herd. In the United States, according to the National Animal Health Monitoring Service, in 89.3% and 20% of dairy and beef operations, respectively, AI is utilized [1]. The Data Retrieval Committee of the International Embryo Transfer Society (IETS) takes great lengths to track use of embryo technologies in the world. The report of the Data Retrieval Committee is published yearly in the December IETS newsletter [2], and the most current data are from the December 2018 IETS Newsletter [2]. The report stated that there were approximately 1,487,343 in vivo−derived (IVD) and in vitro−produced (IVP) embryos collected/produced, which were available for transfer into recipients in 2017 [2]. A total of 495,054 IVD embryos were collected and of these 162,109 were transferred fresh and 244,178 were transferred frozen [2]. Japan did not report production of IVD embryos in 2017, which most likely accounts for the reduction of IVD embryo numbers by approximately 100,000 [2]. There were 992,289 transferrable IVP embryos produced in 2017 with 751,044 of them transferred [2]. Additionally, there were 500,808 IVP embryos transferred fresh and 256,844 transferred frozen [2]. There is a tremendous increase in IVP embryos from previous years [2,3]. As reported in the 2017 IETS Newsletter, there were 666,215 IVP embryos collected via ovum pickup and 326,623 transferred fresh and 121,490 transferred frozen [3]. There has been an increase in the number of micromanipulated embryos as well, with 3935 IVP and 1122 IVD sexed embryos and 2919 IVP and 1681 IVD genotyped embryos [2,3]. There are no current numbers for cloned and transgenic embryos by the Data Retrieval Committee of the IETS but there are previous reports of 105 fresh transfers of cloned embryo [4]. There are a number of companies that produce cloned embryos, but reporting is still an issue.

The United States is currently the largest exporter of cattle embryos, exporting 41.7% of the 26,453 IVD embryos and 65% of the 6133 IVP embryos [2]. The use of embryo technologies continues to increase in other species including swine, caprine, ovine, equine, and wildlife as well. In 2017, 18,652 sheep, 3076 goat, and 20,459 horse IVD embryos were transferred [2]. Additionally, there was an increase in equine IVP embryos with 14,788 of them transferred [2]. Estimations of the true number of IVD and IVP embryos collected and transferred worldwide are five times the current reports [2]. Obviously, there is legitimate concern that due to the level of domestic and international embryo production, that transfer of embryos could facilitate broad distribution of pathogens resulting in occasional transmission.

Semen, oocytes, and IVD, IVP, micromanipulated, cloned, and transgenic embryos are all vulnerable to contamination. Pathogens can take many forms such as bacteria, viruses, protozoans, fungi, and prions, and can result in a variety of negative effects on semen, oocytes, and all types of embryos. Some of these pathogens can be very difficult to detect and may result in just slight decreases in overall embryo production but others can have a significant and often devastating effects on the clinical efficiency of AI and embryo technologies. Donors and recipients can be vectors of disease, and in fact there are a variety of pathogens that pose a serious concern to the embryo and the fetus [5]. There is a recent review of these pathogens, in which bacterial, fungal, protozoan, and viral causes of reproductive dysgenesis in cattle, sheep, goats, pigs, horses, dogs, and cats are summarized [5]. Such review focuses on the importance of immunization and biosecurity protocols to minimize disease transmission and the resultant reproductive losses [5]. It is important to

Reproductive Technologies in Animals. DOI: https://doi.org/10.1016/B978-0-12-817107-3.00016-3

be ever mindful of the myriad of diseases that might result in a decrease in efficiency when utilizing advanced reproductive technologies and to establish appropriate testing and vaccination protocols.

With the application of each of the generations of reproductive technologies, there are environmental changes that have the potential to either enhance or diminish the spread of infectious agents among susceptible animals [6,7]. Often technologies are utilized in concert, as is the case with IVD, and IVP embryos and their transfer, and this is especially the case with IVP micromanipulated embryos, cloned and transgenic embryos. Hence, there are multiple possibilities and multiple routes that a pathogen might gain access to the system. The IETS realizes the need for extensive examination of these technologies and the need to minimize any potential for pathogen transmission. Annually, the research subcommittee of the Health and Safety Advisor Committee (HASAC) of the IETS closely evaluates current research involving embryonic pathogens. This committee determines if changes to the recommended standard operating procedures including handling, processing, and treatment of oocytes and embryos are necessary for the particular pathogen examined [6,7]. The World Organization of Animal Health (OIE) closely assesses the recommendations put forth by the IETS and often adopts them, resulting in alterations or additions to the Terrestrial Animal Health Code [6]. The Terrestrial Animal Health Code provides standards for quality control, risk management for prevention of disease transmission when utilizing reproductive technologies, and provides standards for the improvement of animal health and welfare and veterinary public health worldwide, including standards for safe international trade involving terrestrial animals (mammals, reptiles, birds, and bees) and their products [8]. Embryonic pathogens are placed into official categories from one to four and this is detailed in Chapter 4.8 of the Terrestrial Animal Health Code under "Categorization of Diseases and Pathogenic Agents" [8]. Categorization is currently under review by the IETS HASAC Research subcommittee and some changes may occur in the coming years. As for now, Category 1 diseases are diseases or pathogenic agents for which sufficient evidence has accrued to show that the risk of transmission is negligible provided that the embryos are properly handled between collection and transfer in accordance with the Manual of the IETS [8−12]. The following diseases or pathogenic agents are in category 1: blue tongue virus (BTV) (cattle), bovine spongiform encephalopathy (BSE) (cattle), *Brucella abortus* (cattle), enzootic bovine leukosis, foot-and-mouth disease (cattle), infection with Aujeszky's disease virus (pigs): trypsin treatment required, infectious bovine rhinotracheitis/infectious pustular vulvovaginitis: trypsin treatment required, and Scrapie (sheep) [8]. Category 2 diseases are those for which substantial evidence has accrued to show that the risk of transmission is negligible provided that the embryos are properly handled between collection and transfer, in accordance with processing procedures detailed in the IETS Manual, but for which additional transfers are required to verify existing data [8]. The following diseases are in category 2: BTV (sheep), caprine arthritis/encephalitis, and classical swine fever [8]. Category 3 diseases or pathogenic agents are those for which preliminary evidence indicates that the risk of transmission is negligible provided that the embryos are properly handled between collection and transfer according to the IETS Manual, but for which additional in vitro and in vivo experimental data are required to substantiate the preliminary findings [8]. The following diseases or pathogenic agents are in category 3: atypical Scrapie, bovine immunodeficiency virus, BSE (goats), bovine viral diarrhea virus (BVDV) (cattle), *Campylobacter fetus* (sheep), foot-and-mouth disease (pigs, sheep, and goats), *Haemophilus somnus* (cattle), Maedivisna (sheep), *Mycobacterium paratuberculosis* (cattle), *Neospora caninum* (cattle), ovine pulmonary adenomatosis, porcine reproductive and respiratory disease syndrome, Rinderpest (cattle—eradicated), porcine circovirus type 2 (pigs), and swine vesicular disease [8]. Category 4 diseases and pathogenic agents, as listed in the Terrestrial Animal Health Code 4.8.14, are those for which studies have been done, or are in progress which indicate that no conclusions are yet possible with regard to the level of transmission risk or that the risk of transmission via embryo transfer might not be negligible even if the embryos are properly handled according to the IETS Manual between collection and transfer [8]. The following diseases or pathogenic agents are in category 4: African swine fever, Akabane (cattle), Bovine anaplasmosis, BTV (goats), border disease (sheep), bovine herpesvirus-4, *Chlamydia psittaci* (cattle, sheep), contagious equine metritis, enterovirus (cattle, pigs), equine herpesvirus 1 (equine rhinopneumonitis), equine viral arteritis, *Escherichia coli* 09:K99 (cattle), *Leptospira borgpetersenii* serovar *hardjobovis* (cattle), *Leptospira* sp. (pigs), lumpy skin disease, *Mycobacterium bovis* (cattle), *Mycoplasma* spp. (pigs), ovine epididymitis (*Brucella ovis*), parainfluenza-3 virus (cattle), parvovirus (pigs), Q fever (*Coxiella burnetii*), Scrapie (goats), *Tritrichomonas foetus* (cattle), *Ureaplasma* and *Mycoplasma* spp. (cattle, goats), and vesicular stomatitis (cattle, pigs) [8]. Obviously, the number of reproductive pathogens is numerous and ever changing and growing. An example is the presence of emerging atypical BVDV viruses called HoBi-like viruses (HoBiPeVs), which have been called "atypical pestiviruses, BVDV 3," which are of special concern since they cannot always be detected via testing for BVDV, and BVDV vaccines may have limited efficiency against providing protection against these viruses [13,14]. Additionally, the sources of contamination can be varied and transmission can occur through multiple routes. Transmission may occur transplacentally from the recipient to the embryo or even iatrogenically through lack of sterilization of equipment utilized for transfer of semen and/or embryos.

The donor animal, the recipient, semen, oocyte, embryo, culture environment, washing media, along with animal source proteins utilized in the media, and even animal source hormones used in superstimulation (superovulation) techniques, and liquid nitrogen may all harbor pathogens resulting in contamination of the system [7]. In a study by Pinherio de Oliveria et al., in 2013, the presence of genetic material was assessed in 88 cell cultures samples from eight laboratories [5]. *Mycoplasma*, porcine circovirus 1 (PCV1), bovine leukemia virus (BLV), and/or BVDV were found in the samples [15]. Trypsin samples and 13 fetal calf serum samples from different lots from five of the laboratories were analyzed as well. The percentages of the different contaminants of cell culture detected were 34.1% for *Mycoplasma*, 35.2% for PCV1, 23.9% for BVDV RNA, and 2.3% for BLV. Fetal calf sera and trypsin samples analyzed in this study had BVDV RNA and PCV1 DNA [15]. It is safe to say based on this study that culture, sera, and trypsin used by different laboratories show a high rate of contaminants.

Cryopreservation is utilized universally with advanced reproductive technologies for storage of germplasm. Often, there is a false idea of security when storing germplasm in liquid nitrogen. The thought that pathogens could survive the extremely cold temperatures of $-196°C$ seems somewhat unbelievable. However, some pathogens can survive the freezing process, such as BTV, *Mycoplasma*, BVDV, and bovine herpes virus 1 (BHV-1) [5,7,16]. A study by Bielanski et al., 2009 evaluated the effect of cryopreservation on the infectivity of slow cooled and vitrified IVP embryos that had infective embryo-associated BVDV or BHV-1 [16]. There was a reduction in infective embryo-associated virus (EAV) following both freezing methods, in comparison to embryos that were exposed to virus and not cryopreserved (31% vs 72%, respectively; $P < .001$) [16]. Vitrified embryos were not significantly higher in infective EAV than was seen in the slow cooled (38% vs 22%; $P < .002$) [16]. A higher percentage of BHV-1 infectivity was maintained following cryopreservation (42%), versus that seen with infected BVDV embryos (24%) [16]. So, cryopreservation may decrease infectivity of some pathogens but does not eliminate the threat of infectivity of others [16]. Contamination of liquid nitrogen tanks can be the result of infected biological materials and incidence of poorly sealed containers, or straws, or use of open vitrification systems for cryopreservation of oocytes, and embryos [17]. Appropriately sealed containers, standard operating procedure for cleaning, and maintaining tanks, are a must to decrease the potential disease transmission. Bielanski has thoroughly described standard operating procedures for maintaining liquid nitrogen tanks to prevent transmission of pathogens [18]. Good quality control within the laboratory, effective biosecurity protocols of donors, and recipients including testing for diseases should be implemented when instituting an advanced reproductive technologies program.

16.2 First generation of advanced reproductive technologies

AI, the first generation of advanced reproductive technologies, has always been thought to minimize disease transmission especially, with regards to venereally transmitted diseases such as *Tritrichomonas foetus* (*T. foetus*) and *Campylobacter fetus venerealis* in cattle [19]. However, there are a number of diseases such as BTV, BVDV, BHV-1, and *Mycoplasma* that can be present and infective in semen following cryopreservation [18]. Reliable AI Centers go to great lengths to make sure that all bulls collected are free of diseases at the time of collection. Strict quarantine methods are utilized to minimize any chance of bulls contacting or spreading any disease while at the AI Center [20]. Semen should be processed under National Association of Animal Breeders guidelines for Certified Semen Services (CSS), which represent the minimum requirements for disease control of semen produced for AI Centers and provide a minimum industry standard for health management [20]. These requirements include testing and examination of bulls prior to entry, during an isolation period and semiannually thereafter, with testing continuing throughout their time at the AI Center [20]. Semen straws will have CSS printed on them if collection has been performed in AI Centers that have followed CSS requirements. For CSS AI Center requirements, donor bulls must be negative for tuberculosis 60 days prior to entry via a caudal fold intradermal test [20]. Additionally, bulls must be negative for the following diseases within 30 days prior to entry: brucellosis, leptospirosis (*Leptospira canicola, Leptospira grippotyphosa, Leptospira hardjo, Leptospira pomona, Leptospira icterohaemorrhagiae*), BVDV, and *T. foetus* [20]. Bulls are then held in quarantine/isolation and undergo additional testing for tuberculosis, *B. abortus*, BVDV, leptospirosis-blood test for the same five serotypes as mentioned above, *C. fetus venerealis* and *T. foetus*. [20]. Additionally, BVDV testing of semen is required by CSS due to results reported by Givens et al., in 2009, that discovered that BVDV infections could result in prolonged and/or persistent testicular infections which resulted in the presence of BVDV in the semen [20,21]. Bulls acutely or persistently infected with BVDV will shed virus in their semen and will be viremic [21]. Persistently infected animals shed virus from all bodily secretions so embryos or semen should never be used for embryo transfer or AI [21]. However, with acute infections of BVDV there can be two outcomes, one that the animal does not shed virus in the semen and that a prolonged testicular infections can result [21]. Bulls with prolonged testicular infections will be negative for virus in the blood and negative on immunohistochemistry testing, but shed virus in their semen [21]. Vaccinating with a noncytopathic vaccine may even result in a prolonged testicular infection, so it is important to only vaccinate with cytopathic strains of BVDV [21]. The CSS requirements now include additional BVDV testing prior to release of semen from the AI Center [20]. The testing required for persistent testicular infections include one of the following test methods: (1) test all bulls during the isolation period for BVDV by the serum neutralization test for both types I and II of BVDV. All bulls that test positive must have one negative polymerase chain reaction (PCR) test on processed semen or (2) all bulls must have one negative PCR test on processed semen [20]. Clearly, the importance of appropriate biosecurity and testing methodology needs to be a priority in AI collection centers for bulls and for other species as well, to prevent widespread transmission of diseases via AI.

AI has been around for a long time but synchronization of estrus and timed insemination has made AI much more convenient for the producer. If cows or heifers are cycling when estrus synchronization is implemented, that gives three opportunities to conceive during a 45-day period or four opportunities during a 65-day period, which means more chance for utilizing AI to improve genetics in the herd without economic losses due to a spread out calving season. Additionally, controlled intravaginal drug-releasing devices (CIDRs) have increased the percentage of cows responding to the synchronization program by more than 20% when utilizing timed insemination protocols [22]. The question of CIDRs reuse has been evaluated and conflicting results have been reported [23–25]. In a recent study, there was a significant decrease in serum progesterone concentrations following insertion of reused CIDRs when compared to serum progesterone concentrations following insertion of new CIDRs [25]. This is not in complete agreement with the results of another study that concluded that there was adequate progesterone in the CIDR used for a second time [23]. Hence, reuse of 1.38g CIDRs to aid in synchronization of estrus in cattle is not recommended due to an unreliable amount of progesterone left in the CIDR and that the CIDR may act as a vector and aid in disease transmission [25]. There are bacterial, viral, and protozoal pathogens that infect the vaginal area such as Ureaplasma, BHV-1, and *Trichomonas foetus* (*T. foetus*), respectively [5,19]. Use of double sheaths and/or individual sheaths with plastic sanitary chemise is strongly encouraged to aid in minimizing contamination of the uterus [5,19].

16.3 Second generation of advanced reproductive technologies

In vivo derived embryo production is considered the second generation of advanced reproductive technologies [26]. In vivo embryo production involves superovulation, nonsurgical embryo collection, and nonsurgical transfer of embryos into synchronized recipient [26]. The original techniques utilized to produce IVD embryos were developed more than 60 years ago and Betteridge detailed these techniques in a review [5,26,27]. Reproductive technologies nowadays are utilized extensively throughout the world across all animal production species, and especially in the cattle industry. There is though, the potential for increasing or decreasing the distribution of indigenous or foreign animal pathogens when implementing these techniques [7]. However, IVD embryo production has the reputation of a means to prevent transmission of pathogens that are considered Category 1 diseases [7]. Again, embryos must be properly handled using the IETS processing procedures that were validated with IVD embryos as detailed in the IETS Manual [8–12]. The initial evidence for safety came from "in vitro-in vitro" studies for the Category 1 agents with the exception of BSE [7,28,29]. Additional studies were performed where embryos from infected animals were shown to be free of virus via in vitro assay or through disease-free transfers to recipients, providing greater confidence to the safety of embryo transfer [30–33].

A number of pathogens are known to be transmitted from mother to fetus via the placenta. These include BVDV, *N. caninum*, BTV, BLV, BHV-1, and *Mycobacterium avium* ssp. *Paratuberculosis* (Johne's disease) [5,33–37]. Thus a complete risk assessment should always give due consideration to the entirety of infectious agents capable of establishing transient, latent, or persistent infection of the donor and/or recipient, which can result in fetal disease. Often the disease may spread from the recipient to other animals through horizontal transmission. Otherwise, collection and transfer of embryos, albeit even pathogen-free embryos, could lead to the birth of diseased offspring resulting from subsequent in utero infections. The 2019 official biosecurity recommendations for international movement of embryos are outlined in Chapter 4.8.5 of the Terrestrial Animal Health Code under "Risk Management" [12]. These general strategies were developed to help ensure that the transfer of IVD embryos would not result in the transmission of infectious agents [12]. Three general strategies are available to ensure that transfer of IVD embryos will not result in the transmission of infectious agents originating from donor animals [12]. Each method represents an environmental change that is intended to deter pathogen transmission. The preferred method in cattle is embryo processing with or without trypsin treatment because it is relatively inexpensive and has been substantiated by a considerable amount of research to prevent the spread of a variety of viruses, bacteria and even prions (BSE) [9,12,28,29,38,39]. Other methods include donor testing and recipient quarantine and testing [38]. Testing donors for presence of disease at the time of embryo collection, and again, weeks or months later (while embryos are held in the cryopreserved state), can determine if such animals are or are not a possible pathogen source [7]. The testing of recipients following transfer of embryos and testing of offspring after birth will function as in vivo bioassays utilizing the recipients and offspring as sentinel animals [9]. A lengthy quarantine of animals and testing is necessary to determine the true pathogen source. These methods are very expensive and require multiple tests to determine that accidental exposure did not occur during the quarantine period [7]. Therefore, economically, the use of embryo processing is the preferred method for certifying health of embryos. This method has greatly facilitated international trade of embryos because of the combination of animal welfare, disease control, and economic advantages it provides [38].

16.4 The third generation of advanced reproductive technologies

In vitro embryo production has been described as the third advanced reproductive technology and the second generation of embryo technologies [26]. Most of the essential components for IVP embryo production have been developed and refined over the past 35 years [39−41]. Fundamental procedures that are currently used with in vitro production include superovulation of donor cattle, collection of oocytes via transvaginal follicular aspiration (ovum pickup) or through abattoir ovaries, in vitro maturation (IVM) of oocytes, in vitro capacitation of spermatozoa, in vitro fertilization (IVF), in vitro culture (IVC) of embryos to the blastocyst stage, and embryo transfer to recipients [26,42]. With the advent of intracytoplasmic sperm injection (ICSI), the fertilization process changes within the IVP embryo production. During ICSI, the barrier of the zona pellucida (ZP) is cut through and a single spermatozoon is injected into the oocyte. ICSI embryos are cultured similarly to embryos produced through ordinary IVF protocols, and then at the right stage transferred into recipients. The excellent barrier of the ZP is breached with ICSI, allowing pathogens to gain access to a protected site that normally would not. In IVP embryo production, multiple changes in the oocyte and embryo environments occur, resulting in multiple opportunities for pathogens to enter, amplify, and spread. The sources of pathogens include gametes and their associated fluids, somatic cells used in IVM, IVF, or IVC and the materials of animal origin used to supplement maturation, fertilization or culture media [e.g., fetal bovine serum (FBS), bovine serum albumin, or follicle-stimulating hormone (FSH)] [7,43]. FBS can be a source of adventitious bacteria, fungi, or viruses when adequate quality assurances are not in place [7,22,43,44]. The most common contaminant in FBS is BVDV [43−46]. Studies which surveyed for contaminants in in vitro embryo production laboratories have been conducted in a number of different countries such as Canada, Denmark, France, and the United States [41,47−53]. As expected, levels of bacterial contamination can be quite high from 13% to 68% of samples when abattoir-origin materials are utilized in IVP embryo production systems [41]. Also, low to moderate levels of BVDV and BHV-1 have been found in abattoir-origin materials with a range of 1%−12% of samples testing positive for BVDV and 0%−12% of samples testing positive for BHV-1 [41,53]. Fortunately, the majority of IVP embryo production occurs without utilizing oocytes of abattoir origin or uterine tubal cells and only 1.2% of IVP embryos transferred derived from oocytes of abattoir origin [2]. Additionally, in 2017, no IVD or IVP exported embryos were produced utilizing oocytes of abattoir origin [2]. However, there are other materials from the abattoir that are utilized in reproduction in cattle on a wide basis such as FBS serum and FSH, which still have potential for contamination [41]. Follotropin-V is the only FDA-approved product within the United States for superovulation treatments and it is made from porcine pituitaries [7]. Clinical trials are underway for testing of synthetic FSH products. The use of synthetic products is in fact one more step in the direction of minimizing transmission of disease.

Donors and recipients can harbor diseases. Certain diseases such as BVDV and BHV-1 are widely distributed among populations of cattle and result in severe disease affecting multiple body systems including the respiratory, gastrointestinal, and reproductive systems [52]. Unfortunately, BVDV infections can be asymptomatic [54,55]. Persistently infected cattle are often asymptomatic for long periods of time and serve as reservoirs for production of large quantities of virus [54]. Most of the field isolates of BVDV are noncytopathic, and this can result in unapparent persistent infections in donor animals from which semen, oocytes, and embryos are collected or from which laboratory cell lines are established and used in cocultures with oocytes and embryos [44,56]. All materials (animal products, ovaries, oocytes, follicular fluid, cumulus, uterine tubal cells, FSH, and FBS) of animal origin that have been integral to IVP embryo production in cattle have been shown to contain BVDV when harvested from actively infected animals [52,56]. Each phase of IVP embryo production is ideally suited for replication of BVDV and other viruses such as BHV and/or bacteria. The culture system provides adequate time and substrate for pathogen replication. Hence, a low level of contamination at the beginning of the embryo production system can be effectively amplified, leading to exposure of embryos to a high level of pathogen by the end of the production line. Synthetic media have been developed to minimize contamination and transmission of pathogens through IVP embryo production. However, semen, oocytes, and embryos will always provide a route for transmission of pathogens into culture systems. As with IVD embryos, it was originally hypothesized that the ZP of IVF embryos would protect the developing conceptus from infections and in fact, artificial exposure studies have demonstrated that an intact zona is an effective barrier that is seldom penetrated [57]. Some studies have clarified that the processing procedures recommended for use on IVD embryos are not sufficient for IVP embryos due to the difference in the ZP features [57–59]. Several viruses such as BTV, BHV-1, foot-and-mouth disease virus, BVDV, and bacteria such as *Leptospira* have been shown to adhere more readily to the zona of IVP embryos [52,53,57,60]. The presence of virus within the IVP embryo production system has been shown to result in reduced rates of maturation, fertilization, and development [52,61–69]. Stringfellow et al. reported a reduced development to the blastocyst stage (9%) in BVD-exposed cumulus-oocyte complexes (COCs), when compared to nonexposed COCs (12%) [65]. However, with some viruses, such as Caprine Arthritis Encephalitis Virus the difference in blastocyst rates between control and infected was not as dramatic (36.0% and 34.6%, respectively) [66]. Other studies have shown that BVDV remains associated with in vitro–produced embryos after viral exposure and washings, with or without trypsin treatment [60,67,68]. Infection of susceptible recipients was reported in Bielanski's study where 18 of 35 recipients receiving embryos exposed to the BVDV type 2 strain seroconverted (51%), and only 2 of 11 pregnancies resulted in live offspring [67]. These two offspring were reported to be negative for BVDV, leading to the conclusion that embryos with EAV underwent embryonic death [67]. Additionally, some interesting results were achieved in an attempt to determine whether the amount of BVDV associated with individual embryos would constitute an in utero, infective dose [69,70]. The cumulative results of these in vitro studies, evaluating the risks of transmitting BVDV via embryo transfer are somewhat contradictory [69–71]. On one hand, it is clear that multiple strains of BVDV will remain associated with IVP embryos following washing with trypsin treatment [69–71]. On the other hand, there appears to be innate deterrents to transmission of the EAV to permissive cells in vitro, while reproducing at the best the environment of the uterus and the uterine tube [69–71]. Additionally, media free of animal origin are being produced and utilized, decreasing thus the chances of entry of BVDV into the culture system, although it cannot be ignored the possibility of EAV. Official recommendations for international shipment of in vitro–produced embryos in order to control the spread of diseases are outlined in Chapter 4.9 of the Terrestrial Animal Health Code of the OIE [72]. Impediments to the establishment of universally recognized sanitary precautions for the production of bovine IVP embryos include: the lab-to-lab variability in techniques utilized, the comparative ease with which pathogens adhere to the ZP, and the fact that raw materials such as FBS are collected from abattoirs [73]. It is apparent that mechanical washing and trypsin treatment of IVP embryos do not provide the same level of effectiveness when utilized for cleansing IVD embryos [60,68]. These treatments though have been shown to definitely reduce the level of contamination of IVP embryos following artificial exposure to some agents such as BHV-1 and BVDV [58,60,68]. Thus as is the case with IVD embryos, further studies are necessary in order to complete the risk assessment, which includes evaluating fully if the quantity of embryo-associated BVDV is sufficient to infect recipients and/or their fetuses after transfer into the uterus. Obviously, the strain of the virus seems to have a tremendous effect on the affinity of the virus for the embryos and it is necessary to take this into account when assessing the true risk [74].

16.5   The fourth generation of advanced reproductive technologies

The fourth generation of reproductive technologies includes cloning and transgenic embryo production [26]. These technologies specifically involve maturation and enucleation of oocytes, culture of cell lines from which the nucleus is harvested, insertion of cell nuclei, activation through chemical and/or electrical means, IVC, possibly cryopreservation, and surgical or nonsurgical transfer [7,26]. The same hazards that are associated with the IVP embryo production apply directly to cloned and transgenic embryo production with some additional major concerns. These concerns include breaking or removing the ZP, harvesting oocytes from an abattoir, and culturing cell lines for long periods of time from which the donor nuclei will be recovered [7]. The cell lines utilized may be cultured for 8–16 weeks in order to allow for cellular amplification and even longer if a transgene has been inserted into the nucleus. The amplification procedures open the door to multiple chances for contamination of all types of pathogens, especially of bacterial, fungal, and viral origin. Since animal products are utilized, the chances for viral, bacterial, and fungal contamination increase dramatically. Noncytopathic BVDV and bovine herpes viruses such as BHV-1 can be present in abattoir-origin materials [41,52,53,56]. These viruses can persist and amplify in cultured cell lines, leading to a much greater chance for contamination and transmission of disease than IVD and IVP production.

The presence of viruses such as noncytopathic BVDV may be manifested by recording of poor cloning efficiency and development, low pregnancy rates, early embryonic death, fetal wastage, and neonatal abnormalities. In a report by

Shin et al., in 2000, a high incidence of developmental failure, embryonic death, and fetal resorption, occurred in cloned bovine fetuses derived from cell lines infected with BVDV [75]. A fetal fibroblast cell line was inadvertently infected with a noncytopathic strain of BVDV in this case [75]. In a study by Stringfellow et al., in 2005, fetal fibroblast cell lines were screened for BVDV by veterinary diagnostic labs and revealed that 15 of 39 fetal fibroblast cell lines used in cloning research were positive for BVDV as determined by various assays including reverse transcription PCR [76]. However, it was determined that only 5 of the 39 cell lines were actually infected with BVDV, and furthermore, three of these five lines were determined not to be infected at the very early cryopreserved passage [76]. This evidence leads to the conclusion that the cell lines became infected after culture in media containing contaminated FBS [76]. Sequence comparison of the amplified cDNA from one lot confirmed that FBS was the source of infection for one of these cell lines [76]. Since BVDV was isolated from the remaining two cell lines at the early passages, the fetuses from which they were established could not be ruled out as the source of virus [76]. These studies highlight the greater potential for contamination and resultant disease transmission with cloning and transgenic embryo production. Some risk analyses have estimated the chance of certain diseases such as equine infectious anemia (EIA) to result in disease transmission via transfer of contaminated cloned equine embryos [77]. This risk assessment determined that the plausibility of introducing EIA into the US population of horses via cloned horse embryos was very low [77]. Often, infected embryos have reduced development and pregnancy rates, ultimately resulting in poor cloning efficiency and very low numbers of transferable embryos [76]. However, this is not the case for every pathogen. There are reports of comparable development rates with a BHV-5 infected system [78]. In a study by Silva-Frade et al., BHV-5 infected bovine oocytes, replicated, and suppressed some apoptotic pathways without significantly affecting embryonic development [78]. However, in a follow-up study presumptive zygotes directly infected 1 day after fertilization produced a lower number of IVP embryos, and BHV-5 was transmissible to the embryo during in vitro development [79]. In a more recent study, inactivated Sendai virus was utilized to fuse cloned bovine embryos and actually prevented endoplasmic recticulum stress−associated apoptosis [80]. Blastulation rate and blastocyst quality were improved when Sendai virus fusion was used in the somatic cell nuclear transfer embryos [80]. Sendai virus is a murine paramyxovirus that causes respiratory disease in mice, and it represents a unique condition in which this virus is used in embryo production. The virus is inactivated so that the chances of a pathogenic state are virtually none [80]. Certainly, different viruses have different effects within the system and the severity of the effect is dependent on the stage of embryonic development. Therefore excellent quality control is imperative in order to prevent embryonic pathogens and transmission of pathogens via embryo transfer to recipients. It is of major importance with all generations of embryo technologies and the needs for quality control are increased with each subsequent generation.

16.6 Experimental treatments

Standard operating procedures in laboratories include the addition of antibiotics and antifungals to media to decrease and/or prevent bacterial and fungal contamination in culture. However, viral contamination poses a more complicated threat. A recent study looked at the addition of interferon-tau to IVP embryos in culture with BVDV and BHV-1 [81]. A significant decrease in the amount of BVDV in culture was seen but the same result was not observed for BHV-1 [81]. Other antiviral compounds such as DB606 (2-(4-[2-imidazolinylphenyl)-5-(4-methoxyphenylfuran) have been shown to inhibit the replication of BVDV in bovine uterine tubal epithelial cells, Madin Darby bovine kidney cells, and fetal fibroblast cells, without negatively affecting embryonic development nor future fertility of heifers derived from IVP embryos treated with DB606 [82,83]. Another similar compound, 2-(2-benzimidazolyl)-5-[4-(2-imidazolino) phenyl] furan dihydrochloride (DB772), was analyzed for its activity against diverse pestiviruses (Newcomer) such as isolates of BVDV 2, border disease virus, HoBiPeVs, pronghorn virus, and Bungowannah virus [84]. Various concentrations of DB772 were added to infected cells in vitro, and it was found that DB772 effectively inhibits all pestiviruses studied at concentrations $>0.20\ \mu$M with no evidence of cytotoxicity [84]. New antiviral drugs, called direct-acting antivirals (DAAs) are now on the market and have sustained virologic response rate as high as 60%−100% for hepatitis C virus (HCV) [85]. These DAAs target the nonstructural HCV proteases NS3−4A and NSA5 or inhibit RNA-dependent RNA polymerase, and this could also be very efficacious against emerging pestiviruses such HoBiPeVs [85]. Zika virus, in the same family Flaviviridae as BVDV and HCV [54,86], is responsible for another emerging disease that has looked to antivirals to inhibit the RNA-dependent RNA polymerase to prevent severe repercussions of the virus on the embryo and fetus [86]. Obviously, these new drugs regimens primarily have applications in humans but can have application in animal medicine as well. Antiviral compounds such as DB606, DB772, and DAAs along with interferon-tau show potential for prevention of replication of viruses in cells and the ability to eliminate viruses from infected semen and IVD and IVP embryos, and from fibroblast cell lines utilized for cloning and transgenics [83–87].

16.7 Zoonotic pathogens

Public health concerns can be raised with some pathogens, especially those that are zoonotic. A good example of this is seen with *C. burnetii* (Q fever) which has been detected in flushing media and uterine tissue samples from goats [88]. In a recent study by Alsaleha et al. [88], it was reported that *C. burnetii* was not removed when IVD goat embryos were exposed in vitro and subsequently underwent standard washing procedures. *C. burnetii* is a zoonotic pathogen and recipient contamination might also result in human infection. So appropriate handling of donors, recipients, and embryos is of paramount importance in order to prevent infection to people working with animals.

Conclusion

The advent of sensitive and adequate testing methodologies, to prevent use of contaminated donor and recipient animals, and the availability of effective synthetic media and synthetic hormones have substantially reduced the risk for pathogen transmission via use of advanced reproductive technologies. The use of antivirals within culture systems for prevention of embryonic pathogens is promising. The use of these new modalities might provide a reliable methodology to curtail infectious virus associated with embryos, oocytes, and semen in the future. However, proper immunization protocols of donors and recipients together with effective biosecurity programs and excellent laboratory quality control will always be the first line of defense in prevention of pathogens when utilizing all generations of advanced reproductive technologies.

References

[1] USDA, Health and management practices on U.S. Dairy Operations, Section I: Population estimates–D. Reproduction; 2014. p. 79. <https://www.aphis.usda.gov/animal_health/nahms/dairy/downloads/dairy14/Dairy14_dr_PartIII.pdf>.

[2] Viana J. Statistics of embryo production and transfer in domestic farm animals; Is it a turning point? In 2017 more in vitro-produced than in vivo-derived embryos were transferred worldwide. In: International embryo transfer society data retrieval committee annual report. IETS Newsletter; December 2018. p. 8–25. <https://www.iets.org/pdf/Newsletter/Dec18_IETS_Newsletter.pdf>.

[3] Perry G. 2016 Statistics of embryo production and transfer in domestic farm animals. In: International embryo transfer society data retrieval committee annual report. IETS Newsletter; December 2017. p. 8–23 <https://www.iets.org/pdf/Newsletter/Dec17_IETS_Newsletter.pdf>.

[4] Thibier M. Data retrieval committee annual report. In: Embryo transfer newsletter; December 2006. <www.iets.org/pdf/comm_data/December2006.pdf>.

[5] Givens M, Marley M. Infectious causes of embryonic and fetal mortality. Theriogenology 2008;3:270–85.

[6] Thibier M, Stringfellow D. Health and safety advisory committee (HASAC) of the international embryo transfer society (IETS) has managed critical challenges for two decades. Theriogenology 2003;59(3–4):1067–78.

[7] Gard JA, Stringfellow DA. Shaping the norms that regulate international commerce of embryos. Theriogenology 2014;81:56–66.

[8] Anonymous Terrestrial Animal Health Code Section 4 World Organization for Animal Health (OIE), Paris. Terrestrial Code, 28th Edition, 2019, ISBN of volume I: 978-92-95108-85-1;ISBN of volume II: 978-92-95108-86-8. <https://www.oie.int/index.php?id = 169&L = 0&htmfile = titre_1.4.htm>; 2019.

[9] Stringfellow DA. Recommendations for the sanitary handling of in-vivo-derived embryos. In: Stringfellow DA, Givens MD, editors. Manual of the international embryo transfer society. 4th Edition Savoy, IL: IETS; 2009. p. 65–8.

[10] Thibier M. Embryo transfer: a comparative biosecurity advantage in international movements of germplasm. Rev Sci Tech 2011;30(1):177–88.

[11] Food and Agriculture Organization of the United Nations (FAO) – embryo training manual; <http://www.fao.org/docrep/004/T0117E/T0117E14.htm>.

[12] Terrestrial animal code chapter 4.8.5 of the terrestrial animal code under "risk management"; <www.oie.int>.

[13] Bauermann FV, Ridpath JF, Weiblen R, Flores EF. HoBi-like viruses: an emerging group of pestiviruses. J Vet Diagn Invest 2013;25(1):6–15. Available from: https://doi.org/10.1177/1040638712473103.

[14] Flores EF, Cargnelutti JF, Monteiro FL, Bauermann FV, Ridpath JF, Weiblen R. A genetic profile of bovine pestiviruses circulating in Brazil (1998-2018). Anim Health Res Rev 2018;19(2):134–41. Available from: https://doi.org/10.1017/S1466252318000130.

[15] Pinheiro de Oliveira TF, AugustoFonseca AA, Camargos MF, Oliveira AM, Cottorello ACP, Antonizete dos ReisSouza A, et al. Detection of contaminants in cell cultures, sera and trypsin. Biologicals 2013;41(6):407–14.

[16] Bielanski A, Lalonde A. Effect of cryopreservation by slow cooling and vitrification on viral contamination of IVF embryos experimentally exposed to bovine viral diarrhea virus and bovine herpesvirus-1. Theriogenology 2009;72(7):919–25 Epub 2009 Jul 17.

[17] Bielanski A, Vaita G. Risk of contamination of germplasm during cryopreservation and cryobanking in IVF units. Hum Reprod 2009;24 (10):2457—67 <https://doi.org/10.1093/humrep/dep117>.

[18] Bielanski A. Risk of contamination of germ plasm during cryopreservation and cryobanking. In: Manual of the international embryo transfer society, 4th ed. Stringfellow DA, Givens MD, editors, Savoy, IL: IETS: Appendix C; 2009. p. 131—40.

[19] Edmondson MA, Joiner KS, Spencer JA, Riddell KP, Rodning SP, Gard J, et al. Impact of a killed tritrichomonas foetus vaccine on clearance of the organisms and subsequent fertility of heifers following experimental inoculation. Theriogenology 2017;90:245—51.

[20] National Association of Animal Breeders, Certified Semen Services Guidelines and Minimum Requirements. 2014; p. 1—15. <https://www.naab-css.org/uploads/userfiles/files/CSSMinReq-Jan2014201607-ENG.pdf>.

[21] Givens MD, Riddell KP, Edmondson MA, Walz PH, Gard JA, Zhang Y, et al. Epidemiology of prolonged testicular infections with bovine viral diarrhea virus. Vet Microbiol 2009;139(1—2):42—51.

[22] Lucy MC, Billings HJ, Butler WR, Ehnis LR, Fields MJ, Kesler DJ, et al. Efficacy of an intravaginal progesterone insert and an injection of PGF 2" for synchronizing estrus and shortening the interval to pregnancy in postpartum beef cows, peripubertal beef heifers, and dairy heifers. J Anim Sci 2001;79:982.

[23] Herrmann JA, Wallace RL. The effect of new and reused CIDRs on serum progesterone concentrations in lactating dairy cows. The Bovine Practitioner. 2007;41—7.

[24] Ongubo MN, Rachuony HA, Lusweti FN, Kios DK, Kitilit JK, Musee K, et al. Factors affecting conception rates in cattle following embryo tranfer. Uganda J Agric Sc 2015;16(1):19—27.

[25] Muth-Spurlock AM, Poole DH, Whisnant CS. Comparison of pregnancy rates in beef cattle after a fixed-time AI with once- or twice-used controlled internal drug release devices. Theriogenology. 2016;85(3):447—51 <https://doi.org/10.1016/j.theriogenology.2015.09.019>.

[26] Dyck MK, Zhou C, Tsoil S, Grant J, Dixon WT, Foxcroft GR. Reproductive technologies and the porcine embryonic transcriptome. Anim Reprod Sci 2014;149(1—2):11—18 <https://doi.org/10.1016/j.anireprosci.2014.05.013>.

[27] Betteridge K. Reflections on the golden anniversary of the first embryo transfer to produce a calf. Theriogenology 2000;53:3—10.

[28] Wrathall AE, Brown KFD, Sayers AR, Wells GAH, Simmons MM, Farrelly SSJ, et al. Studies of embryo transfer from cattle clinically affected by bovine spongiform encephalopathy (BSE). Vet Rec 2002;150:365—78.

[29] Wrathall A, Brown K, Sayers A, Wells G, Simmons M, Farrelly S, et al. Studies of embryo transfer from cattle clinically affected by bovine spongiform encephalopathy. (BSE). Vet Rec 2000;150:365—78.

[30] Bielanski A, Hare W. Procedures for design and analysis of research on transmission of infectious disease by embryo transfer. In: Stringfellow DA, Seidel SM, editors. 3. Savoy, IL: International Embryo Transfer Society; 1998. p. 143—9.

[31] Backx A, Heutink R, Van Rooij E, Van Rijn P. Transplacental and oral transmission of wild-type bluetongue virus serotype 8 in cattle after experimental infection. Vet Micro 2009;138(3—4):235—43.

[32] Jimmeénze-Clavero MA, Aguero M, san Miguel E, Mayoral T, Lopez MC, Ruano MJ, et al. High throughput detection of bluetongue virus by a new real-time fluorogenic reverse transcription polymerase chain reaction: application on clinical samples from current mediterranean outbreaks. J Vet Diagn Invest 2006;18(1):7—17.

[33] Wrathall AE, Simmons HA, Van Soom A. Evaluation of risks of viral transmission to recipients of bovine embryos arising from fertilization with virus-infected semen. Theriogenology 2006;65(2):247—74.

[34] Menzies FD, McCullough SJ, McKeown IM, Jess S, Murchie AK, Fallows JG, et al. Evidence for transplacental and contact transmission of bluetongue in cattle. Vet Rec 2008;163:203—9.

[35] Sweeney RW, Whitlock RH, Rosenberger AE. Mycobacterium paratuberculosis isolated from fetuses of infected cows not manifesting signs of disease. Am J Vet Res 1992;53:477—80.

[36] Whittington RJ, Windsor PA. In utero infection of cattle with mycobacterium avium subspecies paratuberculosis: a critical review and meta analysis. Vet J 2009;179(1):60—9.

[37] Fray MD, Prenctice H, Clarke MC, Charleston B. Immunohistochemical evidence for localization of bovine viral diarrhea virus, a single-stranded RNA virus, in ovarian oocytes in the cow. Vet Pathol 1998;35:253—9.

[38] Stringfellow DA. The potential of bovine embryo transfer for infectious disease control. Rev Sci Tech Int Epiz 1985;4 859—866.

[39] Stringfellow DA, Riddell KP, Zurovac O. The potential of embryo transfer for infectious disease control in livestock. N Zealand Vet J 1991;39:8—17.

[40] Gard JA, Givens MD, Stringfellow DA. Bovine viral diarrhea virus (BVDV). Epidemiologic concerns relative to semen and embryos. Theriogenology 2007;68(3):434—42.

[41] Galik PK, Givens MD, Stringfellow DA, Crichton EG, Bishop MD, Eilertsen KJ. Bovine viral diarrhea virus (BVDV) and anti-BVDV antibodies in pooled samples of follicular fluid. Theriogenology 2002;57:1219—27.

[42] Gordon I. Laboratory Production of Cattle Embryos. Wallingford: CAB International; 1994.

[43] Stringfellow DA. Use of materials of animal origin in embryo production schemes: continued caution is recommended. Embryo Transf Newsl 2002;20:11.

[44] Rossi CR, Bridgman RS, Kiesel GK. Viral contamination of bovine fetal lung cultures and bovine fetal serum. Am J Veterinary Res 1980;41:1680—1.

[45] Bolin S, Ridpath JF. Prevalence of bovine viral diarrhea virus genotypes and antibody against those viral genotypes in fetal bovine serum. J Vet Diagn Invest 1998;10:135—9.

[46] Nettleton P, Vilcek S. Detection of pestiviruses in bovine serum. In: Proceedings of the international workshop organized by EDQM, Paris; March 29—30, 2001. p. 69—75.

[47] Bielanski A, Loewen KS, Del Campo MR, Sirard MA, Willadsen S. Isolation of bovine herpesvirus-1 (BVH-1) and bovine viral diarrhea virus (BVDV) in association with the in vitro production of bovine embryos. Theriogenology 1993;40:531−8.

[48] Bielanski A, Stewart B. Ubiquitous microbes isolated from in vitro fertilization (IVF) system. Theriogenology 1996;45:269 (Abstract).

[49] Bielanshi A, Algire J, Lalonde A, Nadin-Davis S. Transmission of bovine viral diarrhea virus (BVDV) via in vitro-fertilized embryos to recipients, but not to their offspring. Theriogenology 2009;71:499−508.

[50] Guerin B, Nibart M, Marquant-Le Guienne B, Humblot P. Sanitary risks related to embryo transfer. Theriogenology 1997;47:33−42.

[51] Marquant-LeGuienne B, Remond M, Cosquer R, Humblot P, Kaiser C, Lebreton F, et al. Exposure of in vitro-produced embryos to foot-and-mouth disease virus. Theriogenology 1998;50:109−16.

[52] Stringfellow DA, Givens MD. Preventing disease transmission through the transfer of in vivo-derived bovine embryos. Livest Prod Sci 2000;62:237−51.

[53] Stringfellow DA, Givens MD, Waldrop JG. Biosecurity issues associated with current and emerging embryo technologies. Reprod Fertil Dev 2004;16:93−102.

[54] Houe H. Epidemiological features and economical importance of bovine virus diarrhea virus (BVDV) infections. Vet Microbiol 1999;64:135−44.

[55] Nettleton P, Vilcek S. Detection of pestiviruses in bovine serum. In: Proceedings of the international workshop organized by EDQM, Paris; March 29−30, 2001. p. 69−75.

[56] Engles M, Ackermann M. Pathogenesis of ruminant herpesvirus infections. Vet Microbiol 1996;53:3−15.

[57] Vanroose G, Nauwynek H, Van Soom H, Ysebarert MT, Charlier G, Van Oostveldt P, et al. Structural aspects of the zona pellucida of in-vitro-produced embryos: a scanning electron and confocal laser scanning microscopic study. Biol Reprod 2000;62:463−9.

[58] Edens MSD, Galik PK, Riddell KP, Givens MD, Stringfellow DA, Loskutoff NM. Bovine herpesvirus-1 associated with single, trypsin-treated embryos was not infective for uterine tubal cells. Theriogenology. 2003;60:1495−504.

[59] D'Angelo M, Visintin JA, Richtzenhain LJ, Gonçalves RF. Evaluation of trypsin treatment on the inactivation of bovine herpesvirus type 1 on in vitro produced pre-implantation embryos. Reprod Domest Anim 2009;44:536−9.

[60] Bielanski A, Jordan L. Washing or washing and trypsin treatment is ineffective for removal of non-cytopathic bovine viral diarrhea virus from bovine oocytes or embryos after experimental viral contamination of an in vitro fertilization system. Theriogenology 1996;46 1467−76.

[61] Bielanski A. Dubuc. In vitro fertilization of ova from cows experimentally infected with a non-cytopathic strain of bovine viral diarrhea virus. Anim Reprod Sci 1995;38:215−21.

[62] Booth PJ, Collins ME, Jenner L, Prentice H, Ross J, Badsberg JH, et al. Noncytopathic bovine viral diarrhea virus (BVDV) reduces cleavage but increases blastocyst yield of in vitro produced embryos. Theriogenology 1998;50:769−77.

[63] Vanroose G, Nanwynek H, Van Soom A, Vanopdenbosch E, De Kruif A. Effects of bovine herpesvirus-1 on bovine viral diarrhea virus on development of in vitro-produced bovine embryos. Mol Reprod Dev 1999;54:255−63.

[64] Dinkin MB, Stallknecht DE, Brackett BG. Reduction of infectious epizootic hemorrhagic disease virus associated with in vitro-produced bovine embryos by non-specific protease. Anim Reprod Sci 2001;65:205−13.

[65] Stringfellow DA, Riddell KP, Galik PK, Damiani P, Bishop MD, Wright JC. Quality controls for bovine viral diarrhea virus-free IVF embryos. Theriogenology. 2000;53:827−39 [PubMed: 10735047].

[66] Lamara A, Fieni F, Mselli-Lakhal L, Chatagnon G, Bruyas JF, Tainturier D, et al. Early embryonic cells from in vivo-produced goat embryos transmit the caprine arthritis-encephalitis virus (CAEV). Theriogenology. 2002;58:1153−63 [PubMed: 12240918]..

[67] Bielanski A, Sapp T, Lutze-Wallace C. Association of bovine embryos produced by in vitro fertilization with a noncytopathic strain of BVDV type II. Theriogenology 1998;49:1231−8.

[68] Trachte E, Stringfellow DA, Riddell KP, Galik PK, Riddell MG, Wright J. Washing and trypsin treatment of in vitro derived bovine embryos exposed to bovine viral diarrhea virus. Theriogenology 1998;50:717−26.

[69] Givens MD, Galik PK, Riddell KP, Brock KV, Stringfellow DA. Quantity and infectivity of embryo-associated bovine viral diarrhea virus and antiviral influence of a blastocyst impede in vitro infection of uterine tubal cells. Theriogenology 1999;52:887−900.

[70] Givens MD, Galik PK, Riddell KP, Stringfellow DA. Uterine tubal cells remain uninfected after culture with in vitro-produced embryos exposed to bovine viral diarrhea virus. Vet Microbiol 1999;70:7−20.

[71] Givens MD, Galik PK, Riddell KP, Brock KV, Stringfellow DA. Replication and persistence of different strains of bovine viral diarrhea virus in an in vitro embryo production system. Theriogenology 2000;54:1093−107.

[72] Collection and processing of oocytes and in vitro produced embryos from livestock and horses. Terrestrial Animal Health Code of the OIE, 2019; 29th Edition: Chapter 4.9.2019: <https://www.oie.int/index.php?id = 169&L = 0&htmfile = chapitre_coll_embryo_invitro.htm>.

[73] Nibart M, Marquant-Le Guienne B, Humbolt P. General sanitary procedures associated with in vitro-production of embryos. In: Stringfellow DA, Seidel SM, editors. Manual of the international embryo transfer society. Savoy, IL: IETS; 1998. p. 67−77.

[74] Waldrop JG, Stringfellow DA, Riddell KP, Galik PK, Riddell MG, Givens MD, et al. Different strains of noncytopathic bovine viral diarrhea virus (BVDV) vary in their affinity for in vivo-derived bovine embryos. Theriogenology 2004;62:45−55.

[75] Shin T, Sneed L, Hill JR, Westhusin ME. High incidence of developmental failure in bovine fetuses derived by cloning bovine viral diarrhea virus-infected cells. Theriogenology 2000;53:243.

[76] Stringfellow DA, Riddell KP, Givens MD, Galik PK, Sullivan E, Dykstra CC, et al. Bovine viral diarrhea virus (BVDV) in cell lines used for somatic cell cloning. Theriogenology 2005;63:1004−13.

[77] Dibaba AB, Habtemariam T, Tameru B, Nganwa D. The risk of introduction of equine infectious anemia virus into USA via cloned horse embryos imported from Canada. Theriogenology 2012;77(2):445–58. Available from: https://doi.org/10.1016/j.theriogenology.2011.08.019.

[78] Silva-Frade C, Gameiro R, Martins Jr A, Cardoso TC. Apoptotic and developmental effects of bovine Herpesvirus type-5 infection on in vitro-produced bovine embryos. Theriogenology 2010;74:1296–303.

[79] Silva-Frade C, Martins Jr A, Borsanelli AC, Cardoso TC. Effects of bovine Herpesvirus Type 5 on development of in vitro-produced bovine embryos. Theriogenology 2010;73:324–31.

[80] Song BS, Kim JS, Yoon SB, Lee KS, Koo DB, Lee DS, et al. Inactivated Sendai-virus-mediated fusion improves early development of cloned bovine embryos by avoiding endoplasmic-reticulum-stress-associated apoptosis. Reprod Fertil Dev 2011;23(6):826–36. Available from: https://doi.org/10.1071/RD10194.

[81] Galik P, Gard J, Givens M, Spencer T, Marley S, Stringfellow D. et al. Effects of ovine interferon-tau on replication of bovine viral diarrhea virus and bovine herpesvirus-1. Reprod Fertil Dev, 2008;1(2):147.

[82] Givens MD, Stringfellow DA, Riddell KP, Galik PK, Carson RL, Riddell MG, et al. Normal calves produced after transfer of in vitro fertilized embryos cultured with an antiviral compound. Theriogenology 2006;65:344–55.

[83] Givens MD, Stringfellow DA, Dykstra CC, Riddell KP, Galik PK, Sullivan E, et al. Prevention and elimination of bovine viral diarrhea virus infections in fetal fibroblast cells. Antivir Res 2004;64:113–18.

[84] Newcomer B, Marley M, Ridpath J, Neil J, Bovkin D, Kumar A, et al. Efficacy of an antiviral compound to inhibit replication of multiple pestivirus species. Antivir Res 2012;96(2):127–9.

[85] Spera AM, Eldin TK, Tosone G, Orlando R. Antiviral therapy for hepatitis C: has anything changed for pregnant/lactating women? World J Hepatol 2016;8(12):557–65. Available from: https://doi.org/10.4254/wjh.v8.i12.557.

[86] Hercik K, Kozak J, Sala M, Dejmek M, Hrebabecky H, Zbornikova E, et al. Adenosine triphosphate analogs can efficiently inhibit the Zika virus RNA-dependent RNA polymerase. Antivir Res 2017;137:131–3. Available from: https://doi.org/10.1016/j.antiviral.2016.11.020.

[87] Walker A, Kimura K, Roberts R. Expression of bovine interferon-tau variants according to sex and age of conceptuses. Theriogenology 2009;72:44–53.

[88] Alsaleha A, Fienia F, Rodolakisb A, Bruyasa J, Rouxa C, Larrata M, et al. Can *Coxiella burnetii* be transmitted by embryo transfer in goats? Theriogenology 2013;80:571–5.

Chapter 17

Reproductive technologies and animal welfare

Fabio Napolitano[1], David Arney[2], Daniel Mota-Rojas[3] and Giuseppe De Rosa[4]

[1]School of Agriculture, Forest, Food and Environmental Sciences, University of Basilicata, Potenza, Italy, [2]Institute of Veterinary Medicine and Animal Sciences, Estonian University of Life Sciences, Tartu, Estonia, [3]Department of Animal Production and Agriculture, Metropolitan Autonomous University (UAM)— Campus Xochimilco (UAM), México City, México, [4]Department of Agricultural Sciences, University of Naples Federico II, Portici, Italy

17.1 Introduction

The reproduction of farm animals is under strict hormonal control both in males and females, which, in turn, in several species such as horses, donkeys, sheep, goats, and buffaloes, is affected by seasonality (i.e., length of the daylight). The reproductive behavior is species-specific and courtship is usually initiated by the male that follows a pattern aiming at stimulating the female and copulate. However, the reproductive cycle in females and the reproductive activities in males are often manipulated in a number of different ways with the aim of improving reproduction efficiency and increase the economical returns of farms. For instance, the cycling females can be treated to induce luteolysis, induce and regulate ovulation and estrus, and increase the number of ovulating follicles. In addition, in noncycling females cycling can be stimulated to end seasonal anestrus, whereas in pregnant females parturition can be induced to synchronize groups of animals. Males can also be manipulated to collect semen from untrained young animals. All these procedures and the management of the reproductive activity of farm animals in general can have consequences on the level of animal welfare.

In general, animal welfare and reproduction are strictly related. Numerous studies have reported that both chronic and acute stress and the associated reduced level of animal welfare may impair the reproductive functions in both males and females. For instance, stress can reduce the secretion of gonadotropin-releasing hormone (GnRH) by the hypothalamus and consequently decrease the production of luteinizing hormone and follicle-stimulating hormone (FSH) by the hypophysis and sex steroids from the gonads [1] induce ovulation failure and the formation of cystic follicles [2]. As a consequence, a decreased libido is detected in males [3] and reduced fertility and lack of embryo implantation are observed in females [4]. Conversely, how reproductive management practices may affect animal welfare has received much less attention so far. In particular, when breeding technologies are applied, three distinct elements have to be considered and the synergic effect they can produce on the animals: handling, restraint, and the application of the breeding technology *per se*. Reproductive technologies have been developed in the last 60 years, and therefore, the animals are unlikely to possess any innate adaptive mechanism to cope and generally perceive them as adverse. However, modern reproductive technologies may also favor animal welfare by minimizing the effect of unfavorable events (e.g., dystocia, mortality). Therefore the present chapter aims at identifying the most common practices and assessing the impact, either positive or negative, of these reproductive technologies on animal welfare.

17.2 Handling and restraining

The application of any reproductive technology to farm animals imply handling (i.e., animals being moved by stockpeople) and often sorting and restraining (i.e., partial or complete inhibition of animal's movements). Handling and sorting are required to select and move the animals in appropriate facilities where the reproductive technologies are then applied, whereas restraining is used to minimize the risk of injuries to animals, provide safety to the operator, and

Reproductive Technologies in Animals. DOI: https://doi.org/10.1016/B978-0-12-817107-3.00017-5

appropriately finalize the reproductive practices. All these procedures can be perceived as aversive by the animals with negative effects on their welfare. In particular, handling and sorting often imply separation from pen-mates, social isolation, and forced interaction with humans, which can determine increased cortisol level and heart rate [5], and induce behavioral alterations such as increased levels of vocalization and orientation toward the home-pen and the companions [6]. Hemsworth [7] postulated that a negative attitude of handlers toward the animals tend to negatively affect their behavior, which is often characterized by negative interactions (such as talking impatiently, yelling, using sticks and slapping) with the handled animals. As a consequence, the animals become more fearful of humans and reactive to handling. This implies reduced animal welfare and stock-people safety, with the onset of a vicious circle where stock-people tend to use increasingly forceful handling toward increasingly fearful animals [7]. Therefore the attitude of stock-people should be targeted in order to improve the quality of their relationship with the animals and increase human safety and animal welfare and productivity [8]. Grandin [9] suggests that handling should be conducted quietly in order to prevent excitement and fright and keep the animals calm, with stock-people moving slowly and being almost completely silent, which makes handling easier and safer. The same author recommends that if the animals get nervous or excited while handled they should be allowed to calm down before starting any veterinary procedure and therefore increasing the success rate of the applied reproductive technology. In addition, as handling procedures are more stressful for isolated animals, several animals should be kept together during the application of the same reproductive technologies [10].

One way to keep the animals calm is to habituate and train them to the procedures they will encounter in their productive life, including handling, sorting, and restraining. Restraining per se can induce increased cortisol levels and altered immune functions [11]. However, animals can be taught to voluntarily enter a restraining device and once trained, they will show low levels of activation of the pituitary-adrenal axis, which may suggest a reduced level of stress [12]. When animals have to be restrained and reproductive technologies imposed to the animals, the restraining facility should be not slippery, as repeated slips in excited animals induce fear and stress, veterinarians and stock-people should move slowly and equipment should be moved slowly, as fast-moving objects can be perceived as a threat, and, if the restraining equipment keep the animal off the ground, the entire body should be supported to avoid discomfort and pain [9]. In addition, when untamed animals (e.g., untrained or kept under extensive management) are treated, the use of blinds blocking the vision of frightened animals may have a calming effect and reduce excitement [13].

17.3 Pain

The control of reproduction in farm animals may involve practices that are painful to the animals. In general, pain caused by reproductive technologies and inducing tissue damages can be classified as acute or chronic. Acute pain is associate with the initial tissue damage, whereas chronic pain is related to the healing process and, albeit less intense, can take several days before attenuation. Both acute and chronic pain are associated with physiological (e.g., cortisol and other stress-related hormones, heart rate, blood pressure) and behavioral alterations (the most obvious include the restlessness observed when the reproductive technology is initially imposed to the animals, and the subsequent increased levels of vocalization while painful procedures are applied). Although the level of pain is often inferred from the degree of behavioral alterations, it should be noted that farm animals are prey and, particularly those living in extensive conditions, may hide their pain to reduce the risk of being localized by predators [14]. The two main methods available for pain reduction in treated animals are the use of less painful reproductive technologies, when available, and the administration of painkillers such as anesthetic and analgesic medications.

17.4 Application of reproductive technologies to females

The following parts of this chapter illustrate the direct effects of the application of the main reproductive technologies on animals with an emphasis on cattle. In particular, adult females, adult males, and newborns may either suffer or benefit from the consequences, which are also summarized in Table 17.1.

17.4.1 Denied expression of natural behavior

It is important to consider impacts not only of the reproductive technologies themselves, such as the denial of a natural mating, but also denial of behavior as a consequence of the isolation of the animal for the practical application of the technology, such as the removal of the animal from the herd in order to inseminate the animal and the consequent social isolation, which is known to affect their behavior [15]. If this period of isolation is extended, through inefficient

TABLE 17.1 Summary of the main reproductive technologies used in cattle and their effects on animals.

Reproductive technology	Effect	Indicator	References
Application of estrus detection devices	Not studied	Not available	Not available
Artificial insemination	Negative/mild	Plasma cortisol level	[23]
Timed insemination	Negative/mild	Abscesses	[25]
Multiple ovulation embryo transfer/donors	Negative/mild	Plasma cortisol level	[30]
Multiple ovulation embryo transfer/recipients	Negative/mild	Malformations, abortion	[31]
Ovum pickup/donors	Negative/mild	Behavior	[48]
Ovum pickup/recipients	Negative/severe	Dystocia	[31]
Ovum pickup/calves	Negative/severe	Malformations, mortality	[31]
Uterine and fetal electronic monitoring/cows	Positive/marked	Reduced dystocia	[67]
Uterine and fetal electronic monitoring/calves	Positive/marked	Reduced mortality	[67]
Cesarean section/cows	Negative/mild	Altered activity budget	[73]
Cesarean section/cows	Positive/marked	Reduced dystocia	[74]
Cesarean section/calves	Positive/marked	Reduced mortality	[74]
Electroejaculation	Negative/severe	Vocalization, plasma cortisol and progesterone levels	[85]
Transrectal ultrasound-guided massage of accessory sex glands	Negative/mild	Heart rate, vocalization, plasma cortisol level	[92]

insemination practice, delay in waiting for the personnel responsible of the practice or any other cause of delay, the cow will be prevented from showing any of the preferred natural behaviors that she may be motivated to express such as feeding, social behavior, and lying. Of these, denial of feeding may be the most likely to be a cause of frustration particularly if, as it often occurs, selected cows for artificial insemination (AI) are isolated immediately after milking when, on the contrary, cows may be expecting to feed.

AI definitely denies natural mating, and whether or not this is a significant frustration of an important behavioral motivation is not well known. Dairy cows do show sexually motivated behavior, and in fact it would be difficult to identify estrus in cows that did not. Mounting of herd-mates prior to estrus is commonly observed (Fig. 17.1), and the cows would not do this unless they were motivated to do so. So, can we conclude that they are also motivated to be mated by a bull? And that if natural mating is denied, then are they missing out on something important? In many dairy farms, cows will never meet a sexually mature bull. Can such cows be said to miss something that they have never known? Well yes, possibly; there are other behaviors that are denied to animals in production systems, that they never have the opportunity to express, but when given the opportunity they will readily do so such as grazing at pasture [16] or nest building by pigs [17]. From an ethical viewpoint, this manifest prevention of a behavioral feature of an animal's life experience may be considered unjustifiable.

FIGURE 17.1 Mounting of herd-mates prior to estrus. *Photo: Aino Nömmeots.*

It is difficult to see how any of the technologies used in estrus detection, such as progesterone testing of milk at normal milking times or activity monitors applied to the leg, neck or ear, if these are attached appropriately and checked frequently, inhibit the natural behavioral motivations of dairy cows. Nevertheless, it is possible that ear tags may get snagged on housing or milking furniture, that incorrectly attached neck monitors might cause chafing, and activity monitors attached to the leg might become caked in dung and make lying uncomfortable, lying known to be a behavior that dairy cows are highly motivated to express [18].

Inducing estrus by the administration of sex hormones will affect the behavior of cows, and while these behavioral changes are natural, the fact that they are being induced implies that they cannot be considered the natural behavioral choices that the cow would have made had there been no intervention. And by inducing estrus before the cow is naturally ready to return into estrus, the cow is deprived of the expression of natural behaviors that she would have expressed in the lost anestrus period.

The seriousness of these deprivations of natural behavior will, as so often, be dependent on the quality of stockmanship. If this is of good quality, then the cows will probably not suffer unduly. From an ethical point of view, the seriousness will depend on the ethical position of the reader.

17.4.2 Estrus detection and artificial insemination

Estrus detection is a key factor in enhancing reproductive performance in the dairy and beef cattle industries. It is usually done by visually checking for signs of this condition, such as the female positioning herself to be mounted [19]. However, several systems using sensors have been developed to detect estrus automatically. These include pedometers and neck-mounted collars that register physical activity and pressure-sensing apparatuses that record standing estrus [20]. No studies have been conducted on the effects of the application of this equipment on cow welfare and it can only be hypothesized that habituation may occur after a period of time, which may vary according to the temperament of the animal.

AI is one of the first reproductive technologies applied to and routinely used in farm animals. It has been estimated that in the early 2000s more than a hundred million cows per year were artificially inseminated [21]. In cattle this technology consists in putting one hand into the rectum of the in-estrus cow to localize and manipulate the cervix while inserting with the other hand the insemination gun through the cervix into the uterus where the semen is released.

The procedure per se is not considered painful and the cortisol level does not increase in inseminated cows as compared with control untreated cows [22]. However, AI is likely to cause discomfort with potential detrimental effects in more nervous cows. In these animals Hays and Vandemark [23] observed a reduced increase of intra-mammary pressure as a possible consequence of the application of a procedure perceived as stressful. The authors explained these results on the basis of a possible production of stress hormones counteracting the effect of oxytocin. However, AI may also have beneficial effects in terms of reduced risk of transmission of venereal diseases to the cows, as the semen is checked before admission to routine use. According to one-welfare approach, AI shows additional positive effects for the stock-people involved in animal handling, as the presence of the bull on the farm may reduce their safety.

The success of AI relies on a prompt and accurate heat detection and this represents one of the main constraints. To overcome this constraint in cattle and buffaloes, various synchronization protocols have been developed (see Ref. [24] for a review) in order to perform a timed artificial insemination (TAI). These protocols involve the use of reproductive hormones (progesterone, GnRH, prostaglandin $F_{2\alpha}$) usually injected intramuscularly or released through an intravaginal device (i.e., progesterone). Thus, with repeated injection of hormones the animals may be inseminated according to the farmers' plan. TAI is also used in embryo transfer (ET) programs. From a welfare point of view the animals are subjected to acute stress due to intramuscular injections and handling (see also previous paragraphs on pain and handling and restraining). Additionally, these injections per se may determine muscle damage with consequent formation of bruises or abscesses [25,26]. Besides, the use of intravaginal devices, which may be in situ 5−7 days, may also cause discomfort and superficial vaginal infection, especially when correct asepsis and hygiene procedures are not followed. However, to date these aspects have never been investigated.

17.4.3 Embryo transfer and multiple ovulation

The development of multiple ovulation embryo transfer (MOET) technique has been used in producing transgenic animals, breeding and rescuing endangered species. MOET in farm animal breeding has increased the reproductive capacity of genetically outstanding dams. This technique is widely used in the conventional dairy cattle breeding industry and to a lesser degree also in other farm animals.

Briefly, in the cow MOET involves transferring embryos from a donor to recipients for the rest of gestation. Prior ovulation a donor is repeatedly injected with FSH in order to induce multiple ovulations. Alternatively, a progesterone-releasing device associated with various estrogens injections are used. After ovulation, the donor is inseminated two or three times to fertilize as many ovulated oocytes as possible. Embryos are collected after 5−7 days at the morula or blastocyst stage, before they attach to the uterine wall. In cattle and buffaloes the collection is performed introducing a catheter into the uterus, where a saline solution is injected to flush the embryos out, which are then assessed for their viability. The estrus of donors and recipients is synchronized within 24 hours in order to transplant an embryo into a uterine environment as similar as possible to the donor. Then, only viable embryos are introduced, through a special pipette, in the uterus of recipients, where the estrus cycle has been synchronized with prostaglandin ($PGF_{2\alpha}$) or other protocols. Alternatively, embryos can be stored in liquid nitrogen for later use. Prior to the collection and transferring of embryos, animals receive a light epidural anesthesia to reduce rectal contractions. Donor cows for MOET are subjected to hormone treatment for superovulation every 5−6 weeks with 6−7 viable embryos produced, resulting in 3−4 pregnancies. In sheep, goats, and pigs both the recovery and transfer of embryos are performed by surgical methods (see Refs. [27] and [28], for a review).

Obviously, if we consider the animals' perspective there are no benefits to either donors or recipients as MOET may increase the risks of pain, suffering, and long-term detrimental effects compared to AI, thus jeopardizing their welfare. In fact, the stress resulting from frequent handling of donors to receive FSH injections (6−8 times) may increase the risk of injuries to both animals and stock-people, especially in animals not used to be restrained such as beef cattle [29]. Additionally, this may result in an increased stress (see also previous paragraph on handling and restraining) as observed in beef heifers, where the animals receiving eight injections had higher cortisol concentrations and lower superovulation response than those treated with two injections of slow FSH release [30].

With regard to the recipient, it has been observed that the use of MOET increased abortion rate, the incidence of congenital malformations, and the birth weight of calves compared to calves produced by AI, but to a lesser extent than in vitro produced calves [31]. This may be due to changes in the reproductive tract or the features of oocytes generating the embryos that can alter embryonic developmental programming (see Ref. [32] for a review). In addition, in a recent study no effect of MOET on postnatal consequences expressed in terms of milk yield and milk quality were found between primiparous cows born from MOET or AI [33], probably due to fact that milk production is less influenced by developmental programming errors [32].

17.4.4 Ovum pickup and in vitro production of embryos

The development of ultrasound-guided transvaginal follicular aspiration for ovum pickup (OPU) and in vitro production (IVP) of embryos may be seen as a natural evolution of MOET. In order to reduce generation interval, OPU is also applied to young animals (i.e., prepubertal females). In farm animals, this technique is mostly used in cattle (e.g., [34], with limited applications also in buffaloes e.g., [35], sheep, e.g., [36], goats, e.g., [37], and pigs, e.g., [38]). To date, in sheep and goats, oocytes are obtained by laparoscopic OPU.

Briefly, when an OPU session is performed, a female is restrained in a chute and receives an epidural anesthesia. Then, the ovary is positioned and held through the rectum by one hand, and at the same time an ultrasound transducer with a surmounted needle is introduced in the vagina to visualize antral follicles, and finally, the vagina and the ovary are punctured to aspirate the oocyte from each individual follicle. The sessions are usually repeated twice per week for a certain number of weeks (see Ref. [34] for a review). In addition, in order to increase the number of oocytes retrieved and, consequently, the number of embryos produced per session, the donor prior to OPU is superstimulated using different hormonal protocols [39–41].

OPU has been considered either a noninvasive [34] or mildly invasive technique [42]. However, the use of OPU has raised concerns both in the public opinion and scientific community. Apart from the stress resulting from denying the expression of mating behavior (see previous paragraph on denied expression of natural behavior), handling and restraining (see previous paragraph on handling and restraining), some detrimental effects on the welfare of donors have been observed. In particular, the use of repeated epidural anesthesia may cause intercoccygeal disc injuries [43]. Behavioral responses to OPU may be more intense during the late than the mid and early OPU sessions [44], as animal awareness of a stressful situation increases throughout the sessions. Repeated ovarian punctures may cause in some donors' hemorrhage and ovarian lesions [43]. Some macroscopic and microscopic changes (i.e., presence of luteal structures and follicles of various sizes, thickening of connective tissue in the ovarian tunica albuginea, accumulation of fibrous tissue around the ovaries) of ovaries [45–48], minor modifications of the endocrine profile [46,47], and irregular cyclic activity [45,49] have also been observed. However, most of these physiological and behavioral responses show a great individual variation [48].

With regard to the recipient, there is a large body of evidence showing that the use of in vitro production of embryos in ruminant species may result in the so-called "large offspring syndrome" (see Ref. [50] for a review). This syndrome is closely related to embryo culture conditions changing the normal pattern of DNA methylation, thus resulting in abnormal embryonic gene expression [51], although the etiology is still not well elucidated [52]. As a consequence, field studies conducted in the Netherlands showed that IVP calves had an increased birth weight, gestation period, perinatal mortality, congenital malformations, and frequency of dystocic calving compared to AI calves [31,53]. These results were confirmed by small-scale experimental studies [54–56], whereas Rasmussen et al. [57] did not observe any effect of IVP embryos on the postnatal development of calves. As suggested by Bonilla et al. [56], these controversial results may be due to confounding factors such as the medium used to produce IVP embryos or the sample size.

The negative effect of IVP may be even more pronounced when a reverse X-sorted semen is used as a result of a modification of embryonic programming that may persist after birth with an effect also on the first lactation, expressed in terms of milk yield and milk quality, when compared with offsprings born after AI [33]. It has been postulated that the sorting process may result in a loss of microRNA or other molecules important for the normal developmental programming of embryos [58].

Conversely, no significant effect of treatment on postpartum efficiency of the dams assessed in terms of incidence of prolapse or metritis, pregnancy rate at first service, services per conception, and milk production was found [56].

17.4.5 Uterine and fetal electronic monitoring

While commonly used in mares, uterine and fetal electronic monitoring (FEM) is much less common in cattle due to the lower economic value of the animals. However, problems around the time of parturition may arise, particularly in heifers with detrimental effects on mothers and calves in the first days after delivery. In order to allow stock-people and veterinarians to attend calving, the moment of parturition has to be precisely predicted. To this aim numerous methods have been described, including the recording of variations in body temperature [59,60], estrogen and progesterone blood levels [61,62], concentration of electrolytes in mammary gland secretions [63], ultrasound monitoring [64], and video monitoring [65]. However, the methods currently available on market to predict the onset of parturition in cattle include sensors monitoring tail movements, sensors monitoring uterine contractions, probes monitoring vaginal temperature and allantochorion expulsion, devices inserted on the lips of the vulva or into the vagina detecting the approaching calf.

One apparatus designed to reduce indices of fetal death by accurately signaling the onset of calving regardless of the fetus' presentation, position, or posture, is the iVET birth monitoring system [66]. All these methods have the potential to decrease the rates of fetal death while also improving the health and welfare of newborn calves and mothers [67].

Similar electronic monitoring devices have been utilized with parturient sows. Studies by Mota-Rojas et al. [68] indicate that FEM is an excellent reproductive technology for assessing fetal suffering and low neonate welfare and monitoring the welfare of sows during farrowing. These are noninvasive uterine and tocographic transducers placed on the dam's abdominal region to monitor the frequency, intensity, and number of uterine contractions during farrowing, while also evaluating the relation of these uterine indicators to the heart rate of the fetuses in order to identify symptoms of acute fetal syndrome. In addition, FEM technology aids in determining the correct dose of uterine contraction stimulators, such as vetrabutine chlorhydrate or oxytocin, without affecting fetal welfare by reducing the volume of blood they receive, which can cause meconium aspiration syndrome or the rupture the umbilical cords of fetuses that have not been expelled [69]. The use of high-tech equipment to monitor the reproductive processes of farm animals—including mares, sows and cows—is growing steadily, because it allows to determine the most appropriate time to intervene with the aim of reducing pain and the duration of parturition, while simultaneously enhancing the welfare of newborns by minimizing fetal suffering, increasing vitality levels, thus reducing the latency to first suckling [70].

However, breeding programs should include calving ease as a preventive measure to effectively contain and possibly lower the incidence of dystocia [71], thus reducing the need for monitoring, which in most cases imposes the application of various devices on or in the body of the animals.

17.4.6   Cesarean section

Double-muscle beef cattle suffer from feto-maternal disproportion. In these cases, as in other cases of dystocia (e.g., feto-pelvic disproportion, incomplete dilatation, uterine torsion, faulty fetal presentation), cesarean section is performed to reduce mortality rates of both calves and cows [72] while maintaining cow fertility. Even if pain relief is usually administered to cows experiencing cesarean partum, after the surgery cows suffer pain due to the section per se and, in case of nonelective cesarean section, also due to failed vaginal parturition attempts. Kolkman et al. [73], observed mild effects, and mostly limited to the first day after parturition, of cesarean section on the behavior of the cows (reduced eating and ruminating; increased lying on the right side) as compared to naturally calving cows. These negative effects, however, could be minimized by using nonsteroidal antiinflammatory drugs administered postpartum [75]. In addition, treated cows showed a normal development of the relationship with their calves even without the stimulation of the cervix and the related oxytocin production consequent to the passage of the fetus [76]. Nevertheless, concerns may arise in the public opinion, when cesarean section is conducted routinely (elective cesarean section) as a necessary intervention performed at or even before cervix dilatation and without any attempt of traction. This is the consequence of selective breeding aimed at increasing the body weight of calves at parturition and their muscle mass for production purposes. It has been argued that re-breeding programs including back-selection should be implemented in order to allow natural parturition [77].

17.5   Male

17.5.1   Denied expression of natural behavior

For most beef bulls, as indeed for most rams, behavior associated with reproduction is usually as natural as it could be expected from an animal part of a production system: copulation is usually natural, bulls and rams identify females in estrus, select females, and mate. It is with the dairy bull that the most significant departure from what might be considered the allowance of the expression of normal behavior is most likely. Dairy bulls are usually kept for collection of their semen and housed on AI units where they are kept separate from other animals of their own kind, or indeed other species. It has been suggested for some time that we should consider two aspects of the natural sexual behavior of bulls: libido (sex drive) and copulation [78]. It is also important, perhaps, to consider as normal sexual behavior, the lack of opportunity to check the estrus status of females: flehmen response, sniffing and licking the perineal region, chin resting on the female and mounting or mock mounting of females if the bulls are kept isolated from female animals. The flehmen response can be encouraged by the administration of urine from estrus females onto the artificial dummy [79], but the full repertoire of sexual behavior cannot be experienced. The experience of receiving solicitation of cows to bulls will also be lost, and likewise, the opportunity to express normal aggression with other males if they are kept isolated from male conspecifics. Libido may be genetic [80], and different animals and breeds may therefore have different strong motivations. But that bulls so have such a motivation is clear to anyone who has seen a beef bull frustratedly bellowing across the hedge to a group of unattainable cows. Whether the act of copulation with a bullock teaser or an artificial dummy can be considered adequate replacement for copulation with a receptive cow to satisfy the motivation to copulate is moot.

But sexual behavior is not everything, and the life experience of dairy bulls kept for semen collection is likely to impose restrictions on a range of aspects of their natural behavior. They are usually high-value animals. Biosecurity is therefore of great importance, and in order to reduce the threat of disease and climatic threats to health and wellbeing, they are less likely to be housed outdoors, and are therefore deprived of the natural behaviors that they would be enabled to express were they allowed access to pasture. Tuomisto et al. [81] have shown that, at least in cold climates, these concerns are not well founded. Because of the risk of injury from aggression, dairy bulls are often kept singly. While reducing excessive aggression is on the whole a good thing, aggression is a natural behavior and, as with all other social behaviors, keeping an animal isolated will remove the animal from expressing these natural social behaviors.

17.5.2 Electroejaculation

Electroejaculation (EE) is a reliable method of obtaining a semen sample from sire males. This method is usually used in young bulls, rams, and bucks not trained to the use of artificial vagina in order to obtain a semen sample for a breeding soundness examination.

This technique involves restraining. In the bull, prior to the insertion of a probe in the rectum, a transrectal massage over the area of accessory sex glands (e.g., seminal vesicles, prostate) is usually performed to sexually excite the animal and relax the anal sphincter. Then, the probe is inserted and electrical pulses (voltage ranging from 8 to 16 V) are released for 1−2 seconds with a period of rest of 0.5−2 seconds until the occurrence of ejaculation. During stimulation the voltage of pulses is gradually increased (see Ref. [82] for more details).

Obviously, EE is regarded as a stressful and painful procedure, which may negatively affect animal welfare. As a consequence, a significant increase in vocalizations and plasma cortisol and progesterone concentrations in bulls subjected to EE is observed [83−85]. Rams during EE show increased vocalizations, heart rate, rectal temperature, cortisol, glycemia, creatine kinase, and granulocyte concentrations [86,87]. Intrarectal or caudal epidural anesthesia has also been used as a means to reduce pain during EE [83,84,88]. In bulls, epidural lidocaine treatment reduced the elevations of cortisol and progesterone and heart rate during EE, although the reductions were not significant [83,84], whereas cortisol and progesterone concentrations were significantly elevated 25 minutes after EE, whether or not bulls received caudal epidural anesthesia [83]. In addition, the treatment with lidocaine does not reduce the behavioral responses of bulls, indicating that epidural lidocaine is only minimally effective in reducing pain and stress, whereas Pagliosa et al. [89] reported that bulls after the use of epidural xylazine showed a reduction in the behavioral signs of discomfort. Although in some European countries this procedure is compulsory, more studies are needed to elucidate whether it works effectively. Finally, it has been observed that a reduction of the electrical stimulation required to induce ejaculation may be attained by the administration of oxytocin and a prostaglandin $F_{2\alpha}$ analog [90,91].

An alternative method to collect semen is the transrectal massage in bulls [82,88,90] or the transrectal ultrasound-guided massage of the accessory sex glands (TUMASG) in small ruminants [92,93]. In bucks, animals subjected to EE emitted more vocalizations and showed a greater increase of cortisol concentration than TUMASG bucks, whereas sperm quality was not different between TUMASG and EE [92]. In the same study heart rate also tended to be greater in EE animals than TUMASG bucks. In wild anesthetized small ruminants, respiratory rate, cortisol and creatine kinase concentration changes were greater with EE than with TUMASG [93]. Therefore it seems that TUMASG may be less stressful and painful than EE without affecting semen quality. However, more studies are needed to draw any firm conclusion about whether TUMASG is an effective procedure to ameliorate the welfare of animals in comparison to EE.

Conclusion

The first important remark about the use of reproductive technologies in farm practices is that only few studies are available about their effects on the animals and further studies are needed to verify whether their routine application may impair a satisfactory welfare state. Based, therefore, on the available literature, we may conclude that most of the reviewed reproductive technologies (e.g., estrus detection and AI) show a mild effect on the animals and in several cases an habituation may occur, although individual more sensitive subjects may perceive even these practices as noxious. Conversely, other technologies may have marked direct detrimental effects in terms of acute or chronic pain (e.g., EE) and/or indirectly induce negative emotional states (e.g., through restraining and isolation). However, some of these technologies have been developed with the primary aim of increasing animal welfare, as directly linked to farm profits. In particular, uterus and fetal monitoring by allowing the prediction of calving may also allow a prompt intervention in case of dystocia. Nevertheless also in these cases, preventive measures, such as the application of appropriate breeding programs including calving ease to reduce the incidence of dystocia and cesarean sections, should be preferred for an effective improvement of animal welfare.

References

[1] Dobson H, Smith R, Royal M, Knight C, Sheldon I. The high-producing dairy cow and its reproductive performance. Reprod Domest Anim 2007;42:17−23. Available from: https://doi.org/10.1111/j.1439-0531.2007.00906.x.

[2] Biran D, Braw-Tal R, Gendelman M, Lavon Y, Roth Z. ACTH administration during formation of preovulatory follicles impairs steroidogenesis and angiogenesis in association with ovulation failure in lactating cows. Domest Anim Endocrinol 2015;53:52−9. Available from: https://doi.org/10.1016/j.domaniend.2015.05.002.

[3] Moberg GP. Effects of environment and management stress on reproduction in the dairy cow. J Dairy Sci 1976;59:1618−24. Available from: https://doi.org/10.3168/jds.S0022-0302(76)84414-1.

[4] Etim NN, Offiong EEA, Udo MD, Williams ME, Evans EI. Physiological relationship between stress and reproductive efficiency 2013;5.

[5] Waiblinger S, Menke C, Korff J, Bucher A. Previous handling and gentle interactions affect behaviour and heart rate of dairy cows during a veterinary procedure. Appl Anim Behav Sci 2004;85:31−42. Available from: https://doi.org/10.1016/j.applanim.2003.07.002.

[6] Napolitano F, Serrapica M, Braghieri A, Claps S, Serrapica F, De Rosa G. Can we monitor adaptation of juvenile goats to a new social environment through continuous qualitative behaviour assessment? PLoS One 2018;13:e0200165. Available from: https://doi.org/10.1371/journal.pone.0200165.

[7] Hemsworth PH. Human−animal interactions in livestock production. Appl Anim Behav Sci 2003;81:185−98. Available from: https://doi.org/10.1016/S0168-1591(02)00280-0.

[8] Breuer K, Hemsworth PH, Barnett JL, Matthews LR, Coleman GJ. Behavioural response to humans and the productivity of commercial dairy cows. Appl Anim Behav Sci 2000;66:273−88. Available from: https://doi.org/10.1016/S0168-1591(99)00097-0.

[9] Grandin T. The importance of measurement to improve the welfare of livestock poultry and fish. Improving animal welfare. A practical approach. Wallingford: CAB International; 2010. p. 1−20.

[10] Arave CW, Albright JL, Sinclair CL. Behavior, milk yield, and leucocytes of dairy cows in reduced space and isolation. J Dairy Sci 1974;57:1497−501. Available from: https://doi.org/10.3168/jds.S0022-0302(74)85094-0.

[11] Minton JE, Coppinger TR, Reddy PG, Davis WC, Blecha F. Repeated restraint and isolation stress alters adrenal and lymphocyte functions and some leukocyte differentiation antigens in lambs. J Anim Sci 1992;70:1126−32. Available from: https://doi.org/10.2527/1992.7041126x.

[12] Grandin T, Rooney MB, Phillips M, Cambre RC, Irlbeck NA, Graffam W. Conditioning of nyala (*Tragelaphus angasi*) to blood sampling in a crate with positive reinforcement. Zoo Biol 1995;14:261−73.

[13] Mitchell KD, Stookey JM, Laturnas DK, Watts JM, Haley DB, Huyde T. The effects of blindfolding on behavior and heart rate in beef cattle during restraint. Appl Anim Behav Sci 2004;85:233−45. Available from: https://doi.org/10.1016/j.applanim.2003.07.004.

[14] Stafford KJ, Mellor DJ. Pain husbandry procedures in livestock and poultry. Improving animal welfare. A practical approach. Wallingford: CAB International; 2010. p. 88−114.

[15] Walker JK, Arney DR, Waran NK, Handel IG, Phillips CJC. The effect of conspecific removal on behavioral and physiological responses of dairy cattle. J Dairy Sci 2015;98:8610−22. Available from: https://doi.org/10.3168/jds.2014-8937.

[16] Motupalli PR, Sinclair LA, Charlton GL, Bleach EC, Rutter SM. Preference and behavior of lactating dairy cows given free access to pasture at two herbage masses and two distances. J Anim Sci 2014;92:5175−84. Available from: https://doi.org/10.2527/jas.2014-8046.

[17] Stolba A, Wood-Gush DGM. The behaviour of pigs in a semi-natural environment. Anim Sci 1989;48:419−25. Available from: https://doi.org/10.1017/S0003356100040411.

[18] Cooper MD, Arney DR, Phillips CJC. Two- or four-hour lying deprivation on the behavior of lactating dairy cows. J Dairy Sci 2007;90:1149−58. Available from: https://doi.org/10.3168/jds.S0022-0302(07)71601-6.

[19] López-Gatius F. Factors of a noninfectious nature affecting fertility after artificial insemination in lactating dairy cows. A review. Theriogenology 2012;77:1029−41.

[20] Miura R, Yoshioka K, Miyamoto T, Nogami H, Okada H, Itoh T. Estrous detection by monitoring ventral tail base surface temperature using a wearable wireless sensor in cattle. Anim Reprod Sci 2017;180:50−7. Available from: https://doi.org/10.1016/j.anireprosci.2017.03.002.

[21] Thibier M, Wagner H-G. World statistics for artificial insemination in cattle. Livest Prod Sci 2002;74:203−12. Available from: https://doi.org/10.1016/S0301-6226(01)00291-3.

[22] Macalay AS, Roussel JD, Seybt SH. Cortisol response in heifers to artificial insemination, natural mating, and no mating at estrus. Theriogenology 1986;117−23.

[23] Hays RL, Vandemark NL. Effect of stimulation of the reproductive organs of the cow on the release of an oxytocin-like substance. Endocrinology 1953;52:634−7.

[24] Colazo MG, Mapletoft RJ. Mapletoft, A review of current timed[HYPHEN]AI (TAI) programs for beef and dairy cattle. Can Vet J. 2014;55:772−80.

[25] Fajt VR, Wagner SA, Pederson LL, Norby B. The effect of intramuscular injection of dinoprost or gonadotropin-releasing hormone in dairy cows on beef quality. J Anim Sci 2011;89:1939—43. Available from: https://doi.org/10.2527/jas.2010-2923.

[26] De Rosa G, Grasso F, Winckler C, Bilancione A, Pacelli C, Masucci F, et al. Application of the welfare quality protocol to dairy buffalo farms: prevalence and reliability of selected measures. J Dairy Sci 2015;98:6886—96. Available from: https://doi.org/10.3168/jds.2015-9350.

[27] Hasler JF. Forty years of embryo transfer in cattle: a review focusing on the journal Theriogenology, the growth of the industry in North America, and personal reminisces. Theriogenology 2014;81:152—69. Available from: https://doi.org/10.1016/j.theriogenology.2013.09.010.

[28] K Vijayalakshmy, J Manimegalai, R Verma. Embryo transfer technology in animals: an overview. J Entomol Zool Stud 2018;6:2215—18.

[29] JF Hasler, D Hockley. Efficacy of hyaluronan as a diluent for a two injection FSH superovulation protocol in Bos taurus beef cows. In: Proceedings of the 17th international congress on animal reproduction (ICAR), Reproduction in Domestic Animals. Vancouver; 2012, p. 459. doi:10.1111/j.1439-0531.2012.02119.x.

[30] Biancucci A, Sbaragli T, Comin A, Sylla L, Monaci M, Peric T, et al. Reducing treatments in cattle superovulation protocols by combining a pituitary extract with a 5% hyaluronan solution: is it able to diminish activation of the hypothalamic pituitary adrenal axis compared to the traditional protocol? Theriogenology 2016;85:914—21. Available from: https://doi.org/10.1016/j.theriogenology.2015.10.041.

[31] van Wagtendonk-de Leeuw AM, Mullaart E, de Roos APW, Merton JS, den Daas JHG, Kemp B, et al. Effects of different reproduction techniques: AI, moet or IVP, on health and welfare of bovine offspring. Theriogenology 2000;53:575—97. Available from: https://doi.org/10.1016/S0093-691X(99)00259-9.

[32] Hansen PJ, Siqueira LGB. Postnatal consequences of assisted reproductive technologies in cattle. Anim Reprod 2017;14:490—6. Available from: https://doi.org/10.21451/1984-3143-AR991.

[33] Siqueira LGB, Dikmen S, Ortega MS, Hansen PJ. Postnatal phenotype of dairy cows is altered by in vitro embryo production using reverse X-sorted semen. J Dairy Sci 2017;100:5899—908. Available from: https://doi.org/10.3168/jds.2016-12539.

[34] Qi M. Transvaginal ultrasound-guided ovum pick-up (OPU) in cattle. J Biomim Biomater Tissue Eng 2013;18:3.

[35] Sá Filho MF, Carvalho NAT, Gimenes LU, Torres-Júnior JR, Nasser LFT, Tonhati H, et al. Effect of recombinant bovine somatotropin (bST) on follicular population and on in vitro buffalo embryo production. Anim Reprod Sci 2009;113:51—9. Available from: https://doi.org/10.1016/j.anireprosci.2008.06.008.

[36] Silva JCB, Okabe WK, Traldi AS. From cattle to sheep: a view of the difficulties and success of commercial in vitro production of sheep embryos. Anim Reprod 2012;9:195—200.

[37] Souza-Fabjan JMG, Pereira AF, Melo CHS, Sanchez DJD, Oba E, Mermillod P, et al. Assessment of the reproductive parameters, laparoscopic oocyte recovery and the first embryos produced in vitro from endangered Canindé goats (Capra hircus). Reprod Biol 2013;13:325—32. Available from: https://doi.org/10.1016/j.repbio.2013.09.005.

[38] Yoshioka K, Uchikura K, Suda T, Matoba S. Production of piglets from in vitro-produced blastocysts by ultrasound-guided ovum pick-up from live donors. Theriogenology 2020;141:113—19. Available from: https://doi.org/10.1016/j.theriogenology.2019.09.019.

[39] Vieira LM, Rodrigues CA, Castro Netto A, Guerreiro BM, Silveira CRA, Moreira RJC, et al. Superstimulation prior to the ovum pick-up to improve in vitro embryo production in lactating and non-lactating Holstein cows. Theriogenology 2014;82:318—24. Available from: https://doi.org/10.1016/j.theriogenology.2014.04.013.

[40] Cavalieri FLB, Morotti F, Seneda MM, Colombo AHB, Andreazzi MA, Emanuelli IP, et al. Improvement of bovine in vitro embryo production by ovarian follicular wave synchronization prior to ovum pick-up. Theriogenology 2018;117:57—60. Available from: https://doi.org/10.1016/j.theriogenology.2017.11.026.

[41] Ribas BN, Missio D, Junior Roman I, Neto NA, Claro I, dos Santos Brum D, et al. Superstimulation with eCG prior to ovum pick-up improves follicular development and fertilization rate of cattle oocytes. Anim Reprod Sci 2018;195:284—90. Available from: https://doi.org/10.1016/j.anireprosci.2018.06.006.

[42] Bols PEJ, Vandenheede JMM, Van Soom A, de Kruif A. Transvaginal ovum pick-up (OPU) in the cow: a new disposable needle guidance system. Theriogenology 1995;43:677—87. Available from: https://doi.org/10.1016/0093-691X(94)00073-4.

[43] McEvoY TG, Thompson H, Dolman DF, Watt RG, Reis A, Staines ME. Effects of epidural injections and transvaginal aspiration of ovarian follicles in heifers used repeatedly for ultrasound-guided retrieval of ova and embryo production. Vet Rec 2002;151:653—8.

[44] Petyim S, Båge R, Madej A, Larsson B. Ovum pick-up in dairy heifers: does it affect animal well-being? Reprod Domest Anim 2007;42:623—32. Available from: https://doi.org/10.1111/j.1439-0531.2006.00833.x.

[45] Boni R, Roelofsen MWM, Pieterse M, Kogut J, Kruip T. Follicular dynamics, repeatability and predictability of follicular recruitment in cows undergoing repeated follicular puncture. Theriogenology 1997;48:277—89. Available from: https://doi.org/10.1016/S0093-691X(97)84075-7.

[46] Petyim S, Bage R, Forsberg M, Rodriguez-Martinez H, Larsson B. The effect of repeated follicular puncture on ovarian function in dairy heifers. J Vet Med Ser A 2000;47:627—40. Available from: https://doi.org/10.1046/j.1439-0442.2000.00327.x.

[47] Petyim S, Bage R, Forsberg M, Rodriguez-Martinez H, Larsson B. Effects of repeated follicular punctures on ovarian morphology and endocrine parameters in dairy heifers. J Vet Med Ser A 2001;48:449—63. Available from: https://doi.org/10.1046/j.1439-0442.2001.00375.x.

[48] Petyim S, Båge R, Madej A, Larsson B. Ovum pick-up in dairy heifers: does it affect animal well-being? Reprod Domest Anim 2007;42:623—32. Available from: https://doi.org/10.1111/j.1439-0531.2006.00833.x.

[49] Carlin SK, Garst AS, Tarraf CG, Bailey TL, McGilliard ML, Gibbons JR, et al. Effects of ultrasound-guided transvaginal follicular aspiration on oocyte recovery and hormonal profiles before and after GnRH treatment. Theriogenology 1999;51:1489—503. Available from: https://doi.org/10.1016/S0093-691X(99)00092-8.

[50] Hill JR. Incidence of abnormal offspring from cloning and other assisted reproductive technologies. Annu Rev Anim Biosci 2014;2:307−21. Available from: https://doi.org/10.1146/annurev-animal-022513-114109.

[51] Hori N, Nagai M, Hirayama M, Hirai T, Matsuda K, Hayashi M, et al. Aberrant CpG methylation of the imprinting control region KvDMR1 detected in assisted reproductive technology-produced calves and pathogenesis of large offspring syndrome. Anim Reprod Sci 2010;122:303−12. Available from: https://doi.org/10.1016/j.anireprosci.2010.09.008.

[52] Li Y, Donnelly CG, Rivera RM. Overgrowth syndrome. Vet Clin North Am Food Anim Pract 2019;35:265−76. Available from: https://doi.org/10.1016/j.cvfa.2019.02.007.

[53] van Wagtendonk-de Leeuw AM, Aerts BJG, den Daas JHG. Abnormal offspring following in vitro production of bovine preimplantation embryos: a field study. Theriogenology 1998;49:883−94. Available from: https://doi.org/10.1016/S0093-691X(98)00038-7.

[54] Bertolini M, Moyer AL, Mason JB, Batchelder CA, Hoffert KA, Bertolini LR, et al. Evidence of increased substrate availability to in vitro-derived bovine foetuses and association with accelerated conceptus growth. Reproduction 2004;128:341−54. Available from: https://doi.org/10.1530/rep.1.00188.

[55] Breukelman SP, Perényi Zs, Ruigh L de, Leeuw AM, van W, Jonker FH, et al. Plasma concentrations of bovine pregnancy-associated glycoprotein (bPAG) do not differ during the first 119 days between ongoing pregnancies derived by transfer of in vivo and in vitro produced embryos. Theriogenology 2005;63:1378−89. Available from: https://doi.org/10.1016/j.theriogenology.2004.07.008.

[56] Bonilla L, Block J, Denicol AC, Hansen PJ. Consequences of transfer of an in vitro-produced embryo for the dam and resultant calf. J Dairy Sci 2014;97:229−39. Available from: https://doi.org/10.3168/jds.2013-6943.

[57] Rasmussen S, Block J, Seidel GE, Brink Z, McSweeney K, Farin PW, et al. Pregnancy rates of lactating cows after transfer of in vitro produced embryos using X-sorted sperm. Theriogenology 2013;79:453−61. Available from: https://doi.org/10.1016/j.theriogenology.2012.10.017.

[58] Bermejo-Álvarez P, Lonergan P, Rath D, Gutiérrez-Adan A, Rizos D. Developmental kinetics and gene expression in male and female bovine embryos produced in vitro with sex-sorted spermatozoa. Reprod Fertil Dev 2010;22:426. Available from: https://doi.org/10.1071/RD09142.

[59] Fujomoto Y, Kimura E, Sawada T, Ishikawa M, Matsunaga H, Mori J. Change in rectal temperature, and heart and respiration rate of dairy cows before parturition. Jpn J Zootech Sci 1988;59:301−5.

[60] Burfeind O, Suthar VS, Voigtsberger R, Bonk S, Heuwieser W. Validity of prepartum changes in vaginal and rectal temperature to predict calving in dairy cows. J Dairy Sci 2011;94:5053−61. Available from: https://doi.org/10.3168/jds.2011-4484.

[61] Shah KD, Nakao T, Kubota H, Maeda T. Peripartum changes in plasma estrone sulfate and estradiol-17β profiles associated with and without the retention of fetal membranes in Holstein-Friesian cattle. J Reprod Dev 2007;53:279−88. Available from: https://doi.org/10.1262/jrd.18080.

[62] Matsas DJ, Nebel RL, Pelzer KD. Evaluation of an on-farm blood progesterone test for predicting the day of parturition in cattle. Theriogenology 1992;37:859−68. Available from: https://doi.org/10.1016/0093-691X(92)90047-U.

[63] Bleul U, Spirig S, Hässig M, Kähn W. Electrolytes in bovine prepartum mammary secretions and their usefulness for predicting parturition. J Dairy Sci 2006;89:3059−65. Available from: https://doi.org/10.3168/jds.S0022-0302(06)72580-2.

[64] Wright IA, White IR, Roussel AJ, Whyte TK, McBean AJ. Prediction of calving date in beef cows by real-time ultrasonic scanning. Veterinary Rec 1988;123:228−9. Available from: https://doi.org/10.1136/vr.123.9.228.

[65] Cangar Ö, Leroy T, Guarino M, Vranken E, Fallon R, Lenehan J, et al. Automatic real-time monitoring of locomotion and posture behaviour of pregnant cows prior to calving using online image analysis. Comput Electron Agric 2008;64:53−60. Available from: https://doi.org/10.1016/j.compag.2008.05.014.

[66] Henningsen G, Marien H, Hasseler W, Feldmann M, Schoon H-A, Hoedemaker M, et al. Evaluation of the iVET® birth monitoring system in primiparous dairy heifers. Theriogenology 2017;102:44−7. Available from: https://doi.org/10.1016/j.theriogenology.2017.07.005.

[67] Paolucci M, Sylla L, Di Giambattista A, Palombi C, Elad A, Stradaioli G, et al. Improving calving management to further enhance reproductive performance in dairy cattle. Vet Res Commun 2010;34:37−40. Available from: https://doi.org/10.1007/s11259-010-9397-y.

[68] Mota-Rojas D, Rosales AM, Trujillo ME, Orozco H, Ramírez R, Alonso-Spilsbury M. The effects of vetrabutin chlorhydrate and oxytocin on stillbirth rate and asphyxia in swine. Theriogenology 2005;64:1889−97. Available from: https://doi.org/10.1016/j.theriogenology.2004.12.018.

[69] Mota-Rojas D, Martínez-Burnes J, Trujillo MaE, López A, Rosales AM, Ramírez R, et al. Uterine and fetal asphyxia monitoring in parturient sows treated with oxytocin. Anim Reprod Sci 2005;86:131−41. Available from: https://doi.org/10.1016/j.anireprosci.2004.06.004.

[70] Mota-Rojas D, Villanueva-García D, Hernández-González R, Roldan-Santiago P, Martínez-Rodríguez R, Mora-Medina P, et al. Assessment of the vitality of the newborn: an overview. Sci Res Essays 2012;7:712−18.

[71] Van Tassell CP, Wiggans GR, Misztal I. Implementation of a sire-maternal grandsire model for evaluation of calving ease in the United States. J Dairy Sci 2003;86:3366−73. Available from: https://doi.org/10.3168/jds.S0022-0302(03)73940-X.

[72] Uystepruyst C, Coghe J, Dorts T, Harmegnies N, Delsemme M-H, Art T, et al. Optimal timing of elective caesarean section in Belgian white and blue breed of cattle: the calf's point of view. Vet J 2002;163:267−82. Available from: https://doi.org/10.1053/tvjl.2001.0683.

[73] Kolkman I, Aerts S, Vervaecke H, Vicca J, Vandelook J, de Kruif A, et al. Assessment of differences in some indicators of pain in double mus-cled Belgian blue cows following naturally calving vs caesarean section. Reprod Domest Anim 2010;45:160−7. Available from: https://doi.org/10.1111/j.1439-0531.2008.01295.x.

[74] Michaux C, Leroy P. Genetic and non-genetic analysis of neonatal mortality in Belgian Blue breed. Stocarstvo 1997;51:439−42. Available from: https://doi.org/10.1136/vr.123.9.228.

[75] Barrier AC, Coombs TM, Dwyer CM, Haskell MJ, Goby L. Administration of a NSAID (meloxicam) affects lying behaviour after caesarean section in beef cows. Appl Anim Behav Sci 2014;155:28−33. Available from: https://doi.org/10.1016/j.applanim.2014.02.015.

[76] Vandenheede M, Nicks B, Désiron A, Canart B. Mother–young relationships in Belgian blue cattle after a caesarean section: characterisation and effects of parity. Appl Anim Behav Sci 2001;72:281–92. Available from: https://doi.org/10.1016/S0168-1591(01)00118-6.

[77] J. De Tavernier, D. Lips, E. Decuypere, J. Van Outryve, Ethical objections to Caesareans: implications on the future of the Belgian White Blue, Third Congress of the European Society for Agricultural and Food Ethics, 2001, 291–294.

[78] Chenoweth PJ, Leroy P. Change this reference to: Sexual Behavior of the Bull: A Review. J Dairy Sci 1983;66:173–9. Available from: https://doi.org/10.3168/jds.S0022-0302(83)81770-6.

[79] Sankar R, Archunan G. Flehmen response in bull: role of vaginal mucus and other body fluids of bovine with special reference to estrus. Behav Process 2004;67:81–6. Available from: https://doi.org/10.1016/j.beproc.2004.02.007.

[80] Petherick JC. A review of some factors affecting the expression of libido in beef cattle, and individual bull and herd fertility. Appl Anim Behav Sci 2005;90:185–205. Available from: https://doi.org/10.1016/j.applanim.2004.08.021.

[81] Tuomisto L, Huuskonen A, Ahola L, Kauppinen R. Different housing systems for growing dairy bulls in Northern Finland – effects on performance, behaviour and immune status. Acta Agric Scand Sect A Anim Sci 2009;59:35–47. Available from: https://doi.org/10.1080/09064700902919074.

[82] Palmer CW, Brito LFC, Arteaga AA, Söderquist L, Persson Y, Barth AD. Comparison of electroejaculation and transrectal massage for semen collection in range and yearling feedlot beef bulls. Anim Reprod Sci 2005;87:25–31. Available from: https://doi.org/10.1016/j.anireprosci.2004.09.004.

[83] Falk AJ, Waldner CL, Cotter BS, Gudmundson J, Barth AD. Effects of epidural lidocaine anesthesia on bulls during electroejaculation. Can Vet J 2001;42:5.

[84] Etson CJ, Waldner CL, Barth AD. Evaluation of a segmented rectal probe and caudal epidural anesthesia for electroejaculation of bulls. Can Vet J 2004;45:6.

[85] Whitlock BK, Coffman EA, Coetzee JF, Daniel JA. Electroejaculation increased vocalization and plasma concentrations of cortisol and progesterone, but not substance P, in beef bulls. Theriogenology 2012;78:737–46. Available from: https://doi.org/10.1016/j.theriogenology.2012.03.020.

[86] Damián J, Ungerfeld R. The stress response of frequently electroejaculated rams to electroejaculation: hormonal, physiological, biochemical, haematological and behavioural parameters: stress response to electroejaculation in rams. Reprod Domest Anim 2011;46:646–50. Available from: https://doi.org/10.1111/j.1439-0531.2010.01722.x.

[87] Abril-Sánchez S, Freitas-de-Melo A, Damián J, Giriboni J, Villagrá-García A, Ungerfeld R. Ejaculation does not contribute to the stress response to electroejaculation in sheep. Reprod Domest Anim 2017;52:403–8. Available from: https://doi.org/10.1111/rda.12922.

[88] Mosure WL, Meyer RA, Gudmundson J, Barth AD. Evaluation of possible methods to reduce pain associated with electroejaculation in bulls. Can Vet J 1998;39:3.

[89] Pagliosa RC, Derossi R, Costa DS, Faria FJC. Efficacy of caudal epidural injection of lidocaine, xylazine and xylazine plus hyaluronidase in reducing discomfort produced by electroejaculation in bulls. J Vet Med Sci 2015;77:1339–45. Available from: https://doi.org/10.1292/jvms.14-0369.

[90] Palmer CW, Amundson SD, Brito LFC, Waldner CL, Barth AD. Use of oxytocin and cloprostenol to facilitate semen collection by electroejaculation or transrectal massage in bulls. Anim Reprod Sci 2004;80:213–23. Available from: https://doi.org/10.1016/j.anireprosci.2003.07.003.

[91] Ungerfeld R, Casuriaga D, Giriboni J, Freitas-de-Melo A, Silveira P, Brandão FZ. Administration of cloprostenol and oxytocin before electroejaculation in goat bucks reduces the needed amount of electrical stimulation without affecting seminal quality. Theriogenology 2018;107:1–5. Available from: https://doi.org/10.1016/j.theriogenology.2017.10.034.

[92] Abril-Sánchez S, Freitas-de-Melo A, Beracochea F, Damián JP, Giriboni J, Santiago-Moreno J, et al. Sperm collection by transrectal ultrasound-guided massage of the accessory sex glands is less stressful than electroejaculation without altering sperm characteristics in conscious goat bucks. Theriogenology 2017;98:82–7. Available from: https://doi.org/10.1016/j.theriogenology.2017.05.006.

[93] Ungerfeld R, López-Sebastián A, Esteso M, Pradiee J, Toledano-Díaz A, Castaño C, et al. Physiological responses and characteristics of sperm collected after electroejaculation or transrectal ultrasound-guided massage of the accessory sex glands in anesthetized mouflons (*Ovis musimon*) and Iberian ibexes (*Capra pyrenaica*). Theriogenology 2015;84:1067–74. Available from: https://doi.org/10.1016/j.theriogenology.2015.06.009.

Index

Printed in the United States
By Bookmasters